D0079653

BRIEF Microbiology LABORATORY THEORY & APPLICATION

SECOND EDITION

Michael J. Leboffe
San Diego City College

Burton E. Pierce

MORTON
PUBLISHING
925 W. Kenyon Avenue, Unit 12
Englewood, CO 80110

www.morton-pub.com

Book Team

Publisher:	Douglas N. Morton
Biology Editor:	David Ferguson
Editorial Assistant:	Rayna Bailey
Production Manager:	Joanne Saliger
Production Assistant:	Will Kelley
Cover Design:	Bob Schram, Bookends, Inc.

Copyright © 2012 by Morton Publishing Company

ISBN 13: 978-0-89582-947-4

Library of Congress Control Number: 2011942014

10 9 8 7 6 5 4 3 2 1

All rights reserved. No part of this publication may be reproduced, stored in a retrieval system, or transmitted, in any form or by any means, electronic, mechanical, photocopying, recording, or otherwise, without the prior written permission of the copyright owners.

Printed in the United States of America

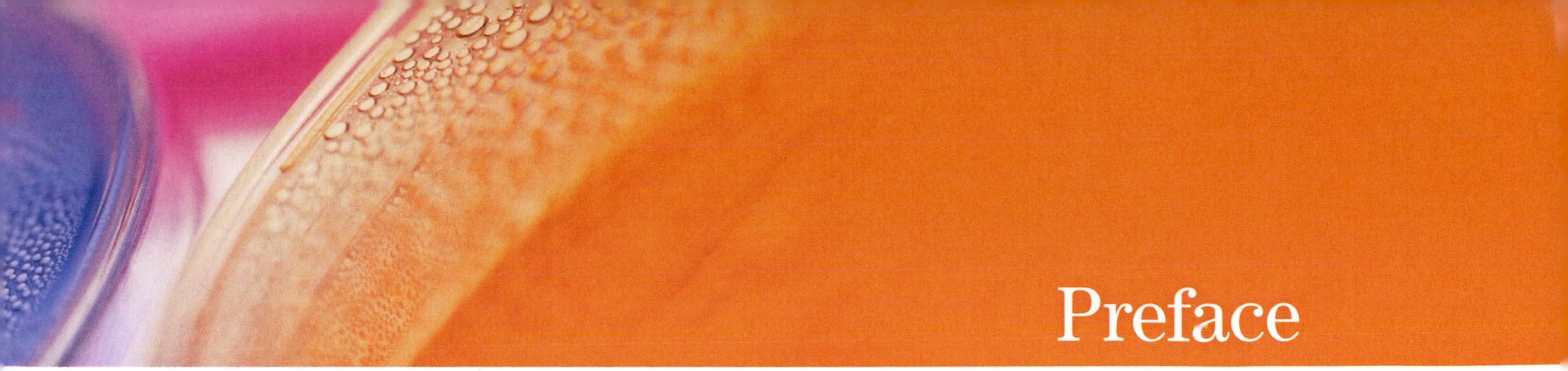

Preface

Welcome to the second edition of *Microbiology Laboratory Theory and Application, Brief Edition*! The response to the first edition was incredibly positive and we are grateful to the adopters for its reception and their support. Many made useful suggestions for improvement, which have been incorporated herein.

We have made a couple of significant changes and many other more subtle changes. Probably the most noticeable change is the use of professionally rendered artwork throughout the manual. While most of these convey the same information as the originals upon which they were based, they are many times more attractive. Beyond replacing all the artwork, we have replaced or added more than 50 new photographs.

The second, more substantive change is reduction in the use of BSL-2 organisms. Where possible, BSL-2 organisms are replaced with equivalent BSL-1 organisms. However, in some instances, a particular exercise required the use of a BSL-2 organism to work, so it is retained with appropriate caution advised. An example of this is *Staphylococcus aureus* in the coagulase test. In a few places, you will see the use of a "surrogate" organism. An example of this is shown in Exercise 4-6 (Hektoen Enteric Agar). The strain of *Providencia stuartii* (BSL-1) produced the same result on the medium as *Shigella flexneri* (BSL-2) and was, therefore, the obvious alternative. However, because one of the purposes of the medium is to isolate *Shigella* species, we decided to retain *S. flexneri* in the recommended organism list with *P. stuartii*, in parentheses, as a surrogate.

Other changes by section are outlined below.

❖ Introduction

Minor changes were made to update the information on BSL categories and strengthen the emphasis on safety. A short treatment of chemical hazards and their standard labeling was added.

❖ Section 1

A new, simple lab comparing the effectiveness of several hand-cleansing agents was added (Exercise 1-1). More importantly, Exercise 1-4 (Common Aseptic Transfers) was rewritten and, we hope, streamlined. Procedural diagrams were also added as was extra emphasis on handling microbes in such a way as to reduce potentially dangerous aerosols. Several new and replacement photos were added. And, we followed the advice of our reviewers to emphasize labeling media *before* inoculation rather than afterward.

❖ Section 2

Several new photos were added to improve this section. Exercise 2-13 (Chemical Germicides) was moved to the end of the section and was rewritten to include a more comprehensive procedure using BSL-1 organisms.

❖ Section 3

Exercise 3-1 (Introduction to the Light Microscope) now includes two activities to supplement the theory and rules of microscope usage. These are the letter "e" slide to familiarize students with the inverted image and the colored threads slide to promote practice at fine focusing. Exercise 3-3 (Eukaryotic Microbes) was totally reorganized to reflect current (but still provisional) classification of these organisms by "Eukaryotic Supergroups." Added to Exercise 3-6 (Gram Stain) is artwork illustrating the differences between Gram-positive and Gram-negative walls. The Acid-Fast Stain (Exercise 3-7) now includes the Ziehl-Neelsen protocol in addition to the Kinyoun method. Throughout, some older photomicrographs were replaced with new ones.

❖ Section 4

These exercises were reordered slightly to reflect a more reasonable approach to the material. The photo in Exercise 4-1 (Phenylethyl Alcohol Agar) was replaced with a better one and the Bile Esculin Test was moved to Section 5.

❖ Section 5

Several exercises, previously in Section 4 and Section 9, were moved to Section 5. Now included in this section are Bile Esculin Test (Exercise 5-10), PYR Test (Exercise 5-17), Bacitracin, Novobiocin, Optochin Tests (Exercise 5-20), and CAMP Test (Exercise 5-22).

❖ Section 6

New artwork and several new photos and/or photomicrographs improve the look of this section. And on the advice of our reviewers to catch up with the 21st century, we have written all of the dilution schemes in this section for digital pipettes.

❖ Section 7

New antibiotics were added to the Kirby-Bauer Test (Exercise 7-2) as were a couple of new photos. The *Morbidity and Mortality Weekly Report* (Exercise 7-3) was updated and includes a photo of the CDC. The Epidemic Simulation (Exercise 7-4) was rewritten to place even greater emphasis on the techniques of safely executing the transfers without aerosol or droplet production.

❖ Section 8

Two exercises were removed: precipitin ring test and the dot blot. The latter was replaced with the Bio-Rad ELISA kit (Exercise 8-6).

❖ Section 9

Exercises 9-1, 9-2, and 9-3 (Identification of Unknown) were rewritten with new or revised identification flowcharts, necessitated by changing organism inventories at standard biological supply houses. An attempt was made to reduce the number of media required for each identification to provide more options for tests, and to reduce the number of BSL-2 organisms used.

❖ Appendices

These still provide useful, but supplemental, materials for those that wish to use them. Appendix A (Biochemical Pathways) had a couple of errors removed and some newer information about ATP yields added. Appendix B (Miscellaneous Transfer Methods) was revised to follow the new style used in Exercise 1-3, as were Appendices C and D (involving pipetting). The alternative procedures appendix (formerly Appendix E) was discontinued. All of the alternative dilution schemes using digital pipettors were fully incorporated in Section 6 procedures.

❖ Acknowledgments

First, we want to thank everyone who has adopted our microbiology titles over the last 16 years. It all started with *A Photographic Atlas for the Microbiology Laboratory* in 1995 (now in its 4th edition). Over the years, it has spawned three other titles: *Exercises for the Microbiology Laboratory* (1996, also in its 4th edition); *Microbiology Laboratory Theory and Application* (2002, now in its 3rd edition); and *Microbiology Laboratory Theory and Application, Brief* (2007, with the 2nd edition resting securely in your hands). Clearly, we have benefitted from suggestions for improvement from innumerable people over the last 16 years and we haven't forgotten you.

We want to thank our colleagues and friends at San Diego City College for their continued support. In alphabetical order, these include Donna DiPaolo, Anita Hettena, Roya Lahijani, David Singer, Minou Spradley, Laura Steininger, and Muu Vu. We would particularly like to thank Debra Reed for her involvement and assistance in running many necessary tests and for her work finding safer alternatives to many of the BSL-2 organisms used in the first edition. We also are grateful to San Diego City College for making lab facilities available through the Civic Center Program. It would be next to impossible to write a laboratory manual without a lab to work in! Lastly, thanks to the Biology 205 students over the years (and in particular the Fall 2011 MW and TTh morning sections) who have performed the labs and have pointed out rough spots that were not apparent to us.

While we have been the grateful beneficiaries of occasional suggestions from adopters, we especially want to recognize the efforts of Janice Smith, Tarrant County College, Fort Worth, Texas; Johana Melendez, Hillsborough Community College, Tampa Bay, Florida; and Diane Doidge, Grand View University, Des Moines, Iowa, for reviewing the first edition and for their helpful suggestions. Reviewing a complex work of this size is not an easy task and we are indebted to them for supplying fresh eyes to scrutinize its contents.

As always, we are indebted to the people at Morton Publishing. It is hard to imagine a better group of people for authors to work with. Thanks to Doug Morton (President) for his vision of producing high quality, yet affordable, texts for students. Thanks also go to Chrissy Morton DeMier (Business Manager), David Ferguson (Biology/Acquisitions Editor), Carter Fenton (Sales and Marketing Manager), Joanne Saliger (Production Manager), Will Kelley (Production Assistant), and Rayna Bailey (Editorial Assistant) for playing their parts in putting their authors in a position to be successful. Each person listed plays a crucial role in the publication of our works, but special recognition is owed to Joanne Saliger, because her efforts in designing the layout are seen by the readers. The quality of a book is mostly in its content, but extraordinary design surely makes the book more palatable. Thanks also go to Bob Schram of Bookends, Inc. for the cover design. Finally, we thank the talented people at Imagineering Art in Toronto, Ontario, Canada, for their artistic renderings. Their inclusion is probably the most noticeable change in this edition.

Contents

Safety and Laboratory Guidelines

Microbiology lab can be an interesting and exciting experience, but at the outset you should be aware of some potential hazards. Improper handling of chemicals, equipment and/or microbial cultures is dangerous and can result in injury or infection. Safety with lab equipment will be addressed when you first use that specific piece of equipment, as will specific examples of chemical safety. Our main concern here is to introduce you to safe handling and disposal of microbes.[1]

Because microorganisms present varying degrees of risk to laboratory personnel (students, technicians, and faculty), people outside the laboratory, and the environment, microbial cultures must be handled safely. Four biosafety levels (BSLs) that provide a set of minimum standards for laboratory practices, facilities, and equipment to be used when handling organisms at each level have been defined. These are described in the U.S. Government publication, *Biosafety in Microbiological and Biomedical Laboratories*, 5th Ed. (2009), and are summarized below and in Table I-1. For complete information, readers are referred to the original document.

BSL-1: Organisms do not typically cause disease in healthy individuals and present a minimal threat to the environment and lab personnel. Standard microbiological practices are adequate. These microbes may be handled in the open, and no special containment equipment is required. Examples include *Bacillus subtilis*, *Escherichia coli*, *Rhodospirillum rubrum*, and *Lactobcillus acidophilus*.

BSL-2: Organisms are commonly encountered in the community and present a moderate environmental and/or health hazard. These organisms are associated with human diseases of varying severity. Individuals may do laboratory work that is not especially prone to splashes or aerosol generation, using standard microbiological practices. Examples include *Salmonella*, *Staphylococcus aureus*, *Clostridium dificile*, and *Borrelia burgdorferi*.

[1] Your instructor may augment or revise these guidelines to fit the conventions of your laboratory.

TABLE I-1 Summary of Recommended Biosafety Levels for Infectious Agents ◆

BSL	Agents	Practices	Safety Equipment (Primary Barriers)	Facilities (Secondary Barriers)
1	Not known to consistently cause disease in healthy adults	Standard microbiological practices	◆ Primary barriers: none required ◆ Personal Protective Equipment (PPE): Laboratory coats and gloves; eye, face protection as needed	Laboratory bench practices and sink required
2	◆ Agents associated with human disease ◆ Routes of transmission include percutaneous injury, ingestion, and mucous membrane exposure	BSL-1 practices plus: ◆ Limited lab access ◆ Biohazard warning signs ◆ "Sharps" precautions ◆ Biosafety manual defining any needed waste decontamination	◆ Primary barriers: Class I or II BSCs or other physical containment devices used for all manipulations of agents that cause splashes or aerosols of infectious materials ◆ PPEs: laboratory coats, gloves, face protection as needed	BSL-1 plus: ◆ Autoclave available
3	Indigenous or exotic agents with potential for aerosol transmission; disease may have serious or lethal consequences	BSL-2 practices plus: ◆ Controlled lab access ◆ Decontamination of all waste ◆ Decontamination of all lab clothing before laundering ◆ Baseline serum	◆ Primary barriers: Class I or II BSCs or other physical containment devices used for all open manipulations of agents ◆ PPEs: protective lab clothing, gloves, respiratory protection as needed	BSL-2 plus: ◆ Physical separation from access corridors ◆ Access to self-closing, double door ◆ Exhausted air not recirculated ◆ Negative airflow into laboratory
4	Dangerous/exotic agents that pose high risk of life-threatening disease, aerosol-transmitted lab infections; or related agents with unknown risk of transmission	BSL-3 practices plus: ◆ Clothing change before entering ◆ Shower on exit ◆ All material decontaminated on exit from facility	◆ Primary barriers: All procedures conducted in Class III BSCs or Class I or II BSCs *in combination with* full-body, air-supplied, positive pressure personnel suit	BSL-3 plus: ◆ Separate building or isolated zone ◆ Dedicated supply and exhaust, vacuum, and decon systems ◆ Other requirements outlined in the text

Source: Reprinted from *Biosafety in Microbiological and Biomedical Laboratories*, 5th edition (Department of Health and Human Services Publication No. (CDC) 21-1112) 2009.

BSL-3: Organisms are of local or exotic origin and are associated with respiratory transmission and serious or lethal diseases. Special ventilation systems are used to prevent aerosol transmission out of the laboratory, and access to the lab is restricted. Specially trained personnel handle microbes in a Class I or II biological safety cabinet (BSC), not on the open bench (see Figure I-1). Examples include *Bacillus anthracis*, *Mycobacterium tuberculolosis*, and West Nile virus.

BSL-4: Organisms have a great potential for lethal infection. Inhalation of infectious aerosols, exposure to infectious droplets, and autoinoculation are of primary concern. The lab is isolated from other facilities, and access is strictly controlled. Ventilation and waste management are under rigid control to prevent release of the microbial agents to the environment. Specially trained personnel perform transfers in Class III BSCs. Class II BSCs may be used as long as personnel wear positive pressure, one-piece body suits with a life-support system. Examples include agents causing hemorrhagic diseases, such as Ebola virus, Marburg virus, and Lassa fever.

The microorganisms used in introductory microbiology courses depend on the institution, objectives of the course, and student preparation. Most introductory courses use organisms that may be handled at BSL-1 and BSL-2 levels so we have followed that practice in designing this set of exercises, though we have replaced many BSL-2 organisms or have recomended BSL-1 substitutes in many of the exercises. Following are general safety rules to reduce the chance of injury or infection to you and to others, both inside and outside the laboratory. Although they

I-1 Biological Safety Cabinet in a BSL-2 Laboratory ◆ In this Class II BSC, air is drawn in from the room and is passed through a HEPA filter prior to release into the environment. This airflow pattern is designed to keep aerosolized microbes from escaping from the cabinet. The microbiologist is pipetting a culture. When the BSC is not in use at the end of the day, an ultraviolet light is turned on to sterilize the air and the work surface.
(San Diego County Public Health Laboratory)

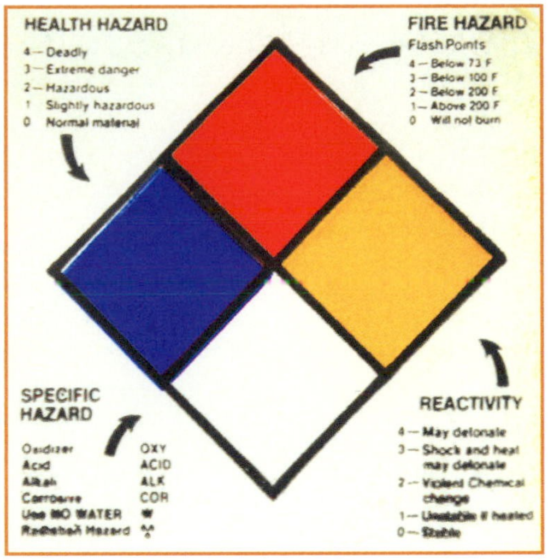

I-2 Chemical Hazards ◆ Be aware of these (or similar) labels on the chemicals you handle. Your instructor will advise you on appropriate safety measures to be taken with each.

represent a mixture of BSL-1 and BSL-2 guidelines, we believe that it is best to err on the side of caution and that students should learn and practice the safest level of standards (relative to the organisms they are likely to encounter) at all times. Please follow these and any other safety guidelines required by your college.

Chemical safety is also important in a microbiology laboratory. Be aware of the hazards presented by the chemicals you are handling. Most will be labeled with a sticker as shown in Figure I-2. Numbers are assigned to the degree of health, fire, and reactivity hazard posed by the chemical. There also is a space to enter specific hazards, such as acid, corrosive, and radioactivity.

Student Conduct

- ◆ To reduce the risk of infection, do not smoke, eat, drink, or bring food or drinks into the laboratory room—even if lab work is not being done at the time.
- ◆ Do not apply cosmetics or handle contact lenses in the laboratory.
- ◆ Wash your hands *thoroughly* with soap and water before working in the lab, after handling living microbes, and before leaving the laboratory at any time. Also, wash your hands after removing gloves.
- ◆ Do not remove any organisms or chemicals from the laboratory.
- ◆ Lab time is precious, so come to lab prepared for that day's work. Figuring out what to do as you go along is likely to produce confusion and accidents.
- ◆ Work carefully and methodically. Do not hurry through any laboratory procedure.

Basic Laboratory Safety

- ◆ Wear protective clothing (i.e., a lab coat) in the laboratory when handling microbes. Remove the coat prior to leaving the lab and autoclave it regularly (Figure I-3).
- ◆ Do not wear sandals or open-toed shoes in the laboratory.
- ◆ Wear eye protection whenever you are heating chemicals, even if you wear glasses or contacts (Figure I-3).
- ◆ Turn off your Bunsen burner when it is not in use. In addition to being a fire and safety hazard, it is an unnecessary source of heat in the room.
- ◆ Tie back long hair, as it is a potential source of contamination as well as a likely target for fire.
- ◆ If you are feeling ill, go home. A microbiology laboratory is not a safe place if you are ill.
- ◆ If you are pregnant, immune compromised, or are taking immunosuppressant drugs, please see the instructor. It may be in your best *long-term* interests to postpone taking this class. Discuss your options with your instructor.
- ◆ If it is your lab's practice to wear disposable gloves while handling microorganisms, be sure to remove them each time you leave the laboratory. The proper method for removal is with the thumb under the cuff of the other hand's glove and turning it inside out without snapping it. Gloves should then be disposed of in the container for contaminated materials. Then, wash your hands.
- ◆ Wear disposable gloves while staining microbes and handling blood products—plasma, serum, antiserum,

I-3 Safety First ◆ This student is prepared to work safely with microorganisms. The lab area is uncluttered, tubes are upright in a test tube rack, and the flame is accessible but not in the way. The student is wearing a protective lab coat, gloves, and goggles, all of which are to be removed prior to leaving the laboratory. Not all procedures require gloves and eye protection. Your instructor will advise you as to the standards in your laboratory.

or whole blood (Figure I-3). Handling blood can be hazardous, even if you are wearing gloves. Consult your instructor before attempting to work with any blood products.

- Use an antiseptic (e.g., Betadine) on your skin if it is exposed to a spill containing microorganisms. Your instructor will tell you which antiseptic you will be using.
- Never pipette by mouth. Always use mechanical pipettors (see Figure C-1, Appendix C).
- Dispose of broken glass or any other item that could puncture an autoclave bag in an appropriate "sharps" or broken-glass container (Figure I-4).
- Use a fume hood to perform any work involving highly volatile chemicals or stains that need to be heated.
- Find the first-aid kit, and make a mental note of its location.
- Find the fire blanket, shower, and fire extinguisher, note their locations, and develop a plan for how to access them in an emergency.
- Find the eye wash basin, learn how to operate it, and remember its location.

Reducing Contamination of Self, Others, Cultures, and the Environment

- Wipe the desktop with a disinfectant (e.g., Amphyl® or 10% chlorine bleach) before *and* after each lab period. Never assume that the class before you disinfected the work area. An appropriate disinfectant will be supplied. Allow the disinfectant to evaporate; do not wipe it dry.
- Never lay down culture tubes on the table; they always should remain upright in a tube holder (Figure I-3). Even solid media tubes contain moisture or condensation that may leak out and contaminate everything it contacts.
- Cover any culture spills with paper towels. Soak the towels immediately with disinfectant, and allow them to stand for 20 minutes. Report the spill to your instructor. When you are finished, place the towels in the container designated for autoclaving.
- Place all nonessential books and papers under the desk. A cluttered lab table is an invitation for an accident that may contaminate your expensive school supplies.

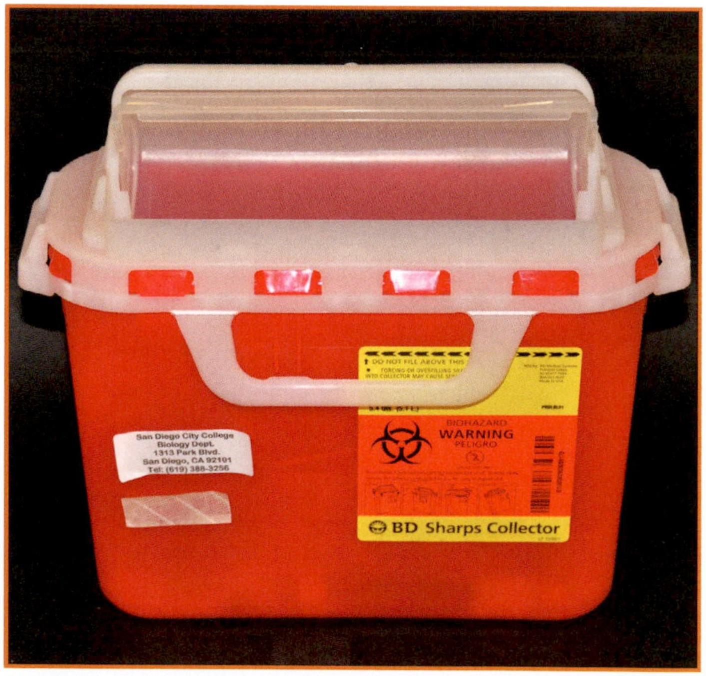

I-4 Sharps Container ◆ Needles, glass, and other contaminated items that can penetrate the skin or an autoclave bag should be disposed of in a sharps container. Do not fill above the dashed black line. Notice the autoclave tape in the lower left. The white stripes will turn black after proper autoclaving. Above the autoclave tape is the address of the institution that produced the biohazardous waste.

- When pipetting microbial cultures, place a disinfectant-soaked towel on the work area. This reduces contamination and possible aerosols if a drop escapes from the pipette and hits the tabletop.

Disposing of Contaminated Materials

In most instances, the preferred method of decontaminating microbiological waste and reusable equipment is the autoclave (Figure I-5).

- Remove all labels from tube cultures and other contaminated *reusable* items and place them in the designated autoclave container. This will likely be an open autoclave pan to enable cleaning the tubes and other items following sterilization.
- Dispose of plate cultures (if plastic Petri dishes are used) and other contaminated nonsharp *disposable* items in the designated autoclave container (Figure I-6). Petri dishes should be taped closed. (Note: To avoid recontamination of sterilized culture media and other items, autoclave containers are designed to be permanently closed, autoclaved, and discarded. Therefore, do not place reusable and nonreusable items in the same container.)
- Dispose of all blood product samples and disposable gloves in the container designated for autoclaving.
- Place used microscope slides of bacteria in a "sharps" container designated for autoclaving, or soak them in disinfectant solution for at least 30 minutes before cleaning or discarding them. Follow your laboratory guidelines for disposing of glass.
- Place contaminated broken glass and other sharp objects (anything likely to puncture an autoclave bag) in a sharps container designated for autoclaving (Figure I-4). Uncontaminated broken glass does not need to be autoclaved, but should be disposed of in a specialized broken glass container.

References

Barkley, W. Emmett, and John H. Richardson. 1994. Chapter 29 in *Methods for General and Molecular Bacteriology*, edited by Philipp Gerhardt, R. G. E. Murray, Willis A. Wood, and Noel R. Krieg. American Society for Microbiology, Washington, DC.

Collins, C. H., Patricia M. Lyne, and J. M. Grange. 1995. Chapters 1 and 4 in *Collins and Lyne's Microbiological Methods*, 7th ed. Butterworth-Heineman, Oxford.

Darlow, H. M. 1969. Chapter VI in *Methods in Microbiology,* Volume 1, edited by J. R. Norris and D. W. Ribbins. Academic Press, Ltd., London.

Fleming, Diane O., and Debra L. Hunt (Eds.). 2000. *Laboratory Safety—Principles and Practices*, 3rd ed. American Society for Microbiology, Washington, DC.

Koneman, Elmer W., Stephen D. Allen, William M. Janda, Paul C. Schreckenberger, and Washington C. Winn, Jr. 1997. *Color Atlas and Textbook of Diagnostic Microbiology*, 5th ed. Lippincott-Raven Publishers, Philadelphia and New York.

Power, David A., and Peggy J. McCuen. 1988. Pages 2 and 3 in *Manual of BBL™ Products and Laboratory Procedures*, 6th ed. Becton Dickinson Microbiology Systems, Cockeysville, MD.

I-5 An Autoclave ◆ Media, cultures, and equipment to be sterilized are placed in the basket of the autoclave. Steam heat at a temperature of 121°C (produced at atmospheric pressure plus 15 psi) for 15 minutes is effective at killing even bacterial spores. Some items that cannot withstand the heat, or have irregular surfaces that prevent uniform contact with the steam, are sterilized by other means.

I-6 An Autoclave Bag ◆ Nonreusable items (such as plastic Petri dishes) are placed in an autoclave bag for decontamination. Petri dishes should be taped closed. Do not overfill or place sharp objects in the bag. Notice the autoclave tape at the middle right. The white stripes will turn black after proper autoclaving. At the lower right is the address of the institution that produced the biohazardous waste.

Chosewood, L. Casey, and Deborah E. Wilson (Eds.). December 2009. U.S. Department of Health and Human Services, *Biosafety in Microbiological and Biomedical Laboratories*, 5th ed. Department of Health and Human Services Publication No. (CDC) 21-1112.

A Word About Experimental Design

Like most sciences, microbiology has descriptive and experimental components. Here we are concerned with the latter. Science is a philosophical approach to finding answers to questions. In spite of what you may have been taught in grade school about *THE* "Scientific Method," science can approach problems in many ways, rather than in any *single* way. The nature of the problem, personality of the scientist, intellectual environment at the time, and good, old-fashioned luck all play a role in determining which approach is taken. Nevertheless, in experimental science, one component that is always present is a **control** (or controls).

A controlled experiment is one in which all **variables** except one—the **experimental variable**—are maintained without change. This is the only way the results can be considered reliable. By maintaining all variables except one, other potential sources of an observed event can be eliminated. Then (presumably), a **cause and effect relationship** between the event and the experimental variable can be established. That is, if the event changes when the experimental variable changes, we provisionally link that variable and the event. Alternatively, if there is no observed change, we can eliminate the experimental variable from involvement with the event.

Throughout science experimentation—and this book—you will see the word *control.* Controls are an essential and integral part of all experiments. As you work your way through the exercises in this book, pay attention to all the ways controls are used to improve the reliability of the procedure and your confidence in the results.

Microbiological experimentation often involves tests that determine the ability of an organism to use or produce some chemical, or to determine the presence or absence of a specific organism in a sample. Ideally, a positive result in the test indicates that the microbe has the ability or is present in the sample, and a negative result indicates a lack of that ability or absence in the sample. The tests we run, however, have limitations and occasionally may give **false positive** or **false negative** results. An inability to detect small amounts of the chemical or organism in question would yield a false negative result and would be the result of inadequate **sensitivity** of the test (Figure I-7). An inability to discriminate between the chemical or organism in question and similar chemicals or organisms would yield a false positive result

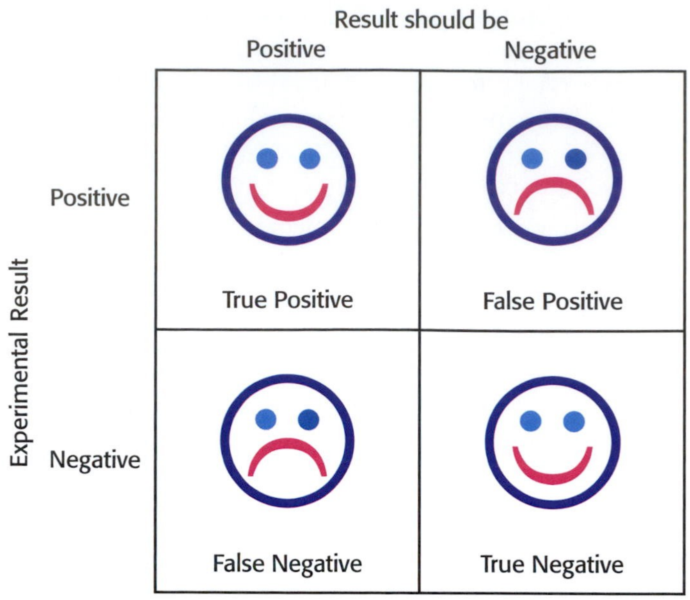

I-7 Limitations of Experimental Tests ◆ Ideally, tests should give a positive result for specimens that are positive, and a negative result for specimens that are negative. False positive and false negative results do occur, however, and these are attributed to inadequate specificity and inadequate sensitivity, respectively, of the test system.

and would be the result of inadequate **specificity** of the test (Figure I-7). Sensitivity and specificity can be quantified using the following equations:

$$\text{Sensitivity } 4 \quad \frac{\text{True Positives}}{\text{True Positives} + \text{False Negatives}}$$

$$\text{Specificity } 4 \quad \frac{\text{True Negatives}}{\text{True Positives} + \text{False Negatives}}$$

The closer sensitivity and specificity are to a value of one, the more useful the test is. As you perform the tests in this book, be mindful of each test's limitations, and be open to the possibility of false positive and false negative results.

References

Forbes, Betty A., Daniel F. Sahm, and Alice. S. Weissfeld. 1998. Chapter 5 in *Bailey and Scott's Diagnostic Microbiology*, 10th ed. Mosby-Year Book, St. Louis.

Lilienfeld, David E., and Paul D. Stolley. 1994. Page 118 in *Foundations of Epidemiology*, 3rd ed. Oxford University Press, New York.

Mausner, Judith S., and Shira Kramer. 1985. Pages 217–220 in *Epidemiology: An Introductory Text*, 2nd ed. W.B. Saunders Company, Philadelphia.

SECTION 1

Fundamental Skills for the Microbiology Laboratory

A necessary skill for safely working in a laboratory, handling foods, and just living in a world full of microbes, is effective hand washing. In Exercises 1-1 and 1-2, you will have the opportunity to compare hand cleansing agents and evaluate your hand-washing technique to correct any deficiencies you observe.

Bacterial and fungal **cultures** are grown and maintained on or in solid and liquid substances called **media**. Preparation of these media involves weighing ingredients, measuring liquid volumes, calculating proportions, handling basic laboratory glassware, and operating a pH meter and an autoclave. In Exercise 1-3 you will learn and practice these fundamental skills by preparing a couple of simple growth media. When you have completed the exercise, you will have the skills necessary to prepare almost any medium if given the recipe.

A third fundamental skill necessary for any microbiologist is the ability to transfer microbes from one place to another without contaminating the original culture, the new medium, or the environment (including the microbiologist). This **aseptic** (sterile) transfer technique is required for virtually all procedures in which living microbes are handled, including isolations, staining, and differential testing. Exercises 1-4 through 1-6 present descriptions of common transfer and inoculation methods. Less frequently used methods are covered in Appendices B through D.

EXERCISE 1-1

A Comparison of Hand-Cleansing Agents[1]

Theory

The concept of good hand hygiene has gone from a controversial beginning (in the early 1800s) to an accepted practice that is still problematic. Current studies designed to test the efficacy of various agents often have subjects wash for unrealistic lengths of time (that is, longer than workers routinely wash on the job), test artificially contaminated hands (or not), and use different standards of evaluation, making comparison difficult. We are still left with the question, "What works best?"

While it has been identified as an important, easily performed behavior that minimizes transfer of pathogens to others, uniform compliance with hand-washing standards has been difficult to achieve. Heavy workloads, skin reactions to the agent (e.g., plain or antimicrobial soap, iodine compounds, alcohol), skin dryness due to frequent washing, and many other factors contribute to noncompliance (see Boyce and Pittet, 2002, and The Joint Commission, 2009). Alcohol-based hand rubs have, in many instances, replaced conventional hand-washing agents because they are more effective than soap and water (disputed), require less time, produce fewer skin reactions, and have been shown to result in a higher level of compliance by health care workers. Examples of agents that can be tested in this exercise are shown in Figure 1-1.

[1] Thanks to reviewers Johana Meléndez-Santiago and Janice Yoder Smith for their suggestions on developing this laboratory exercise.

Application

Effective hand washing to minimize direct person-to-person transmission, and indirect contact of pathogens by health care professionals and food handlers is essential. It also is critical to laboratorians handling pathogens to minimize transmission to others, inoculation of oneself, and contamination of cultures.

In This Exercise

You will evaluate the relative effectiveness of a variety of hand-washing agents.

Materials

Per Student

- ✓ one Nutrient Agar or Trypticase Soy Agar plate
- ✓ access to a sink and paper towels
- ✓ a permanent marking pen

Per Table

- ✓ Bottles of as many as five of the following or suitable replacements:
  - antibacterial liquid soap (0.15% triclosan)

1-1 Hand-Washing Agents ◆ From left to right are: Betadine® scrub (10% povidone iodine), alcohol based hand sanitizer (65% ethyl alcohol), 0.13% benzalkonium chloride-based hand sanitizer, antibacterial soap (0.15 triclosan), and Hibiclens® (4% chlorhexidine gluconate). As written, this exercise uses these agents, but others can be substituted, including tap water.

- hand sanitizer (65% ethyl alcohol)
- Hy5 brand of alcohol-free hand sanitizer (0.13% benzalkonium chloride)
- Betadine® scrub (10% povidone iodine)
- Hibiclens® (4% w/v chlorhexidine gluconate)
- tap water
- Note: Students should not ingest or get any of these products in their eyes. Follow package inserts for treatments. Students with allergies or sensitivities to the active ingredients in any of these products should be assigned an alternate product.

Procedure

Lab One

1 Label the base of a Nutrient Agar or Trypticase Soy Agar plate with your name. Divide the plate in half and label one half "Before" and the other half "After." Then, number 1 through 5 on both sides of the dividing line so the numbers line up. That is, number 1 on the "Before" side should be opposite number 1 on the "After" side, and so on. Use the entire diameter of the plate for the five numbers and spread them out as much as you can. You will be pressing your fingers at each number in Steps 3 and 6.

2 Rub your left hand fingertips (all five of them) on the area to be sampled. (Good choices are the floor, shoe soles, tabletop near a sink, backpack, and your forehead or feet. Be creative!

3 Then, one at a time, gently press your thumb tip next to the "Before" number 1, your index finger next to the "Before" number 2, and so on. Your middle, ring, and pinky fingers are numbered 3, 4, and 5, respectively.

4 Wash your hands with soap and water.

5 Record the hand-cleansing agents you will be using on the Data Sheet.

6 You will now sample the same site as in Step 2 with the fingertips of your right hand, but this time one at a time. Then, you will rub that fingertip in a few milliliters of the hand-cleansing agent in in your left palm. If the agent requires rinsing and drying, perform this at a sink. Agents that only require air-drying can be done at your table.

a. Place a quarter-sized drop of the first hand-cleansing agent in the palm of your left hand. Rub your right thumb on the area to be sampled. Then, rub your thumb into the agent for 10 seconds.[2] (If you don't have a watch, sing the "Happy Birthday Song" once to yourself.) Once your thumb is dry, gently press it next to the number 1 on the "After" side. Wash both hands with soap and water, and dry them.

b. Repeat with each finger of the right hand and the remaining four hand-cleansing agents. Be sure to press each finger onto the agar next to its correct number and to wash your hands in between exposures.

7 Tape the lid on the plate and incubate at 25°C until the next lab period.

Lab Two

1 Examine your plates for growth. (Note: if you see fuzzy growth, get permission from your instructor before you open it. It is probably a fungus and the lid should be left on. Otherwise, it is beneficial to interpretation if you remove the plate's lid.) Each "blob" is probably the product of a single microbial species that has reproduced from one or a few cells to such an extent that it is now visible to the naked eye. This is called a "colony." Different microbial species often produce distinctly different colonies.

2 Fill in the table provided on the Data Sheet as you evaluate the relative amount of growth and microbial diversity (based on colony differences) on the "Before" and "After" sides of the plate for each agent. Recognize that you are comparing "apples and oranges" with respect to their intended uses. That is, the products tested range from surgical scrubs to over-the-counter soaps intended for household use. A surgical scrub may have better antimicrobial activity than hand soap, but that doesn't make it "better" for household use!

[2] The recommended time for hand washing is 15-20 seconds, but because you are only washing the fingertips and concentrating on those, that time has been reduced in order to see representative results.

References

Boyce, John M., and Didier Pittet, 2002. Centers for Disease Control and Prevention. *Guideline for Hand Hygiene in Health-Care Settings: Recommendations of the Healthcare Infection Control Practices Advisory Committee and the HICPAC/SHEA/APIC/IDSA Hand Hygiene Task Force.* MMWR 2002, 51 (No. RR-16): 1-45.

The Joint Commission, 2009. *Measuring Hand Hygiene Adherence: Overcoming the Challenges.* http://www.jointcommission.org/assets/1/18/hh_monograph.pdf.

NAME ______________________ DATE ______________________

LAB SECTION ______________________ I WAS PRESENT AND PERFORMED THIS EXERCISE (INITIALS) ________

A Comparison of Hand-Cleansing Agents

OBSERVATIONS AND INTERPRETATIONS

Record your results in the table below. For **relative amount of growth**, use this qualitative scale for evaluation:

+++ means "a lot of contamination"
++ means "moderate contamination"
+ means "little contamination"
0 means "no contamination"

There is no absolute cutoff between any of these categories, and what you call "moderate contamination" might be called "little contamination" by another student. Just try to be consistent within your evaluation. For **diversity**, look for different colors, shapes, textures, sizes, and anything else that is indicative of a different organism growing and record the number of different organisms present. **Interpretation** would be relative effectiveness of the agent (excellent, good, fair, poor).

Number on Plate	Hand-washing Agent	"Before"		"After"		Interpretation
		Relative Growth	Diversity	Relative Growth	Diversity	
1						
2						
3						
4						
5						

DATA SHEET 1-1

Questions

1 Which agent seemed to be most effective at removing microbes from the fingertips?

2 Which agent seemed to be least effective at removing microbes from the fingertips.

3 What other variables besides the cleansing agent might be responsible for any differences noted between fingers or students?

4 Did you or any other students notice some microbes persisting more than others after washing? What factor(s) might account for this?

Name ______________________ Date ______________________

Lab Section ______________________ I was present and performed this exercise (initials) ________

5 Do some research and fill in the table. Your answer to "Antibacterial Effect" should address the structural and/or functional part of the cell affected, as well as if the agent is bactericidal (kills bacteria) or bacteriostatic (stops bacterial growth, but doesn't kill them). Under "Other Antimicrobial Effects," note any antiviral, antiprotozoal, and/or antifungal properties of the agent.

Active Ingredient	Antibacterial Effect	Other Antimicrobial Effects
Antibacterial Soap with Triclosan		
Isopropanol/Ethanol Hand Sanitizer		
Benzalkonium Chloride Hand Sanitizer		
Povidone Iodine Scrub		
Chlorhexidine Gluconate		

DATA SHEET 1-1

EXERCISE 1-2

Glo Germ™ Hand Wash Education System

Theory

If you didn't perform Exercise 1-1, read the theory for it. Then, continue with this one.

Now that you have compared various hand-cleansing agents, you will examine your individual skill at hand washing. The Glo Germ™ Hand Wash Education System was developed to train people to wash their hands more effectively. The lotion (a powder also is available) contains minute plastic particles (artificial germs) that fluoresce when illuminated with ultraviolet (UV) radiation but are invisible with normal lighting. Initially the hands are covered with the lotion, but the location and density of the germs is unknown because of the normal room lighting. After washing, a UV lamp is shined on the hands. Wherever the "germs" remain, hand washing was not effective. This provides immediate feedback to the washers as to the effectiveness of their hand washing and provides information about where they have to concentrate their efforts in the future.

Application

Effective hand washing to minimize direct person-to-person transmission, and indirect contact of pathogens by health care professionals and food handlers is essential. It also is critical to laboratorians handling pathogens to minimize transmission to others, inoculation of self, and contamination of cultures.

In This Exercise

You will cover your hands with nontoxic, synthetic fluorescent "germs" and compare the degree of contamination before and after hand washing to evaluate your hand-washing technique and demonstrate the difficulty in removing hand contaminants.

Materials[1]

Per Student Group

- ✓ one bottle of Glo Germ™ lotion-based simulated germs
- ✓ one ultraviolet penlight

[1] Available from Glo Germ™, PO Box 189, Moab, UT 84532. 1-800-842-6622 (USA). Online: http://www.glogerm.com/

Procedure

1. Shake the lotion bottle well.
2. Have your lab partner apply 2–3 drops of gel on the palms of both of your hands. Be careful not to get the gel on your clothing, or in your eyes or mouth.
3. Rub your hands together, thoroughly covering your hand surfaces, including the backs and between the fingers (Figure 1-2). Spread the lotion up to the wrists on both sides. Also, scratch the palms with all your fingernails.
4. Have your lab partner shine the UV light on your hands to see the extent of coverage with the lotion. *Do not look directly at the lamp*. This works best in an area with limited ambient light. Do *not* handle the light yourself because you will contaminate it with the artificial germs.
5. Have your lab partner turn on warm water at a sink for you. Then wash your hands with soap and warm water as thoroughly as you can for at least 20 seconds. (If you don't have a watch handy, sing the "Happy Birthday Song" to yourself twice.) Use a fingernail brush if you have one. When you are finished, have your lab partner turn off the water and hand you a fresh paper towel. Dry your hands.
6. Have your lab partner shine the UV light on your hands once more. *Do not look directly at the lamp*. Examine the hand surfaces contaminated by the artificial germs. Turn off the lamp.

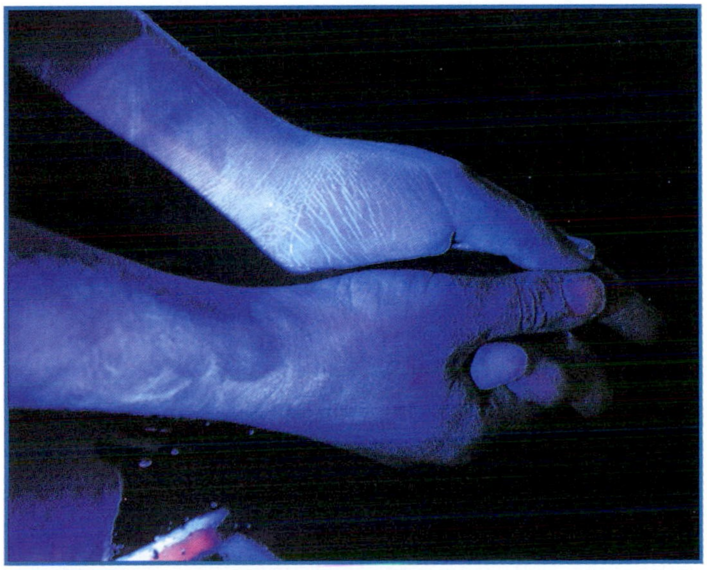

1-2 Hands Covered With Glo Germ™ Prior to Washing ◆ Shown are properly prepared hands covered with the fluorescent Glo Germ™ lotion prior to washing. Note the thorough coverage, including the back of the hands and under the fingernails.

7 Now that you know where the artificial germs remain, wash your hands once more to remove as many as possible. As before, have your lab partner turn the water on and off for you.

8 Repeat the experiment with your lab partner, but with the roles reversed.

9 Record your results on the Data Sheet, and answer the questions.

10 After recording your results, shine the UV lamp once more on your Data Sheet and pen/pencil to see how much of the lotion was transferred to these. *Do not look directly at the lamp.*

References

Glo Germ™. Package insert for the Glo Germ™ Hand Wash Education System. Glo Germ™, PO Box 189, Moab, UT 84532.

NAME ______________________ DATE ______________________

LAB SECTION ______________________ I WAS PRESENT AND PERFORMED THIS EXERCISE (INITIALS) ________

DATA SHEET 1-2

Glo Germ™ Hand Wash Education System

OBSERVATIONS AND INTERPRETATIONS

Record the degree of hand contamination before and after washing in the table below. Use this qualitative scale for evaluation:

+++ means "a lot of contamination"
++ means "moderate contamination"
+ means "little contamination"
0 means "no contamination"

There is no absolute cutoff between any of these categories, and what you call "moderate contamination" might be called "little contamination" by another student. Just try to be consistent within your evaluation.

Body Region	Left Hand		Right Hand	
	Before Washing	After Washing	Before Washing	After Washing
Palm				
Back of Hand				
Fingers				
Between Fingers				
Tops of Fingernails				
Under Fingernails				
Front of Wrist				
Back of Wrist				

Questions

1 What areas were cleaned most thoroughly with your washing technique?

2 What areas were most difficult for you to clean with this washing technique?

3 In general, were your two hands cleaned an equal amount, or was one cleaned more than the other? What could account for any differences?

4 How do your answers to Questions 1, 2, and 3 compare to your lab partner's answers? Why might they differ?

5 Why were you instructed to have your lab partner turn the water on and off and operate the UV lamp rather than you doing these actions yourself?

6 Why might it be advisable to modify the procedure and use the UV light to check your hands prior to application of lotion and the paper towels prior to drying?

7 Shine the UV lamp on this Data Sheet and your pen/pencil. Using the same qualitative scale as before, how much lotion was transferred to this Data Sheet and your writing instrument? What does this tell you about the ease of transferring the "unseen" by contact?

Basic Growth Media

To cultivate microbes, microbiologists use a variety of growth media. Although these media may be formulated from scratch, they more typically are produced by rehydrating commercially available powdered media. Media that are routinely encountered in the microbiology laboratory range from the widely used, general-purpose growth media, to the more specific selective and differential media used in identification of microbes. In Exercise 1-3 you will learn how to prepare simple general growth media.

EXERCISE **1-3**

Nutrient Broth and Nutrient Agar Preparation

❖ Theory

Nutrient Broth and Nutrient Agar are common media used for maintaining bacterial cultures.[1] To be of practical use, they have to meet the diverse nutrient requirements of routinely cultivated bacteria. As such, they are formulated from sources that supply carbon and nitrogen in a variety of forms—amino acids, purines, pyrimidines, monosaccharides to polysaccharides, and various lipids. Generally, these are provided in digests of plant material (phytone) or animal material (peptone, beef extract, and others). Because the exact composition and amounts of carbon and nitrogen in these ingredients are unknown, general growth media are considered to be **undefined**. They are also known as **complex media**.

In most classes (because of limited time), media are prepared by a laboratory technician. Still, it is instructive for novice microbiologists to at least gain exposure to what is involved in media preparation. Your instructor will provide specific instructions on how to execute this exercise using the equipment in your laboratory.

❖ Application

Microbiological growth media are prepared to cultivate microbes. These general growth media are used to maintain bacterial stock cultures.

❖ In This Exercise

You will prepare 1-liter batches of two general growth media: Nutrient Broth and Nutrient Agar. Over the course of the semester, a laboratory technician will probably do this for you, but it is good to gain firsthand appreciation for the work done behind the scenes!

❖ Materials

Per Student Group

✓ one 2-liter Erlenmeyer flask for each medium made

✓ three or four 500.0 mL Erlenmeyer flasks and covers (can be aluminum foil)

[1] This is a true statement but a bit misleading. The implication is that these will support growth of most bacteria. In fact, recent estimates indicate we have yet to find appropriate media to cultivate over 90% of bacteria in the environment!

- ✓ stirring hot plate
- ✓ magnetic stir bars
- ✓ all ingredients listed below in the recipes (or commercially prepared dehydrated media)
- ✓ sterile Petri dishes (1 liter of agar will require 50 plates filled with 20.0 mL)
- ✓ test tubes (16 mm × 150 mm) and caps (1 liter of medium will make 143 tubes filled with 7.0 mL)
- ✓ balance
- ✓ weighing paper or boats
- ✓ spatulas
- ✓ disinfectant

Medium Recipes

Nutrient Broth

♦ beef extract	3.0 g
♦ peptone	5.0 g
♦ distilled or deionized water	1.0 L
pH 6.6–7.0 at 25°C	

Nutrient Agar

♦ beef extract	3.0 g
♦ peptone	5.0 g
♦ agar	15.0 g
♦ distilled or deionized water	1.0 L
pH 6.6–7.0 at 25°C	

Preparation of Medium

Day One

To minimize contamination while preparing media, clean the work surface with a disinfectant, turn off all fans, and close any doors that might allow excessive air movements. Note: 100 mL volumes may be prepared by student groups to reduce materials required.

Nutrient Agar Tubes

1. Weigh the ingredients on a balance (Figure 1-3). If using dehydrated media, weigh the amount given on the bottle.
2. Suspend the ingredients in 1 liter of distilled or deionized water in the 2-liter flask, mix well, and boil until fully dissolved (Figure 1-4). *Be careful not to allow the ingredients to boil over—it can happen quickly.*
3. Dispense 7.0 mL portions into test tubes and cap loosely (Figure 1-5). If your tubes are smaller than those listed in Materials, adjust the volume to fill 20% to 25% of the tube. Fill to approximately 50% for agar deeps.

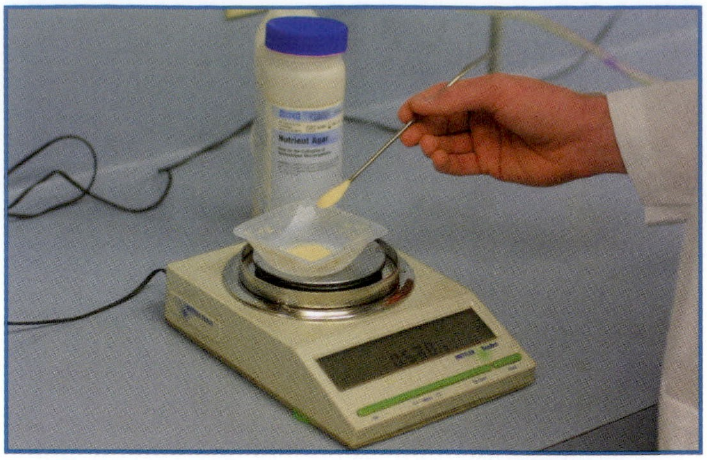

1-3 Weighing Medium Ingredients ♦ Solid ingredients are weighed with an analytical balance. A spatula is used to transfer the powder to a tared weighing boat. Shown here is dehydrated Nutrient Agar, but the weighing process is the same for any powdered ingredient.

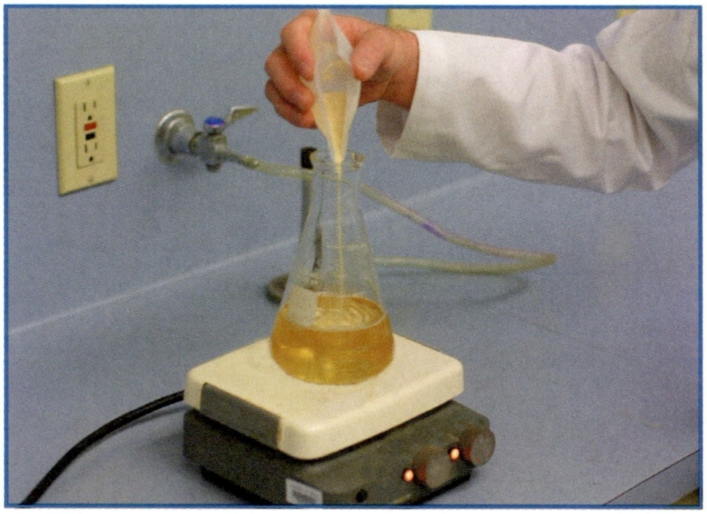

1-4 Mixing the Medium ♦ The powder is added to a flask of distilled or deionized water on a hot plate. A magnetic stir bar mixes the medium as it is heated to a boil to dissolve the powder. *Do not allow it to boil over.*

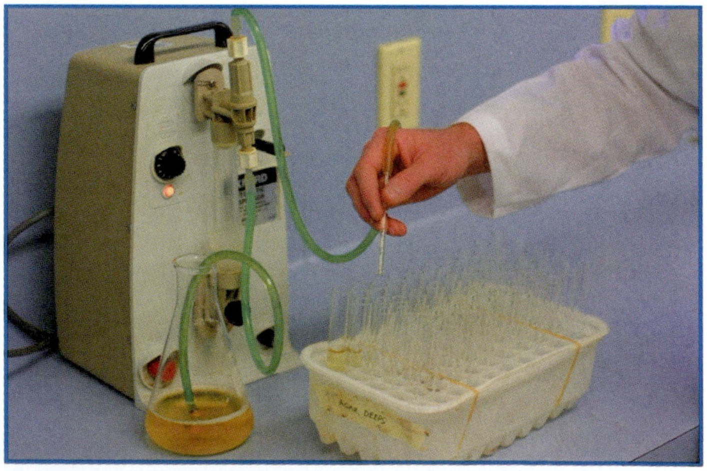

1-5 Dispensing the Medium into Tubes ♦ An adjustable pump can be used to dispense the appropriate volume (usually 7–10 mL) into tubes. Then, loosely cap the tubes.

1-6 Autoclaving the Tubed Medium ◆ The basket of tubes is sterilized for 15 minutes at 121°C in an autoclave. When finished, the tubes are cooled.

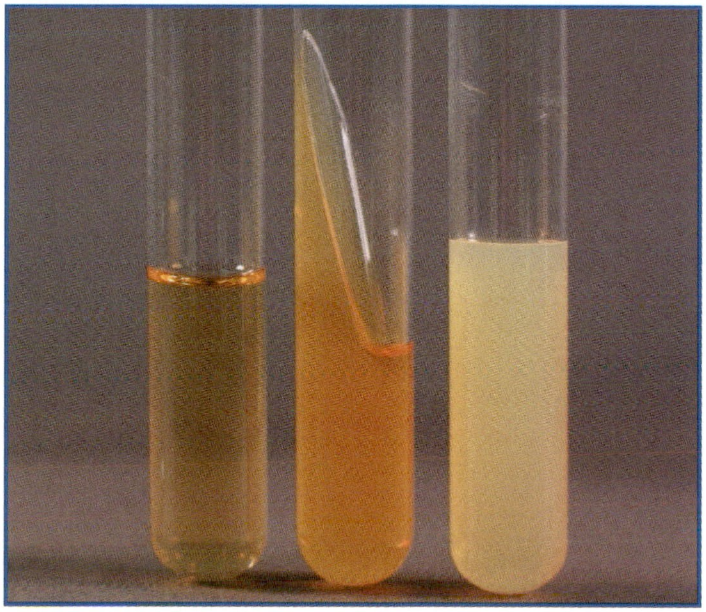

1-7 Tubed Media ◆ From left to right: a broth, an agar slant, and an agar deep tube. The solid media are liquid when they are removed from the autoclave. Agar deeps are allowed to cool and solidify in an upright position, whereas agar slants are cooled and solidified on an angle.

4 Sterilize the medium by autoclaving for 15 minutes at 121°C (Figure 1-6).

5 After autoclaving, cool to room temperature with the tubes in an upright position for agar deep tubes. Cool with the tubes on an angle for agar slants (Figure 1-7).

6 Incubate the slants and/or deep tubes at 35±2°C for 24 to 48 hours.

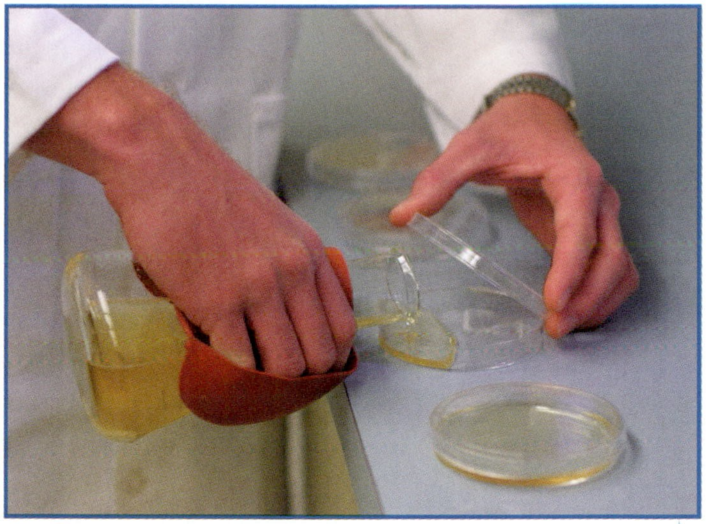

1-8 Pouring Agar Plates ◆ Agar plates are made by pouring sterilized medium into sterile Petri dishes. The lid is used as a shield to prevent airborne contamination. Once poured, the dish is gently swirled so the medium covers the base. Plates are then cooled and dried to eliminate condensation.

Nutrient Agar Plates

1 Weigh the ingredients on a balance (Figure 1-3). If using dehydrated media, weigh the amount given on the bottle.

2 Suspend the ingredients in 1 liter of distilled or deionized water in the 2-liter flask, mix well, and boil until fully dissolved (Figure 1-4). *Be careful not to allow the ingredients to boil over—it can happen quickly.*

3 Divide the mixture into three or four 500.0 mL flasks. Smaller flasks are easier to handle when pouring plates. Don't forget to add a magnetic stir bar and to cover each flask before autoclaving.

4 Autoclave for 15 minutes at 121°C to sterilize the medium (Figure 1-6).

5 Remove the sterile agar from the autoclave, and allow it to cool to 50°C while you are stirring it on a hot plate.

6 Dispense approximately 20.0 mL into sterile Petri plates (Figure 1-8). *Be careful! The flask will still be hot, so wear an oven mitt.* While you pour the agar, shield the Petri dish with its lid to reduce the chance of introducing airborne contaminants. If necessary, *gently* swirl each plate so the agar completely covers the bottom; do not swirl the agar up into the lid. Allow the agar to cool and solidify before moving the plates.

7 Store these plates on a counter top for 24 hours to allow them to dry prior to use.

Nutrient Broth

1. Weigh the ingredients on a balance (Figure 1-3). If using dehydrated media, weigh the amount given on the bottle.
2. Suspend the ingredients in 1 liter of distilled or deionized water in the 2-liter flask. Agitate and heat slightly (if necessary) to dissolve them completely (Figure 1-4). *Be careful not to allow the ingredients to boil over—it can happen quickly.*
3. Dispense 7.0 mL portions into test tubes and cap loosely (Figure 1-5). As with agar slants, if your tubes are smaller than those recommended in Materials, add enough broth to fill them approximately 20% to 25%.
4. Sterilize the medium by autoclaving for 15 minutes at 121°C (Figure 1-6).

Day Two

1. Examine the tubes and plates for evidence of growth.
2. Record your observations on the Data Sheet.

Reference

Zimbro, Mary Jo, and David A. Power. 2003. Pages 404–405 and 408 in *DIFCO™ & BBL™ Manual—Manual of Microbiological Culture Media*. Becton, Dickinson and Company, Sparks, MD.

NAME ______________________ DATE ______________________

LAB SECTION ______________________ I WAS PRESENT AND PERFORMED THIS EXERCISE (INITIALS) ________

Nutrient Broth and Nutrient Agar Preparation

OBSERVATIONS AND INTERPRETATIONS

Record the number of each medium type you prepared, then record the number of apparently sterile ones. Calculate your percentage of successful preparations for each. In the last column, speculate as to probable/possible sources of contamination.

Body Region	Total Number Prepared	Number of Sterile Preparations	Percentage of Successful Preparations	Probable Sources of Contamination (if any)
Nutrient Agar Tubes (Slant or Deep)				
Nutrient Agar Plates				
Nutrient Broths				

Questions

1 Which medium was most difficult to prepare without contamination? Why do think this might be so?

2 For each of the following types of contamination, suggest the most likely point in preparation (or later) at which the contaminant was introduced.

a. Growth in all broth tubes.

b. Growth in one broth tube.

c. Growth only on the surface of a plate.

d. Growth throughout the agar's thickness on a plate.

e. Growth only in the upper 1 cm of agar in an agar deep tube.

f. All plates in a batch have the same type and density of contaminants.

g. Only a few plates in a batch are contaminated, and each looks different.

Aseptic Transfers and Inoculation Methods

As a microbiology student, you will be required to transfer living microbes from one place to another aseptically (i.e., without contamination of yourself, others, the culture, the sterile medium, or the surroundings). While you won't be expected to master all transfer methods right now, you will be expected to perform most of them over the course of the semester. Refer back to this section as needed.

To prevent contamination of the sample, inoculating instruments (Figure 1-9) must be sterilized prior to use. Inoculating loops and needles are sterilized immediately before use in an incinerator or Bunsen burner flame. The mouths of tubes or flasks containing cultures or media are also incinerated at the time of transfer by passing their openings through a flame. Instruments that are not conveniently or safely incinerated, such as Pasteur pipettes, cotton applicators, glass pipettes, and digital pipettor tips are sterilized inside wrappers or containers by autoclaving prior to use.

Aseptic transfers are not difficult; however, a little preparation will help assure a safe and successful procedure. Before you begin, you will need to know where the sample is coming from, its destination, and the type of transfer instrument to be used. These exercises provide step-by-step descriptions of routine transfer methods. Certain less routine transfer methods are discussed in Appendices B through D.

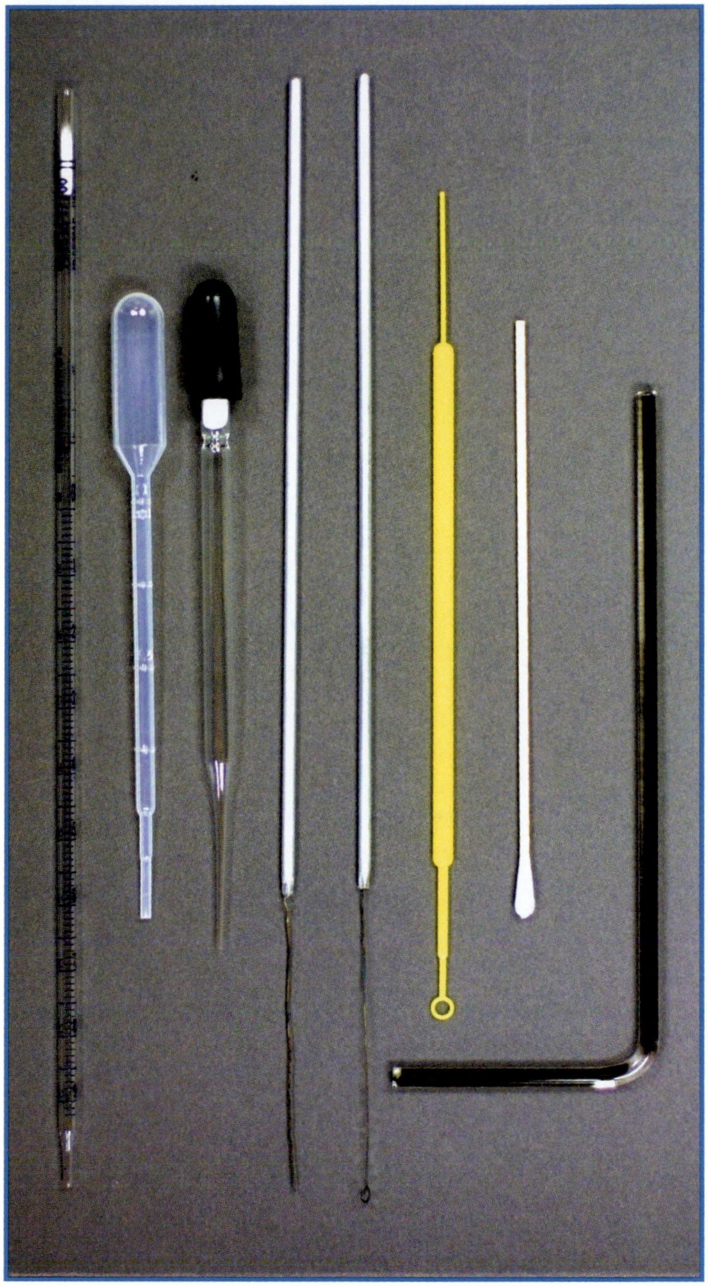

1-9 Inoculating Instruments ◆ Any of several different instruments may be used to transfer a microbial sample, the choice of which depends on the sample source, its destination, and any special requirements imposed by the specific protocol. Shown here are several examples of transfer instruments. From left to right: serological pipette (see Appendix C), disposable transfer pipette, Pasteur pipette, inoculating needle, inoculating loop, disposable inoculating needle/loop, cotton swab (see Appendix B and Exercise 1-4), and glass spreading rod (see Exercise 1-5). (Note: The glass rod is not an inoculation instrument, but it is used to spread an inoculum introduced to an agar plate by another instrument. As such, it is an instrument used in the inoculation process.)

EXERCISE 1-4

Common Aseptic Transfers and Inoculation Methods

Theory

A medium that contains living microbes is called a **culture**. If a culture contains a single species it is said to be a **pure culture**. It is essential to transfer microbes from their pure culture to a sterile medium **aseptically**, that is, without contamination of yourself, others, the environment, the source culture, or the medium being inoculated. In other words, you want your pure culture to stay pure, your new culture to be pure, and the surroundings to remain uninoculated. Following are general techniques and practices that improve your chances of successfully making an aseptic transfer. Equally important are the techniques related to safety. It has been demonstrated that while "…the causative incident for most LAIs (laboratory acquired infections) is unknown…A procedure's potential to release microorganisms into the air as aerosols and droplets is the most important operational risk factor…" (Chosewood and Wilson, 2009). In other words, limiting aerosol production is a *safety issue and not an issue of keeping pure cultures pure*. Please adhere to these practices for your safety and the safety of those around you.

- **Minimize the potential of contamination.** Do not perform any transfers over your books and papers because you may inadvertently and unknowingly contaminate them with droplets or aerosols that settle. Put them safely away. Some labs advise performing transfers over a disinfectant-soaked paper towel.
- **Be organized.** Arrange all media in advance and clearly label them with your name, the date, the medium, and the inoculum (Figure 1-10). Tubes may be labeled with tape, paper held on with a rubber band, or by writing directly on the glass (this option presumes you will clean the writing off before you dispose of the culture for decontamination). You *should* write directly on the base (not the lid) of plastic Petri plates because they are disposable. Be sure not to place any labels in such a way as to obscure or obstruct your view of the tube's or plate's interior.
- **Place all media tubes in a test tube rack when not in use whether they are sterile or not.** Tubes should never be laid on the table surface as they may leak (Figure 1-11).
- **Take your time.** Work efficiently, but do not hurry. You are handling potentially dangerous microbes. Working at a frenzied pace leads to carelessness and accidents.
- **Never hold a tube culture by its cap.** Caps are generally loose to allow aeration and are not secure enough to be used as a handle.
- **Hold the inoculating loop or needle like a pencil in your dominant hand and relax** (Figure 1-12)!
- **Adjust your Bunsen burner so its flame has an inner and outer cone** (Figure 1-13).

Application

Media come in many forms, each with specific applications. **Broths** are used to grow microbes when fresh cultures or large numbers of cells are required. Broths of differential media are also used in microbial identification (Section 5). **Agar slants** are generally used to grow stock cultures that can be refrigerated after incubation and maintained for several weeks. In addition, many differential media are

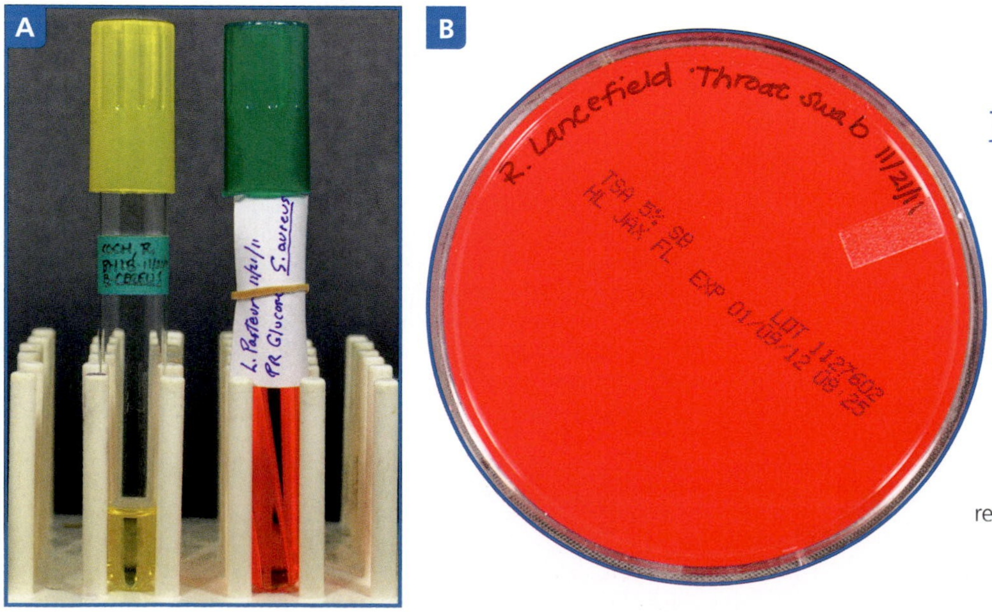

1-10 Label The Media ◆ To avoid confusion after-the-fact, it is best to label sterile media prior to inoculating it. **A** Tubed media can be labeled with tape, paper labels, or directly on the glass with a marking pen. Labels must be removed when tubes are put in the autoclave bin for sterilization. **B** Plastic petri dishes should be labeled on their base, not on their lid, as the lid may get separated from its base during reading or rotated from its correct orientation. Because most labs use disposable Petri dishes, the labels do not have to be removed prior to autoclaving.

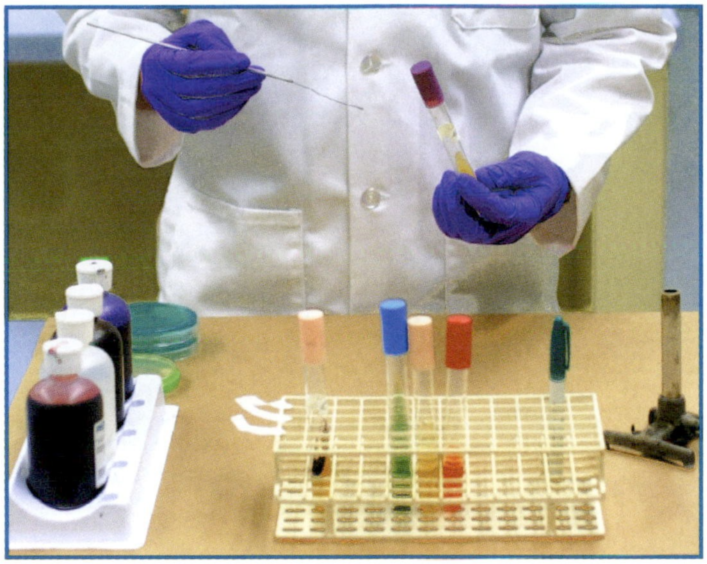

1-11 Microbiologist at Work ◆ Materials are neatly positioned and not in the way. To prevent spills, culture tubes are stored upright in a test tube rack. They are never laid on the table. The microbiologist is relaxed and ready for work. Notice he is holding the loop like a pencil, not gripping it like a dagger.

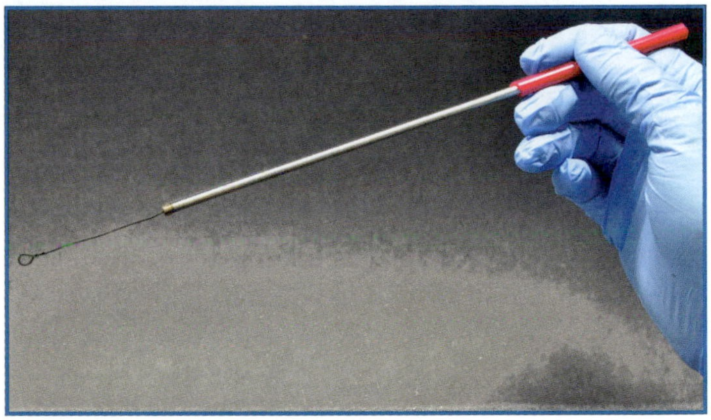

1-12 Hold the Loop Like a Pencil ◆ Holding the loop as shown puts the hand in a convenient position to hold tube caps by the "pinky" finger.

agar slants. **Plated media** are typically used for obtaining isolation of species (Exercises 1-5 and 1-6), differential testing, and quantifying bacterial densities (Exercise 6-1). In all cases, using these media requires aseptic inoculation in which a portion of an existing pure culture is transferred to a sterile medium to start a new pure culture. (A streak plate is an exception to this. See Exercise 1-5.)

Transfers can be made between all forms of media—slants, broths, and plates—depending on the intended use of the new culture. The following is organized into transfers from broth culture to sterile broth, agar slant culture to sterile agar slant, and plate culture to sterile broth. If you can do these, then you have the skills to transfer between most any combination of media. (Inoculation of sterile agar plates is covered in Exercises 1-5 and 1-6 and will complete your skill set.)

The instruments usually used for transfers are either **inoculating loops** or **inoculating needles**. (Pipettes are also sometimes used and these are covered in Appendices C and D.) For simplicity, the following instructions only refer to inoculating loops, but the same apply to inoculating needles.

Transfer from a Broth Culture to a Sterile Broth

As you read these instructions, also follow the Procedural Diagram in Figure 1-14 to get a summary view of the process.

1. Label the sterile broth tube with your name, the date, and the organism you are inoculating it with.
2. Make sure your loop is a closed circle. If it isn't, pinch it closed. Hold it like a pencil, and then flame it from base to tip as shown in Figure 1-15. Be sure the entire wire reaches orange hot at some point.

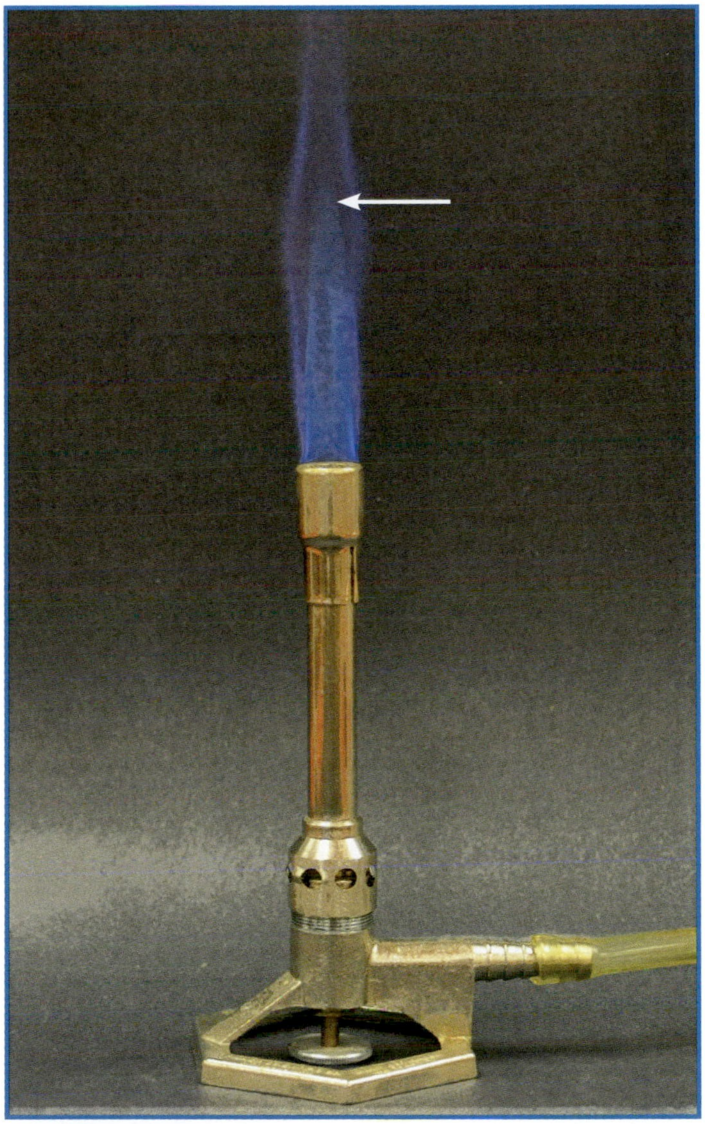

1-13 Bunsen Burner Flame ◆ When properly adjusted, a Bunsen burner produces a flame with two cones. Sterilization of inoculating instruments is done in the hottest part of the flame—the tip of the inner cone (**arrow**). Heat-fixing bacterial smears on slides and incinerating the mouths of open glassware items may be done in the outer cone.

1 Label the sterile broth tube with your name, the date, the medium, and the organism's name

2 Hold the loop like a pencil in your dominant hand

3 Flame the loop and wire from base to tip

4 Pick up the culture tube in your free hand and suspend the organisms in the broth

5 Remove the tube's cap with the little finger of your loop hand

6 Hold the tube on an angle

7 Flame the tube's lip by passing it quickly through the flame two or three times (do not splash the broth out of the open tube)

8 Move the tube over the loop until the loop is in the broth

9 Remove the tube from over the loop being careful not to catch the loop on the tube's lip (the loop should have a film of broth in it)

10 Keeping the loop hand still, flame the tube's lip as before and replace the cap

11 Return the tube to the test tube rack

12 Pick up the sterile broth tube in your free hand

13 Remove the tube's cap with the little finger of your loop hand

14 Hold the tube on an angle

15 Flame the tube's lip by passing it quickly through the flame two or three times

16 Move the tube over the loop until the loop is in the broth

17 Mix the inoculum in the sterile medium

18 Withdraw the tube so the loop is out of the broth, then remove the film in the loop by tapping it on the glass

19 Keeping the loop hand still, flame the tube's lip as before and replace the cap

20 Return the tube to the test tube rack

21 Flame the loop and wire from base to tip

22 Incubate the inoculated culture at the assigned temperature for the assigned time

◆ Actions involving the broth pure culture are in **BLUE**

◆ Actions involving the sterile broth are in **RED**

1-14 Procedural Diagram for an Aseptic Transfer from a Nutrient Broth Pure Culture to a Sterile Nutrient Broth Tube ◆ This is a summary of the procedure. Make every effort to keep your loop hand as still as possible throughout the transfer. Details can be found in the text.

3 Suspend the bacteria in the broth culture with a vortex mixer prior to transfer (Figure 1-16). Be sure not to mix so vigorously that broth gets into the cap or that you lose control of the tube. Start slowly, and then gently increase the speed until the tip of the vortex reaches the bottom of the tube. Alternatively, the culture may be agitated by drumming your fingers along the length of the tube several times (Figure 1-17). Again, be careful not to splash broth into the cap or lose control of the tube.

4 Loosen the cap of the culture tube (this is especially important if you are using a screw-cap tube). Move the culture tube to your loop hand. Remove and hold the tube's cap with the little finger of your loop hand (Figure 1-18). Moving the tube and not the loop prevents excessive movement of the loop that might result in aerosols or droplets.

5 Hold the tube on an angle to minimize the opportunity for airborne contamination (Figure 1-19).

6 Flame the lip of the tube by passing it quickly through the flame two or three times (Figure 1-20). Do not wave it through the flame so fast that broth sloshes out of the open end.

7 This step is important to minimize the production of aerosols. Hold the loop hand still and move the tube up the wire until the loop's tip is in the broth. Continue holding the loop hand still while you remove the tube from over the loop (Figure 1-21). *Be careful not to catch the loop on the lip of the tube or you will produce contaminated droplets and aerosols.* At this point, there should be a visible film of broth in your loop (Figure 1-22). (If there isn't, replace the cap, flame the loop, let it cool, and then pinch the loop so it is a closed circle. Then, start over.)

8 While holding your loop hand still, flame the tube and replace the cap. Set the tube in the rack and pick up the sterile broth tube in your free hand.

9 Repeat the process of flaming the lip of the sterile broth tube while holding the loop hand still. Don't forget you have living microbes exposed to the environment on the loop at this point and careless movements can spread them.

10 Insert the loop into the sterile broth and mix by gently swirling the loop.

11 Before you remove the loop from the tube, tap the loop's face on the inside of the tube to remove the film within the loop (Figure 1-23). Be persistent in this—don't give up until you are successful at its removal. If you were to flame the loop with the film present, you would produce aerosols.

12 Keeping your loop hand still, flame the tube, replace the cap, and set it in the rack.

13 Flame the loop from base to tip until it is uniformly orange hot.

Transfer from Agar Slant Culture to Sterile Agar Slant

As you read these instructions, also follow the Procedural Diagram in Figure 1-24 to get a summary view of the process. This transfer has a lot in common with the broth-to-broth transfer covered above and you are referred back to it at relevant points.

1 Label the sterile agar slant with your name, the date, the medium, and the inoculum.

2 Hold the loop like a pencil and then flame it from base to tip as in a broth transfer (Figure 1-15).

3 Loosen and remove the cap on the culture tube as in a broth transfer (Figure 1-18).

4 Hold the culture tube on an angle with the agar surface facing upward.

5 Flame the tube's lip as in a broth transfer (Figure 1-20).

6 Move the culture tube up the wire of the loop, then gently touch the loop's tip to the growth on the agar's surface (Figure 1-25). You don't need to dig into the agar, nor do you need to scoop up a glob of growth. Just touch the loop to the growth and pick up the smallest amount you can see with your naked eye.

7 Holding your loop hand still, carefully remove the tube from over the wire, flame the culture tube's lip, and replace its cap. Place the culture in the test tube rack.

8 Pick up the sterile agar slant, remove the cap, and flame the tube as before.

9 Move the tube over the wire so the loop is near the bottom of the slant.

10 Touch the tip of the loop to the agar. Then, as you withdraw the tube, move the loop back and forth (Figure 1-26). Be careful not to cut the agar with the loop. This is called a **fishtail inoculation** or **fishtail streak** because you are seeding the agar surface in a wavy pattern resembling the movement of a fish tail.

11 Be careful not to catch the loop on the tube's lip as you remove it. Then, keeping the loop hand still, flame the tube's lip, replace its cap, and put it in the test tube rack.

12 Flame the loop from base to tip as before.

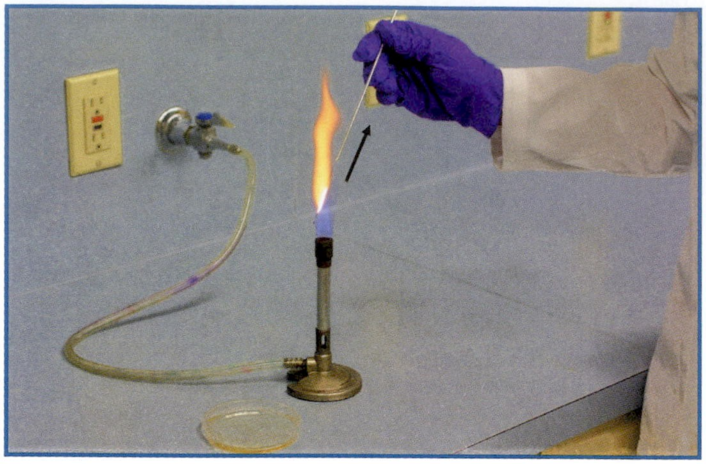

1-15 Flaming Loop ◆ Incineration of an inoculating loop's wire is done by passing it through the tip of the flame's inner cone. Begin at the wire's base and continue to the end, making sure that all parts are heated to a uniform orange color. Allow the wire to cool before touching it or placing it on/in a culture. The former will burn you; the latter will cause aerosols of microorganisms.

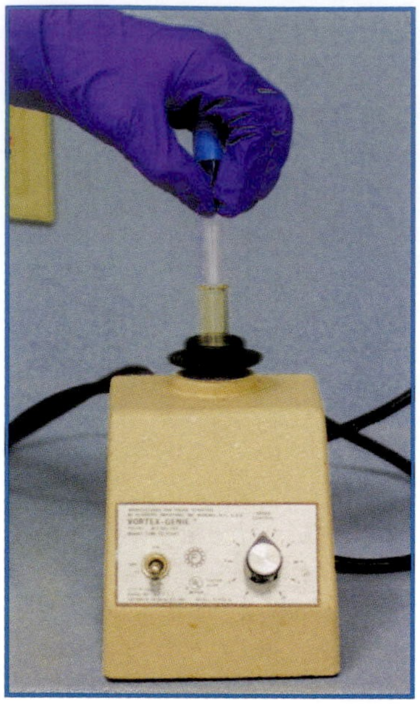

1-16 Vortex Mixer ◆ Bacteria are suspended in a broth with a vortex mixer. The switch on the left has three positions: on (up), off (middle), and touch (down). The rubber boot is activated when touched only if the "touch" position is used; "on" means the boot is constantly vibrating. On the right is a variable speed knob. Caution must be used to prevent broth from getting into the cap or losing control of the tube and causing a spill.

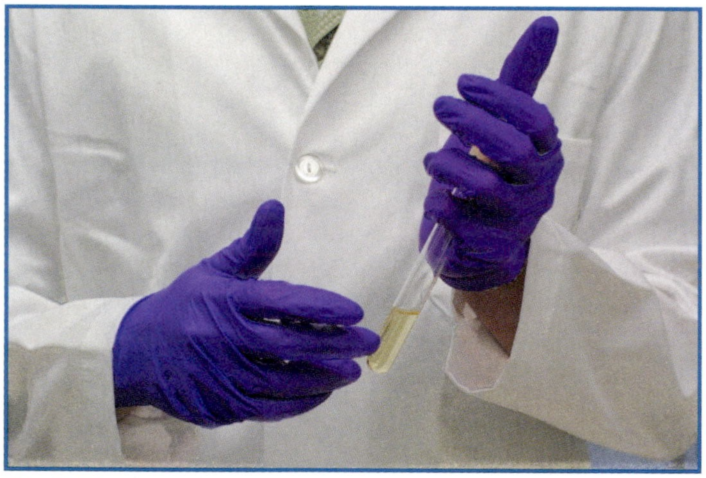

1-17 Mixing Broth by Hand ◆ A broth culture always should be mixed prior to transfer. Tapping the tube with your fingers gets the job done safely and without special equipment.

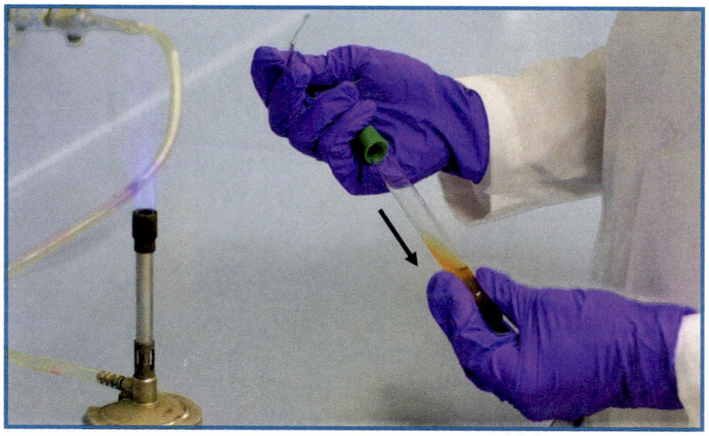

1-18 Removing the Tube Cap ◆ The loop is held in the dominant hand and the tube in the other hand. Remove the tube's cap with your little finger of your loop hand by pulling the tube away with the other hand; keep your loop hand still. Hold the cap in your little finger during the transfer. When replacing the cap, move the tube back to the cap to keep your loop hand still. The replaced cap doesn't have to be on firmly at this time—just enough to cover the tube.

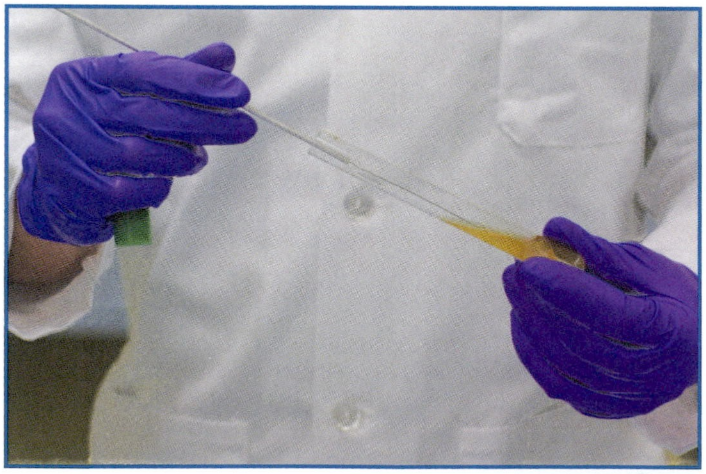

1-19 Holding the Tube at an Angle ◆ The tube is held at an angle to minimize the chance that airborne microbes will drop into it. Notice that the tube's cap is held in the loop hand.

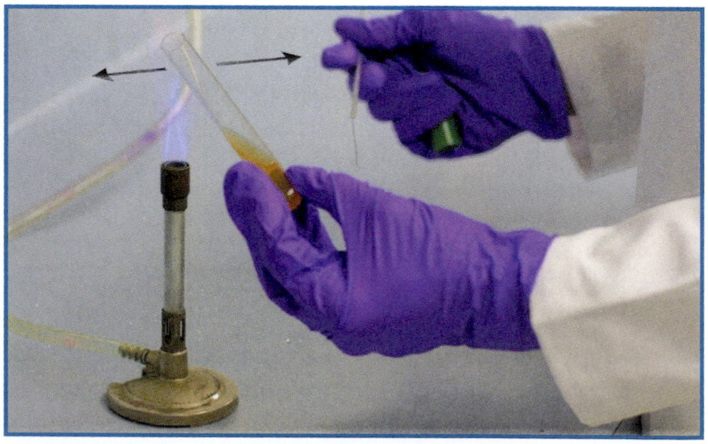

1-20 Flaming the Tube ◆ The tube's mouth is passed quickly through the flame a couple of times to sterilize the tube's lip and the surrounding air. Notice that the tube's cap is held in the loop hand.

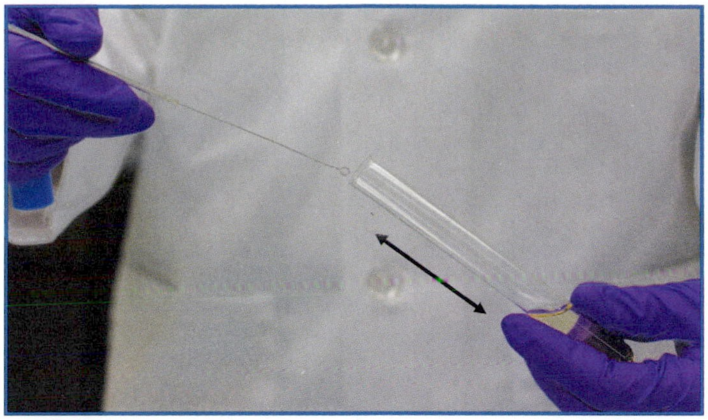

1-21 Loop In/Out of Broth ◆ The open tube is held at an angle to minimize airborne contamination. When placing a loop into a broth tube or removing it, keep the loop hand still and move the tube. *Be careful not to catch the loop on the tube's lip when removing it.* This produces aerosols that can be dangerous or produce contamination.

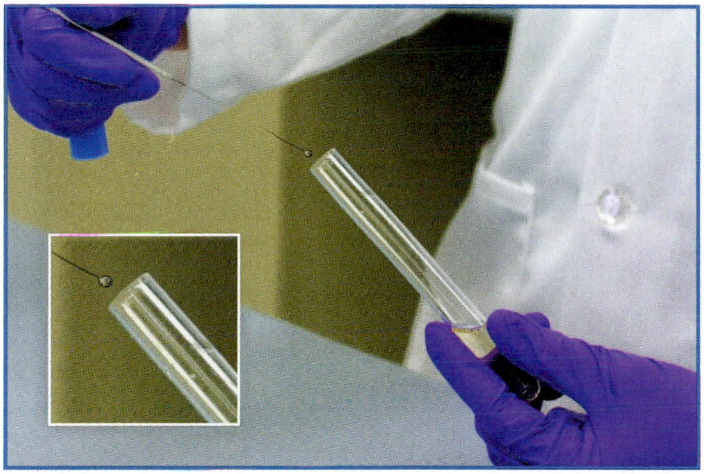

1-22 Removing the Loop from Broth ◆ Notice the film of broth in the loop (see inset). Be careful not to catch the loop on the lip of the tube when removing it. This would produce aerosols that can be dangerous or produce contamination.

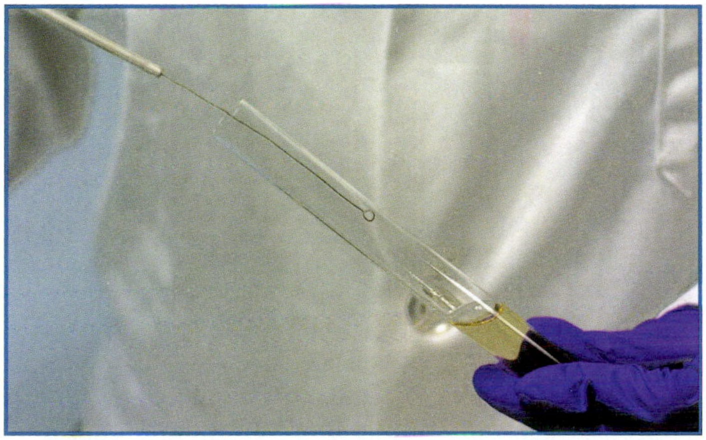

1-23 Removing Excess Broth from Loop ◆ Before removing it from the tube, tap the face of the loop to the glass to remove excess broth. Failing to do so will result in splattering and aerosols when sterilizing the loop in a flame.

Transfers from Plate Culture to Sterile Broth

As you read these instructions, also follow the Procedural Diagram in Figure 1-27 to get a summary view of the process.

1. Flame the loop from base to tip (Figure 1-15).
2. Lift the lid of the Petri dish and use it as a shield from airborne contamination (Figure 1-28).
3. Touch the loop to an uninoculated portion of the plate to cool it. (Loop wires can get very hot if a series of transfers is made in a short period of time. Placing a hot wire on growth may cause the growth to spatter and create aerosols. If you aren't doing a lot of successive transfers, this is probably unnecessary.) Gently touch the loop to some growth on the agar surface and collect the smallest amount you can see. As with the slant, you don't need to dig into the agar or scoop up a glob of growth.
4. Remove the loop and replace the lid.
5. Pick up the sterile broth and continue with Step #9 in the broth-to-broth transfer above.

In This Exercise

You now have the procedures for obtaining growth from a broth culture, an agar slant culture, and a plated culture. You also have the procedures for inoculating a sterile broth and a sterile agar slant. These can be performed in any combination necessary. Today, you will begin by testing your baseline dexterity by using sterile agar and sterile broth as the "culture" tubes and the tubes to be inoculated. After a little practice with these "blanks," you'll do transfers with real cultures and sterile media in the following combinations: agar slant culture to sterile agar slant and sterile broth, broth culture to sterile agar slant and sterile broth, and plate culture to sterile broth.

Materials

Per Student

- ✓ inoculating loop
- ✓ Bunsen burner
- ✓ four sterile Nutrient Broth tubes
- ✓ three sterile Nutrient Agar slants
- ✓ marking pens and labeling tape (or materials for the preferred labeling method in your laboratory)
- ✓ vortex mixer (optional)

Per Lab Group

- ✓ Nutrient Agar (NA) slant culture of *Staphylococcus epidermidis*

1. Label the sterile agar slant with your name, the date, the medium, and the organism's name
2. Hold the loop like a pencil in your dominant hand
3. Flame the loop and wire from base to tip
4. Pick up the culture slant in your free hand
5. Remove the culture tube's cap with the little finger of your loop hand
6. Hold the tube on an angle with the agar surface facing upward
7. Flame the tube's lip by passing it quickly through the flame two or three times
8. Move the tube up the loop's wire until the loop is over some growth
9. Touch the sterile loop to the growth on the agar surface and pick up a small amount of growth
10. Remove the tube from over the loop being careful not to catch the loop on the tube's lip
11. Keeping the loop hand still, flame the tube's lip as before, and replace the cap
12. Return the culture slant to the test tube rack
13. Pick up the sterile agar slant in your free hand
14. Remove the tube's cap with the little finger of your loop hand
15. Hold the tube on an angle with the agar surface facing upward
16. Flame the tube's lip by passing it quickly through the flame two or three times
17. Move the tube upward over the wire so the loop is at the base of the agar
18. Touch the agar with the tip of the loop
19. Slowly remove the tube from over the loop as you move the loop in a zigzag pattern (do not cut the agar)
20. Remove the tube completely from over the loop without catching the loop on the tube's lip
21. Keeping the loop hand still, flame the tube's lip as before and replace the cap
22. Return the inoculated tube to the test tube rack
23. Flame the loop and wire from base to tip
24. Incubate the inoculated culture at the assigned temperature for the assigned time

- Actions involving the agar slant pure culture are in **BLUE**
- Actions involving the sterile agar slant are in **RED**

1-24 Procedural Diagram for an Aseptic Transfer from a Nutrient Agar Slant Pure Culture to a Sterile Nutrient Agar Slant ◆
This is a summary of the procedure. Make every effort to keep your loop hand as still as possible throughout the transfer. Details can be found in the text.

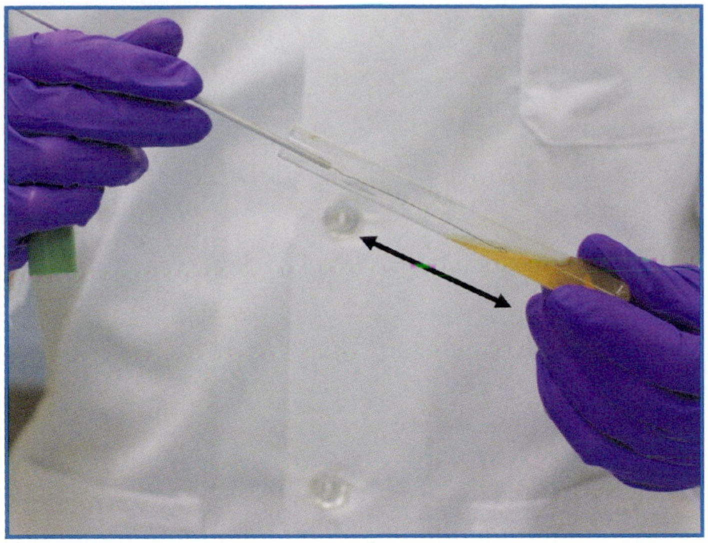

1-25 A Loop and an Agar Slant ◆ When placing a loop into a slant tube or removing it, the loop hand is kept still while the tube is moved. Hold the tube so the agar is facing upward.

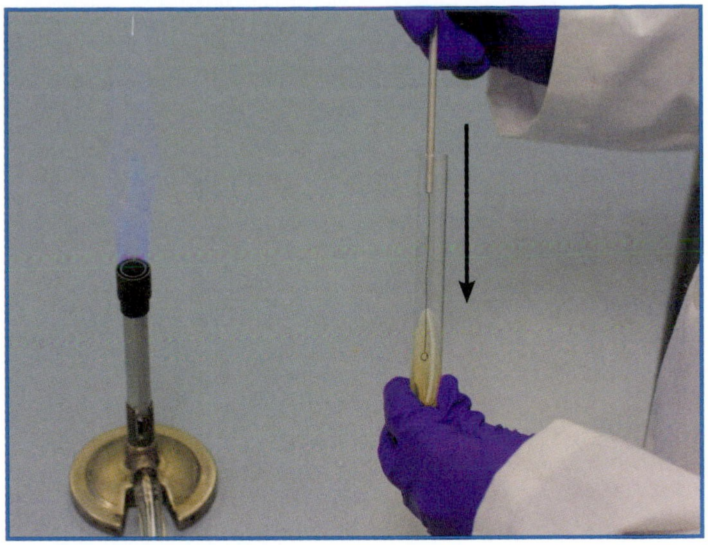

1-26 Fishtail Inoculation of a Slant ◆ Begin at the base of the slant surface and gently move the loop back and forth as you withdraw the tube. Be careful not to cut the agar. After completing the transfer, sterilize the loop.

✓ Nutrient Broth (NB) culture of *Staphylococcus epidermidis*
✓ Nutrient Agar plate of *Staphylococcus epidermidis*

Procedure

Lab One

1 Label one Nutrient Broth (NB) and one Nutrient Agar (NA) tube with your name and the word "sterile."

2 Using the sterile NB tube and the sterile NA tube, practice making transfers between them in all possible combinations. Your lab partner has the same assignment using a second set of tubes. Work with her/him and alternate transferring and evaluating each other's technique. (Not only will you learn by doing and evaluating another person, alternating will allow the tubes to cool.)

3 Incubate the sterile tubes at 35±2°C until the next lab period.

4 Once you have been "cleared" to make transfers using real cultures, perform the following:

a. Label a sterile Nutrient Agar slant and a sterile Nutrient Broth with your name, the medium in the tube (either NA or NB), the source of inoculum (NA slant), and the organism. Then, aseptically transfer from the *S. epidermidis* NA slant culture to the sterile NA slant and the sterile NB.

b. Label a sterile NA slant and a sterile NB with your name, the medium in the tube, the source of inoculum (NB), and the organism. Then, transfer from the *S. epidermidis* NB to the sterile NA slant and the sterile NB.

c. Label a sterile NB with your name, the medium in the tube, the source of inoculum (NA plate), and the organism. Then, transfer from the *S. epidermidis* NA plate culture to the sterile NB. Choose a well-isolated colony and touch the center with the loop as in Figure 1-28.

d. Incubate these five tubes at 35±2°C until the next lab period.

Lab Two

1 Remove your cultures and "sterile" tubes from the incubators and examine them for growth. Record your observations and answer the questions on the Data Sheet.

2 Your instructor may ask you to save your cultures for later use. If so, put them in the refrigerator. Otherwise, remove the labels and dispose of them in the appropriate autoclave container.

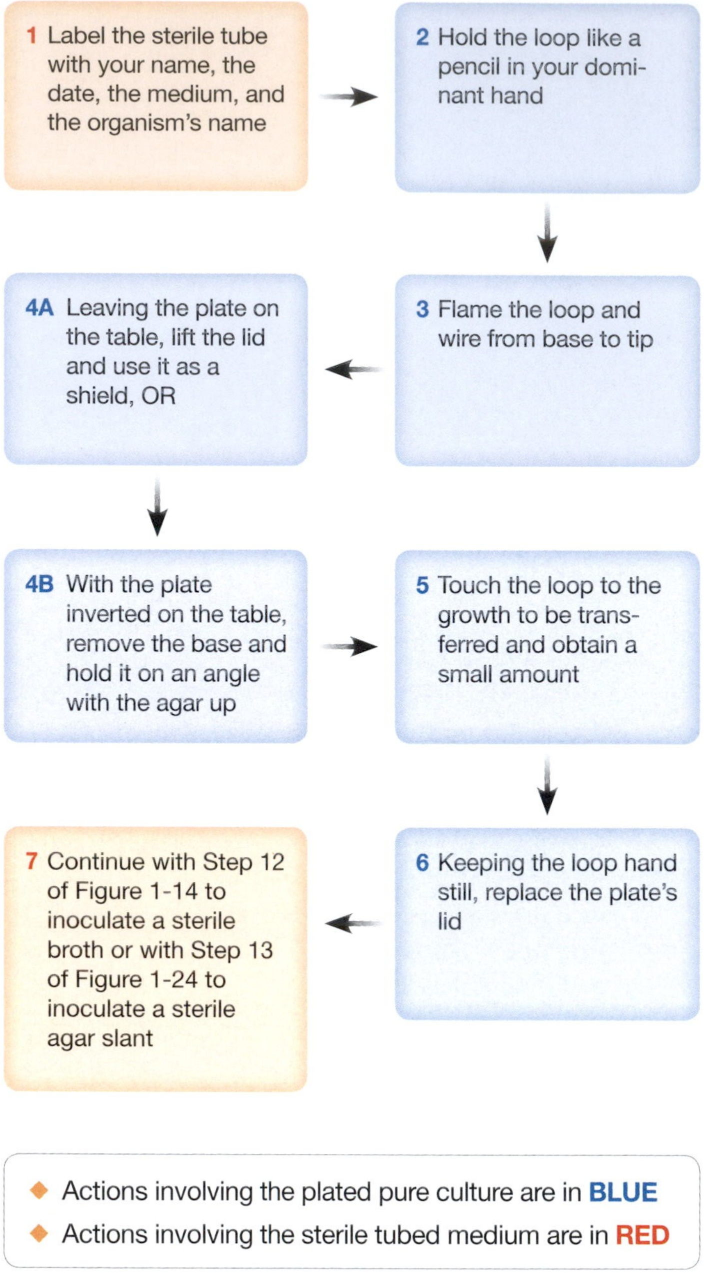
1 Label the sterile tube with your name, the date, the medium, and the organism's name
2 Hold the loop like a pencil in your dominant hand
3 Flame the loop and wire from base to tip
4A Leaving the plate on the table, lift the lid and use it as a shield, OR
4B With the plate inverted on the table, remove the base and hold it on an angle with the agar up
5 Touch the loop to the growth to be transferred and obtain a small amount
6 Keeping the loop hand still, replace the plate's lid
7 Continue with Step 12 of Figure 1-14 to inoculate a sterile broth or with Step 13 of Figure 1-24 to inoculate a sterile agar slant
◆ Actions involving the plated pure culture are in BLUE
◆ Actions involving the sterile tubed medium are in RED

1-27 Procedural Diagram for an Aseptic Transfer from a Nutrient Agar Plate Pure Culture to a Sterile Tubed Medium ◆ This is a summary of the procedure. Make every effort to keep your loop hand as still as possible throughout the transfer. Details can be found in the text. Inoculation of a sterile broth is the same as in Figure 1-14, whereas inoculation of a sterile slant is the same as in Figure 1-24.

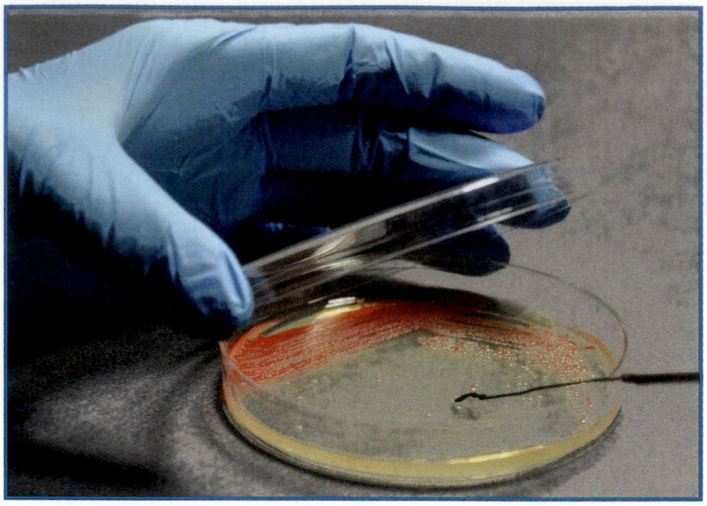

1-28 "Picking" a Colony for Transfer ◆ Touch the tip of the loop to an isolated colony and get a small amount of growth. Use the lid as a shield from airborne contamination.

References

Barkley, W. Emmett, and John H. Richardson. 1994. Chapter 29 in *Methods for General and Molecular Bacteriology*. American Society for Microbiology, Washington, DC.

Chosewood, L. Casey, and Deborah E. Wilson (Editors). December 2009. Page 4 in *Biosafety in Microbiological and Biomedical Laboratories*, 5th ed. U.S. Department of Health and Human Services Publication No. (CDC) 21-1112.

Claus, G. William. 1989. Chapter 2 in *Understanding Microbes—A Laboratory Textbook for Microbiology*. W. H. Freeman and Company, New York, NY.

Darlow, H. M. 1969. Chapter VI in *Methods in Microbiology*, Volume 1. Edited by J. R. Norris and D. W. Ribbins. Academic Press, Ltd., London.

Fleming, Diane O. 1995. Chapter 13 in *Laboratory Safety—Principles and Practices*, 2nd ed. Edited by Diane O. Fleming, John H. Richardson, Jerry J. Tulis, and Donald Vesley. American Society for Microbiology, Washington, DC.

Koneman, Elmer W., Stephen D. Allen, William M. Janda, Paul C. Schreckenberger and Washington C. Winn, Jr. 1997. Chapter 2 in *Color Atlas and Textbook of Diagnostic Microbiology*, 5th ed. Lippincott-Raven Publishers, Philadelphia.

Murray, Patrick R., Ellen Jo Baron, Michael A. Pfaller, Fred C. Tenover, and Robert H. Yolken. 1995. *Manual of Clinical Microbiology*, 6th ed. American Society for Microbiology, Washington, DC.

Power, David A. and Peggy J. McCuen. 1988. *Manual of BBL™ Products and Laboratory Procedures*, 6th ed. Becton Dickinson Microbiology Systems, Cockeysville, MD.

NAME ______________________ DATE ______________________

LAB SECTION ______________________ I WAS PRESENT AND PERFORMED THIS EXERCISE (INITIALS) ________

DATA SHEET 1-4

Common Aseptic Transfers and Inoculation Methods

OBSERVATIONS AND INTERPRETATIONS

Describe the appearance of growth on/in each medium. Growth on a solid medium could be described by color and amount (abundant, sparse, absent). Draw representative samples of each growth type. Growth in broth can be described by its degree of cloudiness (turbidity), using a qualitative scoring system:

+++ means "very turbid"
++ means "somewhat turbid"
+ means "barely turbid"
0 means "not turbid"

Source	Medium Inoculated	
	Nutrient Broth	Nutrient Agar Slant
"Sterile" Media		
Staphylococcus epidermidis on NA Slant		
Staphylococcus epidermidis in NB		
Staphylococcus epidermidis on Sterile NA Plate		

Questions

1 Considering the cultures used to inoculate each medium in this exercise, how many different microbial types should you expect to see in/on each medium?

2 You were asked to describe differences in the appearance of growth in each culture, if present. In which medium was this the most difficult to determine? What made this difficult?

Name ______________________ Date ______________________

Lab Section ______________________ I was present and performed this exercise (initials) ________

3 Which medium was most difficult for you to transfer from? Which medium was most difficult for you to inoculate? Explain your difficulties. (Note: There are no current answers to these questions. They are based on evaluation of your personal experiences.)

4 Did you get growth on/in the sterile NB and NA slant tubes you practiced with? If not, congratulations! If so, where did you see it and what might have been its source(s)?

5 Did you notice a difference in density (turbidity) of growth in NB tubes inoculated from NB and NA slants? Suggest possible reasons why a difference might occur.

6 Did you notice a difference in density of growth on NA slants inoculated from NA slants and NB? Suggest possible reasons why a difference might occur.

EXERCISE 1-5

Streak Plate Methods of Isolation

❖ Theory

A microbial culture consisting of two or more species is said to be a **mixed culture**, whereas a **pure culture** contains only a single species. Obtaining isolation of individual species from a mixed sample is generally the first step in identifying an organism. A commonly used **isolation technique** is the **streak plate** (Figure 1-29).

In the streak plate method of isolation, a bacterial sample (always assumed to be a mixed culture) is streaked over the surface of a plated agar medium. During streaking, the cell density decreases, eventually leading to individual cells being deposited separately on the agar surface. Cells that have been sufficiently isolated will grow into **colonies** consisting only of the original cell type. Because some colonies form from individual cells and others from pairs, chains, or clusters of cells, the term **colony-forming unit** (CFU) is a more correct description of the colony origin.

Several patterns are used in streaking an agar plate, the choice of which depends on the source of inoculum and microbiologist's preference. Although streak patterns range from simple to more complex, all are designed to separate deposited cells (CFUs) on the agar surface so individual cells (CFUs) grow into isolated colonies. A quadrant streak is generally used with samples suspected of high cell density, whereas a simple zigzag pattern may be used for samples containing lower cell densities.

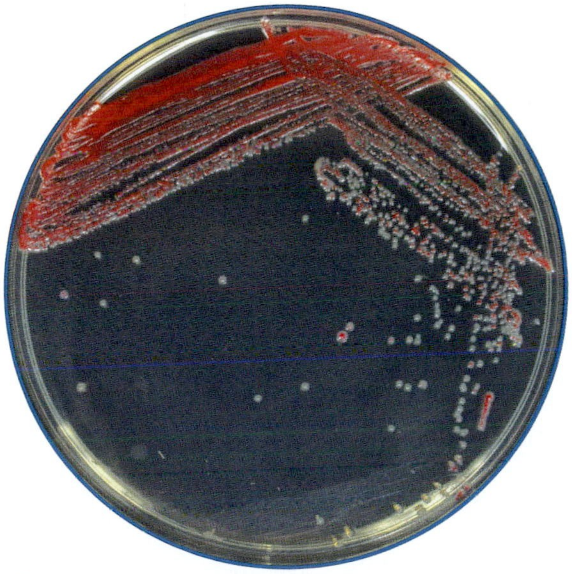

1-29 Streak Plate of *Serratia marcescens* ◆ Note the decreasing density of growth in the four streak patterns. The red colonies are *Serratia marcescens*, the light ones are *Enterobacter aerogenes*. Cells from an individual colony may be transferred to a sterile medium to start a pure culture. Note: the plate was photographed against a dark background.

❖ Application

The identification process of an unknown microbe relies on obtaining a pure culture of that organism. The streak plate method produces individual colonies on an agar plate. A portion of an isolated colony then may be transferred to a sterile medium to start a pure culture.

Following are descriptions of the three different streak techniques.

Inoculation of Agar Plates Using the Quadrant Streak Method

This inoculation pattern is usually performed as the initial streak for isolation of two or more bacterial species in a mixed culture with suspected high cell density.

1. Label the plate to be inoculated with your name, the date, and source of inoculation. Label the plate's base, not the lid.
2. Obtain the sample of mixed culture with a sterile loop.
3. You have two options at this point. Use whichever is more comfortable for you or is required by your instructor.
   a. Leave the sterile agar plate on the table and lift the lid slightly, using it as a shield from airborne contamination (as in Figure 1-28).
   or,
   b. Place the plate lid down on the table (Figure 1-30A). Then remove the base and hold it in the air on an angle (Figure 1-30B).
4. Starting at the edge of the plate lightly drag the loop back and forth across the agar surface as shown in Figure 1-31A. Streaking should be done with the edge of the loop's tip, not the entire face of the loop. Be careful not to cut the agar surface.
5. Remove the loop and replace the lid.
6. Flame your loop from base to tip.
7. Rotate the plate a little less than 90°.
8. Let the loop cool for a few moments (or you can touch an open part of the agar), then perform another streak with the sterile loop beginning at one end of the first streak pattern (Figure 1-31B). Intersect the first streak only two or three times.
9. Flame the loop, then repeat with a third streak beginning in the second streak (Figure 1-31C).

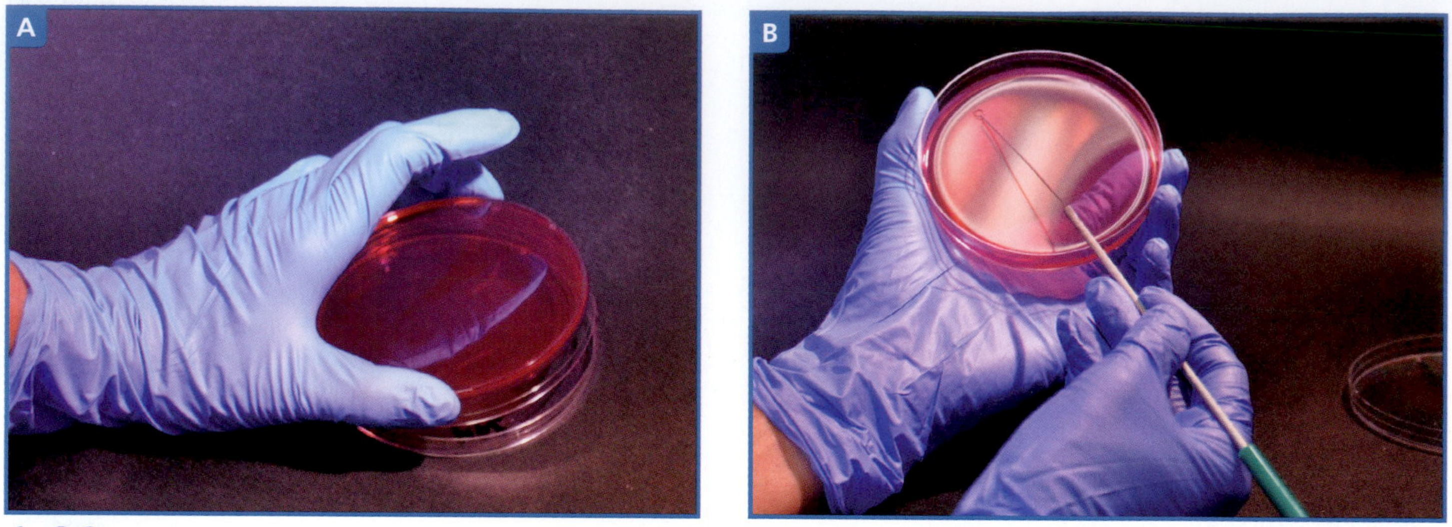

1-30 **Streak Plate Inoculation** ◆ **A** Some microbiologists prefer to hold the Petri dish in the air when performing a streak plate. To do this, place the plate lid down on the table and lift the base from it, holding it on an angle. **B** Perform the streak as described in the text and as shown in Figure 1-31.

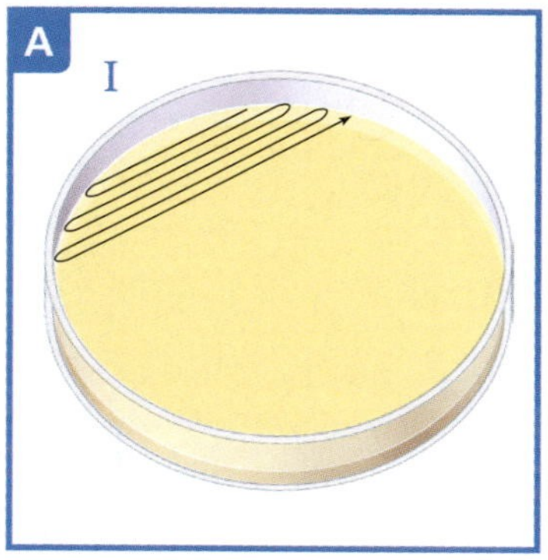

1-31A **Beginning the Streak Pattern** ◆ Label the plate's base. Then, streak the mixed culture back and forth in one quadrant of the agar plate. Do not cut the agar with the loop. Flame the loop, then proceed.

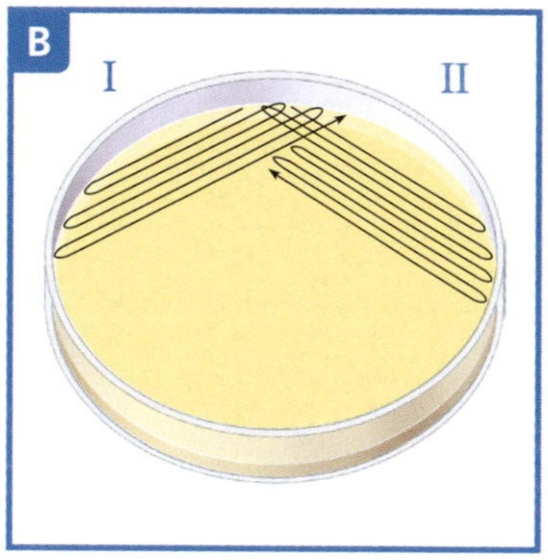

1-31B **Streaking Again** ◆ Rotate the plate nearly 90° and touch the agar in an uninoculated region to cool the loop. Streak again using the same wrist motion. Flame the loop.

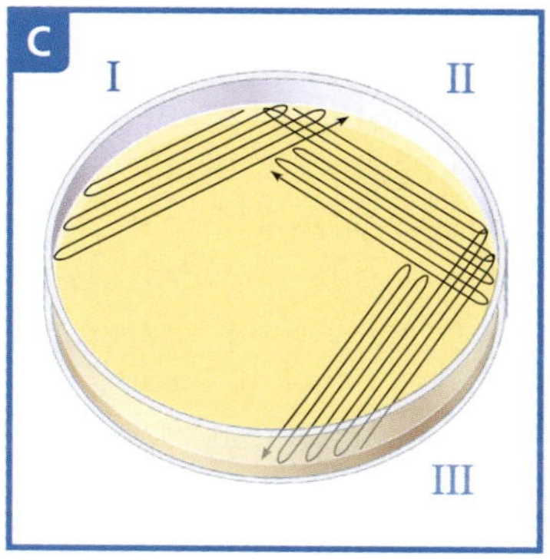

1-31C **Streaking Yet Again** ◆ Rotate the plate nearly 90° and streak again using the same wrist motion. Be sure to cool the loop prior to streaking. Flame again.

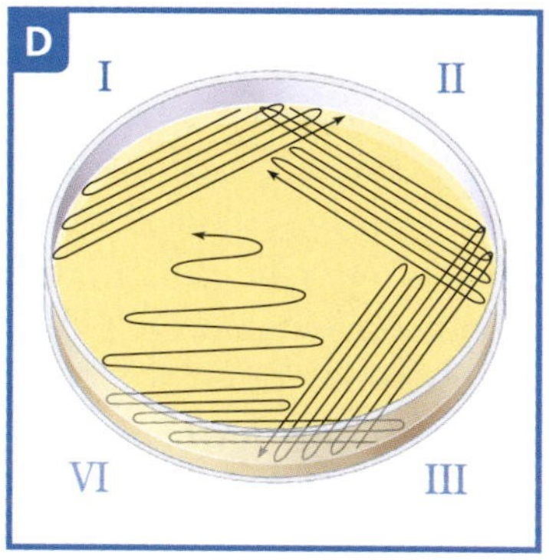

1-31D **Streaking Into the Center** ◆ After cooling the loop, streak one last time into the center of the plate. Flame the loop and incubate the plate in an inverted position for the assigned time at the appropriate temperature.

10 Flame the loop, then perform a fourth streak beginning in the third streak and extending into the middle of the plate. Be careful not to enter any streaks but the third (Figure 1-31D).

11 Flame the loop.

12 Incubate the plate in an inverted position for the assigned time at the appropriate temperature.

Zigzag Inoculation of Agar Plates Using a Cotton Swab or Loop

This inoculation pattern is usually performed when the sample does not have a high cell density. A cotton swab would generally be used when collecting a sample directly from a patient or an environmental surface. A loop would be used if the sample is a liquid obtained from an environmental source, such as pond water, for example.

1 Label the base of the plate with your name, date, and sample.

2 Hold the swab (or loop) comfortably in your dominant hand and lift the lid of the Petri dish with the other. Use the lid as a shield to protect the agar from airborne contamination. Alternatively, the plate can be held in the air as shown in Figure 1-30B.

3 Lightly drag the cotton swab (or loop) across the agar surface in a zigzag pattern. Be careful not to cut the agar surface (Figure 1-32).

4 Replace the lid.

5 Dispose of the swab according to your lab's practices (generally in a sharps container). If a loop was used, flame it.

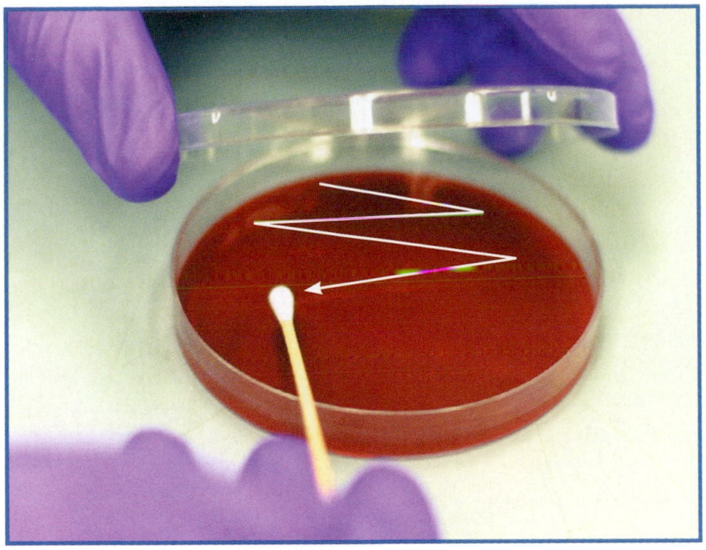

1-32 Zigzag Inoculation ◆ Use the cotton swab to streak the agar surface to get isolated colonies after incubation. Roll the swab as you inoculate, but be careful not to cut the agar. Properly dispose of the swab in a biohazard container.

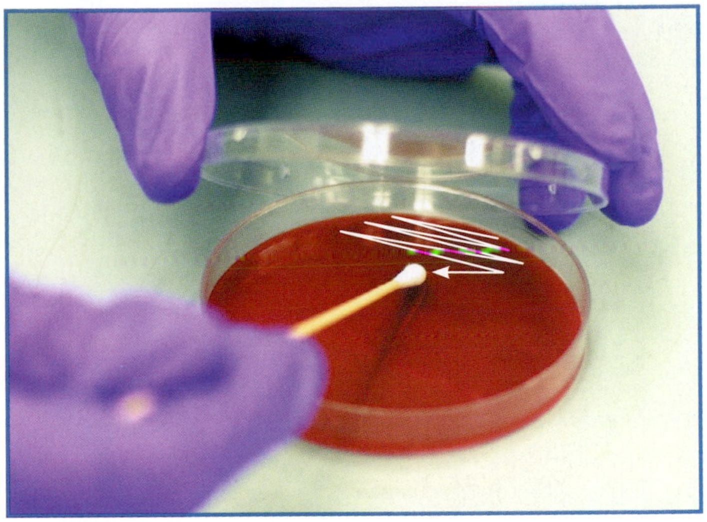

1-33 Inoculation in Preparation for a Quadrant Streak ◆ If the sample is expected to have a high density of organisms, streak one edge of the plate with the swab. Then continue with the quadrant streak, using a loop. Be careful not to cut the agar. Properly dispose of the swab in a biohazard container.

6 Incubate the plate in an inverted position for the assigned time at the appropriate temperature.

Inoculation of Agar Plates with a Cotton Swab in Preparation for a Quadrant Streak Plate

This inoculation pattern is usually performed as the initial streak for isolation of a mixed culture obtained directly from a patient or environmental surface with suspected high cell density.

1 Label the plate's base with your name, date and sample.

2 Hold the swab comfortably in your dominant hand and lift the lid of the Petri dish with the other. Use the lid as a shield to protect the agar from airborne contamination (Figure 1-28).

3 Lightly drag the cotton swab back and forth across the agar surface in one quadrant of the plate (Figure 1-33). This replaces the first streak as shown in Figure 1-31A.

4 Dispose of the swab according to your lab's practices (generally in a sharps container).

5 Further streaking is performed with a loop as shown in Figures 1-31B through 1-31D.

6 Incubate the plate in an inverted position for the assigned time at the appropriate temperature.

In This Exercise

You will learn how to isolate individual organisms from a mixed culture, the first step in producing a pure culture.

Two related streaking techniques will be used, the choice of which is determined by the anticipated cell density of the sample.

Materials

Per Student Pair

- ✓ inoculating loop (one per student)
- ✓ four Trypticase Soy Agar (TSA) or Nutrient Agar (NA) plates
- ✓ two sterile cotton swabs in sterile distilled water
- ✓ sterile microtube
- ✓ two sterile transfer pipettes
- ✓ broth cultures of:
  - *Micrococcus luteus*
  - *Staphylococcus epidermidis*

Procedure

Lab One

Quadrant Streak for Isolation

1 Using a pencil, practice quadrant-streaking the plate on the Data Sheet before trying it with living bacteria. Hint: Keep your wrist relaxed.

2 Each student should label one Nutrient Agar or Trypticase Soy Agar plate with their name, the date, and the two organisms used.

3 Per student pair: Add three drops of *M. luteus* to the sterile microtube. Properly dispose of the transfer pipette.

4 Add three drops of *S. epidermidis* to the microtube. Mix the two species by drawing the entire six drops up into the pipette and dispensing back into the tube several times. Do not dispense so forcefully as to produce bubbles and aerosols.

5 Each student should transfer one drop of the mixture to her/his plate. Properly dispose of the pipette.

6 Using a flamed loop, each student should perform a quadrant streak beginning with the drop of inoculum. Follow the diagrams in Figure 1-31. Begin the streaking process immediately; do not allow the drop to soak into the agar. Avoid digging into or cutting the agar, which ruins the plate and may create dangerous aerosols.

Zigzag Streak for Isolation

1 Each student should use a sterile cotton swab to sample an environmental source (see Appendix B), then do a simple zigzag streak on a Trypticase Soy Agar or Nutrient Agar plate (Figure 1-32). Dispose of the swab in an autoclave container. Label the plate with your name, the date, and the sample source.

Incubation

1 Tape the four plates together, invert them, and incubate them at 25°C for 24 to 48 hours.

Lab Two

1 After incubation, examine the plates for isolation.

2 Compare your streak plates with your lab partner's plates and critique each other's technique. Remember, a successful streak plate is one that has isolated colonies; the pattern doesn't have to be textbook quality—it's just that textbook quality provides you with a greater chance of getting isolation.

References

Collins, C. H., and Patricia M. Lyne. 1995. Chapter 6 in *Collins and Lyne's Microbiological Methods*, 7th ed. Butterworth-Heineman.

Delost, Maria Dannessa. 1997. Chapter 1 in *Introduction to Diagnostic Microbiology*. Mosby, Inc., St. Louis, MO.

Forbes, Betty A., Daniel F. Sahm, and Alice S. Weissfeld. 2002. Chapter 1 in *Bailey & Scott's Diagnostic Microbiology*, 11th ed. Mosby-Yearbook, St. Louis, MO.

Koneman, Elmer W., Stephen D. Allen, William M. Janda, Paul C. Schreckenberger and Washington C. Winn, Jr. 1997. Chapter 2 in *Color Atlas and Textbook of Diagnostic Microbiology*, 5th ed. J. B. Lippincott Company, Philadelphia, PA.

Power, David A., and Peggy J. McCuen. 1988. Pages 2 and 3 in *Manual of BBL™ Products and Laboratory Procedures*, 6th ed. Becton Dickinson Microbiology Systems, Cockeysville, MD.

Name ______________________ Date ______________________

Lab Section ______________________ I was present and performed this exercise (initials) ________

Streak Plate Methods of Isolation

RESULTS AND INTERPRETATIONS

1 Using your pencil, perform a quadrant streak on the "practice plate" below. Rotate the page between streaks as you would an agar plate.

2 Examine the streak plate made from the mixture of two organisms. Have your lab partner write a critique of your isolation technique in the space below. The following should be addressed: Did you obtain isolation? Were the first three streaks near the edges of the plate? Did any streaks intersect streaks they shouldn't have? Was the whole surface of the agar used? Was the agar cut by the loop? Is there evidence of growth not occurring on a streak line? How could this happen?

3 Did you achieve isolation using the quadrant streak? If so, in which streak (1, 2, 3, or 4) did it occur? If you did not achieve isolation, what might you do differently next time to improve your results?

4 Most colonies on streak plates grow from isolated colony-forming units (CFUs). On rare occasions, however, a colony can be a mixture of two different organisms. If a culture is started from this colony (thinking it is pure), correct identification will be next to impossible because the extra organism could confound the identifying test results. How could you verify the purity of a colony? (The answers may vary depending on what experience you have had prior to performing this exercise.) If you found the colony to be a mixture of organisms, what could you do to purify it?

5 Examine the environmental sample. Was the zigzag streak method appropriate to the cell density in that sample?

EXERCISE 1-6

Spread Plate Method of Isolation

❖ Theory

The spread plate technique is a method of isolation in which a diluted microbial sample is deposited on an agar plate and spread uniformly across the surface with a glass rod. With a properly diluted sample, cells (CFUs) will be deposited far enough apart on the agar surface to grow into individual colonies.

❖ Application

After incubation, a portion of an isolated colony can be transferred to a sterile medium to begin a pure culture. The spread plate technique also has applications in quantitative microbiology (see Section 6). Following is a description of the spread plate technique.

Spread Plate Technique

1. Label the plate base with your name, date, organism, and any other relevant information.
2. Arrange the alcohol beaker, Bunsen burner, and agar plate as shown in Figure 1-34. This arrangement minimizes the chances of catching the alcohol on fire.
3. Lift the plate's lid and use it as a shield to protect from airborne contamination.
4. Using an appropriate pipette, deposit the designated inoculum volume on the agar surface. (Please see Appendices C and D for use of pipettes.) From this point, the remainder of steps should be completed within about 15 seconds to prevent the inoculum from soaking into the agar.
5. Properly dispose of the pipetting instrument used to inoculate the medium, because it is contaminated. Each lab has its own specific procedures and your instructor will advise you what to do.
6. Remove the glass spreading rod from the alcohol and pass it through the flame to ignite the alcohol (Figure 1-35). Remove the rod from the flame and allow the alcohol to burn off completely. Do not leave the rod in the flame; the combination of the alcohol and brief flaming are sufficient to sterilize it. *Be careful not to drop any flaming alcohol on the work surface. Be especially careful not to drop flaming alcohol back into the alcohol beaker. If a fire should start, extinguish it according to your lab's guidelines. Notify your instructor for help!*
7. After the flame has gone out on the glass rod, lift the lid of the plate and use it as a shield from airborne contamination. Then touch the rod to the agar surface away from the inoculum to cool it.
8. To spread the inoculum, hold the plate lid with the base of your thumb and index finger and use the tip of your thumb and middle finger to rotate the base (Figure 1-36). At the same time, move the rod in a back-and-forth motion across the agar surface. After

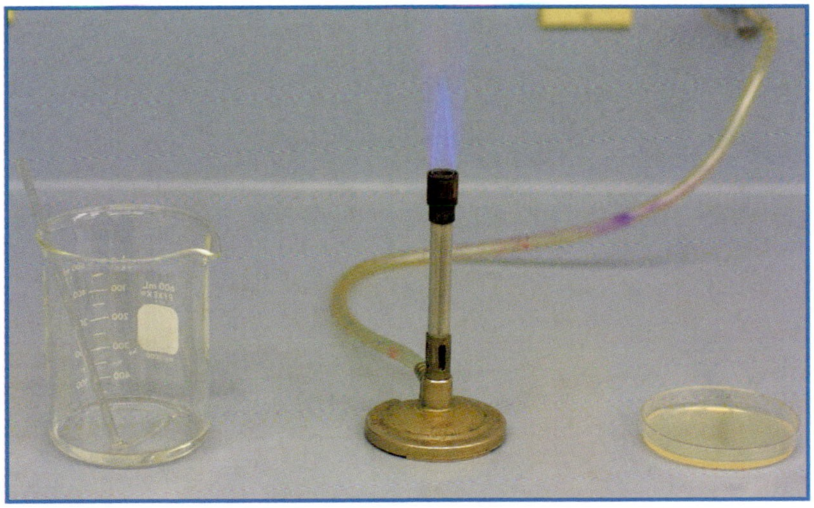

1-34 Spread Plate Set-up ◆ The spread plate technique requires a Bunsen burner, a beaker with alcohol, a glass spreading rod (shown in the beaker), and the plate. Position these components in your work area as shown: isopropyl alcohol, flame, and plate. This arrangement reduces the chance of accidentally catching the alcohol on fire.

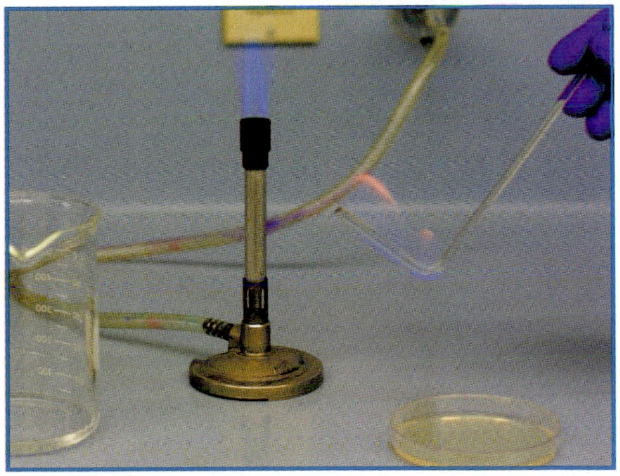

1-35 Flaming the Glass Rod ◆ Remove the glass spreading rod from the alcohol and pass it through the flame to ignite the alcohol. Remove the rod from the flame and allow the alcohol to burn off completely. Do not leave the rod in the flame; the combination of the alcohol and brief flaming are sufficient to sterilize it. *Be careful not to drop any flaming alcohol on the work surface or back into the alcohol beaker.*

a couple of turns, do one last turn with the rod next to the plate's edge. Alternatively, place the plate on a rotating platform and spread the inoculum (Figure 1-37).

9 Remove the rod from the plate and replace the lid.

10 Return the rod to the alcohol in preparation for the next inoculation. There is no need to flame it again.

11 Incubate the plate in an inverted position at the appropriate temperature for the assigned time. (If you plated a volume of inoculum greater than 0.5 mL, wait a few minutes and allow it to soak in before inverting the plate.)

1-36 Spreading Plate Inoculum ◆ After the flame has gone out on the rod, lift the lid of the plate and use it as a shield from airborne contamination. Then, touch the rod to the agar surface away from the inoculum in order to cool it. To spread the inoculum, hold the plate lid with the base of your thumb and index finger, and use the tip of your thumb and middle finger to rotate the base. At the same time, move the rod in a back-and-forth motion across the agar surface. After a couple of turns, do one last turn with the rod next to the plate's edge.

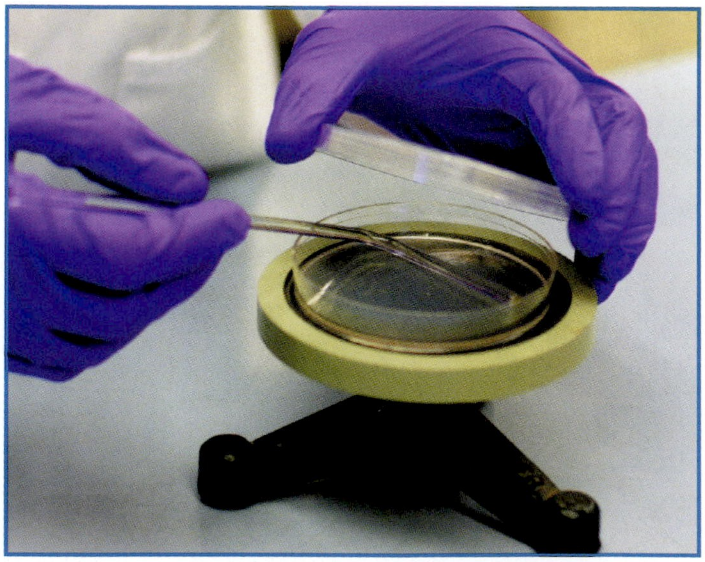

1-37 Inoculating with a Turntable ◆ A turntable makes it easier to rotate the plate during the spread plate technique.

In This Exercise

You will perform a spread plate inoculation. In the context of this exercise, it is used as an isolation procedure, but it also can be used in quantifying cell densities in broth samples (as in Exercise 6-1).

Materials

Per Student Pair

- ✓ inoculating loop (each student)
- ✓ six sterile plastic transfer pipettes
- ✓ 500.0 mL beaker with 50.0 mL of isopropyl alcohol
- ✓ glass spreading rod
- ✓ Bunsen burner and striker
- ✓ four Nutrient Agar plates
- ✓ one sterile microtube
- ✓ four capped microtubes with about 1.0 mL sterile distilled or deionized water (dH_2O)
- ✓ vortex mixer (optional)
- ✓ broth cultures of:
  - *Micrococcus luteus*
  - *Staphylococcus epidermidis*

Procedure

Lab One

1 Using a different pipette for each, transfer a few drops of *M. luteus* and *S. epidermidis* to a sterile microtube. Use the second pipette to mix well by gently drawing and dispensing the mixture in and out of the tube a couple of times. Do not spray the mixture!

2 Label the four microtubes containing sterile dH_2O "A," "B," "C," and "D."

3 Label the four Nutrient Agar plates "A," "B," "C," and "D."

4 Transfer a loopful of the mixture from the microtube to Tube A and mix well with the loop or a vortex mixer. If using a vortex mixer, be sure to cap the tube.[1]

5 Transfer a loopful of the mixture in Tube A to Tube B and mix well with the loop or a vortex mixer.

6 Transfer a loopful of the mixture in Tube B to Tube C and mix well with the loop or a vortex mixer.

[1] Because this is not a quantitative procedure, it is not necessary to flame the loop between transfers.

7 Transfer a loopful of the mixture in Tube C to Tube D and mix well with the loop or a vortex mixer.

8 Using a sterile transfer pipette, place a couple of drops of sample from Tube A on Plate A. Spread the inoculum with a glass rod as described in "Spread Plate Technique" and as shown in Figures 1-34 through 1-37. Let the plate sit for a few minutes.

9 Repeat Step 8 for Tubes B, C, and D and Plates B, C, and D, respectively.

10 Tape the four plates together (be sure they are facing the same direction), invert them, and incubate them at 25°C for 24 to 48 hours.

Lab Two

1 After incubation, examine the plates for isolation. *M. luteus* produces yellow colonies, and *S. epidermidis* produces light colonies.

2 Fill in the Data Sheet and answer the questions.

References

Clesceri, WEF, Chair; Arnold E. Greenberg, APHA; Andrew D. Eaton, AWWA ; and Mary Ann H. Franson. 1998. Pages 9–38 in *Standard Methods for the Examination of Water and Wastewater*, 20th ed. Joint publication of American Public Health Association, American Water Works Association and Water Environment Federation. APHA Publication Office, Washington, DC.

Downes, Frances Pouch, and Keith Ito. 2001. Page 57 in *Compendium of Methods for the Microbiological Examination of Foods*. American Public Health Association. Washington DC.

Gerhard, Philipp, R. G. E. Murray, Willis A. Wood, and Noel R. Kreig. 1994. Pages 255–257 in *Methods for General and Molecular Bacteriology*. American Society for Microbiology, Washington, DC.

NAME ______________________ DATE ______________________

LAB SECTION ______________________ I WAS PRESENT AND PERFORMED THIS EXERCISE (INITIALS) ________

Spread Plate Method of Isolation

OBSERVATIONS AND INTERPRETATIONS

Record your observations in the table below.

Organism	Plate(s) With Isolation	Comments
M. luteus		
S. epidermidis		

Questions

1 On which plate did you obtain isolation with *E. coli*? How about *S. marcescens*? Do you have reason to suspect that they should become isolated on the same dilution plate? Why or why not?

2 Once you obtained isolation at a particular dilution, did you continue to have isolation on subsequent dilution plates? Is this what you would expect? Why or why not?

3 What is the consequence of not spreading the inoculum adequately over the agar surface?

4 To get isolated colonies on a plate, only about 300 cells can be in the inoculum. What will happen if the cell density of the inoculum significantly exceeds this number?

5 Suppose you have two organisms in a mixture and Organism A is 1000 times more abundant than Organism B. Will you (without counting on good luck!) be able to isolate Organism B using the spread plate technique? Explain your answer.

SECTION

2

Microbial Growth

The microbial world is extraordinarily diverse. Every species demonstrates a unique combination of characteristics, a few of which can be easily observed. In this section we illustrate some of those characteristics and factors that affect them.

Allowing for variables, such as the growth medium and incubation conditions, much can be determined about an organism by simply looking at the colonies it produces. Distinguishing different growth patterns is an important skill—one that you can use as you progress through the semester. Note the growth characteristics of all the organisms provided for your laboratory exercises, and jot them down or even sketch them. If and when you are given an unknown species, you will find your records very useful.

You will begin this section with an exercise intended to awaken you to the vast microbial population living all around us. Then you will examine some microbial growth characteristics and cultivation methods. Next you will look at some environmental factors affecting microbial growth, including pH, oxygen, temperature, and osmotic pressure. Finally, you will examine some physical and chemical microbial control agents and systems.

Diversity and Ubiquity of Microorganisms

Microorganisms are found everywhere that other forms of life exist. They can be isolated from soil, bodies of water, even from the air. As unwanted parasites or colonizers, some microorganisms cause diseases or infections. Most, however, are harmless saprophytes that live in, on, or around plants and animals and decompose dead organic matter. In so doing, they perform the essential function of nutrient recycling in ecosystems.

In this section, you will grow microorganisms from seemingly uninhabited sources. Then you will learn to identify the various growth characteristics that these "invisible" cohabitants produce when they are cultivated in broth and on solid media.

EXERCISE **2-1**

Ubiquity of Microorganisms

❖ Theory

In the literature on microorganisms, a frequently encountered phrase is "ubiquitous in nature." This means that the organism being considered can be found just about everywhere. More specifically, the organism likely could be isolated from soil, water, plants, and animals (including humans). Although the word *ubiquitous* doesn't apply to every species, it does apply to many, and certainly to microorganisms as a group.

Many microorganisms are **free-living—**they do not reside on or in a specific plant or animal **host** and are not known to cause disease. They are **nonpathogenic**. Other microorganisms are **pathogens** and generally are associated with their host (or hosts). Even many of the **commensal** or **mutualistic** strains inhabiting our bodies are **opportunistic pathogens**. That is, they are capable of producing a disease state if introduced into a suitable part of the body. Any area, including sites outside the host organism, where a microbe resides and serves as a potential source of infection is called a **reservoir**.

❖ Application

This exercise is designed to demonstrate the ubiquitous nature of microorganisms and the ease with which they can be cultivated. (It should be noted that although we can find living microorganisms virtually everywhere and confirm their presence by cultivation, molecular techniques developed over the last two or three decades demonstrate that the organisms successfully grown in the lab represent a minute fraction of those still uncultivable.)

❖ In This Exercise

Today, you will work in small groups to sample and culture several locations in your laboratory. Your instructor may have other locations outside of the lab to sample as well. Remember that even relatively "harmless" bacteria, when cultivated on a growth medium, are in sufficient numbers to constitute a health hazard. Treat them with care.

❖ Materials

Per Student Group

- ✓ eight Nutrient Agar plates
- ✓ one sterile cotton swab

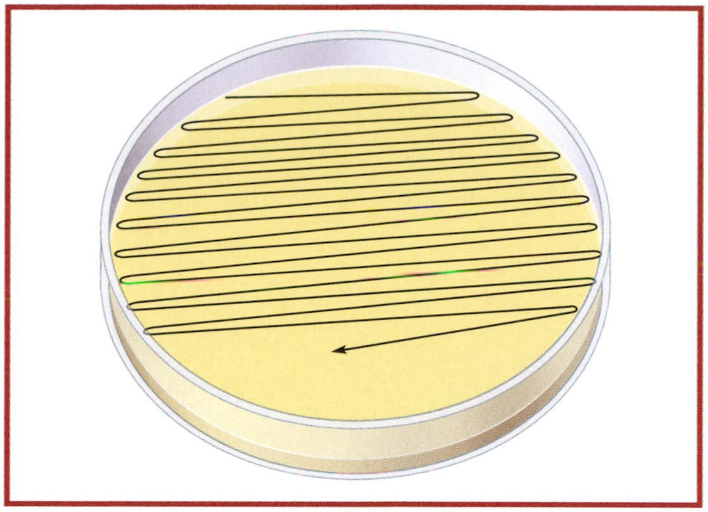

2-1 **Simple Streak Pattern on Nutrient Agar** ◆ Roll the swab as you inoculate the plate. Do not press so hard that you cut the agar.

Procedure

Lab One

1. Number the plates 1 through 8.
2. Open plate number 1 and expose it to the air for 30 minutes or longer. Set it aside and out of the way of the other plates.
3. Use the cotton swab to sample your desk area and then streak plates 2 and 3 in the pattern shown in Figure 2-1. (Pressing very lightly, roll the swab on the agar as you streak it.)
4. Cough several times on the agar surface of plate 4.
5. Rub your hands together, and then touch the agar surface of plate 5 lightly with your fingertips. (A light touch is sufficient; touching too firmly will crack the agar.)
6. Remove the lid of plate 6 and vigorously scratch your head above it. (Keep this plate away from plate 1 to avoid cross-contamination.)
7. Leave plates 7 and 8 covered; do not open them.
8. Label the base of each plate with the date, type of exposure it has received, and the name of your group.
9. Invert all plates and incubate them for 24 to 48 hours at the following temperatures:

Plates 1, 2, and 8:	25°C
Plates 3, 4, 5, 6, and 7:	37°C

Lab Two

1. Using the plate diagrams on the Data Sheet, draw the growth patterns on each of your agar plates. Be sure to label them according to incubation time, temperature, and source of inoculum.
2. Save these plates in a refrigerator for use in Exercise 2-2.

References

Forbes, Betty A., Daniel F. Sahm, and Alice S. Weissfeld. 2002. Chapter 10 in *Bailey & Scott's Diagnostic Microbiology*, 11th ed. Mosby, St. Louis.

Holt, John G. (Ed.). 1994. *Bergey's Manual of Determinative Bacteriology*, 9th ed. Williams and Wilkins, Baltimore.

Winn, Washington C., et al. 2006. *Koneman's Color Atlas and Textbook of Diagnostic Microbiology*, 6th ed. Lippincott Williams & Wilkins, Baltimore.

Varnam, Alan H., and Malcolm G. Evans. 2000. *Environmental Microbiology*. ASM Press, Washington, DC.

NAME ______________________________ DATE ______________________________

LAB SECTION ______________________ I WAS PRESENT AND PERFORMED THIS EXERCISE (INITIALS) ______

DATA SHEET 2-1

Ubiquity of Microorganisms

OBSERVATIONS AND INTERPRETATIONS

1 Use the circles below as Petri dishes. Then, for each plate, choose two different colonies and draw each as seen from above and from the side. Label the plates according to incubation time, temperature, and source of inoculum. Also include other useful colony information, such as color and relative abundance.

2 Save the plates for Exercise 2-2.

Questions

1 What was the purpose of incubating the unopened plates? Be specific. What is an appropriate name for these plates?

Name ______________________ Date ______________________

Lab Section ______________________ I was present and performed this exercise (initials) ________

DATA SHEET 2-1

2 If growth appears on both unopened plates, what are some likely explanations? What if growth appears on only one plate? How does growth on the unopened plates affect the reliability (your interpretation) of the other plates?

3 Why were the specific types of exposure (air, hair, tabletop, etc.) chosen for this exercise?

4 Why were you asked to incubate the plates at two different temperatures? Be specific. What is the likely source (reservoir) of organisms that grew best at 37°C, and how do you think they survive at room temperature without nutrients?

5 Did you get different appearing colonies on plates 2 and 3? If so, explain why.

6 The plates you are using for this lab will be autoclaved eventually to completely sterilize them. The measures taken to disinfect the tabletops (the source of the organisms on plates 2 and 3) are not as extreme. Why?

EXERCISE 2-2

Colony Morphology

Theory

When a single bacterial cell is deposited on a solid nutrient medium, it begins to divide. One cell makes two, two make four, four make eight . . . one million make two million, and so on. Eventually a **colony** appears where the original cell was deposited. Color, size, shape, and texture of microbial growth are determined by the genetic makeup of the organism, but are also greatly influenced by environmental factors including nutrient availability, temperature, and incubation time.

The basic categories of colony morphology include colony shape, margin (edge), elevation, texture, and color.

1. *Shape* may be described as **circular**, **irregular**, or **punctiform** (tiny).
2. The *margin* may be **entire** (smooth, with no irregularities), **undulate** (wavy), **lobate** (lobed), **filamentous**, or **rhizoid** (branched like roots).
3. *Elevations* include **flat, raised, convex, pulvinate** (very convex), and **umbonate** (raised in the center).
4. *Texture* may be **moist**, **mucoid**, or **dry**.
5. *Pigment production* (color), is another useful characteristic and may be combined with optical properties such as **opaque**, **translucent**, **shiny**, or **dull**.

Application

Recognizing different bacterial growth morphologies on agar plates is a useful step in the identification process. Once purity of a colony has been confirmed by an appropriate staining procedure, cells can be cultivated and maintained on sterile media for a variety of purposes.

In This Exercise

Today you will be viewing colony characteristics on the plates saved from Exercise 2-1 and (if available) prepared streak plates provided by your instructor. Colony characteristics may be viewed with the naked eye or with the assistance of a colony counter (Figure 2-2). Figures 2-4 through 2-33 show a variety of bacterial colony forms and characteristics. Where applicable, contrasting environmental factors are indicated.

Materials

Per Student Group

- ✓ colony counter (optional)
- ✓ dissecting microscope (optional)
- ✓ metric ruler
- ✓ plates from Exercise 2-1
- ✓ (optional) streak plate cultures of any of the following:
  - *Micrococcus luteus*
  - *Corynebacterium xerosis*
  - *Lactobacillus plantarum*
  - *Mycobacterium smegmatis* (BSL-2)
  - *Bacillus subtilis*
  - *Proteus mirabilis* (BSL-2)

Procedure

1 Using the terms in Figure 2-3, describe the colonies on your plates. Measure colony diameters (in mm) with a ruler and include them with your descriptions on the Data Sheet. (Note: Remember that many microorganisms are opportunistic pathogens, so be sure to handle the plates carefully. Do not open plates containing fuzzy growth, as a fuzzy appearance suggests fungal growth containing spores that can spread easily and contaminate the laboratory and other cultures. If you are in doubt, check with your instructor.)

2 Unless you have been instructed to save today's cultures for future exercises, discard all plates in an appropriate autoclave container.

2-2 The Colony Counter ◆ Subtle differences in colony shape and size can best be viewed on the colony counter. The transmitted light and magnifying glass allow observation of greater detail, however, colony color is best determined with reflected light. The grid in the background is used as a counting aid. Each big square is a square centimeter.

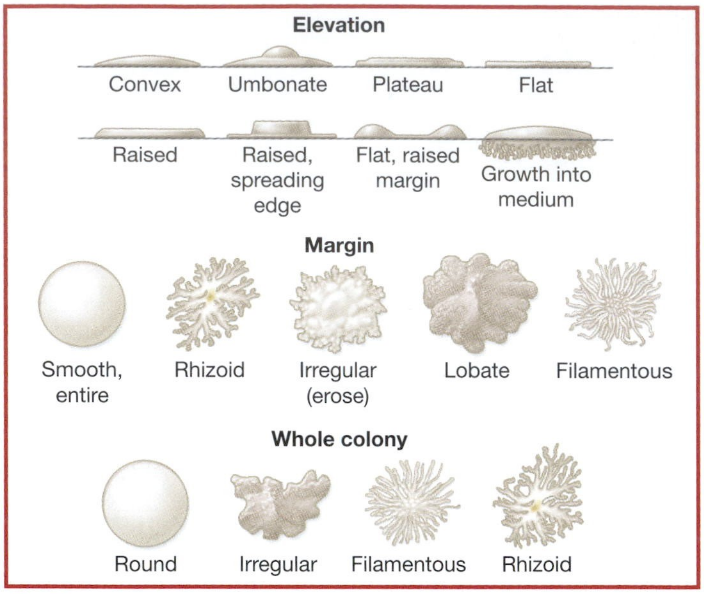
Elevation
Convex
Umbonate
Plateau
Flat
Raised
Raised, spreading edge
Flat, raised margin
Growth into medium
Margin
Smooth, entire
Rhizoid
Irregular (erose)
Lobate
Filamentous
Whole colony
Round
Irregular
Filamentous
Rhizoid

2-3 A Sampling of Bacterial Colony Features ◆ These terms are used to describe colonial morphology. Descriptions also should include color, surface characteristics (dull or shiny), consistency (dry, butyrous-buttery, or moist) and optical properties (opaque or translucent).

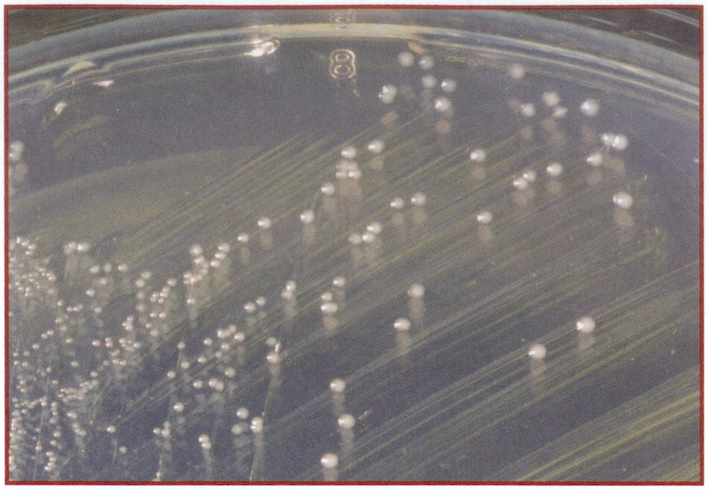

2-4 *Enterococcus faecium* Grown on Nutrient Agar ◆ The colonies are white, circular, convex, and have an entire margin. *E. faecium* (formerly known as *Streptococcus faecium*) is found in human and animal feces.

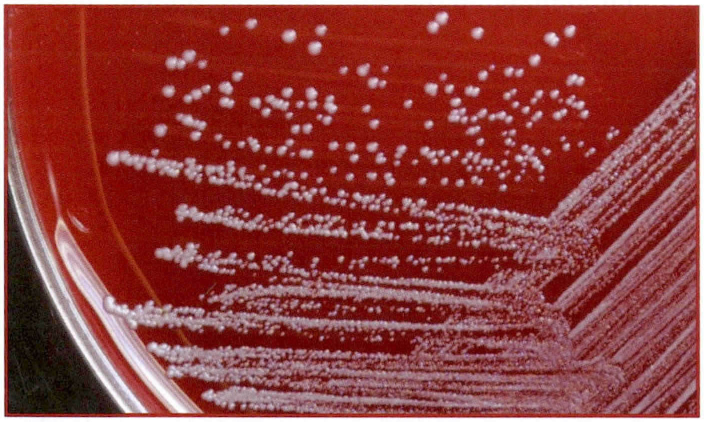

2-5 *Staphylococcus epidermidis* Grown on Sheep Blood Agar ◆ The colonies are white, raised, circular, and entire. *S. epidermidis* is an opportunistic pathogen.

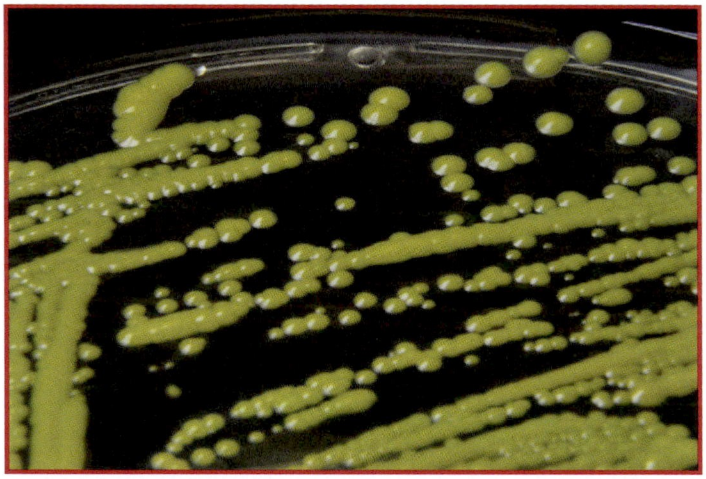

2-6 *Micrococcus luteus* Grown on Brain–Heart Infusion Agar ◆ These colonies are yellow, smooth, and convex with a regular margin. They range in size from 1 to 3 mm. *M. luteus* is common in soil, dust, and on human skin.

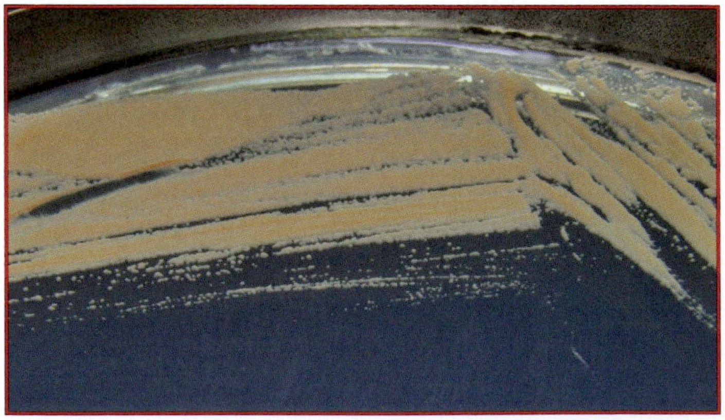

2-7 *Kocuria rosea* Grown on Brain–Heart Infusion Agar ◆ Pink, punctiform (these are less than 1 mm in diameter), smooth, regular colonies typify *K. rosea*, an inhabitant of water, dust, and salty foods.

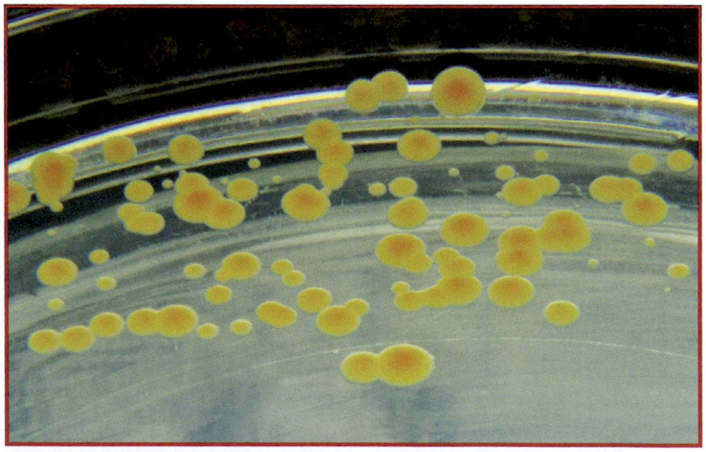

2-8 *Sarcina aurantiaca* Grown on Brain–Heart Infusion Agar ◆ *S. aurantiaca* produced yellow-orange, smooth, convex, regular colonies on BHIA. These are between 1 to 3 mm in diameter.

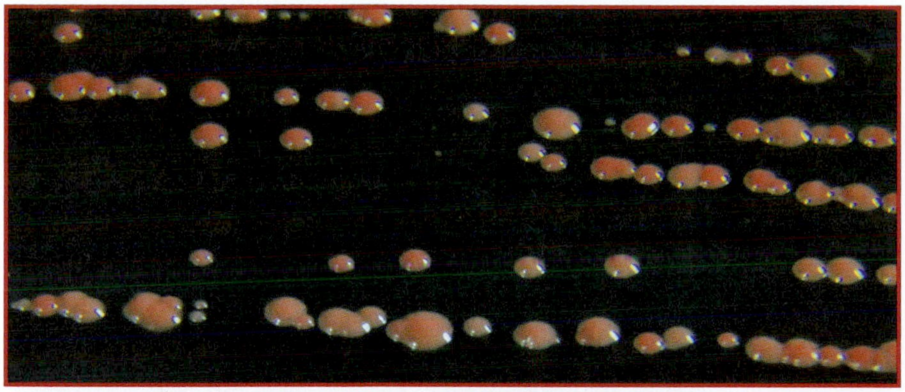

2-9 ***Rhodococcus rhodochrous* Grown on Brain–Heart Infusion Agar** ◆ These colonies are reddish-pink, smooth, and convex with a regular margin. They are about 1 mm in diameter. *Rhodococcus* species are soil organisms.

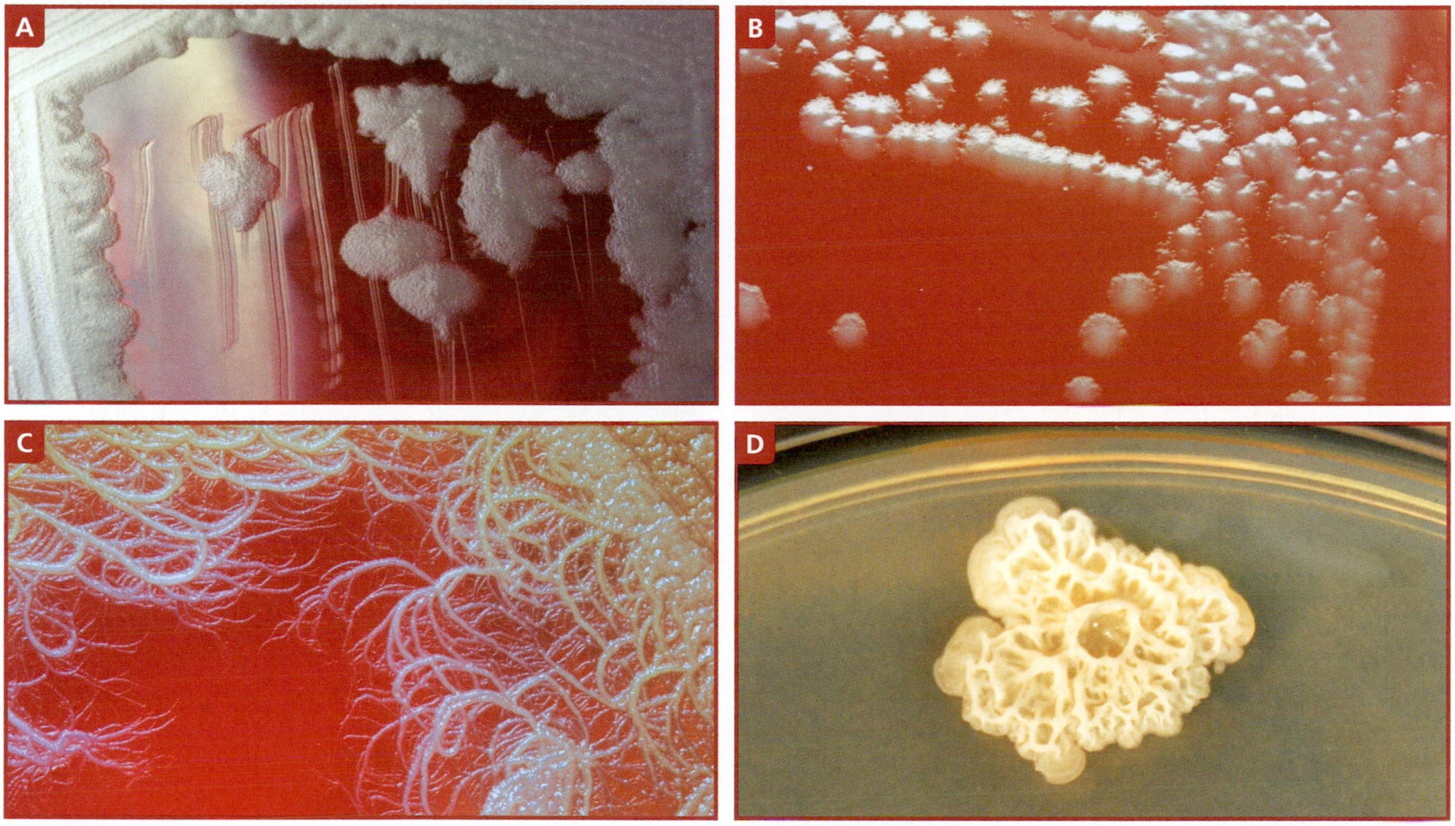

2-10 Comparison of Four *Bacillus* Species ◆ **A** *B. cereus* grown on Blood Agar produces distinctively large (up to 7 mm), gray, granular, irregular colonies. They often produce a "mousy" smell. Also note the extensions of growth along the streak line. **B** *B. anthracis* colonies resemble *B. cereus*, but are usually smaller and adhere to the medium more tenaciously. **C** *B. mycoides* produces rapidly spreading, rhizoid colonies. **D** This unknown *Bacillus* isolated as a laboratory contaminant produced wrinkled, irregular colonies with an undulate (wavy) margin.

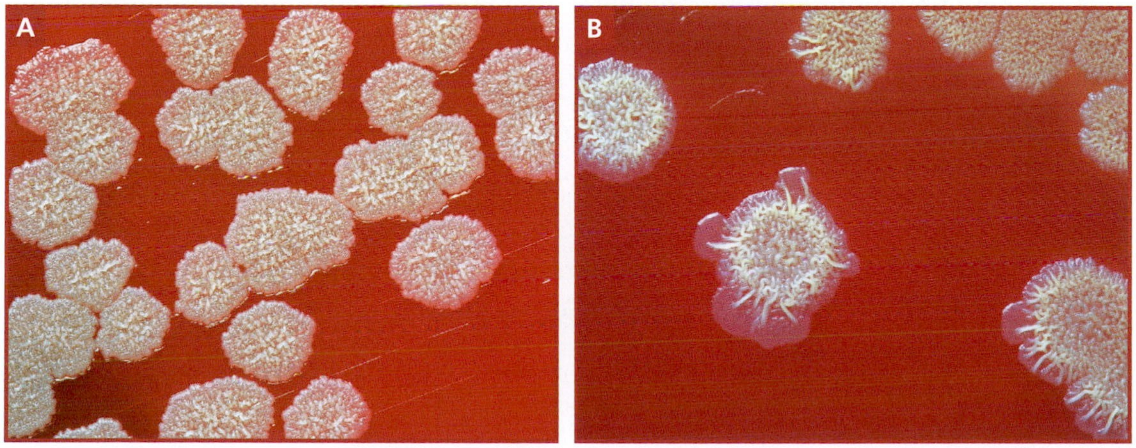

2-11 Effect of Age on Colony Morphology ◆ **A** Close-up of *Bacillus subtilis* on Sheep Blood Agar after 24 hours of incubation. **B** Close-up of *B. subtilis* on Sheep Blood Agar after 48 hours. Note the wormlike appearance.

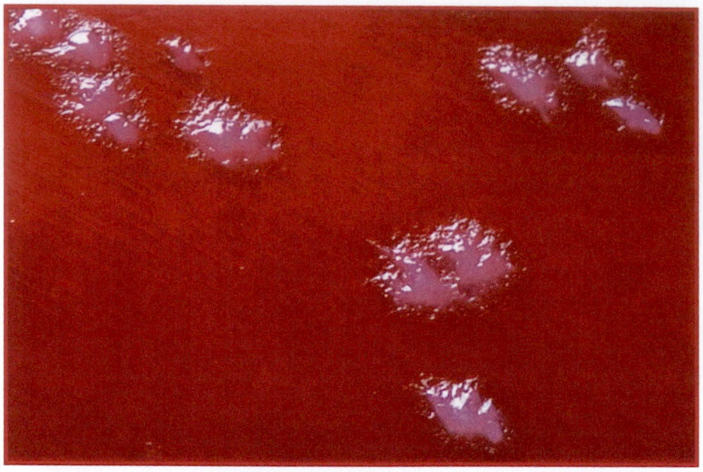

2-12 ***Clostridium sporogenes*** **on Sheep Blood Agar** ◆ These irregularly circular colonies have a raised center and a flat, spreading edge of tangled filaments (reminiscent of the mythological creature Medusa, which had snakes for hair). They vary in size from 2 to 6 mm.

2-13 ***Providencia stuartii*** **Grown on Nutrient Agar** ◆ The colonies are shiny, buff, and convex. *P. stuartii* is a frequent isolate in urine samples obtained from hospitalized and catheterized patients. *P. stuartii* is highly resistant to antibiotics.

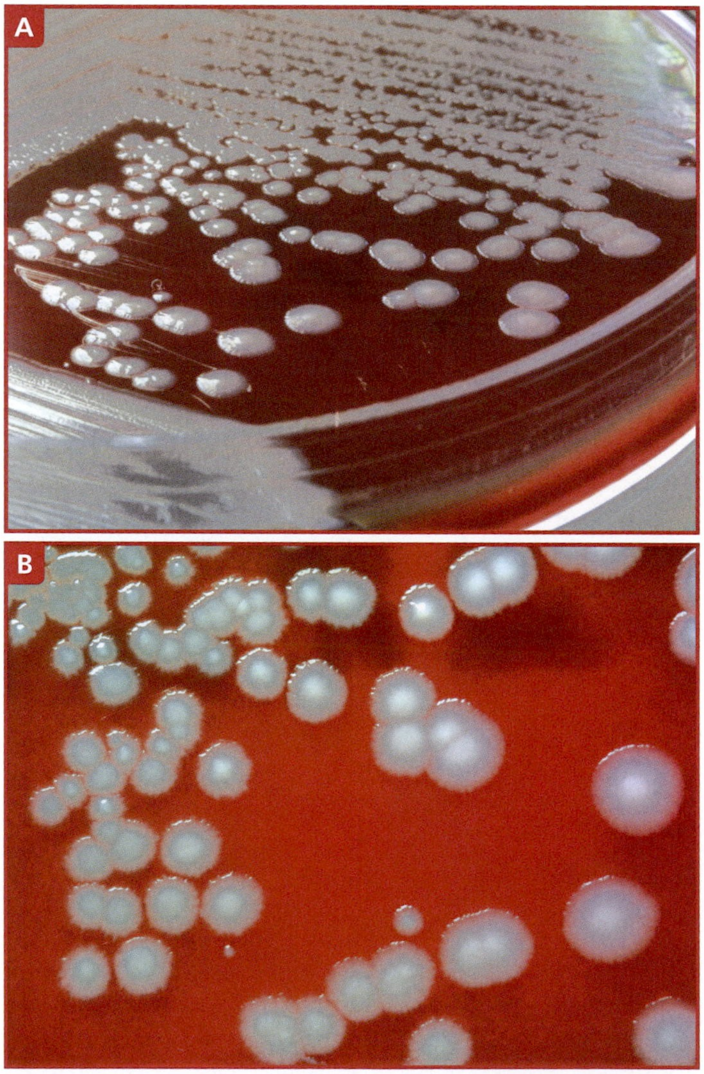

2-15 ***Alcaligenes faecalis*** **Colonies on Sheep Blood Agar** ◆ The colonies of this opportunistic pathogen are umbonate with an opaque center and a spreading edge. **A** Side view. Note the raised center. **B** Close-up of the *A. faecalis* colonies showing spreading edge.

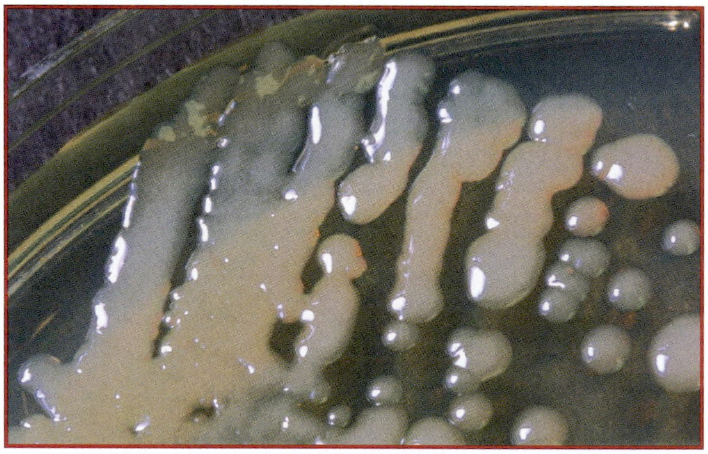

2-14 ***Klebsiella pneumoniae*** **Grown on Nutrient Agar** ◆ The colonies are mucoid, raised, and shiny. While it is a normal inhabitant of the human intestinal tract, it is associated with community-acquired pneumonia and nosocomial urinary tract infections.

2-16 ***Citrobacter koseri*** **Grown on Sheep Blood Agar** ◆ The colonies are round, smooth, and opaque with a regular margin. This species is also able to partially hemolyze red blood cells (α-hemolytic), as evidenced by the greening around each colony. They range in size from 1 to 2 mm.

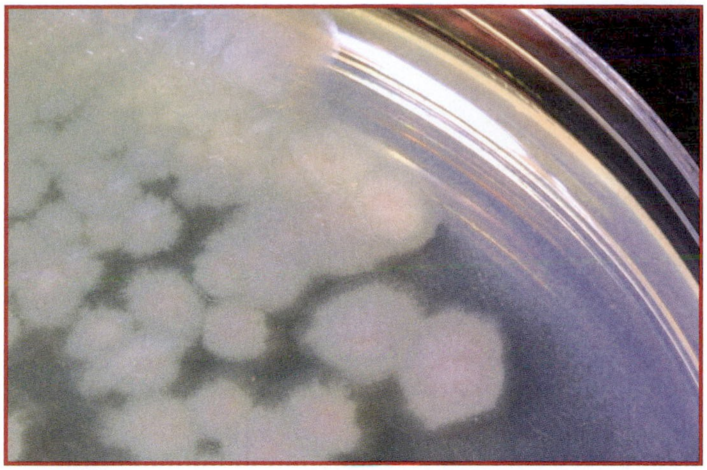

2-17 ***Erwinia amylovora*** **Colonies Grown on Nutrient Agar** ◆ Note the irregular shape and spreading edges. *E. amylovora* is a plant pathogen.

2-18 ***Rhizobium leguminosarum*** **Grown on Brain–Heart Infusion Agar** ◆ The colonies are convex, circular, and filamentous. They are translucent at the edges and about 5 mm in diameter. *R. leguminosarum* is capable of producing root nodules (tumors) in many legumes and subsequently fixing atmospheric nitrogen.

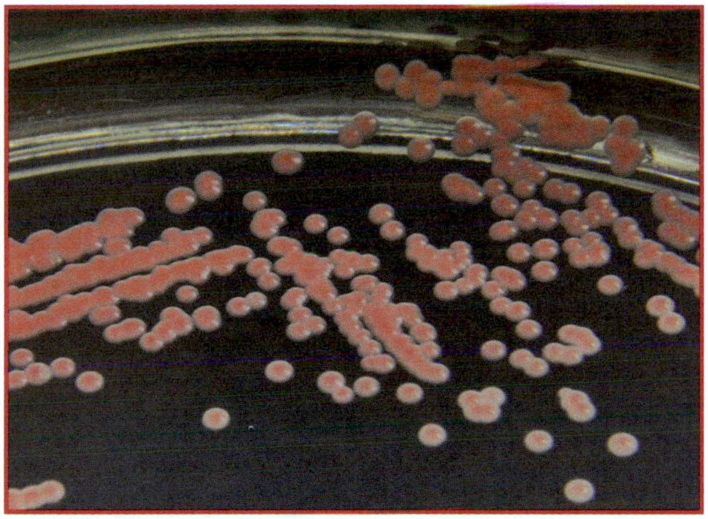

2-19 ***Deinococcus radiodurans*** **Grown on Trypticase Soy Agar** ◆ These small (between 1 and 2 mm in diameter), round, convex, and regular colonies took 36 hours to develop the orange color. This species is highly resistant to ionizing radiation.

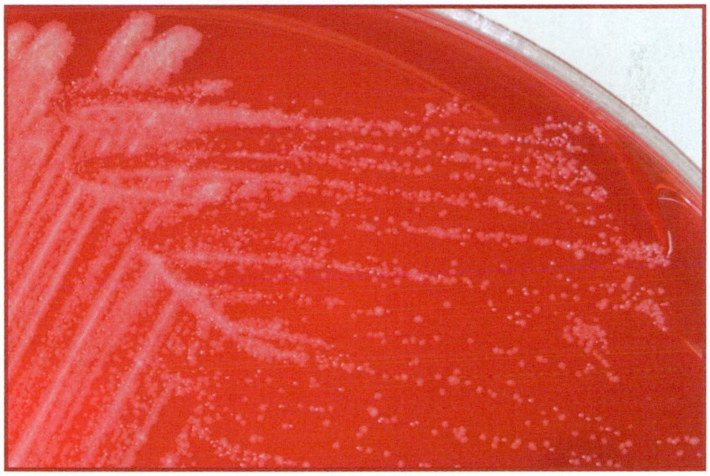

2-20 ***Mycobacterium smegmatis*** **Grown on Sheep Blood Agar** ◆ The colonies of this slow growing relative of *M. tuberculosis* are punctiform.

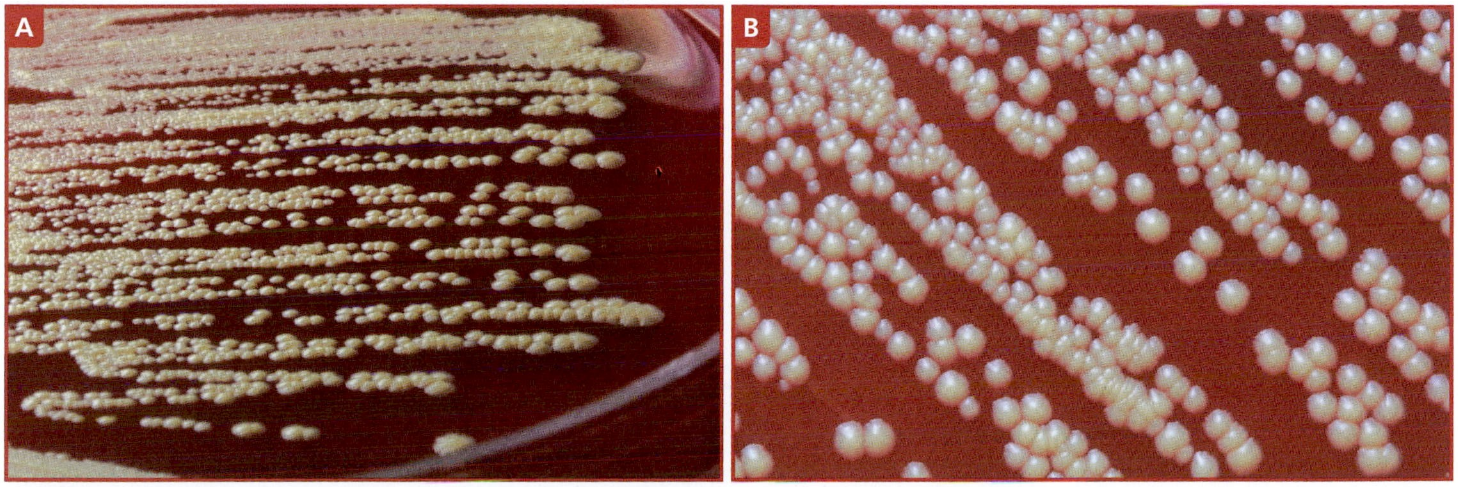

2-21 ***Corynebacterium xerosis*** **Grown on Sheep Blood Agar** **A** As seen in this view from the side, the colonies are dull, buff, and convex. **B** Close-up of circular *C. xerosis* colonies. *C. xerosis* is rarely an opportunistic pathogen.

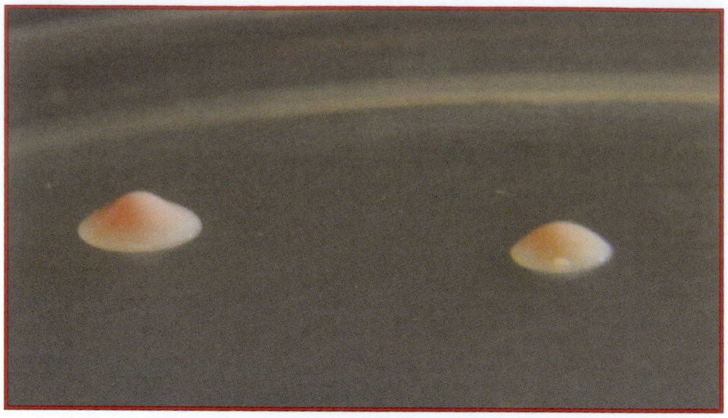

2-22 Umbonate Colony of an Anaerobic Lab Contaminant ◆ The colony on the left is truly umbonate. The one on the right is getting there. Their diameters are about 3 mm.

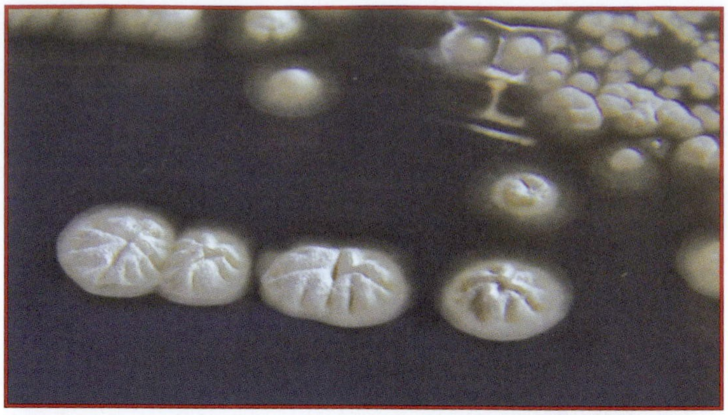

2-23 *Streptomyces griseus* Grown on Brain–Heart Infusion Agar ◆ These colonies are circular and ridged with a granular appearance. At a later stage of development, they produce yellow reproductive spores. Growth of streptomycetes is associated with an "earthy" smell. This one plate fragranced the entire incubator!

2-24 Swarming Growth Pattern ◆ Members of the genus *Proteus* will swarm at certain intervals and produce a pattern of concentric rings due to their motility. This is a photograph of *P. vulgaris* demonstrating swarming behavior on DNase agar.

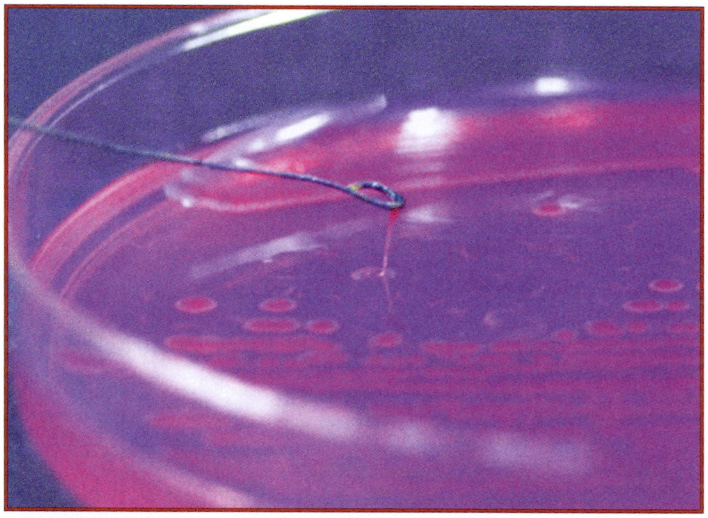

2-25 Mucoid Colonies ◆ *Pseudomonas aeruginosa* grown on Endo agar demonstrates a mucoid texture. *P. aeruginosa* is found in soil and water, and can cause infections in burn patients.

2-26 Butyrous Colony of an Unknown Soil Isolate ◆ This 12 mm colony was found on a Glycerol Yeast Extract plate inoculated with a diluted soil sample. It was almost liquid in composition, something that is indicated by its contact with the yellow colony to its right.

2-27 *Chromobacterium violaceum* Grown on Sheep Blood Agar ◆ *C. violaceum* produces shiny, purple, convex colonies. It is found in soil and water, and rarely produces infections in humans.

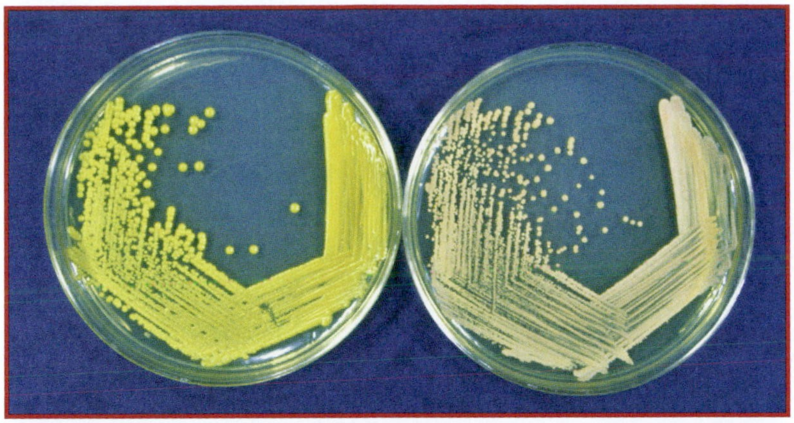

2-28 Pigment Production ◆ Closely related species may look very different, as seen in these plates of *Micrococcus luteus* (left) and *Kocuria rosea*, formerly *Micrococcus roseus* (right).

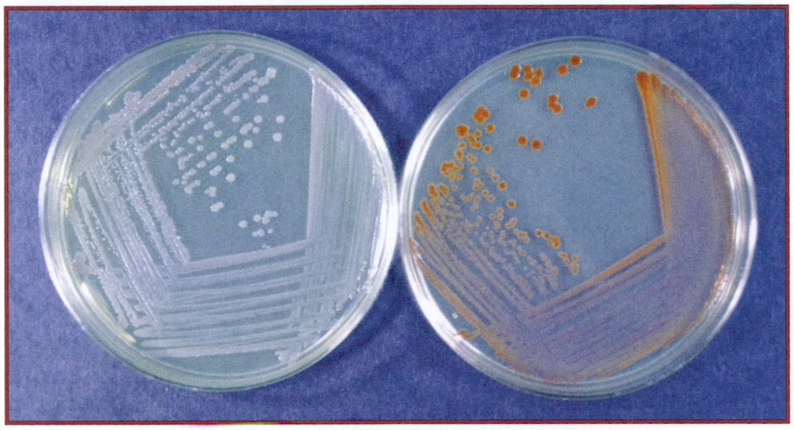

2-29 Influence of Temperature on Pigment Production ◆ Pigment production may be influenced by temperature. *Serratia marcescens* produces less orange pigment when grown at 37°C (left) than when grown at 25°C (right).

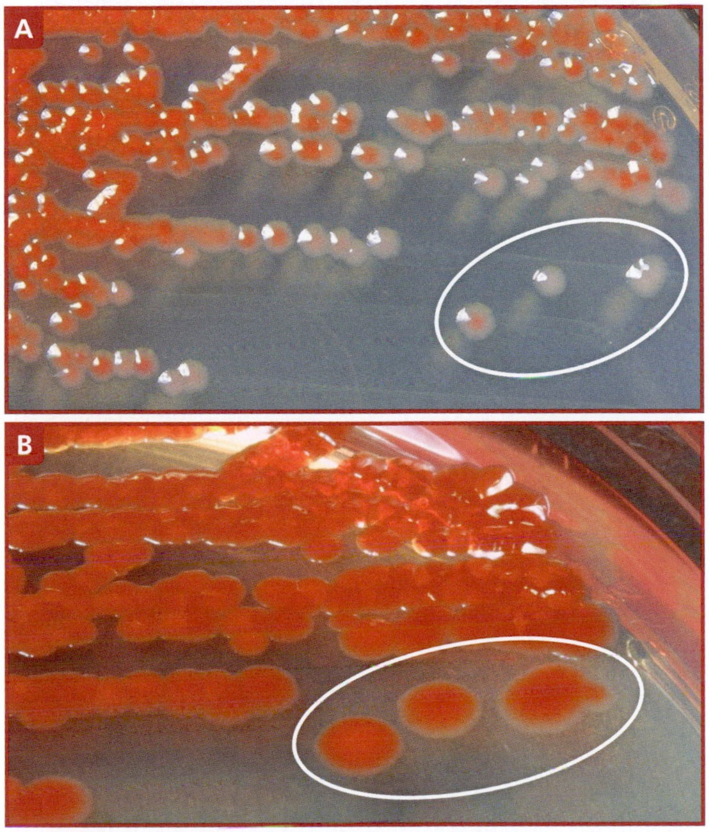

2-30 Influence of Age on Pigment Production ◆ **A** *Serratia marcescens* grown on Sheep Blood Agar after 24 hours. **B** The same plate of *S. marcescens* after 48 hours. Note in particular the change in the 3 colonies in the lower right (encircled).

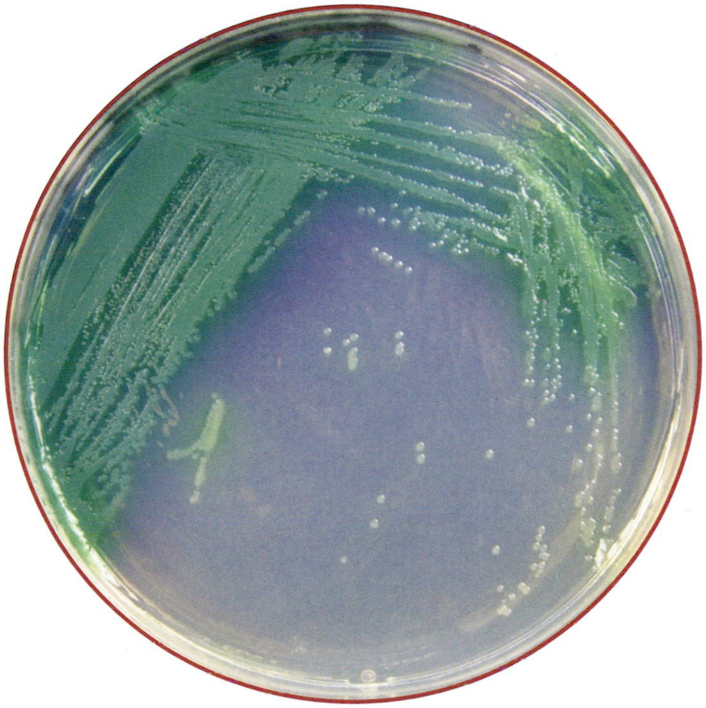

2-31 Diffusible Pigment of *Pseudomonas aeruginosa* ◆ The blue-green pigment pyocyanin often makes *P. aeruginosa* easy to identify.

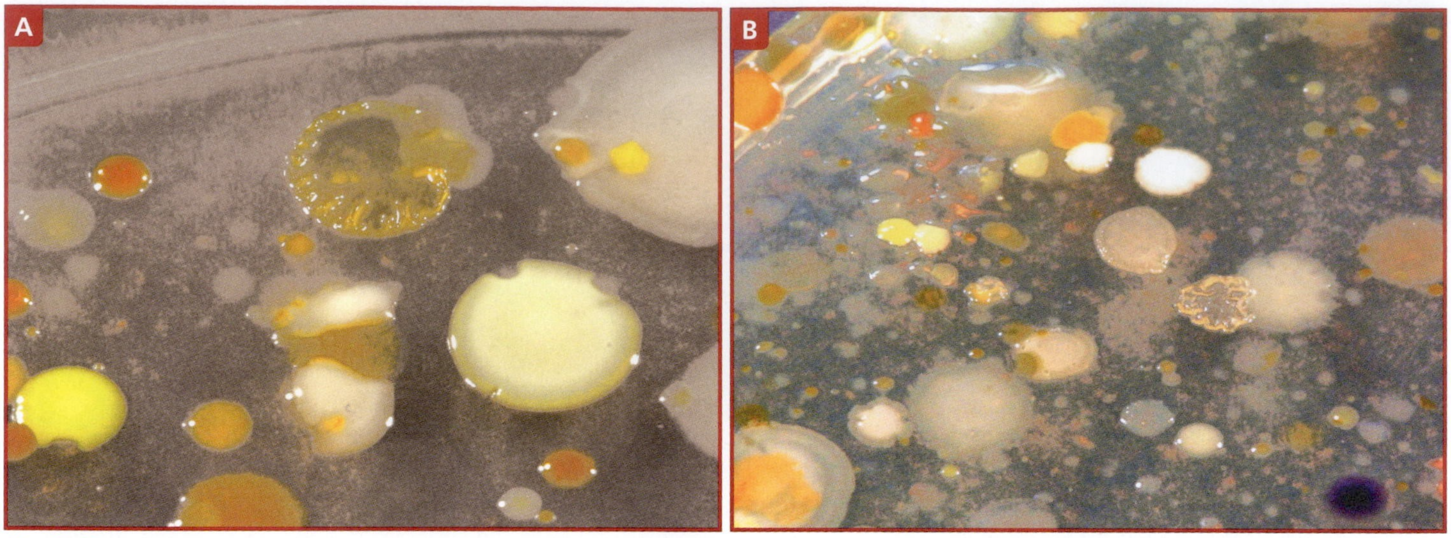

2-32 **Two Mixed Soil Cultures on Nutrient Agar** ◆ These plates show the morphological diversity present in two soil samples.

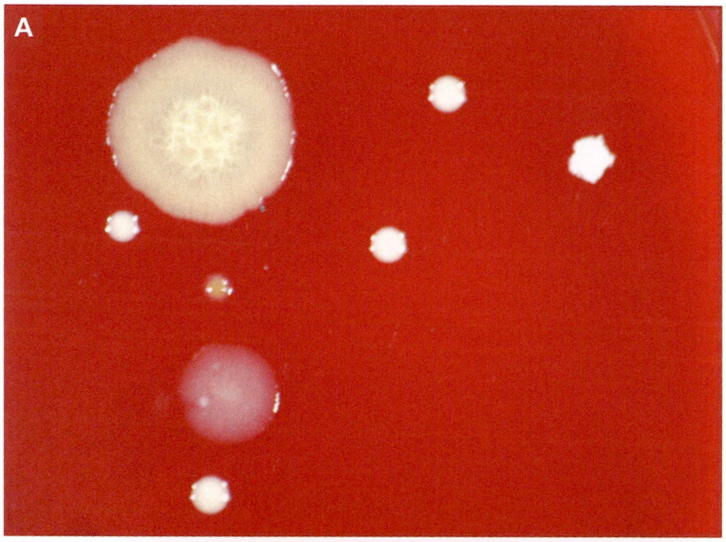

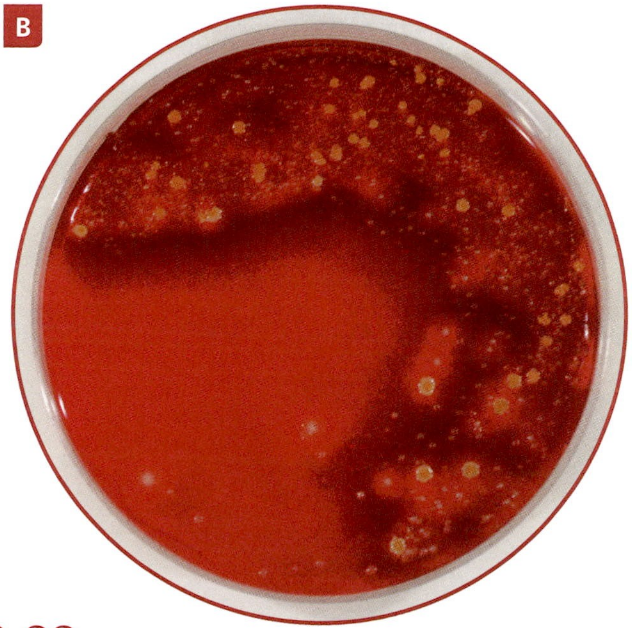

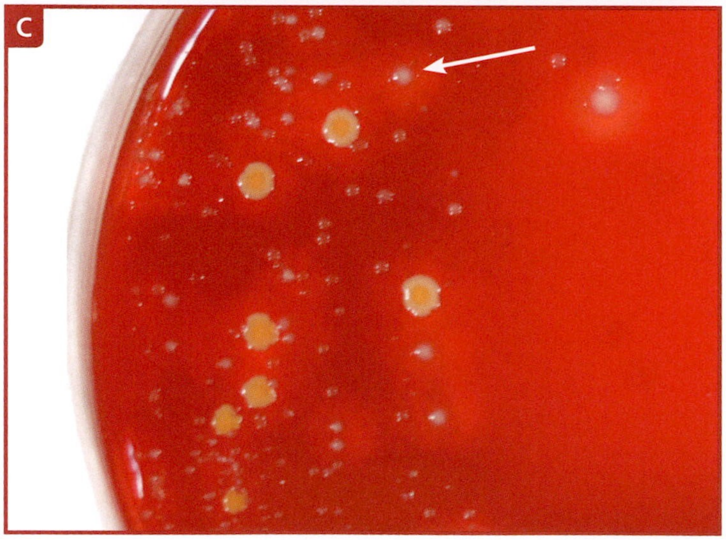

2-33 **Three Throat Cultures on Sheep Blood Agar** ◆
A There are probably five different species on this plate. **B** Note the α-hemolysis (darkening of the agar; see page 341 for more information) shown by much of the growth. **C** This is a close-up of the same plate as in **B**. Note the weak β-hemolysis of the white colony in the upper right (arrow). White growth with β-hemolysis is characteristic of *Staphylococcus aureus*. For information about hemolytic reactions on Blood Agar Plates, see Exercise 5-21.

References

Claus, G. William. 1989. Chapter 14 in *Understanding Microbes—A Laboratory Textbook for Microbiology*. W. H. Freeman and Co., New York.

Collins, C. H., Patricia M. Lyne, and J. M. Grange. 1995. Chapter 6 in *Collins and Lyne's Microbiological Methods,* 7th ed. Butterworth-Heineman, Oxford, England.

Forbes, Betty A., Daniel F. Sahm, and Alice S. Weissfeld. 2002. *Bailey & Scott's Diagnostic Microbiology,* 11th ed. Mosby-Yearbook, St. Louis.

Winn, Washington C., et al. 2006. *Koneman's Color Atlas and Textbook of Diagnostic Microbiology,* 6th ed. Lippincott Williams & Wilkins, Baltimore.

NAME ______________________ DATE ______________________

LAB SECTION ______________________ I WAS PRESENT AND PERFORMED THIS EXERCISE (INITIALS) ________

Colony Morphology

OBSERVATIONS AND INTERPRETATIONS

Using the terms in Figure 2-3, describe and sketch representative colonies (including side views) on your plates, including those from Exercise 2-1. Use a colony counter if necessary. Measure colony diameters, and include them with your descriptions. Use a separate paper as needed.

Organism/Plate	Colony Description and Sketch

Organism/Plate	Colony Description and Sketch

Questions

1 A description of colony morphology provides important information about an organism. What other information should you include when describing physical growth characteristics?

2 Three critical aspects of a description of bacterial growth are colony size, color, and shape. At least three other important factors—not physical descriptions—typically are included when describing bacterial growth. What do you suppose they are and why are they important?

EXERCISE 2-3

Growth Patterns on Slants

Theory

Agar slants are useful primarily as media for cultivation and maintenance of stock cultures. Organisms cultivated on slants, however, do display a variety of growth characteristics. We offer these more as an item of interest than one of diagnostic value.

Most of the organisms you will see in this class produce **filiform** growth (dense and opaque with a smooth edge). All of the organisms in Figure 2-34 are filiform and **pigmented**. Most species in the genus *Mycobacterium* produce **friable** (crusty) growth (Figure 2-35—*Mycobacterium phlei*). Some motile organisms produce growth with a **spreading edge** (Figure 2-35—*Alcaligenes faecalis*). Still others produce **translucent** or **transparent** growth (Figure 2-35—*Lactobacillus plantarum*).

Growth patterns of virtually all microorganisms can be influenced by environmental factors. Two important factors influencing bacterial growth are incubation time and temperature. Figure 2-36 illustrates the effect of temperature on pigment production in *Serratia marcescens*. See also Environmental Factors Affecting Microbial Growth on page 77.

Application

Although not definitive by themselves, growth characteristics on slants can provide useful information when attempting to identify an organism.

2-34 Pigment Production on Slants ◆ From left to right, *Staphylococcus epidermidis* (white), *Pseudomonas aeruginosa* (green), *Chromobacterium violaceum* (violet), *Serratia marcescens* (red/orange), *Kocuria rosea* (rose), and *Micrococcus luteus* (yellow).

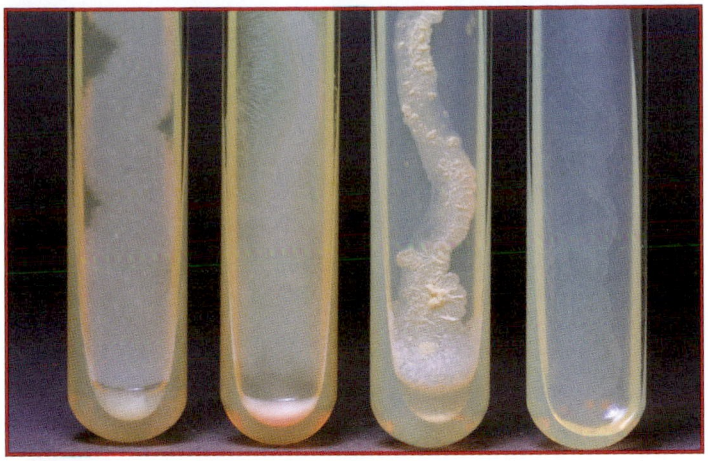

2-35 Growth Texture on Slants ◆ From left to right, *Bacillus* spp., *Alcaligenes faecalis*, *Mycobacterium phlei*, *Lactobacillus plantarum*.

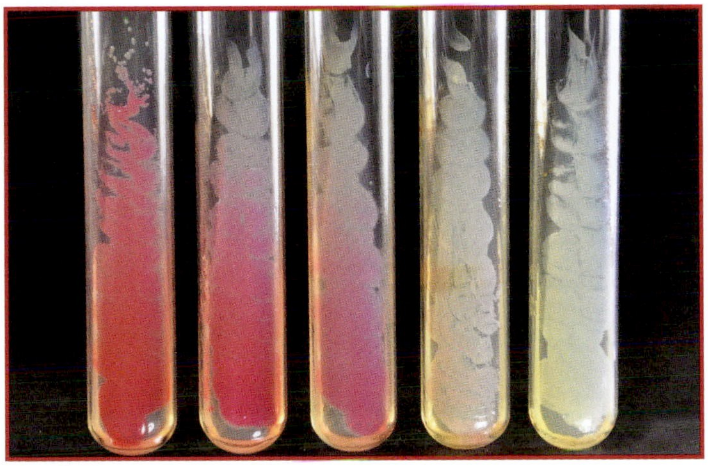

2-36 Influence of Temperature on Pigment Production ◆ *Serratia marcescens* was grown for 48 hours on Trypticase Soy Agar slants at five different temperatures. From left to right: 25°C, 30°C, 33°C, 35°C, and 37°C. A difference of 2°C makes the difference between being pigmented or not!

In This Exercise

In Exercise 2-2, you examined growth on agar plates. Today you will be looking at the growth patterns of six organisms on prepared agar slants. If your instructor had you save your plates from Exercise 2-2, you will compare the growth patterns on the two different types of media.

Materials

Per Student Group

✓ Tryptic Soy Agar or Brain–Heart Infusion Agar Slant cultures of:
- *Micrococcus luteus*
- *Corynebacterium xerosis*
- *Lactobacillus plantarum*
- *Bacillus subtilis*

Optional

- *Mycobacterium smegmatis* (BSL-2)
- *Proteus mirabilis* (BSL-2)

✓ Uninoculated tube of the medium used for cultures.

Procedure

Lab One

1. Examine the slants and describe the different growth patterns on the Data Sheet. Include a sketch of a representative portion of each.
2. (Optional) Compare the slant growth to that on the agar plates from Exercise 2-2.
3. (Optional) Save the tubes for comparison with broths in Exercise 2-4.

References

Claus, G. William. 1989. Chapter 17 in *Understanding Microbes—A Laboratory Textbook for Microbiology*. W. H. Freeman and Co., New York.

Name ______________________ Date ______________________

Lab Section ______________________ I was present and performed this exercise (initials) ________

DATA SHEET 2-3

Growth Patterns on Slants

OBSERVATIONS AND INTERPRETATIONS

In the chart below, describe the growth on your slants, including shape, margin, texture, and color. Draw a picture if necessary, and include other information about the conditions of incubation, such as time, temperature, and the medium used. Use additional paper if necessary.

Organism	Growth Description
Control	

Organism	Growth Description

Questions

1 List some reasons why growth characteristics are more useful on agar plates than on agar slants.

2 Why are agar slants better suited than agar plates to maintain stock cultures?

3 Match the following:

_____ Filiform	1. Produces colored growth
_____ Spreading edge	2. Smooth texture with solid edge
_____ Transparent	3. Solid growth seeming to radiate outward
_____ Friable	4. Almost invisible or easy to see light through
_____ Pigmented	5. Rough texture with a crusty appearance

EXERCISE **2-4**

Growth Patterns in Broth

Theory

Microorganisms cultivated in broth display a variety of growth characteristics. Some organisms float on top of the medium and produce a type of surface membrane called a **pellicle**. Others sink to the bottom as **sediment**. Some bacteria produce **uniform fine turbidity** (Figure 2-37), and others appear to clump in what is called **flocculent** growth (Figure 2-38). The ability to produce pigments, although not specifically a growth pattern, is nonetheless an identifiable characteristic of some bacteria (Figure 2-39).

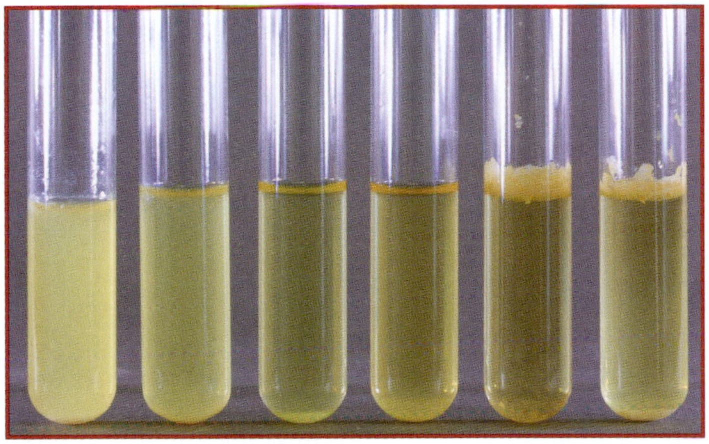

2-37 Growth Patterns in Broth ◆ From left to right in pairs (by type of organism): *Enterobacter aerogenes* and *Citrobacter diversus*—motile members of *Enterobacteriaceae* (uniform fine turbidity), *Enterococcus faecalis* and *Staphylococcus aureus*—nonmotile Gram-positive cocci (sediment), *Mycobacterium phlei* and *Mycobacterium smegmatis* (relatives of *Mycobacterium tuberculosis*)—nonmotile with a waxy cell wall (pellicle).

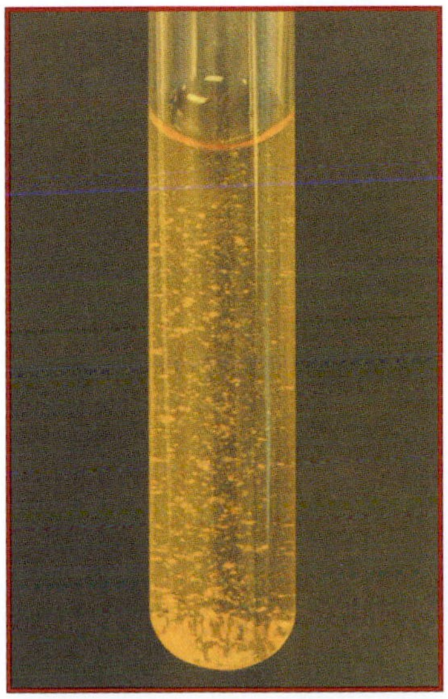

2-38 Flocculence in Broth ◆ This is a *Streptococcus* species from a throat culture demonstrating flocculence in Todd-Hewitt Broth.

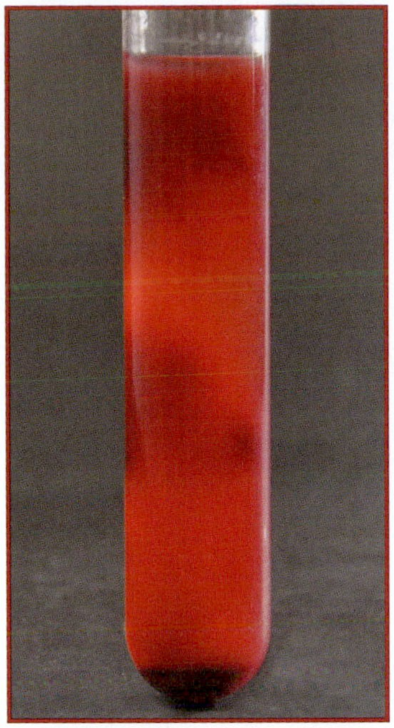

2-39 Pigments in Broth ◆ *Rhodospirillum rubrum* has a red color due to carotenoid pigments. It grows as a photoheterotroph in the presence of light and the absence of oxygen.

Application

Bacterial genera, and frequently individual species within a genus, demonstrate characteristic growth patterns in broth that provide useful information when attempting to identify an organism.

In This Exercise

Today, you will be examining the growth characteristics of six different bacteria in broth. If you were instructed to save plates or slants from previous exercises, you will compare the growth patterns on the different types of media.

Materials

Per Student Group

✓ Brain–Heart Infusion (BHI) Broth cultures of:
- *Micrococcus luteus*
- *Corynebacterium xerosis*
- *Lactobacillus plantarum*
- *Bacillus subtilis*

Optional
- *Mycobacterium smegmatis* (BSL-2)
- *Proteus mirabilis* (BSL-2)

✓ (optional) your slants from Exercise 2-3

✓ one sterile uninoculated BHI Broth control

Procedure

Examine the tubes and compare them with the control. Describe the different growth patterns on the Data Sheet.

References

Claus, G. William. 1989. Chapter 17 in *Understanding Microbes—A Laboratory Textbook for Microbiology*. W. H. Freeman and Co., New York.

Forbes, Betty A., Daniel F. Sahm, and Alice S. Weissfeld. 2002. *Bailey & Scott's Diagnostic Microbiology,* 11th ed. Mosby-Yearbook, St. Louis.

Name ______________________ Date ______________

Lab Section ______________ I was present and performed this exercise (initials) ______

Growth Patterns in Broth

OBSERVATIONS AND INTERPRETATIONS

Examine the tubes and enter the descriptions below. Draw a picture if necessary, and include other information about the conditions of incubation, such as time, temperature, and the medium used. Use a separate paper if necessary.

Organism	Growth Description
Control	

Organism	Growth Description

Questions

1 What factors besides physical growth characteristics are important when recording data about an organism? Why?

2 Match the following:

_____ Flocculent	1. Evenly cloudy throughout
_____ Sediment	2. Growth at top around the edge
_____ Ring	3. Growth on the bottom
_____ Pellicle	4. Membrane at the top
_____ Uniform fine turbidity	5. Suspended chunks or pieces

Environmental Factors Affecting Microbial Growth

Bacteria have limited control over their internal environments. Whereas many eukaryotes have evolved sophisticated internal control mechanisms, bacteria are almost completely dependent on external factors to provide conditions suitable for their existence. Minor environmental changes can dramatically change a microorganism's ability to transport materials across the membrane, perform complex enzymatic reactions, and maintain critical cytoplasmic pressure.

One way to observe bacterial responses to environmental changes is to artificially manipulate external factors and measure growth rate. Understand that growth rate, as it is used in this manual, is synonymous with reproductive rate. Optimal growth conditions, as might be expected, result in faster growth and greater cell density (as evidenced by greater turbidity) than do less than optimal conditions.

In this series of laboratory exercises, you will examine the effects of oxygen, temperature, pH, osmotic pressure, and the availability of nutritional resources on bacterial growth rate. You also will learn some methods for cultivating anaerobic bacteria. When possible, you will attempt to classify organisms based on your results.

EXERCISE 2-5

Evaluation of Media

❖ Theory

Living things are composed of compounds from four biochemical families:

1. proteins,
2. carbohydrates,
3. lipids, and
4. nucleic acids.

Even though all organisms share this fundamental chemical composition, they differ greatly in their ability to make these molecules. Some are capable of making them out of the simple carbon compound carbon dioxide (CO_2). These organisms, called **autotrophs,** require the least "assistance" from the environment to grow. The remaining organisms, called **heterotrophs**, require preformed organic compounds from the environment.

Some heterotrophs are metabolically flexible and require only a few simple organic compounds from which to make all their biochemicals. Others require a greater portion of their organic compounds from the environment. An organism that relies heavily on the environment to supply ready-made organic compounds is referred to as **fastidious.** Heterotrophs vary greatly in their dependence on the environment to supply organic compounds and, as such, range from highly fastidious to **nonfastidious**. Autotrophs are less fastidious than the most nonfastidious heterotrophs.

Successful cultivation of a microbe in the laboratory requires an ability to satisfy its nutritional needs. The absence of a single required chemical resource prevents its growth. In general, the more fastidious the organism, the more ingredients a medium must have. **Undefined** media (also known as complex media) are composed of extracts from plant or animal sources and are rich in nutrients. Even though the exact composition of the medium and the amount of each ingredient are unknown, undefined media are useful in growing the greatest variety of culturable microbes. A **defined** medium (or chemically defined medium) is one in which the amount and identity of every ingredient is known. Defined media typically support a narrower range of organisms.

❖ Application

The ability of a microbiologist to cultivate a microorganism requires some knowledge of its metabolic needs. One quick way to make this determination is to transfer it to a variety of media containing different nutritional components and observe how well it grows.

In This Exercise

Today you will evaluate the ability of three media—Brain–Heart Infusion Broth, Nutrient Broth, and Glucose Salts Medium—to support bacterial growth. You will do this by visually comparing the density of growth (turbidity) between organisms in the various broths. It is important to make your inoculations as uniform as possible. Make sure that your loop is fully closed so it will hold enough broth to produce a film across its opening (similar to a toy loop for blowing bubbles). Inoculate each medium with a single loopful of broth.

Materials

Per Student Group

✓ five tubes each of:
- Brain–Heart Infusion Broth
- Nutrient Broth
- Glucose Salts Medium

✓ Broth cultures of:
- *Escherichia coli*
- *Lactococcus lactis*
- *Moraxella catarrhalis*
- *Staphylococcus epidermidis*

Medium Recipes

Nutrient Broth

Beef extract	3.0 g
Peptone	5.0 g
Distilled or deionized water	1.0 L

Brain–Heart Infusion Broth

Calf Brains, Infusion from 200 g	7.7 g
Beef Heart, Infusion from 250 g	9.8 g
Proteose peptone	10.0 g
Dextrose	2.0 g
Sodium chloride	5.0 g
Disodium phosphate	2.5 g
Distilled or deionized water	1.0 L

Glucose Salts Medium

Glucose	5.0 g
Sodium chloride	5.0 g
Magnesium sulfate	0.2 g
Ammonium dihydrogen phosphate	1.0 g
Dipotassium phosphate	1.0 g
Distilled or deionized water	1.0 L

Procedure

Lab One

1. Inoculate each medium with a single loopful of each organism. Leave one of each tube uninoculated.
2. Incubate all tubes (including the uninoculated ones) at 35 ± 2°C for 24–48 hours.

Lab Two

1. Mix all tubes well and examine them for turbidity. Score relative amounts of growth using "0" for no growth, and "1," "2," and "3" for successively greater degrees of growth.
2. Record your results on the Data Sheet. (Note: Results may vary. Record what you see, not what you expect.)

References

Delost, Maria Dannessa. 1997. Page 144 in *Introduction to Diagnostic Microbiology: A Text and Workbook*. Mosby, St. Louis.

Forbes, Betty A., Daniel F. Sahm, and Alice S. Weissfeld. 2002. *Bailey & Scott's Diagnostic Microbiology,* 11th ed. Mosby-Yearbook, St. Louis.

Zimbro, Mary Jo, and David A. Power, eds. 2003. *Difco™ and BBL™ Manual—Manual of Microbiological Culture Media.* Becton Dickinson and Co., Sparks, MD.

NAME ______________________ DATE ______________________

LAB SECTION ______________ I WAS PRESENT AND PERFORMED THIS EXERCISE (INITIALS) ______

DATA SHEET 2-5

Evaluation of Media

OBSERVATIONS AND INTERPRETATIONS

1 On the top line of the data table below, write "defined" or "undefined" ("complex") for each medium.

2 Score the relative amount of growth in each tube. Compare the five tubes for each medium with each other, as well as the amount of growth for each organism in the various media. Use "0" for no growth, "3" for abundant growth, and "1" and "2" for degrees of growth in between.

Organism	Nutrient Broth	Brain–Heart Infusion Broth	Glucose Salts Medium	Interpretation (relative fastidiousness)
Defined/Undefined				
Uninoculated Broth				
Ability of medium to support a wide range of microorganisms (good, fair, poor)				

Questions

1 What does comparing growth of the four organisms in a given medium tell you? (That is, are you gaining information primarily about the organism or about the medium?)

2 What does comparing growth of a given organism in the three media tell you? (Again, are you gaining information primarily about the organism or about the medium?)

3 Evaluate the media.

a. Which medium supports growth of the widest range of organisms?

b. Which medium supports the fewest organisms?

c. Is there a correlation between your answers to Questions 3a and 3b, and the terms "defined medium" and "undefined medium"? If so, what is it? If not, attempt to explain the relationship you observe.

4 Evaluate the organisms.

a. Which organism appears to be most fastidious? How can you tell?

b. Which organism appears to be least fastidious? How can you tell?

5 What is the biochemical basis for the spectrum of fastidiousness seen in the microbial world? (That is, why are some organisms fastidious and others are nonfastidious?)

Aerotolerance

Most microorganisms can survive within a range of environmental conditions, but not surprisingly, tend to produce growth with the greatest density in the areas where conditions are most favorable. One important resource influencing microbial growth is oxygen. Some organisms require oxygen for their metabolic needs. Some other organisms are not affected by it at all. Still other organisms cannot even survive in its presence. This ability or inability to live in the presence of oxygen is called **aerotolerance**.

Most growth media are sterilized in an autoclave during preparation. This process not only kills unwanted microbes, but removes most of the free oxygen from the medium as well. After the medium is removed from the autoclave and allowed to cool, the oxygen begins to diffuse back in. In tubed media (both liquid and solid) this process creates a gradient of oxygen concentrations, ranging from **aerobic** at the top, nearest the source of oxygen, to **anaerobic** in the lower regions. Because of microorganisms' natural tendency to proliferate where the oxygen concentration best suits their metabolic needs, differing degrees of population density will develop in the medium over time that can be used to visually examine their aerotolerance.

Obligate (strict) aerobes, organisms that require oxygen for respiration, grow at the top where oxygen is most plentiful. **Facultative anaerobes** grow in the presence *or* absence of oxygen. When oxygen is available, they respire aerobically. When oxygen is not available, they either respire anaerobically (reducing sulfur or nitrate instead of oxygen) or ferment an available substrate. Refer to Appendix A and Section Five for more information on anaerobic respiration and fermentation. Where an oxygen gradient exists, facultative anaerobes grow throughout the medium but appear denser near the top. **Aerotolerant anaerobes**, organisms that don't require oxygen and are not adversely affected by it, live uniformly throughout the medium. Aerotolerant anaerobes are fermentative even in the presence of free oxygen. **Microaerophiles**, as the name suggests, survive only in environments containing lower than atmospheric levels of oxygen. Some microaerophiles called **capnophiles** can survive only if carbon dioxide levels are elevated. Microaerophiles will be seen somewhere near the middle or upper middle region of the medium. Finally, **obligate (strict) anaerobes** are organisms for which even small amounts of oxygen are lethal and, therefore, will be seen only in the lower regions of the medium, depending on how far into the medium the oxygen has diffused.

EXERCISE 2-6

Fluid Thioglycollate Medium

❖ Theory

Fluid Thioglycollate Medium is prepared as a basic medium (as used in this exercise) or with a variety of supplements, depending on the specific needs of the organisms being cultivated. As such, this medium is appropriate for a broad variety of aerobic and anaerobic, fastidious and nonfastidious organisms. It is particularly well adapted for cultivation of strict anaerobes and microaerophiles.

Key components of the medium are yeast extract, pancreatic digest of casein, dextrose, sodium thioglycollate, L-cystine, and resazurin. Yeast extract and pancreatic digest of casein provide nutrients; sodium thioglycollate and L-cystine reduce oxygen to water; and resazurin (pink when oxidized, colorless when reduced) acts as an indicator. A small amount of agar is included to slow oxygen diffusion and stabilize growth.

As mentioned in the introduction to aerotolerance, oxygen removed during autoclaving will diffuse back into the medium as the tubes cool to room temperature. This produces a gradient of concentrations from fully aerobic at the top to anaerobic at the bottom. Thus, fresh media will appear clear to straw-colored with a pink region at the top where the dye has become oxidized (Figure 2-40). Figure 2-41 demonstrates some basic bacterial growth patterns in the medium as influenced by the oxygen gradient.

❖ Application

Fluid Thioglycollate Medium is a liquid medium designed to promote growth of a wide variety of fastidious micro-organisms. It can be used to grow microbes representing all levels of oxygen tolerance; however, it generally is associated with the cultivation of anaerobic and micro-aerophilic bacteria.

❖ In This Exercise

Today you will be inoculating Fluid Thioglycollate Medium to determine the aerotolerance categories of three bacteria. Your instructor may have additional organisms or may have you collect environmental samples to test.

Materials

Per Student Group

✓ four Fluid Thioglycollate Medium tubes
✓ fresh cultures of:
- *Alcaligenes faecalis*
- *Clostridium sporogenes*
- *Staphylococcus epidermidis*

Medium Recipe

Fluid Thioglycollate Medium

Yeast extract	5.0 g
Pancreatic digest of casein	15.0 g
Dextrose	5.5 g
Sodium chloride	2.5 g
Sodium thioglycollate	0.5 g
L-cystine	0.5 g
Agar	0.75 g
Resazurin	0.001 g
Distilled or deionized water	1.0 L

Procedure

Lab One

1. Obtain four Fluid Thioglycollate tubes and label them with your name, the date, medium, and organism.
2. Using your loop, inoculate three broths with the organisms provided. (Note: When inoculating Thioglycollate Broth, it helps to dip the loop all the way to the bottom of the tube and gently mix the broth with the loop as you remove it. Finish mixing by gently rolling the tube between your hands.) Do not inoculate the fourth tube.
3. Incubate the tubes at 35 ± 2°C for 24 to 48 hours.

2-40 Aerobic Zone in Thioglycollate Medium ◆
Note the pink region in the top (oxidized) portion of the broth. The bottom (reduced) portion of the medium remains colorless.

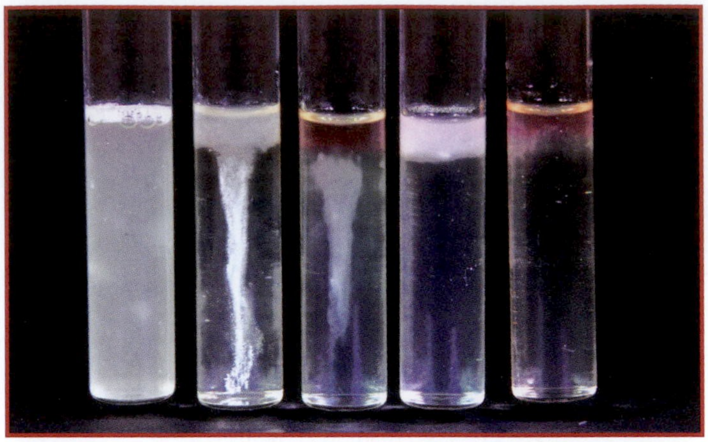

2-41 Growth Patterns in Thioglycollate Medium ◆
Growth patterns of a variety of organisms are shown in these Fluid Thioglycollate Broths. Pictured from left to right are: aerotolerant anaerobe, facultative anaerobe, strict anaerobe, strict aerobe, and microaerophile. Compare these tubes with the uninoculated broth in Figure 2-40.

Lab Two

1. Check the control tube for growth to assure sterility of the medium. Note any changes that may have occurred as a result of incubation, especially in the colored region at the surface.
2. Using the control as a comparison, examine and note the location of the growth in all tubes.
3. Enter your observations and interpretations on the Data Sheet.

References

Allen, Stephen D., Christopher L. Emery, and David M. Lyerly. 2003. Chapter 54 in *Manual of Clinical Microbiology*, 8th ed. Edited by Patrick R. Murray, Ellen Jo Baron, James H. Jorgensen, Michael A. Pfaller, and Robert H. Yolken. ASM Press, American Society for Microbiology, Washington, DC.

Forbes, Betty A., Daniel F. Sahm, and Alice S. Weissfeld. 2002. Chapter 10 in *Bailey & Scott's Diagnostic Microbiology*, 11th ed. Mosby, St. Louis.

Winn, Washington C., et al. 2006. *Koneman's Color Atlas and Textbook of Diagnostic Microbiology*, 6th ed. Lippincott Williams & Wilkins, Baltimore.

Zimbro, Mary Jo and David A. Power, eds. 2003. *Difco™ and BBL™ Manual—Manual of Microbiological Culture Media*. Becton Dickinson and Company, Sparks, MD.

Name ______________________ Date ______________________

Lab Section ______________________ I was present and performed this exercise (initials) ______

DATA SHEET 2-6

Fluid Thioglycollate Medium

OBSERVATIONS AND INTERPRETATIONS

Draw a diagram of each broth showing the location of growth. Indicate the amount of growth in the broth.
"0" = no growth "3" = abundant growth "1" and "2" = degrees of growth in between

Organism	Location of Growth in Medium	Aerotolerance Category
Control		

Questions

1 Why is there a colored band at the surface of Fluid Thioglycollate Medium? Which is more desirable, a thick or a thin band?

2 Where would you expect to see growth of a strict aerobe? Anaerobe? Microaerophile? Facultative anaerobe?

3 Why is it important that this medium be fresh? Which type of organism (aerobe, anaerobe, microaerophile, facultative anaerobe) would most likely be affected negatively by the use of old media? Which would most likely be affected positively?

EXERCISE 2-7

Anaerobic Jar

Theory

The GasPak® Anaerobic System by BBL™ is a plastic jar in which to create anaerobic, microaerophilic, or CO_2-conditions depending on the specific needs of the bacteria being cultivated. The components required for anaerobic growth include a chemical gas generator packet (envelope) containing sodium borohydride and sodium bicarbonate, and a paper indicator strip saturated with methylene blue to confirm the absence of oxygen (Figure 2-42). Methylene blue is colorless when reduced and blue when oxidized. Also included in the packet is a small amount of palladium to act as a catalyst for the reaction that will produce the necessary conditions inside the jar.

After the inoculated media (typically agar plates) are placed inside the jar, the opened gas generator envelope is placed inside along with the anaerobic indicator strip. Water is added to the envelope and the jar lid immediately is fastened down. The sodium borohydride and sodium bicarbonate in the envelope react with the water to produce hydrogen and carbon dioxide gases. The palladium catalyzes a reaction between the hydrogen and free oxygen in the jar to produce water, as shown in Figure 2-43. Removal of free oxygen produces anaerobic conditions in the jar within approximately an hour, as evidenced by a white indicator strip and moisture on the inside of the jar.

Figure 2-44 illustrates some typical growth patterns under aerobic and anaerobic conditions.

2-42 The Anaerobic Jar ◆ Note the sachet and the white indicator strip inside the jar. The sachet has performed properly, reducing the oxygen level within the jar to less than 1%, as evidenced by the indicator strip. If the indicator were blue, it would mean free oxygen remained in the jar and the resulting growth would be in question relative to its ability to survive in anaerobic conditions.

$$2H_2 + O_2 \xrightarrow{Pd} 2H_2O$$

2-43 Conversion of H_2 and O_2 to Water Using a Palladium Catalyst ◆

Application

This procedure provides a means of cultivating anaerobic and microaerophilic bacteria.

In This Exercise

Today each group will inoculate two Nutrient Agar plates with three organisms. (Your instructor may have additional organisms for you to test.) One of the plates will go into the jar, and the second plate will not. Both plates will be incubated at 35 ± 2°C for 24 to 48 hours. Following incubation, the growth on the two plates will be examined and compared.

Materials

Per Class

✓ one anaerobic jar with gas generator packet[1]

Per Group

✓ two Nutrient Agar plates

✓ fresh broth cultures of:

- *Alcaligenes faecalis*
- *Clostridium sporogenes*
- *Staphylococcus epidermidis*

[1] Available from Becton Dickinson Microbiology Systems, Sparks, MD http://www.bd.com

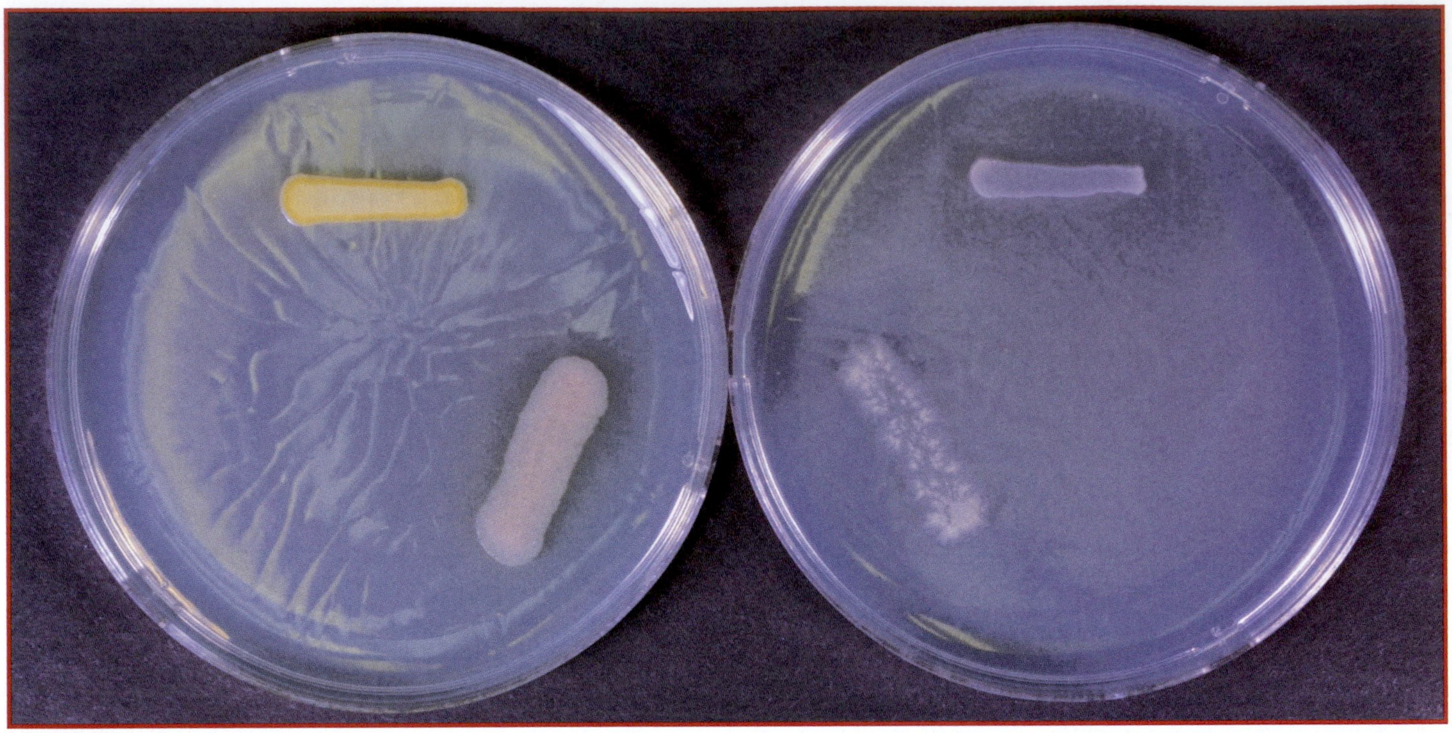

2-44 Plates Incubated Inside and Outside the Anaerobic Jar ◆ Both Nutrient Agar plates were inoculated with *Staphylococcus aureus* (top), *Pseudomonas aeruginosa* (right), and *Clostridium sporogenes* (left). The plate on the left was incubated aerobically outside the jar; the plate on the right was incubated inside the anaerobic jar. Note the relative amounts of growth of the three organisms.

Procedure

Lab One

1. Obtain two Nutrient Agar plates. Using a marking pen, divide the bottom of each plate into three sectors.
2. Label each plate with your name, the date, and organism by sector.
3. Using your loop, inoculate the sectors of both plates with the organisms provided. (Note: Inoculate with single streaks about 1 cm long.) Tape the lids in place.
4. Place one plate in the anaerobic jar in an inverted position.
5. When all groups have placed their plates in the jar, discharge the packet as follows (or follow the instructions for your system):
   a. Stick the methylene blue strip on the wall of the jar.
   b. Open the packet and add 10 mL of distilled water.
   c. Place the open packet in the jar with the label facing inward.
   d. Immediately close the jar.
6. Place the second plate and the anaerobic jar in the $35 \pm 2°C$ incubator for 24 to 48 hours. Be sure the plate is inverted.

Lab Two

1. Examine and compare the growth on the plates. (Note: Density of growth will be the most useful basis for comparison.)
2. Record your results and interpretations in the chart provided on the Data Sheet.

References

Allen, Stephen D., Christopher L. Emery, and David M. Lyerly. 2003. Chapter 54 in *Manual of Clinical Microbiology,* 8th ed. Edited by Patrick R. Murray, Ellen Jo Baron, James H. Jorgensen, Michael A. Pfaller, and Robert H. Yolken. ASM Press, American Society for Microbiology, Washington, DC.

Forbes, Betty A., Daniel F. Sahm, and Alice S. Weissfeld. 2002. Chapter 10 in *Bailey & Scott's Diagnostic Microbiology,* 11th ed. Mosby, St. Louis.

Winn, Washington C., et al. 2006. *Koneman's Color Atlas and Textbook of Diagnostic Microbiology,* 6th ed. Lippincott Williams & Wilkins, Baltimore.

Zimbro, Mary Jo, and David A. Power, eds. 2003. *Difco™ and BBL™ Manual—Manual of Microbiological Culture Media.* Becton Dickinson and Company, Sparks, MD.

NAME ________________________________ DATE ________________________________

LAB SECTION ________________________ I WAS PRESENT AND PERFORMED THIS EXERCISE (INITIALS) ______

Anaerobic Jar

OBSERVATIONS AND INTERPRETATIONS

Indicate the amount of growth in the broth.
"0" = no growth "3" = abundant growth "1" and "2" = degrees of growth in between

Organism	Growth on Aerobic Plate	Growth on Anaerobic Plate	Aerotolerance Category

Questions

1 If, after incubation, you observed that the methylene blue indicator strip inside the jar was blue, what would you guess the internal environment to be—aerobic or anaerobic? How would you expect the growth on the plate inside the jar to differ from the plate incubated outside the jar?

DATA SHEET 2-7

2 Which of the three organisms would be most *negatively* affected by the conditions described in question 1?

3 An alternative to the anaerobic jar is a candle jar, in which a candle is placed in the jar, lit, and the lid closed to enable the flame to use the available oxygen. Typically, in this system, not all of the oxygen is used. Which types of organisms would most likely benefit from this environment?

EXERCISE 2-8

The Effect of Temperature on Microbial Growth

Theory

Bacteria have been discovered living in habitats ranging from –10 degrees Celsius to more than 110 degrees Celsius. The temperature range of any single species, however, is a small portion of this overall range. As such, each species is characterized by a minimum, maximum, and optimum temperature—collectively known as its **cardinal temperatures** (Figure 2-45). Minimum and maximum temperatures are, simply, the temperatures below and above which the organism will not survive. Optimum temperature is the temperature at which an organism shows the greatest growth over time—its highest growth rate.

Organisms that grow only below 20 degrees Celsius are called **psychrophiles**. These are common in ocean, Arctic, and Antarctic habitats where the temperature remains permanently cold with little or no fluctuation. Organisms adapted to cold habitats that fluctuate from about 0 degrees to above 30 degrees Celsius are called **psychrotrophs.** Bacteria adapted to temperatures between 15 degrees and 45 degrees Celsius are known as **mesophiles**.

Most bacterial residents in the human body, as well as numerous human pathogens, are mesophiles. **Thermophiles** are organisms adapted to temperatures above 40 degrees Celsius. Typically, they are found in composting organic material and in hot springs. Thermophiles that will not grow at temperatures below 40 degrees are called **obligate thermophiles**; those that will grow below 40 degrees are known as **facultative thermophiles.** Bacteria isolated from hot ocean floor ridges living between 65 and 110 degrees Celsius are called **extreme thermophiles.** Extreme thermophiles grow best above 80 degrees Celsius. Figure 2-46 illustrates bacterial temperature ranges and classifications.

Application

This is a qualitative procedure designed for observing the effect of temperature on bacterial growth. It allows an estimation of the cardinal temperatures for a single species.

In This Exercise

Today you will examine the growth characteristics of four organisms at five different temperatures. In addition, you will observe the influence of temperature on pigment production.

Materials

Per Class

✓ Five incubating devices set at 10°C, 20°C, 30°C, 40°C, and 50°C. These devices may be any combination of the following:
- refrigerator
- incubator
- hot water bath
- cold water bath

Per Student Group

✓ 20 sterile Nutrient Broths

✓ two Trypticase Soy Agar (TSA) plates

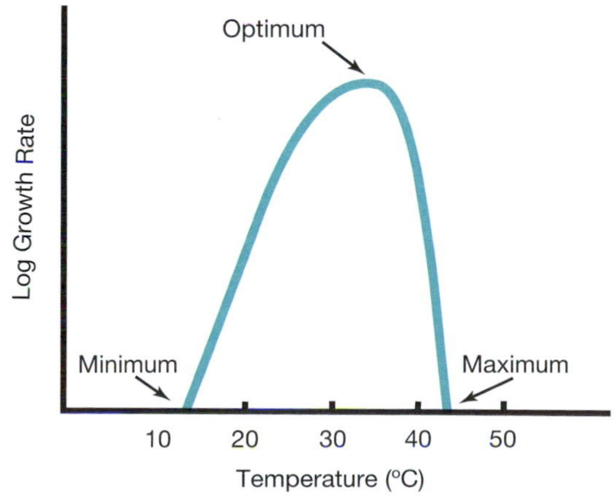

2-45 Typical Growth Range of a Mesophile ◆ The "minimum" and "maximum" are temperatures beyond which no growth takes place. The "optimum" is the temperature at which growth rate is highest.

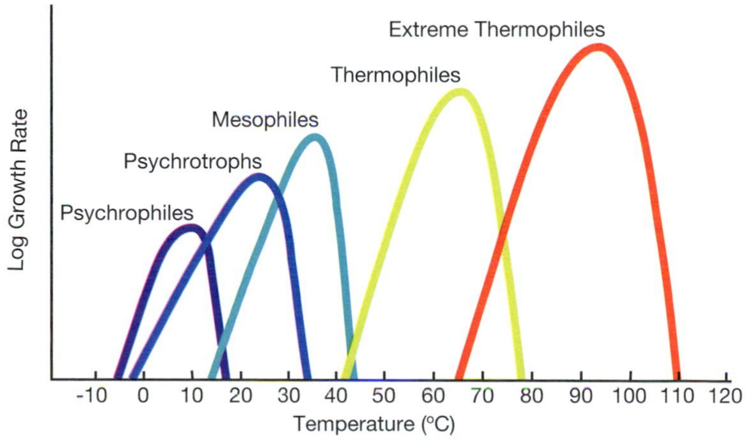

2-46 Thermal Classifications of Bacteria ◆ Refer to the text for a description of each category.

- ✓ four sterile transfer pipettes
- ✓ fresh Nutrient Broth cultures of:
  - ♦ *Escherichia coli*
  - ♦ *Serratia marcescens*
  - ♦ *Geobacillus stearothermophilus*
  - ♦ *Pseudomonas fluorescens*

Procedure

Lab One

1. Obtain 20 Nutrient Broths—one broth for each organism at each temperature. Label them accordingly. Also obtain two TSA plates and label them 20°C and 35°C, respectively.
2. Mix each culture thoroughly before making the following transfers. Using a sterile pipette, transfer a *single drop* of each broth culture to its appropriate Nutrient Broth tube. (Note: Because you will be comparing the amount of growth in the Nutrient Broth tubes, you must be sure to begin by transferring the same volume of culture to each one. Use the same pipette for all transfers with a single organism. Alternatively, your instructor may have you use a digital pipette.)
3. Using a simple zigzag pattern (as in Figure 2-1), inoculate each plate with *Serratia marcescens*.
4. Incubate all tubes in their appropriate temperatures for 24 to 48 hours. Incubate the plates in the 20°C and 35°C incubators in an inverted position.

Lab Two

1. Clean the outside of all tubes with a tissue, and place them in a test tube rack organized into groups by organism.
2. Shake each broth gently until uniform turbidity is achieved.
3. Compare all tubes in a group to each other. Rate each as 0, 1, 2, or 3, according to its turbidity (0 is clear and 3 is highly turbid). Record these in the Broth Data chart on the Data Sheet.
4. Examine the plates incubated at different temperatures, compare the growth characteristics, and enter your results in the Plate Data chart on the Data Sheet.
5. Using the data from the Broth Data chart, determine the cardinal temperatures and classification of each of the four organisms.
6. On the graph paper provided on the Data Sheet, plot the data (numeric values versus temperature) for the four organisms.

References

Forbes, Betty A., Daniel F. Sahm, and Alice S. Weissfeld. 2002. Chapters 2 and 10 in *Bailey & Scott's Diagnostic Microbiology,* 11th ed. Mosby-Yearbook, St. Louis.

Holt, John G., Ed. 1994. *Bergey's Manual of Determinative Bacteriology,* 9th ed. Williams and Wilkins, Baltimore.

Moat, Albert G., John W. Foster, and Michael P. Spector. 2002. Pages 597–601 in *Microbial Physiology,* 4th ed. Wiley-Liss, New York.

Prescott, Lansing M., John P. Harley, and Donald A. Klein. 2005. Chapter 6 in *Microbiology,* 6th ed., WCB McGraw-Hill, Boston.

Varnam, Alan H., and Malcolm G. Evans. 2000. *Environmental Microbiology.* ASM Press, Washington, DC.

White, David. 2000. Pages 384–387 in *The Physiology and Biochemistry of Prokaryotes,* 2nd ed. Oxford University Press, New York.

Winn, Washington C., et al. 2006. *Koneman's Color Atlas and Textbook of Diagnostic Microbiology,* 6th ed. Lippincott Williams & Wilkins, Baltimore.

Name ______________________ Date ______________________

Lab Section ______________________ I was present and performed this exercise (initials) ______

DATA SHEET 2-8

The Effect of Temperature on Microbial Growth

OBSERVATIONS AND INTERPRETATIONS

1 Record your numeric values for each organism at each temperature.

Broth Data						
Organism	10°C	20°C	30°C	40°C	50°C	Classification

Enter your observations as 0, 1, 2, or 3 (0 is clear and 3 is very turbid).

2 Record the growth characteristics of the *Serratia marcescens* incubated at two temperatures.

Plate Data	
Incubation Temperature	Description of Growth
20°C	
35°C	

Questions

1. Using the data from the chart, determine the cardinal temperatures for each of the four organisms. Circle the optimum temperature for each organism. Use brackets to designate the range for each.

2. Plot the numeric values versus temperature on the graph paper provided for this Data Sheet.

3. Why is it not advisable to connect the data points for each organism in your graph?

4. In what way(s) could you adjust incubation temperature to grow an organism at less than its optimal growth rate?

5. Why do different temperatures produce different growth rates?

NAME ______________________ DATE ______________________

LAB SECTION ______________________ I WAS PRESENT AND PERFORMED THIS EXERCISE (INITIALS) ______

DATA SHEET 2-8

EXERCISE 2-9

The Effect of pH on Microbial Growth

❖ Theory

The conventional means of expressing the concentration (or activity) of hydrogen ions in a solution is "pH." The term pH, which stands for "pondus hydrogenii" (variably defined as hydrogen power or hydrogen potential), was invented in 1909 by the Danish biochemist, Søren Peter Lauritz Sørensen. The 0–14 pH range he developed is a logarithmic scale designed to simplify acid and base calculations that otherwise would be expressed as molar values. Sørensen's formula for the calculation of pH is expressed as follows:

$$pH = -\log [H^+]$$

For example, an aqueous solution containing 10^{-6} moles of disassociated hydrogen ions per liter would be converted using Sørensen's formula as follows:

$$pH = -\log [H^+]$$
$$pH = -\log 10^{-6} M\ H^+$$
$$pH = 6$$

Pure water contains 10^{-7} moles of hydrogen ions per liter and has a pH of 7. As hydrogen ions increase, the solution becomes more acidic and the pH decreases (see Table 2-1).

Bacteria live in habitats throughout the pH spectrum; however, the range of most individual species is small. Like temperature and salinity, pH tolerance is used as a means of classification. The three major classifications are

1. **acidophiles**: organisms adapted to grow well in environments below about pH 5.5
2. **neutrophiles:** organisms that prefer pH levels between 5.5 and 8.5
3. **alkaliphiles**: organisms that live above pH 8.5.

Under normal circumstances, bacteria maintain a near-neutral internal environment regardless of their habitat; pH changes outside an organism's range may destroy necessary membrane potential (in the production of ATP) and damage vital enzymes beyond repair. This **denaturing** of cellular enzymes may be as minor as conformational changes in the proteins' tertiary structure, but usually is lethal to the cell.

Acids from carbohydrate fermentation and alkaline products from protein metabolism are sufficient to disrupt microbial enzyme integrity when grown in vitro. This is

TABLE 2-1 pH Scale

Solution Classification Acidity/Alkalinity	pH	H+ Concentration in Moles/Liter	Common Examples	Organismal Classification
↑	0	10^{0}	Nitric acid	↑
	1	10^{-1}	Stomach acid	
	2	10^{-2}	Lemon juice	
	3	10^{-3}	Vinegar, cola	
	4	10^{-4}	Tomatoes, orange juice	
	5	10^{-5}	Black coffee	Acidophiles
Acidic	6	10^{-6}	Urine	
Neutral	7	10^{-7}	Pure water	Neutrophiles
Alkaline	8	10^{-8}	Seawater	
	9	10^{-9}	Baking soda	Alkaliphiles
	10	10^{-10}	Soap, milk of magnesia	
	11	10^{-11}	Ammonia	
	12	10^{-12}	Lime water [$Ca(OH)_2$]	
	13	10^{-13}	Household bleach	
↓	14	10^{-14}	Drain cleaner	↓

why buffers made from weak acids such as hydrogen phosphate are added to bacteriological growth media. In solution, buffers are able to alternate between weak acid ($H_2PO_4^-$) and conjugate base ($HPO_4 2^-$) to maintain H^+/OH^- equilibrium.

$$H^+ + HPO_4^{2-} \rightarrow H_2PO_4^-$$

$$OH^- + H_2PO_4^- \rightarrow HPO_4^{2-} + H_2O$$

Application

This is a qualitative procedure used to estimate the minimum, maximum, and optimum pH for growth of a bacterial species.

In This Exercise

Today you will cultivate and observe the effects of pH on four organisms. Then you will classify them based on your results.

Materials

Per Student Group

- ✓ five of each pH adjusted Nutrient Broth as follows: pH 2, pH 4, pH 6, pH 8, and pH 10
- ✓ four sterile transfer pipettes
- ✓ fresh Nutrient Broth cultures of:
  - *Lactobacillus plantarum*
  - *Lactococcus lactis*
  - *Staphylococcus saprophyticus*
  - *Alcaligenes faecalis*

Medium Recipe

pH-Adjusted Nutrient Broth

♦ Beef extract	3.0 g
♦ Peptone	5.0 g
♦ Distilled or deionized water	1.0 L
♦ NaOH or HCl as needed to adjust pH	

Procedure

Lab One

1. Obtain five tubes of each pH broth—one of each pH per organism (20 tubes total). Label them accordingly.
2. Mix each culture thoroughly before making the following transfers. Using a sterile pipette, transfer a *single drop* of each broth culture to its appropriate Nutrient Broth tube. (Note: Because you will be comparing the amount of growth in the Nutrient Broth tubes, you must be sure to begin by transferring the same volume of culture to each tube. Use the same pipette for all transfers with a single organism. Alternatively, your instructor may have you use a digital pipette.)
3. Incubate all tubes at 35 ± 2°C for 48 hours.

Lab Two

1. Clean the outside of all tubes with a tissue and place them in a test tube rack organized into groups by organism.
2. Shake each broth gently until uniform turbidity is achieved.
3. Compare all tubes in a group to each other. Rate each one as 0, 1, 2, or 3 according to its turbidity (0 is clear and 3 is highly turbid). Enter your observations in the chart on the Data Sheet. (Note: Some color variability may exist between the different pH broths; therefore, base your conclusions solely on turbidity, not color.)
4. Determine the range and classification of each test organism. Record the information on the Data Sheet.
5. On the graph paper provided on the Data Sheet, plot the data (numeric values versus pH) of the four organisms.

References

Forbes, Betty A., Daniel F. Sahm, and Alice S. Weissfeld. 2002. Chapter 10 in *Bailey & Scott's Diagnostic Microbiology,* 11th ed. Mosby, St. Louis.

Holt, John G., ed. 1994. *Bergey's Manual of Determinative Bacteriology,* 9th ed. Williams and Wilkins, Baltimore.

Varnam, Alan H., and Malcolm G. Evans. 2000. *Environmental Microbiology.* ASM Press, Washington, DC.

Winn, Washington C., et al. 2006. *Koneman's, Color Atlas and Textbook of Diagnostic Microbiology,* 6th ed. Lippincott Williams & Wilkins, Baltimore.

Name ______________________ Date ______________________

Lab Section ______________________ I was present and performed this exercise (initials) ______

The Effect of pH on Microbial Growth

OBSERVATIONS AND INTERPRETATIONS

Record the numeric values for each organism at each pH.

Organism	pH 2	pH 4	pH 6	pH 8	pH 10	Classification

Enter visual readings as 0, 1, 2, or 3 (0 is clear and 3 is very turbid).

Questions

1 Circle the pH optimum for each organism. Place brackets around the range. Is there any overlap between species?

2 Account for the inability of organisms to grow outside their pH ranges. Why, for instance, are alkaliphiles able to survive at high pHs when neutrophiles cannot?

3 Where is the pH optimum relative to the pH range for each organism? Do you see any parallels between these data and the data produced in Exercise 2-8? Explain.

4 Plot the data (numeric values versus pH) for each organism on the graph paper provided.

5 Why is it not advisable to connect the data points for each organism in your graph?

Name ______________________________ Date ______________________________

Lab Section ______________________ I was present and performed this exercise (initials) ______

Data Sheet 2-9

EXERCISE 2-10

The Effect of Osmotic Pressure on Microbial Growth

Theory

Water is essential to all forms of life. It is not only the principal component of cellular cytoplasm, but also an essential source of electrons and hydrogen ions. Prokaryotes, like plants, require water to maintain cellular **turgor pressure**. Whereas eukaryotic cells burst with a constant influx of water, prokaryotes require water to prevent shrinking of the cell membrane, resulting in separation from the cell wall—an occurrence known as **plasmolysis**.

Many bacteria regulate turgor pressure by transporting in and maintaining a relatively high cytoplasmic potassium or sodium ion concentration, thereby creating a concentration gradient that promotes inward **diffusion** of water. For bacteria living in saline habitats, the job of maintaining turgor pressure is continuous because of the constant efflux of water.

Irrespective of a cell's efforts to control its internal environment, natural forces will cause water to move through its semipermeable membrane from an area of low **solute** concentration to an area of high solute concentration. In a solution in which solute concentration is low, water concentration is high, and vice versa. Therefore, water moves through a cell membrane from where its concentration is high to where its concentration is low. This process is called **osmosis**, and the force that controls it is called **osmotic pressure.**

Osmotic pressure is a quantifiable term and refers, specifically, to the ability of a solution to *pull water toward itself* through a semipermeable membrane. If a bacterial cell is placed into a solution that is **hyposmotic** (a solution having low osmotic pressure), there will be a *net* movement of water into the cell. If an organism is placed into a **hyperosmotic** solution (a solution having high osmotic pressure), there will be a net movement of water out of the cell. For a bacterial cell in an **isosmotic** solution (a solution having osmotic pressure equal to that of the cell), water will tend to move in both directions equally; that is, there is no net movement (Figure 2-47).

Bacteria constitute a diverse group of organisms and, as such, have evolved many adaptations for survival. Microorganisms tend to have a distinct range of salinities that are optimal for growth, with little or no survival outside that range. For example, some bacteria called **halophiles** grow optimally in NaCl concentrations of 3% or higher. **Extreme halophiles** are organisms with specialized cell membranes and enzymes that require salt concentrations from 15% up to about 25% and will not survive where salinity is lower (Figure 2-48). Except for a few **osmotolerant** bacteria, which will grow over a wide range of salinities, most bacteria live where NaCl concentrations are less than 3%.

Application

This is a qualitative procedure used to demonstrate bacterial tolerances to NaCl.

In This Exercise

You will be growing three microorganisms at a variety of NaCl concentrations to determine the tolerance range of salinities and optimum salinity for each organism.

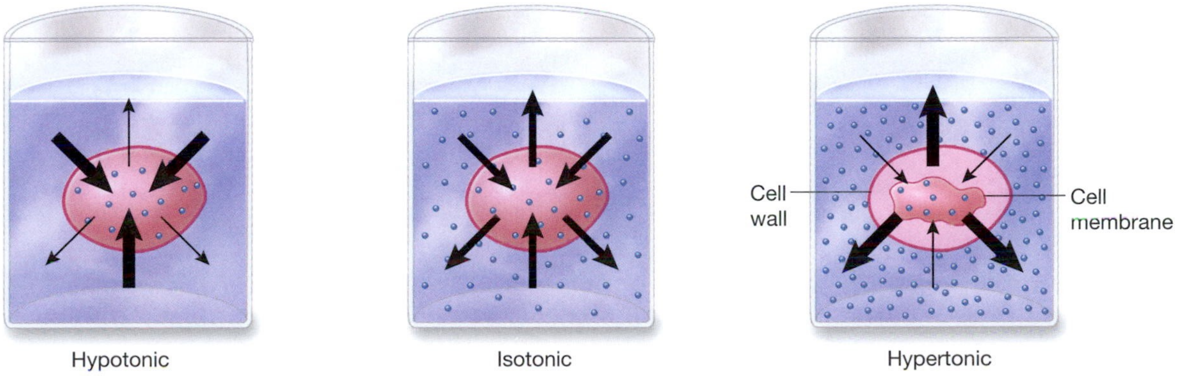

Cell wall
Cell membrane
Hypotonic
Isotonic
Hypertonic

2-47 The Effect of Osmotic Pressure on Bacterial Cells ◆ This osmosis diagram illustrates the movement of water into and out of cells. The labels refer to the osmotic pressure of the solution outside the cell. In a hyposmotic environment, the cell has greater osmotic pressure, so the net movement of water (arrows) will be into the cell. In an isosmotic environment, there is no net movement because the osmotic pressure of the cell and that of the environment are equal. (Actually, the water is moving equally in both directions.) In a hyperosmotic environment, the osmotic pressure of the environment is greater, so the net movement is outward and results in plasmolysis. Note the shrinking membrane (CM) and the rigid cell wall (CW) in the hyperosmotic solution.

2-48 **A Saltern in San Diego Bay** ◆ Salterns are low pools of saltwater used in the harvesting of salt. As water evaporates, the saltwater becomes saltier and saltier, until only salt remains. This can then be purified and sold. The colors in the pools result from different communities of halophilic microorganisms that are associated with different salinities as the pools dry out.

Materials

Per Student Group

✓ four tubes each of saline medium prepared with 0%, 5%, 10%, 15%, 20%, and 25% NaCl

✓ three sterile transfer pipettes

✓ fresh cultures of:

- *Escherichia coli*
- *Halobacterium*
- *Staphylococcus epidermidis*

Medium Recipe

Modified Halobacterium Broth

◆ Sodium chloride	0 g, 50 g, 100 g, 150 g, 200 g, or 250 g
◆ Magnesium sulfate, heptahydrate	20.0 g
◆ Trisodium citrate, dihydrate	3.0 g
◆ Potassium chloride	2.0 g
◆ Casamino acids	5.0 g
◆ Yeast extract	5.0 g
◆ Deionized water	1.0 L

Adjust pH to 7.2 using 5 M of concentrated HCl.

Procedure

Lab One

1 Obtain four tubes of each *Halobacterium* Broth—one of each concentration per organism plus a control. Label them accordingly.

2 Mix each culture thoroughly before making the following transfers. Using a sterile pipette, transfer a *single drop* of each broth culture to its appropriate *Halobacterium* Broth tube. (Because you will be comparing the amount of growth in the *Halobacterium* Broth tubes, take care to begin by transferring the same volume of culture to each tube. Use the same pipette for all transfers with a single organism. Alternatively, your instructor may have you use a digital pipette.)

3 Incubate all tubes at 35 ± 2°C for 48 hours.

Lab Two

1 Clean the outside of all tubes with a tissue, and place them in a test tube rack organized into groups by organism.

2 Shake each broth gently until uniform turbidity is achieved.

3 Compare all tubes in a group to each other and to the control. Rate each one as 0, 1, 2, or 3 according to its turbidity (0 is clear and 3 is very turbid.) (Note: The different broths may show some color variability. Check the uninoculated controls and, thus, base your conclusions solely on turbidity, not color.)

4 Enter your results on the Data Sheet.

5 On the graph paper provided with the Data Sheet, plot the numeric value of turbidity versus the salt concentration for all three organisms.

References

Forbes, Betty A., Daniel F. Sahm, and Alice S. Weissfeld. 2002. Chapter 10 in *Bailey & Scott's Diagnostic Microbiology,* 11th ed. Mosby, Inc., St. Louis.

Hauser, Juliana T. 2006. *Techniques for Studying Bacteria and Fungi.* Carolina Biological Supply Company, 2700 York Road, Burlington, NC.

Holt, John G., Ed. 1994. *Bergey's Manual of Determinative Bacteriology,* 9th ed. Williams and Wilkins, Baltimore.

Koneman, Elmer W., Stephen D. Allen, William M. Janda, Paul C. Schreckenberger, and Washington C. Winn, Jr. 1997. *Color Atlas and Textbook of Diagnostic Microbiology,* 5th ed. J. B. Lippincott Co., Philadelphia.

Moat, Albert G., John W. Foster, and Michael P. Spector. 2002. Pages 582–587 in *Microbial Physiology,* 4th ed. Wiley-Liss, New York.

Varnam, Alan H., and Malcolm G. Evans. 2000. *Environmental Microbiology.* ASM Press, Washington, DC.

White, David. 2000. Pages 388–394 in *The Physiology and Biochemistry of Prokaryotes,* 2nd ed. Oxford University Press, New York.

Name ______________________ Date ______________________

Lab Section ______________________ I was present and performed this exercise (initials) ______

Effect of Osmotic Pressure on Microbial Growth

OBSERVATIONS AND INTERPRETATIONS

Record the relative growth in each medium, using 0, 1, 2, and 3 to represent relative growth.

Organism	NaCl Concentration					
	0%	5%	10%	15%	20%	25%
Control						

Questions

1 *Circle* the optimum salinity for each organism.

2 Place *brackets* around the range.

3 Which organism(s) exhibit(s) the greatest tolerance range?

4 Use your textbook or other available reference to look up the habitat of each organism. Do the habitats make sense in light of the tolerance ranges you obtained? Explain your answers.

5 Plot your data ("number" values vs. salt concentration) on the graph paper provided.

6 Why is it not advisable to connect the data points for each organism in your graph?

Name ______________________ Date ______________________

Lab Section ______________________ I was present and performed this exercise (initials) ______

Data Sheet 2-10

Control of Pathogens: Physical and Chemical Methods

Every patient in a hospital or other clinical setting has the right to expect that he or she will not contract a disease or infection while in that institution's care. Every person donating blood at a blood bank or mobile center has the right to expect that all materials and surfaces they come in contact with will be free of pathogens. Workers in health clinics, hospitals, medical laboratories, and public health laboratories have the right to assume that reasonable precautions have been and are being taken to protect their safety while in the workplace.

These are only a few of the many reasons why the importance of understanding and use of pathogen control systems cannot be overemphasized. Fortunately, with relatively few exceptions, the above-described conditions exist in this and other developed countries largely because of the dedication of thousands of employees and the oversight of dozens of international, governmental, and private organizations such as the World Health Organization (WHO), Centers For Disease Control and Prevention (CDC), Food and Drug Administration (FDA), Environmental Protection Agency (EPA), American Public Health Association (APHA), and Association of Official Analytical Chemists (AOAC). These and many other federal and private organizations are responsible for the proper testing, registration, and classification of the substances or systems used to prevent the spread of pathogens.

These substances or systems, both chemical and physical, are referred to broadly as germicides. Some germicides are specific in nature and typically include the name of the target pathogen, such as "tuberculocide," "virucide," or "sporocide." Most germicides are broad-spectrum and, thus, target a wide variety of pathogens. Although some overlap occurs, germicidal systems fall into three categories: decontamination, disinfection, or sterilization.

1 **Decontamination** is the lowest level of control and is defined as "reduction of pathogenic microorganisms to a level at which items are safe to handle without protective attire." Decontamination usually includes physical cleaning with soaps or detergents, and removal of all (ideally) or most organic and inorganic material. Proper cleaning of all instruments and surfaces is considered the critical first step toward disinfection or sterilization because, to be fully effective, a disinfectant or sterilant must come in direct contact with all pathogens present. Materials left to dry on a surface or apparatus can actually shield pathogens from a disinfecting or sterilizing agent or otherwise neutralize it.

2 **Disinfection** is the next level of control and is divided into three sublevels—low, medium, and high—based on effectiveness against specific control pathogens or their surrogates. All sublevels kill large numbers, if not all, of the targeted pathogens but typically do not kill large numbers of spores. Some high-level disinfectants are called chemical sterilants because they have the ability to kill all vegetative cells and spores.

Disinfectants typically are liquid chemical agents but can also be solid or gaseous. Other disinfection methods include dry heat, moist heat, and ultraviolet light. Disinfectants that are designed to reduce or eliminate pathogens on or in living tissue are called antiseptics. For obvious safety reasons, antiseptics are subject to additional testing to minimize the risks of side effects. Some antiseptics are considered drugs and, therefore, are regulated by the FDA.

3 **Sterilization** is the complete elimination of viable organisms including spores and, as such, is the highest level of pathogen control. Sterilization can be achieved by some chemicals, some gases, incineration, dry heat, moist heat, ethylene oxide gas, ionizing radiation (gamma, X-ray, and electron beam), low-temperature plasma (utilizing a combination of chemical sterilants and ultraviolet radiation in a vacuum chamber), or low-temperature ozone (utilizing bottled oxygen, water, and electricity in a chamber to produce a lethal level of ozone).

In this unit you will examine both physical and chemical means of pathogen control. The following exercises illustrate the germicidal effects of UV radiation, disinfection, antisepsis, and steam (moist heat) sterilization. (For more information on microbial control, refer to Exercise 1-2 GloGerm, and to Exercise 7-2 Antimicrobial Susceptibility Test.)

EXERCISE 2-11

Steam Sterilization[1]

Theory

Of the many methods or agents that have been developed for sterilizing surgical and dental instruments, microbiological media, infectious waste, and other materials not harmed by moisture or heat, steam is still the most effective and most common. The device used most commonly for this purpose is called a steam sterilizer, or **autoclave**. Autoclaves are relatively safe, easy to operate, and, if used properly, effective at killing all microbial vegetative cells and spores.

Under atmospheric pressure, water boils at 100ºC (212ºF). At pressures above atmospheric pressure, water must be heated above 100ºC before it will boil. Much like home pressure cookers, which create pressure and high temperatures to shorten cooking times, autoclaves use super-heated steam under pressure to kill heat-resistant organisms. Examples of heat-resistant organisms include members of the spore-producing genera—*Bacillus*, *Geobacillus*, and *Clostridium*.

In the microbiology laboratory, sterilizing temperature usually is set at between 121° and 127°C (250° and 260°F); however, sterilizing time can vary according to the size and consistency of the material being sterilized.[2] At a minimum, to be sure that all vegetative cells and spores have been killed, items being processed must reach optimum temperature for at least 15 minutes. This includes items deep inside the autoclave container that may be partially insulated from the steam by surrounding items. Understandably, larger loads take longer to process than smaller loads do. (Certain sensitive applications, such as microbiological media preparation, in which formula integrity must be maintained and when specific growth inhibiting ingredients are included, lower times and temperatures are acceptable.)

To maintain laboratory safety and comply with laws regarding infectious waste disposal, sterilizers must be checked regularly for operating effectiveness. Special thermometers placed in an autoclave can record the maximum temperature reached inside the chamber but do not measure how low the temperature dips during the normal cycling of the heating elements. Specialized color-coded autoclave tape can be a fairly good indicator that sterilization is complete, but the only way, with certainty, to determine that sterilization has been achieved is by using a device called a **biological indicator** (Figure 2-49).

Biological indicators, as the name suggests, are test systems that contain something living. A typical biological indicator that is particularly useful for testing autoclaves is one that contains **bacterial spores**. Bacterial spores, the dormant form of an organism, are highly resistant to both chemical and physical means of control. Therefore, if an autoclave kills the spores in the test system, it is safe to assume that it will destroy other microbes as well. This is important not only for safety reasons but is of legal importance as well. Public health and safety agencies maintain compliance with hazardous waste disposal regulations by requiring regular testing of autoclaves used to process biohazardous material.

A typical system, and the one selected for today's lab, includes a small heat-resistant plastic vial containing a glass ampule of sterile fermentation broth. Also inside the vial, but outside of the ampule is a strip of filter paper containing bacterial spores. The vial is placed inside the autoclave and heated at 121°C for 15 minutes. After autoclaving, the vial is cooled and crushed with a special device that breaks the inner ampule without damaging the plastic vial. Breaking the ampule allows the fermentation

2-49 Autoclave Biological Indicators ◆ From left to right are: positive control, test vial, and negative control. The positive control was not autoclaved, but the ampule was broken to allow growth. The negative control was not autoclaved and the ampule was not broken. The vial in the center was autoclaved and the ampule was broken. The lack of color change suggests that the organism in the test vial has been killed and the autoclave is working properly.

[1] Steam sterilization is not yet considered reliable at inactivating prion proteins, such as those that cause Bovine Spongiform Encephalopathy, the so-called Mad Cow Disease, and its variant, Creutzfeldt-Jakob disease.

[2] In clinics, hospitals, or other locations where surgical instruments are being processed, the World Health Organization (WHO) recommends a minimum processing time and temperature of 134°C for 18 minutes.

broth to come in contact with the bacterial spores in the filter paper. The vials are then incubated at 55°C for 48 hours. If the spores have been killed in the autoclave, incubation will produce no growth. If they have not been killed, they will germinate and ferment the substrate in the broth. A pH-indicating dye, included in the broth, will reveal any acid produced (during fermentation) with a distinctive color change. No color change during incubation is, thus, an indication that sterilization is complete, the spores have been killed, and the autoclave is operating properly. (For more information on fermentation, refer to Section 5 and Appendix A.)

Application

Biological indicators are available in many forms and commonly used to test the efficiency of steam sterilizers.

In This Exercise

You are going to use bacterial spores and their resistance to steam sterilization to test the effectiveness of your lab's autoclave. This procedure is written for a product called B/T Sure Biological Indicator[3], but can be applied to other brands as well. B/T Sure Biological Indicators include a pH indicator that turns the broth from purple to yellow under the acidic conditions produced by fermentation.

Materials

Per Class

- ✓ one (or more) steam autoclave(s)
- ✓ incubator set at 55°C

Per Student Group

- ✓ four B/T Sure Biological Indicators
- ✓ one autoclave pan

Procedure

1. Obtain four B/T Sure Biological Indicator vials and label them #1, #2, #3, and #4 with a Sharpie® or other permanent marker. Do not label the vial with tape or paper, as this will insulate it from the steam.
2. Place vial #1 on its side uncovered in an autoclave pan.
3. Place vial #2 inside a container (or multiple containers), as well insulated as can be achieved with materials provided by your instructor. We recommend placing the vial inside a screw-capped test tube inside other tubes—two, three, or four layers deep. Place this vial in the autoclave pan with vial #1.
4. Do nothing with vials #3 and #4 as yet.
5. Place the autoclave pan in the autoclave.
6. Follow your instructor's guidelines to add water to the chamber, set the temperature at 121°C (250°F), and set the timer at 15 minutes. When instructed to do so, close and start the autoclave.
7. When autoclaving is complete, all of the steam has been vented, and the machine has been allowed to cool slightly, remove your pan (while protecting your hands with appropriate gloves) and allow the contents to cool to room temperature.
8. Keeping the vials in an upright position, use the crushing device to squeeze vials #1, #2, and #3 (one at a time) until you hear the glass ampule inside the vial break.
9. Place these vials along with vial #4, again in an upright position, into the 55°C incubator for 48 hours.
10. After incubation, examine the vials for color changes.
11. Using Table 2-2 as a guide, record your results on the Data Sheet.

[3] B/T Sure Biological Indicators are available from Barnstead/Thermolyne, Inc., 2555 Kerper Blvd., Dubuque, IA 52004-0797, 800/553-0039.

References

McDonnell, Gerald E. 2007. *Antisepsis, Disinfection, and Sterilization: Types, Action, And Resistance.* ASM Press, American Society for Microbiology, Washington, DC.

Widmer, Andreas F., and Reno Frei. 2007. Chapter 7 in *Manual of Clinical Microbiology*, 9th ed. Edited by Patrick R. Murray, Ellen Jo Baron, James H. Jorgensen, Marie Louise Landry, and Michael A. Pfaller. ASM Press, American Society for Microbiology, Washington, DC.

TABLE 2-2 Autoclave Biological Indicator Test Results and Interpretations ◆ (Assume the ampules inside the vials have been crushed.)

Broth Color	Interpretation
Purple	No fermentation/acid production in the medium. The organism is dead.
Yellow	Fermentation/acid production in the medium. The organism is alive.

Name ______________________ Date ______________________

Lab Section ______________________ I was present and performed this exercise (initials) ______

Steam Sterilization

OBSERVATIONS AND INTERPRETATIONS

Examine all vials and record your results in the table below.

Indicator Vial	Color Result	Interpretation
Vial #1		
Vial #2		
Vial #3		
Vial #4		

Questions

1 Why are the bacterial spores placed on the paper strip and not directly in the fermentation broth?

2 What is the purpose of the unautoclaved/unbroken vial?

3 How would you interpret the following combinations of results produced in this experiment?

a. Uncovered vial—purple, covered vial—purple, positive control—purple, negative control—purple?

b. Uncovered vial—purple, covered vial—purple, positive control—yellow, negative control—purple?

c. Uncovered vial—purple, covered vial—yellow, positive control—yellow, negative control—purple?

d. Uncovered vial—yellow, covered vial—yellow, positive control—yellow, negative control—purple?

e. Uncovered vial—yellow, covered vial—yellow, positive control—yellow, negative control—yellow?

4 What changes would you make to avoid repeating the faulty scenarios illustrated in question 3?

EXERCISE 2-12

The Lethal Effect of Ultraviolet Radiation on Microbial Growth

Theory

Ultraviolet radiation (UV light) is a type of **electromagnetic energy.** Like all electromagnetic energy, UV travels in waves and is distinguishable from all others by its **wavelength**. Wavelength is the distance between adjacent wave crests and is typically measured in nanometers (nm) (Figure 2-50).

Ultraviolet light is divided into three groups categorized by wavelength:

UV-A, the longest wavelengths, ranging from 315 to 400 nm

UV-B, wavelengths between 280 and 315 nm

UV-C, wavelengths ranging from 100 to 280 nm (These wavelengths are most detrimental to bacteria. Bacterial exposure to UV-C for more than a few minutes usually results in irreparable DNA damage and death of the organism. For a discussion on the mutagenic effects of UV and DNA repair, refer to Exercise 10-5.)

Application

Ultraviolet light is commonly used to disinfect laboratory work surfaces.

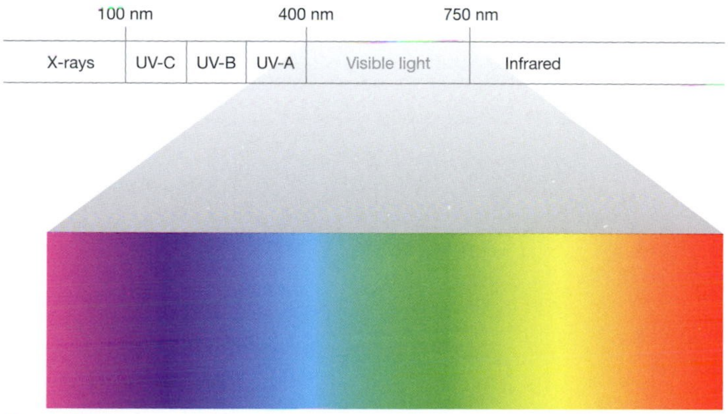

2-50 Electromagnetic Spectrum ◆ The shortest and highest energy wavelengths are those of gamma rays, starting at about at 10^{-5} nm. Radio waves, at the other end of the spectrum, can be one kilometer or longer. Between about 100 nm and 400 nm (just above visible light) is the sliver known as ultraviolet light.

In This Exercise

Today you will compare the effect of UV exposure on two cultures—*Bacillus subtilis* and *Escherichia coli*. Because of the large number of plates to be treated, the work will be divided among six groups of students. Refer to Table 2-3 for assignments.

Materials

Per Student Group

✓ ultraviolet lamp with appropriate shielding
✓ cardboard to cover plates
✓ two Tryptic Soy Agar (TSA) plates (four for group one)
✓ stopwatch or electronic timer
✓ sterile cotton swabs (two per group; four for group 1)
✓ 48-hour Nutrient Broth cultures of:
- *Bacillus subtilis*
- *Escherichia coli*

Procedure

Lab One

1. Enter your group number and exposure time (from Table 2-3) on the Data Sheet.
2. Obtain two TSA plates and label the bottom of each with the name of the organism to be inoculated and your group number. Draw a line to divide the plates in half, and label the sides "A" and "B."
3. Dip a sterile cotton swab into the broth of one culture and wipe the excess on the inside of the tube. Inoculate the appropriate plate by spreading the organism over the entire surface of the agar. Do this by streaking the plate surface completely three times, rotating it one-third turn between streaks. When incubated, this will form a bacterial lawn.
4. Repeat step 3 with the other organism and plate.
5. Place a paper towel on the table next to the UV lamp and soak it with disinfectant.
6. Place your plates under the UV lamp and set the covers, open side down, on the disinfectant-soaked towel. Cover the B half of the plates with the cardboard as shown in Figure 2-51.
7. Turn on the lamp for the prescribed time. ***Caution: Be sure the protective shield is in place and do not look at the light while it is on!***

TABLE **2-3** Group Assignments by Number

Organism	No UV	5 minutes	10 minutes	15 minutes	20 minutes	25 minutes	30 minutes
B. subtilis	1	1	2	3	4	5	6
E. coli	1	1	2	3	4	5	6

8 Immediately replace the covers, invert the plates, and incubate them at 35 ± 2°C for 24 to 48 hours.

2-51 Plates Shielded for UV Exposure ◆

Lab Two

1 Remove your plates from the incubator and observe for growth. Side B should be covered with a bacterial lawn. If this is not the case, see your instructor.

2 Record the growth on side A of each plate in the table on the Data Sheet. Enter "0" if you observe no growth, "1" for poor growth, "2" for moderate growth, and "3" for abundant growth.

3 Using the results from other groups, complete the class data chart on the Data Sheet.

4 On the graph paper provided with the Data Sheet, construct a graph representing growth versus UV exposure time for all three organisms.

References

Lewin, Benjamin. 1990. *Genes IV*. Oxford University Press, Cambridge, MA.

Varnam, Alan H., and Malcolm G. Evans. 2000. *Environmental Microbiology*. ASM Press, Washington, DC.

NAME ______________________ DATE ______________________

LAB SECTION ______________________ I WAS PRESENT AND PERFORMED THIS EXERCISE (INITIALS) ______

DATA SHEET 2-12

The Lethal Effect of Ultraviolet Radiation on Microbial Growth

OBSERVATIONS AND INTERPRETATIONS

Enter your class data below. Score the relative amount of growth on each plate. Use the numbers 0, 1, 2, or 3 (0 is no growth; 3 is abundant growth).

Organism	No UV	5 minutes	10 minutes	15 minutes	20 minutes	25 minutes	30 minutes
B. subtilis							
E. coli							

Questions

1 The purpose of this exercise is to demonstrate the comparative effect of UV on two bacterial populations. This could have been accomplished without the cardboard cover. Why was the cover used?

2 This is not a quantitative exercise. Keeping this in mind, can you see a general trend between bacterial death and UV exposure time?

3 Which organism survived the longest exposure? Why?

4 Why were you told to remove the plate covers prior to exposing them to UV?

5 Using the graph paper provided, construct a single graph of growth (0, 1, 2, 3) versus UV exposure time for the two organisms.

NAME ______________________ DATE ______________________

LAB SECTION ______________________ I WAS PRESENT AND PERFORMED THIS EXERCISE (INITIALS) ______

EXERCISE 2-13

Chemical Germicides: Disinfectants and Antiseptics

Theory

Chemical germicides are substances designed to reduce the number of pathogens on a surface, in a liquid, or on or in living tissue. Germicides designed for use on surfaces (floors, tables, sinks, countertops, surgical instruments, etc.) or liquids are called **disinfectants**. Germicides designed for use on or in living tissue are called **antiseptics**.

Before a new substance can be registered by either the FDA or EPA and allowed on the market, it must be tested and classified according to its effectiveness against pathogens. The Use-Dilution Test, published by the Association of Official Analytical Chemists (AOAC), is one of many commonly used tests for this purpose.

The Use-Dilution Test is a standard procedure used to measure the effectiveness of disinfectants specifically against *Staphylococcus aureus*, *Salmonella enterica* serovar Cholerasuis, and *Pseudomonas aeruginosa*. In the standard procedure, glass beads or stainless steel cylinders coated with living bacteria are exposed to varying concentrations (dilutions) of test germicides, then transferred to a growth medium. After a period of incubation, the medium is examined for growth. If a solution is sufficient to prevent microbial growth at least 95% of the time, it meets the required standards and is considered a usable dilution of that germicide for a specific application. Today's exercise is an adaptation of this method.

Application

This procedure is used to test the effectiveness of germicides against *Salmonella* spp., *Staphylococcus aureus*, and *Pseudomonas aeruginosa*.

In This Exercise

Today you will examine the effectiveness of four germicides—two common household disinfectants and two over-the-counter antiseptics. The disinfectants selected for the exercise are household bleach and Lysol® Brand II Disinfectant. The antiseptics are hydrogen peroxide and isopropyl alcohol. For safety reasons, the organisms used for the test are *Staphylococcus epidermidis* and *Escherichia coli*.

You will first coat the beads with bacteria, expose them to three concentrations of your germicide, and then use them to inoculate sterile Nutrient Broth. If, during exposure to the germicide, all of the bacteria on the bead are killed the broth inoculated with that bead will remain clear. If any of the bacteria survive the germicide exposure, they will reproduce during incubation and make the broth turbid. With these results, you will determine the effective concentration (dilution) of your germicide.

The tasks for the exercise are divided among eight groups of students. Each group will be responsible for one organism and three dilutions of one germicide. Refer to Table 2-4 for your assignments.

Finally, this is an interesting exercise with moderate amount of work involved. If you do not hurry and are careful to use aseptic technique, you will be rewarded with reliable data at the end.

Materials

Per Class (See Table 2-4)

Disinfectants

- ✓ 0.01%, 0.1%, and 1.0% household bleach
- ✓ 25%, 50%, and 100% Lysol® Brand II Disinfectant

Antiseptics

- ✓ 0.03%, 0.3%, and 3% hydrogen peroxide (3% is full strength as purchased at the pharmacy)
- ✓ 10%, 30%, and 50% isopropyl alcohol (70% is full strength as purchased at the pharmacy)

Per Student Group

- ✓ 100 mL flask of sterile deionized water
- ✓ three concentrations of one germicide (listed above)
- ✓ five sterile 60 mm Petri dishes
- ✓ one sterile glass 100 mm Petri dish containing filter or bibulous paper
- ✓ one container of sterile ceramic or glass beads[1]
- ✓ sterile transfer pipette

[1] Sterilized #8 seed beads from a craft store will work for this purpose.

TABLE 2-4 Group Assignments

Germicide	*Staphylococcus aureus*	*Escherichia coli*
Bleach	Group 1	Group 2
Lysol	Group 3	Group 4
Hydrogen Peroxide	Group 5	Group 6
Isopropyl Alcohol	Group 7	Group 8

- ✓ seven sterile Nutrient Broth tubes
- ✓ needle-nose forceps (or appropriate device for aseptically picking up beads)
- ✓ small beaker with alcohol (for flaming forceps)
- ✓ fresh broth cultures of (one per group):
  - *Staphylococcus epidermidis*
  - *Escherichia coli*

Procedure[2]

Timing is important in this procedure. Read through it and make a plan before you begin so your transfers and soaking times are done uniformly and are consistent with those of other groups.

Lab One

1 Enter the name of your organism here:

_______________________________________.

2 Enter the name of your germicide here:

_______________________________________.

3 Obtain all of the necessary items for your group as listed in Materials.

4 Place the materials properly on your workspace as shown in Figure 2-52. Label all seven broths with your group name or number. Also number three of the broths 1–3 also as shown in the diagram. Label the last four broths controls #1, #2, #3, and #4 as shown in the diagram.

5 Label three of the 60 mm plates with the name and concentration of your germicide. Label the other two plates "#4 sterile water" and "#5 sterile water."

6 Add approximately 15 mL of each germicide concentration to its respective plate. Add approximately 15 mL sterile water into each of the other two plates.

7 Mix your bacterial culture until uniform turbidity is achieved.

8 Aseptically transfer one loopful of culture broth to control #1.

9 Alcohol-flame your forceps and aseptically drop four beads into the broth culture. (Note: Store your forceps in the beaker of alcohol between transfers. When it is time to make a transfer, remove the forceps and burn off the alcohol. When finished, return the forceps to the beaker.)

10 After 1 minute, decant the broth into a beaker of disinfectant. Remove as much of the broth as possible without losing the beads in the disinfectant.

11 Dispense the beads onto the sterile filter paper. This can be done by tapping the mouth of the tube on the paper. If this doesn't work, remove the beads with a sterile inoculating loop.

12 Using sterilized forceps spread the beads apart on the paper and allow them to dry. Do not roll them around as this may remove bacteria.

13 After 10 minutes, place one bead in each of the three germicide plates. Mark the time here:

_______________________________________.

14 Place the fourth bead in plate #4 (sterile water).

15 Place a sterile bead in plate #5 (sterile water).

16 After 10 minutes from the time marked above, remove the beads from the solutions, in the same order as they were added, and place them in their respective Nutrient Broths. Mix the broths immediately to disperse any residual disinfectant on the beads.

17 Incubate all seven broths at $35 \pm 2°C$ for 48 hours.

Lab Two

1 Remove all broth tubes from the incubator. Gently mix the controls and examine them for evidence of growth. Enter your results on the Data Sheet. Control #1 should have produced growth (turbidity) and Control #2 should not have produced growth. If both of these conditions have been met, you may proceed. If not, see your instructor.

2 Using controls #1 and #2 as comparisons, examine broths containing beads exposed to germicide (1, 2, and 3). Using "G" to indicate growth and "NG" to indicate no growth, enter your results in both the individual data chart and the class data chart on the Data Sheet.

3 Again using controls #1 and #2 as comparisons, examine controls #3 and #4. Using "G" to indicate growth and "NG" to indicate no growth, enter your results on the Data Sheet.

4 Answer the questions on the Data Sheet.

[2] This procdure has been modified from its original form and is to be used for instructional purposes only.

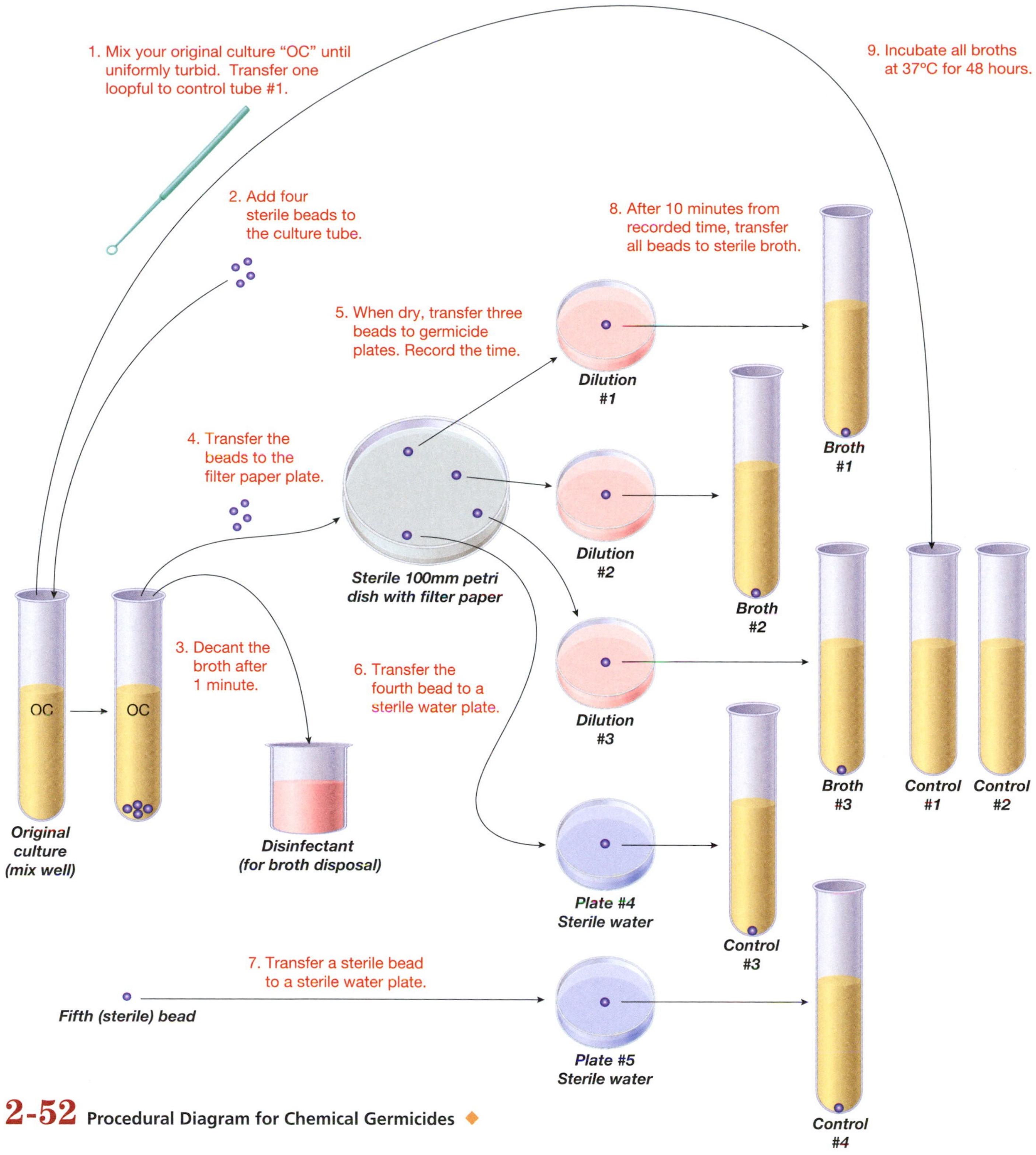
1. Mix your original culture "OC" until uniformly turbid. Transfer one loopful to control tube #1.
2. Add four sterile beads to the culture tube.
3. Decant the broth after 1 minute.
4. Transfer the beads to the filter paper plate.
5. When dry, transfer three beads to germicide plates. Record the time.
6. Transfer the fourth bead to a sterile water plate.
7. Transfer a sterile bead to a sterile water plate.
8. After 10 minutes from recorded time, transfer all beads to sterile broth.
9. Incubate all broths at 37°C for 48 hours.
OC
OC
Original culture (mix well)
Disinfectant (for broth disposal)
Sterile 100mm petri dish with filter paper
Dilution #1
Dilution #2
Dilution #3
Plate #4 Sterile water
Plate #5 Sterile water
Fifth (sterile) bead
Broth #1
Broth #2
Broth #3
Control #1
Control #2
Control #3
Control #4

2-52 Procedural Diagram for Chemical Germicides ◆

References

McDonnell, Gerald E. 2007. *Antisepsis, Disinfection, and Sterilization: Types, Action, And Resistance.* ASM Press, American Society for Microbiology, Washington, DC.

Widmer, Andreas F., and Reno Frei. 2007. Chapter 7 in *Manual of Clinical Microbiology,* 9th ed. Edited by Patrick R. Murray, Ellen Jo Baron, James H. Jorgensen, Marie Louise Landry and Michael A. Pfaller. ASM Press, American Society for Microbiology, Washington, DC.

NAME ______________________ DATE ______________________

LAB SECTION ______________________ I WAS PRESENT AND PERFORMED THIS EXERCISE (INITIALS) ______

Chemical Germicides: Disinfectants and Antiseptics

DATA SHEET 2-13

OBSERVATIONS AND INTERPRETATIONS

1 Enter your individual data below.

Controls	Result
#1	
#2	
#3	
#4	

G = Growth, NG = No Growth

Germicide/Concentrations	Growth
/	
/	
/	

G = Growth, NG = No Growth

2 Enter the class data below.

Organism	Household Bleach			Hydrogen Peroxide			Lysol® Brand II Disinfectant			Isopropyl Alcohol		
	0.01%	0.1%	1%	0.03%	0.3%	3%	25%	50%	100%	10%	30%	50%
S. epidermidis												
E. coli												

G = Growth, NG = No Growth

Questions

1 Compare your results with the class data. Which germicide was most effective and at what concentration? Which was least effective? Defend your choices.

2 Which organism seemed to be most resistant to the germicides?

3 Four controls were used in this exercise. List the specific purposes of each.

SECTION

3

Microscopy and Staining

Microbiology as a biological discipline would not be what it is today without microscopes and cytological stains. Our ability to visualize, sometimes in great detail, the form and structure of microbes too small or transparent to be seen otherwise is attributable to developments in microscopy and staining techniques. In this section you will learn (or refine) your microscope skills. Then you will learn simple and more sophisticated bacterial staining techniques.

The earliest microscopes used visible light to create images and were little more than magnifying glasses. Today, more sophisticated compound light microscopes (Figure 3-1) are used routinely in microbiology laboratories. The various types of light microscopy include bright-field, dark-field, fluorescence, and phase contrast microscopy (Figure 3-2). Although each method has specific applications and advantages, the one used most commonly in introductory classes and clinical laboratories is bright-field microscopy. Many research applications use electron microscopy because of its ability to produce higher quality images of greater magnification.

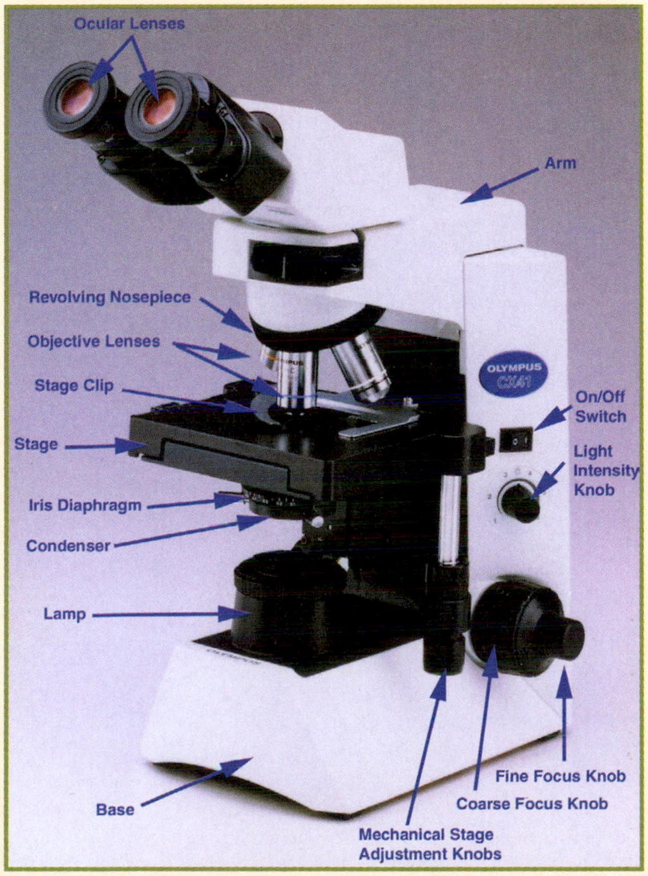
Ocular Lenses
Arm
Revolving Nosepiece
Objective Lenses
Stage Clip
Stage
Iris Diaphragm
Condenser
Lamp
Base
OLYMPUS
On/Off Switch
Light Intensity Knob
Fine Focus Knob
Coarse Focus Knob
Mechanical Stage Adjustment Knobs

3-1 A Binocular Compound Microscope ◆ A quality microscope is an essential tool for microbiologists. Most are assembled with exchangeable component parts and can be customized to suit the user's specific needs.

Photograph courtesy of Olympus America Inc.

EXERCISE 3-1

Introduction to the Light Microscope

❖ Theory

Bright-field microscopy produces an image made from light that is transmitted through a specimen (Figure 3-2A). The specimen restricts light transmission and appears "shadowy" against a bright background (where light enters the microscope unimpeded). Because most biological specimens are transparent, the contrast between the specimen and the background can be improved with the application of stains to the specimen (Exercises 3-4 through 3-9 and 3-11). The "price" of the improved contrast is that the staining process usually kills cells. This is especially true of bacterial-staining protocols.

Image formation begins with light coming from an internal or an external light source (Figure 3-3). It passes through the **condenser** lens, which concentrates the light and makes illumination of the specimen more uniform. **Refraction** (bending) of light as it passes through the **objective lens** from the specimen produces a magnified **real image.** This image is magnified again as it passes through the **ocular lens** to produce a **virtual image** that appears below or within the microscope. The amount of magnification that each lens produces is marked on the lens (Figures 3-4A through 3-4C). Total magnification of the specimen can be calculated by using the following formula:

$$\text{Total Magnification} = \text{Magnification by the Objective Lens} \times \text{Magnification by the Ocular Lens}$$

The practical limit to magnification with a light microscope is around 1300×. Although higher magnifications are possible, image clarity is more difficult to maintain as the magnification increases. Clarity of an image is called **resolution** (Figure 3-5). The **limit of resolution** (or **resolving power**) is an actual measurement of how far apart two points must be for the microscope to view them as being separate. Notice that resolution improves as the limit of resolution is made smaller.

The best limit of resolution achieved by a light microscope is about 0.2 μm. (That is, at its absolute best, a light microscope cannot distinguish between two points closer together than 0.2 μm.) For a specific microscope, the actual limit of resolution can be calculated using the following formula:[1]

$$D = \frac{\lambda}{NA_{condenser} + NA_{objective}}$$

[1] Different equations have been developed to determine approximate limit of resolution, each made from different assumptions. The one used here assumes the $NA_{Objective} \geq NA_{Condenser}$.

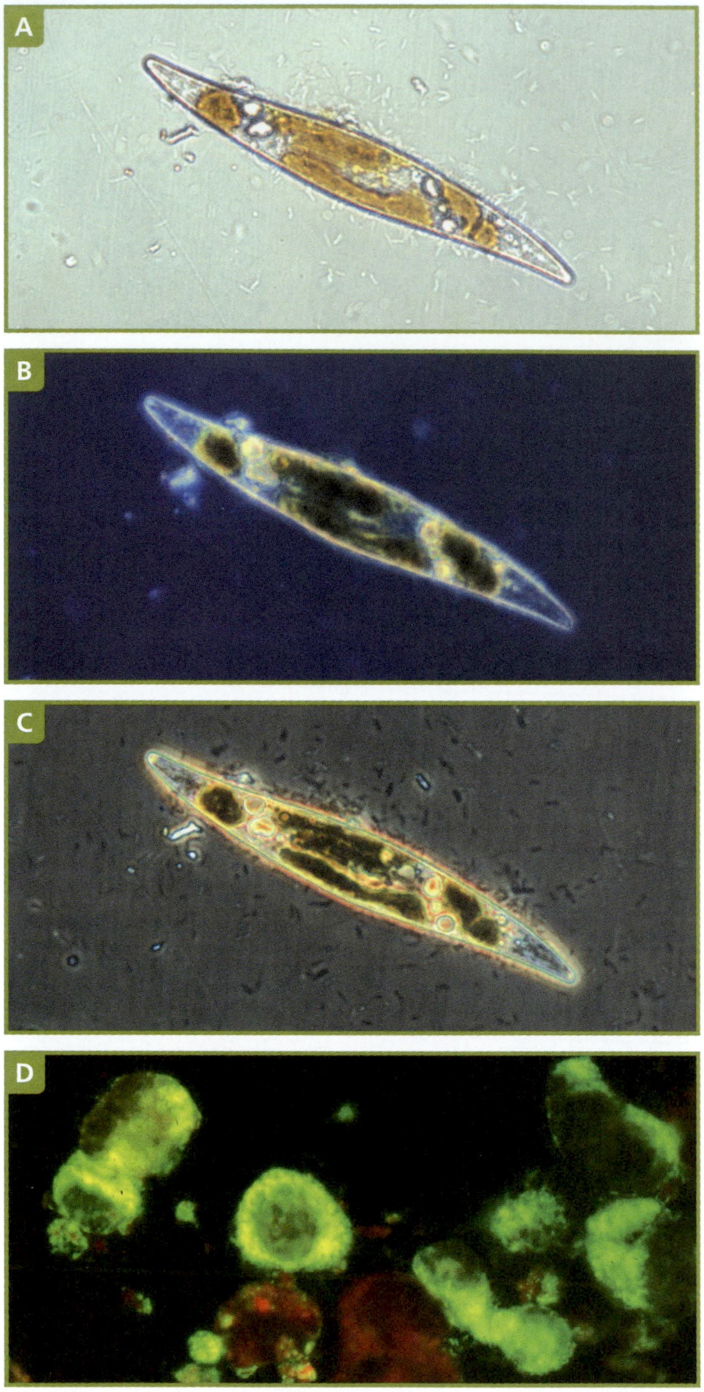

3-2 Types of Light Microscopy ◆ **A** This is a bright-field micrograph of an entire diatom (called a "whole mount"). Because of its thickness, the entire organism will not be in focus at once. Continually adjusting the fine focus to clearly observe different levels of the organism will give a sense of its three-dimensional structure. The bright rods around the diatom are bacteria. **B** This is a dark-field micrograph of the same diatom. Notice that dark field is especially good at providing contrast between the organism's edge and its interior and the background. Notice also that the bacteria are not visible, though this would not always be the case. **C** This phase contrast image of the same diatom shows different details of the interior than what is seen in the other two micrographs. Different parts of the interior and its detail are visible than what is seen in the other two micrographs. Also, notice the bacteria are dark. **D** This is a fluorescence micrograph of *Mycobacterium kansasii*. The apple green is one of the characteristic colors of fluorescence microscopy.

where D is the minimum distance at which two points can be resolved, λ is the wavelength of light used, and $NA_{condenser}$ and $NA_{objective}$ are the numerical apertures of the condenser lens and objective lenses, respectively. Because numerical aperture has no units, the units for D are the same as the units for wavelength, which typically are in nanometers (nm).

Numerical aperture is a measure of a lens's ability to "capture" light coming from the specimen and use it to make the image. As with magnification, it is marked on the lens (Figures 3-4A, 3-4B, and 3-4D). Using immersion oil between the specimen and the oil objective lens increases its numerical aperture and, in turn, makes its limit of resolution smaller. (If necessary, oil also may be placed between the condenser lens and the slide.) The result is better resolution.

The light microscope may be modified to improve its ability to produce images with contrast without staining, which often distorts or kills the specimen. In **dark-field microscopy** (Figure 3-2B), a special condenser is used so only the light reflected off the specimen enters the objective. The appearance is of a brightly lit specimen against a dark

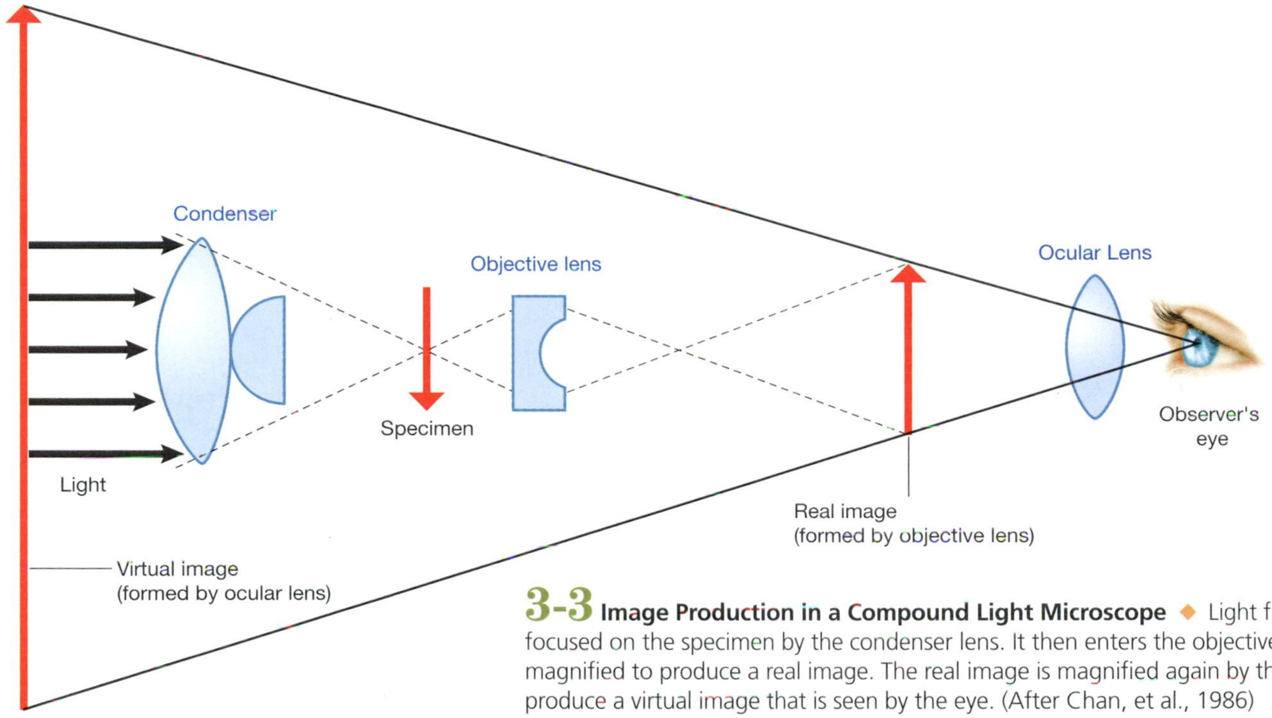

3-3 Image Production in a Compound Light Microscope ◆ Light from the source is focused on the specimen by the condenser lens. It then enters the objective lens, where it is magnified to produce a real image. The real image is magnified again by the ocular lens to produce a virtual image that is seen by the eye. (After Chan, et al., 1986)

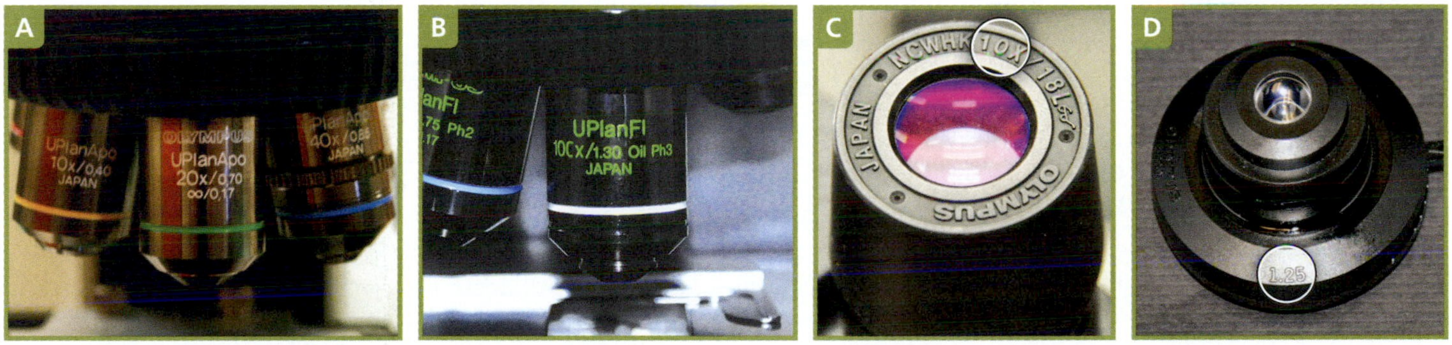

3-4 Markings of Magnification and Numerical Aperture on Microscope Components ◆ **A** Three plan apochromatic objective lenses on the nosepiece of a light microscope. *Plan* means the lens produces a flat field of view. *Apochromatic* lenses are made in such a way that chromatic aberration is reduced to a minimum. From left to right, the lenses magnify 10X, 20X, and 40X, and have numerical apertures of 0.40, 0.70, and 0.85. The 20X lens has other markings on it. The mechanical tube length is the distance from the nosepiece to the ocular and has become standardized at 160 mm. However, this 20X lens has been corrected so the light rays are made parallel, effectively creating an infinitely long mechanical tube length (∞). This allows insertion of accessories into the light path without decreasing image quality. The thickness of cover glass to be used is also given (0.17±0.01 mm). Also notice the standard colored rings for each objective: yellow for 10X, green for 20X (or 16X), and light blue for 40X (or 50X). **B** This is an oil immersion lens. It is the only lens constructed in such a way as not to be damaged by oil and, as such, is the only one with which oil is to be used. Oil lenses are indicated by black and white rings. This particular lens is also constructed for phase contrast microscopy. It indicates it is to be used with the #3 setting on the phase condenser. **C** A 10X ocular lens. **D** A condenser (removed from the microscope) with a numerical aperture of 1.25. The lever at the right is used to open and close the iris diaphragm and adjust the amount of light entering the specimen.

3-5 Resolution and Limit of Resolution ◆ The headlights of most automobiles are around 1.5 m apart. As you look at the cars in the foreground of the photo, it is easy to see both headlights as separate objects. The automobiles in the distance appear smaller (but really aren't) as does the apparent distance between the headlights. When the *apparent* distance between automobile headlights reaches about 0.1 mm, they blur into one because that is the limit of resolution of the human eye.

background, often with better resolution than that of the bright-field microscope.

Phase contrast microscopy (Figure 3-2C) uses special optical components to exploit subtle differences in the refractive indices of water and cytoplasmic components to produce contrast. Light waves that are in phase (that is, their peaks and valleys exactly coincide) reinforce one another, and their total intensity (because of the summed amplitudes) increases. Light waves that are out of phase by exactly one-half wavelength cancel each other and result in no intensity—that is, darkness. Wavelengths that are out of phase by any amount will produce some degree of cancellation and result in brightness that is less than maximum but more than darkness. Thus, contrast is provided by differences in light intensity that result from differences in refractive indices in parts of the specimen that put light waves more or less out of phase. As a result, the specimen and its parts appear as various levels of darks and lights.

Fluorescence microscopy (Figure 3-2D) uses a fluorescent dye that emits fluorescence when illuminated with ultraviolet radiation. In some cases, specimens possess naturally fluorescing chemicals and no dye is needed.

Application

Light microscopy (used in conjunction with cytological stains) is used to identify microbes from patient specimens or the environment. It also may be used to visually examine a specimen for the presence of more than one type of bacteria, or for the presence of other cell types that indicate tissue inflammation or contamination by a patient's cells.

In this Exercise

Today you will become familiar with the operation and limitations of your light microscope. You also will examine two practice slides to learn about microscope functioning.

Materials

- ✓ compound light microscope
- ✓ lens paper
- ✓ nonsterile cotton swabs
- ✓ lens-cleaning solution or 95% ethanol
- ✓ letter "e" slide
- ✓ colored threads slide

Instructions for Using the Microscope

Proper use of the microscope is essential for your success in microbiology. Fortunately, with practice and by following a few simple guidelines, you can achieve satisfactory results quickly. Because student labs may be supplied with a variety of microscopes, your instructor may supplement the following procedures and guidelines with instructions specific to your equipment. Refer to Figure 3-1 as you read the following (if working independently), or follow along on your microscope as your instructor guides you. (Note: This is a thorough treatment of microscope use, and not all parts may be immediately relevant to your laboratory. Refer back to this exercise as necessary.)

Transport

1. Carry your microscope to your workstation using both hands—one hand grasping the microscope's arm and the other supporting the microscope beneath its base.
2. Gently place the microscope on the table.

Cleaning

1. Lens paper is used for gently cleaning the condenser and objective lenses. Light wiping is usually enough. If that still doesn't clean the lens, call your instructor.
2. To clean an ocular, moisten a cotton swab with cleaning solution and gently wipe in a spiral motion starting at the center of the lens and working outward. Follow with a dry swab in the same pattern.

Operation

1. Raise the substage condenser to a couple of millimeters below its maximum position nearly even with the stage and open the iris diaphragm.
2. Plug in the microscope and turn on the lamp. Adjust the light intensity slowly to its maximum.

3 Using the nosepiece ring, move the scanning objective (usually 42) or low power objective (102) into position. Do not rotate the nosepiece by the objectives as this can damage the objective lenses and cause them to unscrew from the nosepiece.

4 Place a slide on the stage in the mechanical slide holder and center the specimen over the opening in the stage.

5 If using a binocular microscope, adjust the distance between the two oculars to match your own inter-pupillary distance.

6 Adjust the iris diaphragm and condenser position to produce optimum illumination, contrast, and image. (As a rule, use the maximum light intensity combined with the smallest aperture in the iris diaphragm that produces optimum illumination. Remember: This is bright-field microscopy, so don't close down the iris diaphragm too much.)

7 Use the coarse-focus adjustment knob to bring the image into focus. (Note: For most microscopes, the distance from the nosepiece opening to the focal plane of each lens has been standardized at 45 mm. This makes the lenses parfocal and gives the user an idea of where to begin focusing.) Bring the image into sharpest focus using the fine-focus adjustment knob. Then observe the specimen with your eyes relaxed and slightly above the oculars to allow the images to fuse into one. If you are using a monocular microscope, keep both eyes open anyway to reduce eye fatigue.

8 If you are using a binocular microscope, adjust the oculars' focus to compensate for differences in visual acuity of your two eyes. Close the eye with the adjustable ocular and bring the image into focus with the coarse- and fine-focus knobs. Then, using only the eye with the adjustable ocular, focus the image using the ocular's focus ring.

9 Scan the specimen to locate a promising region to examine in more detail.

10 If you are observing a nonbacterial specimen, progress through the objectives until you see the degree of structural detail necessary for your purposes. You will have to adjust the fine focus (but not coarse focus because the lenses are parfocal—all that should be necessary after changing lenses is a slight adjustment) and illumination for each objective. Before advancing to the next objective, be sure to position a desirable portion of the specimen in the center of the field or you will risk "losing" it at the higher magnification.

11 If you are working with a bacterial smear, you will have to use the oil immersion lens.

12 To use the oil immersion lens, work through the low (10×), then high dry (40×) objectives, adjusting the fine focus and illumination for each. Before advancing to the next objective, be sure to position a desirable portion of the specimen in the center of the field or you risk "losing" it at the higher magnification.

When the specimen is in focus under high dry, rotate the nosepiece to a position midway between the high dry and oil immersion lenses. Then place a drop of immersion oil on the specimen. *Be careful not to get any oil on the microscope or its lenses, and be sure to clean it up if you do.* Rotate the oil lens so its tip is submerged in the oil drop, pass through it, and then return the oil lens into the oil. This minimizes the occurrence of air bubbles. (Note: Do not move the stage down to add oil to the slide or the specimen will no longer be in focus. On a properly adjusted microscope, the oil and the high dry lenses have the same focal plane. Therefore, when a specimen is in focus on high dry, the oil lens, although longer, will also be in focus and won't touch the slide when rotated into position.) Focus and adjust the illumination to maximize the image quality.

13 When you are finished, lower the stage (or raise the objective) and remove the slide. Dispose of the freshly prepared slides in a jar of disinfectant or a sharps container; return permanent slides to storage.

Storage

When you are finished for the day:

1 Move the scanning objective into position.

2 Center and lower the mechanical stage.

3 Lower the light intensity to its minimum, then turn off the light.

4 Wrap the electrical cord according to your particular lab rules.

5 Clean any oil off the lenses, stage, etc. Be sure to use only cotton swabs or lens paper for cleaning any of the optical surfaces of the microscope (see "Cleaning," page 128).

6 Return the microscope to its appropriate storage place.

Procedure

1 Get out your microscope and record the magnifications and numerical aperture values in the chart on the Data Sheet.

2 Clean your microscope lenses as outlined in "Instructions for Using the Microscope."

3 Plug in the microscope and position the scanning objective over the stage. Make condenser and lamp adjustments appropriate for scanning power.

4 Pick up the letter "e" slide and examine it without the microscope. Record the orientation of the letter when the slide label is on the left. Record the appearance of the letter on your Data Sheet.

5 Place the slide on the stage in the same position as you examined it with your naked eyes. Now, center the "e" in the field and examine it with the scanning objective. Record its orientation as viewed with the microscope on your Data Sheet.

6 Now, move the *stage* to the right and record the direction the *image* moves.[1]

7 Position the "e" in the center of the field again. Move the *stage* toward you and record on your Data Sheet the direction the *image* moves. Then remove the slide from the microscope.

8 Examine the colored thread slide without the microscope (Figure 3-6). See if you can tell where in the stack of three threads each color resides. That is, is the red thread on the top, bottom, or middle? Do the same for the yellow and blue threads. Record your observations on the Data Sheet.

9 Now, place the slide on the microscope and determine the order of the threads using the low and high power objectives. Record your observations on the Data Sheet.

[1] If your microscope doesn't have a mechanical stage, move the slide with your hands in the appropriate direction.

3-6 Challenge of the Threads ◆ Even with the microscope, determining the order of threads from top to bottom is a challenge. This will require patience and use of the fine focus! Making it worse, not all the slides will be the same. Good luck!

References

Abramowitz, Mortimer. 2003. *Microscope Basics and Beyond.* Olympus America Inc., Scientific Equipment Group, Melville, NY.

Ash, Lawrence R., and Thomas C. Orihel. 1991. Pages 187–190 in *Parasites: A Guide to Laboratory Procedures and Identification.* American Society for Clinical Pathology (ASCP) Press, Chicago.

Bradbury, Savile, and Brian Bracegirdle. 1998. Chapter 1 in *Introduction to Light Microscopy.* BIOS Scientific Publishers Limited, Oxford, United Kingdom.

Forbes, Betty A., Daniel F. Sahm, and Alice S. Weissfield. 2002. Pages 119–121 in *Bailey & Scott's Diagnostic Microbiology,* 11th ed. Mosby, St. Louis.

NAME ______________________ DATE ______________________

LAB SECTION ______________________ I WAS PRESENT AND PERFORMED THIS EXERCISE (INITIALS) ________

Introduction to the Light Microscope

DATA AND CALCULATIONS

Record the relevant values off your microscope and perform the calculations of total magnification for each lens.

Lens System	Magnification of Objective Lens	Magnification of Ocular Lens	Total Magnification	Numerical Aperture
Scanning				
Low Power				
High-Dry				
Oil Immersion				
Condenser Lens				

Sketch your observations of the letter "e" slide in the chart below. Be sure the slide is right side up with the label at the left.

Appearance of the "e" with the Naked Eye	Appearance of the "e" Under the Microscope	When the stage moves to the right, the image moves to the…	When the stage moves toward you, the image moves…

Record your observations of the colored thread slide below. Check it under low and high power and see if your answer changes.

Low Power (usually the 10× objective)	Top:
	Middle:
	Bottom:
High Power (usually the 40× objective)	Top:
	Middle:
	Bottom:

Questions

1 Why aren't the magnifications of *both* ocular lenses of a binocular microscope used to calculate total magnification?

2 What is the total magnification for each lens setting on a microscope with 15× oculars and 4×, 10×, 45×, and 97× objectives lenses?

3 Assuming that all other variables remain constant, explain why light of shorter wavelengths will produce a clearer image than light of longer wavelengths.

4 Why is wavelength the main limiting factor on limit of resolution in light microscopy?

Name ______________________ Date ______________________

Lab Section ______________________ I was present and performed this exercise (initials) ________

5 On a given microscope, the numerical apertures of the condenser and low power objective lenses are 1.25 and 0.25, respectively. You are supplied with a filter that selects a wavelength of 520 nm.

a. What is the limit of resolution on this microscope?

b. Will you be able to distinguish two points that are 300 nm apart as being separate, or will they blur into one? Explain your answer.

6 On the same microscope as in Question #5, the high dry objective lens has a numerical aperture of 0.85.

a. What is the limit of resolution on this microscope?

b. Will you be able to distinguish two points that are 300 nm apart as being separate, or will they blur into one? Explain your answer.

7 Calculate the limit of resolution for the oil lens of your microscope. Assume an average wavelength of 500 nm.

8 Examine Figure 3-3 and explain the results you observed with the letter "e" slide.

9 With which objective was it easier to determine the sequence of colored threads?

10 Why should closing the iris diaphragm improve your ability to determine thread order?

11 What does the Colored Thread slide demonstrate about specimens you will be observing later in the class?

EXERCISE 3-2

Calibration of the Ocular Microscope

❖ Theory

An **ocular micrometer** is a type of ruler installed in the microscope eyepiece, composed of uniform but unspecified graduations (Figure 3-7). As such, it must be calibrated before any viewed specimens can be measured. The device used to calibrate ocular micrometers is called a **stage micrometer**. As illustrated in Figure 3-8, a stage micrometer is a type of microscope slide containing a ruler with 10 µm and 100 µm graduations. (Other measuring instruments may be used in place of a stage micrometer, as shown in Figure 3-9.)

When the stage micrometer is placed on the stage, it is magnified by the objective being used; therefore, the size of the graduations (relative to the ocular micrometer divisions) increases as magnification increases. Consequently, the *value* of ocular micrometer divisions decreases as magnification increases. For this reason, calibration must be done for each magnification.

As shown in Figure 3-9, the stage micrometer is placed on the stage and brought into focus such that it is superimposed by the ocular micrometer. Then the first (left) line of the ocular micrometer is aligned with one of the marks on the stage micrometer. (The line chosen on the stage micrometer depends on the power of the lens being calibrated. Lower powers use the large graduations; higher powers use the smaller graduations on the left. Figure 3-9 illustrates proper alignment with the scanning objective.)

Notice in Figure 3-9 that line 25 of the ocular micrometer and the eighth major line of the stage micrometer are perfectly aligned. This indicates that 25 ocular micrometer divisions (also called ocular units, or OU) span a distance of 800 µm because the stage micrometer lines are 100 µm apart. Notice also that line 47 of the ocular micrometer is aligned with the fifteenth major stage micrometer line. This means that 47 ocular units span 1500 µm. These values have been entered for you in Table 3-1.

To determine the value of an ocular unit on a given magnification, divide the distance (as measured with the stage micrometer) by the corresponding number of ocular units.

$$\frac{800\ \mu m}{25\ \text{ocular units}} = \frac{32\ \mu m}{OU}$$

$$\frac{1500\ \mu m}{47\ \text{ocular units}} = \frac{32\ \mu m}{OU}$$

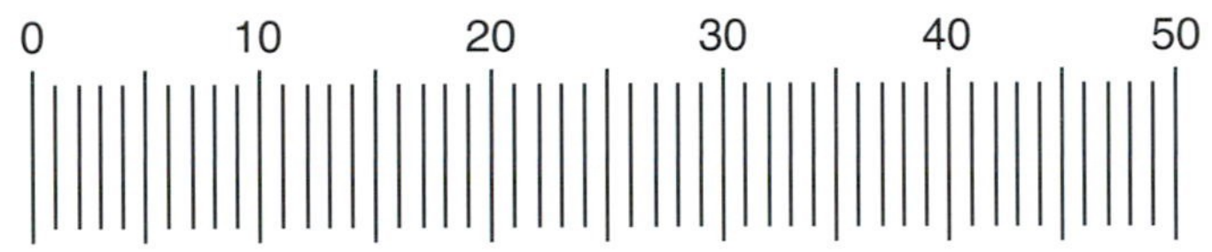

3-7 An Ocular Micrometer ◆ The ocular micrometer is a scale with uniform increments of unknown size. It has to be calibrated for each objective lens on the microscope.

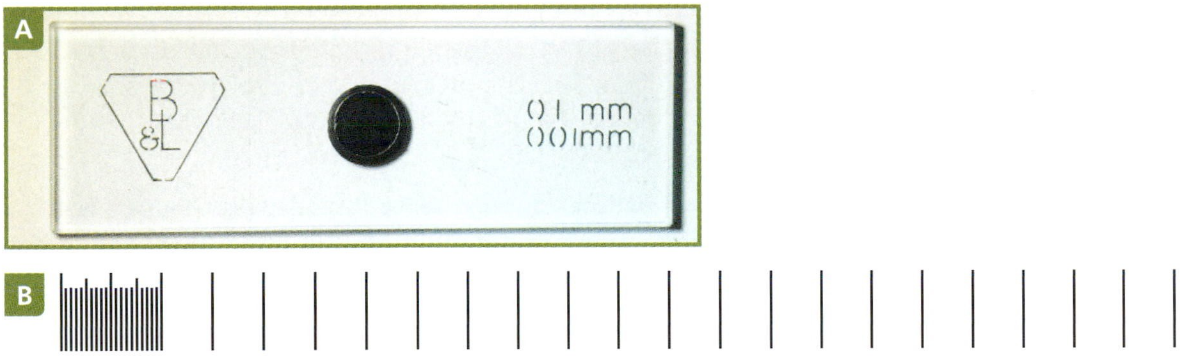

3-8 A Stage Micrometer ◆ **A** The stage micrometer is a microscope slide with a microscopic ruler engraved into it (not visible in the dark center of the slide). The markings on this micrometer indicate that the major increments are 0.1 mm (100 µm) apart. There is also a section of the scale that is marked off in 0.01 mm (10 µm) increments. **B** This drawing represents what the stage micrometer on the slide in **A** looks like. The micrometer is 2200 µm long. The major divisions are 100 µm apart. The 200 µm at the left are divided into 10 µm increments.

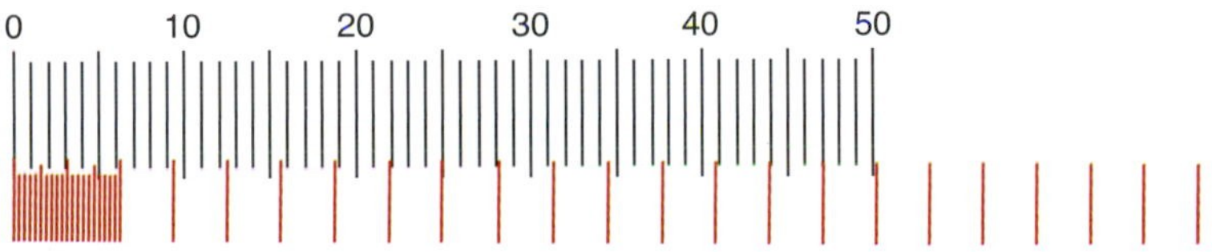

3-9 What They Look Like in Use ◆ When properly aligned, the ocular micrometer scale is superimposed over the stage micrometer scale (shown in red for contrast). Notice that they line up at their left ends.

TABLE **3-1** **Sample Data from Figure 3-9** ◆

Stage Micrometer	Ocular Micrometer
800 µm	25 OU
1500 µm	47 OU

As shown in Table 3-1, it is customary to record more than one measurement. Each measurement is calculated separately. If the calculated ocular unit values differ, use their arithmetic mean as the calibration for that objective lens.

As mentioned previously, each magnification must be calibrated. Because of its short working distance, calibrating the oil immersion lens may be difficult to accomplish using the stage micrometer. It also may be difficult because the distance between stage micrometer lines is too large for accurate measurement when magnified 1000×. If this is the case, its value can be calculated using the calibration value of one of the other lenses. Refer to Table 3-2 for the total magnifications for each objective lens on a typical microscope. Notice that the magnification of the oil immersion lens is 10 times greater than the low power lens. This means that objects viewed on the stage (stage micrometer *or* specimens) appear 10 times larger when changing from low power to oil immersion. But, because the magnification of the ocular micrometer does not change, an ocular division now covers only one-tenth the distance. Thus, the size of an ocular unit using the oil immersion lens can be calculated by dividing the calibration for low power by 10.

Ocular micrometer values can be calculated for any lens using values from any other lens, and they provide a good check of measured values. As calculated from Figure 3-9, 32 µm/OU was the calibration for the scanning objective. For practice, calculate the low, high dry, and oil immersion calibrations. Write the values in the Table 3-2.

Once you have determined the ocular unit values for each objective lens, use the ocular micrometer as a ruler to measure specimens. For instance, if you determine that under the scanning objective a cell is 5 ocular units long, the cell's actual length would be determined as follows (using the sample values from Table 3-2):

Cell Dimension = Ocular Units × Calibration

Cell Dimension = 5 Ocular Units × 32 µm/OU

Cell Dimension = 160 µm

Be sure to include the proper units in your answer!

TABLE **3-2** **Total Magnifications for Different Objective Lenses of a Typical Microscope and the Calibrations of the Ocular Micrometer for the Scanning Objective** ◆

Calculate the remaining calibratons as described in the text.

Power	Total Magnification	Calibration (µm/OU)
Scanning	40×	32
Low Power	100×	
High Dry Power	400×	
Oil Immersion	1000×	

❖ Application

The ability to measure microbes is useful in their identification and characterization.

❖ In This Exercise

This lab exercise involves calibrating the ocular micrometer on your microscope. Actual measurement of specimens will be done in subsequent lab exercises as assigned by your instructor.

❖ Materials

Per Student

- ✓ compound microscope equipped with an ocular micrometer
- ✓ stage micrometer

Procedure

Following is the general procedure for calibrating the ocular micrometer on your microscope. Your instructor will notify you of any specific details unique to your laboratory.

1. Check your microscope and determine which ocular has the micrometer in it.
2. Move the scanning objective into position.
3. Place the stage micrometer on the stage and position it so its image is superimposed by the ocular micrometer and the left-hand marks line up.
4. Examine the two micrometers and, as described above, record two or three points where they line up exactly. Record these values on the Data Sheet and calculate the value of each ocular unit.
5. Change to low power and repeat the process.
6. Change to high dry power and repeat the process.

7 Change to the oil immersion lens (but make sure there is enough room for the oil lens to "clear" the surface of the stage micrometer) and repeat the process. If this cannot be done (either because the stage micrometer lines are too far apart or the slide is too thick for the oil lens to be rotated into position), calculate the calibration from the value of another lens.

8 Compute average calibrations for each objective lens and record these on the Data Sheet.

9 As long as you keep this microscope throughout the term, you may use the calibrations you recorded without recalibrating the microscope.

References

Abramoff, Peter, and Robert G. Thompson. 1982. Pages 5 and 6 in *Laboratory Outlines in Biology—III*. W. H. Freeman and Co., San Francisco.

Ash, Lawrence R., and Thomas C. Orihel. 1991. Pages 187–190 in *Parasites: A Guide to Laboratory Procedures and Identification*. American Society for Clinical Pathology (ASCP) Press, Chicago.

NAME ______________________ DATE ______________________

LAB SECTION ______________________ I WAS PRESENT AND PERFORMED THIS EXERCISE (INITIALS) ________

Calibration of the Ocular Micrometer

DATA AND CALCULATIONS

Record two or three values where the ocular micrometer and the stage micrometer line up for the scanning, low, high dry, and oil immersion (if possible) objective lenses. Then calculate the calibration for each. Be sure to include proper units in your calibrations.

Scanning Objective Lens

Stage Micrometer (μm)	Ocular Micrometer (OU)	Calibration

Low Power Objective Lens

Stage Micrometer (μm)	Ocular Micrometer (OU)	Calibration

High Dry Objective Lens

Stage Micrometer (μm)	Ocular Micrometer (OU)	Calibration

Oil Immersion Objective Lens

Stage Micrometer (μm)	Ocular Micrometer (OU)	Calibration

Calculate the average value for each calibration and record them in the chart below. If necessary, use one of the values to calculate the calibration of the oil lens. Be sure to include proper units in your answer.

Average Calibrations for My Microscope

Objective Lens	Average Calibration
Scanning	
Low Power	
High Dry Power	
Oil Immersion	

EXERCISE 3-3

Microscopic Examination of Eukaryotic Microbes

Theory

Cells are divided into two major groups—the **prokaryotes**[1] and the **eukaryotes**—based on size and complexity. These and other differences are summarized in Table 3-3 and shown in Figure 3-10. The prokaryotes are further divided into two domains—the *Archaea* and the *Bacteria*. The eukaryotes belong to a single domain and are divided into five "supergroups" until kingdom associations can be reasonably determined. Figure 3-11 provides a phylogenetic tree of these groups based on RNA comparisons.

In this lab you will observe representative protozoans, algae, and fungi. Some specimens are on commercially prepared slides, whereas others need to be prepared as wet mounts (Figure 3-12) of living cultures. In a wet mount, a drop of water is placed on the slide and the organisms are introduced into it. Or, if the organism is already in a liquid medium, then a drop of medium is placed on the slide. A cover glass is placed over the preparation to flatten the drop and keep the objective lens from getting wet. A stain may or may not be applied to add contrast.

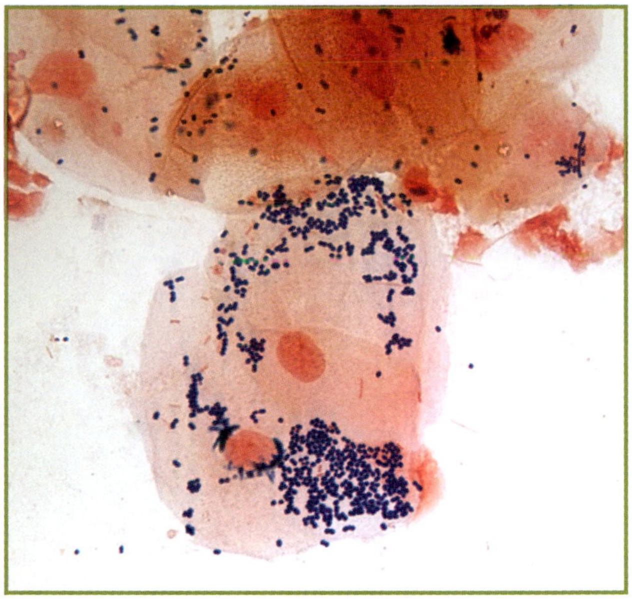

3-10 Prokaryotic and Eukaryotic Cells (Gram stain, Oil Immersion) ◆ This is a direct smear specimen taken from around the base of the teeth below the gum line. The large, pink cells are human epithelial cells and are eukaryotic (notice the prominent, darker pink nuclei). The small, purple cells are bacteria. Typically, prokaryotic cells range in size from 1 to 5 μm, whereas eukaryotic cells are in the 10 to 100 μm range.

Eukaryotic Supergroups

Only a brief introduction to the eukaryotic supergroups (Excavata, Archaeplastida, Chromalveolata, and Unikonta) will be given here, with an emphasis on microscopic species. Rhizaria is not covered because its members are not usually encountered in beginning microbiology classes.

Supergroup Excavata

These are unicellular species that usually have a feeding groove *excavated* from one side of the cell and possess one or more flagella. As shown in Figure 3-11, excavates branched from the other eukaryotic supergroups very early on. Many species have a life cycle in which there is an active feeding stage called the **trophozoite** and an inert

[1] The validity of the term "prokaryote" has recently been called into question. We continue to use it because there is no consensus on its validity and it remains a convenient term.

TABLE 3-3 Summary of Major Prokaryotic and Eukaryotic Features ◆

Character	Prokaryotes	Eukaryotes
Organismal groups	*Bacteria* and *Archaea*	Plants, animals, fungi, and protists
Typical size	1–5 μm	10–100 μm
Membrane-bound Organelles (including a nucleus)	Absent	Present
Ribosomes	70S (30S and 50S subunits)	80S (40S and 60S subunits)
Microtubules	Absent	Present
Flagellar movement	Rotary	Whip-like
DNA	Single, circular molecule called a chromosome	Two to many linear molecules; each is a chromosome
Introns	Absent (sometimes in *Archaea*)	Common
Mitotic division	Absent	Present

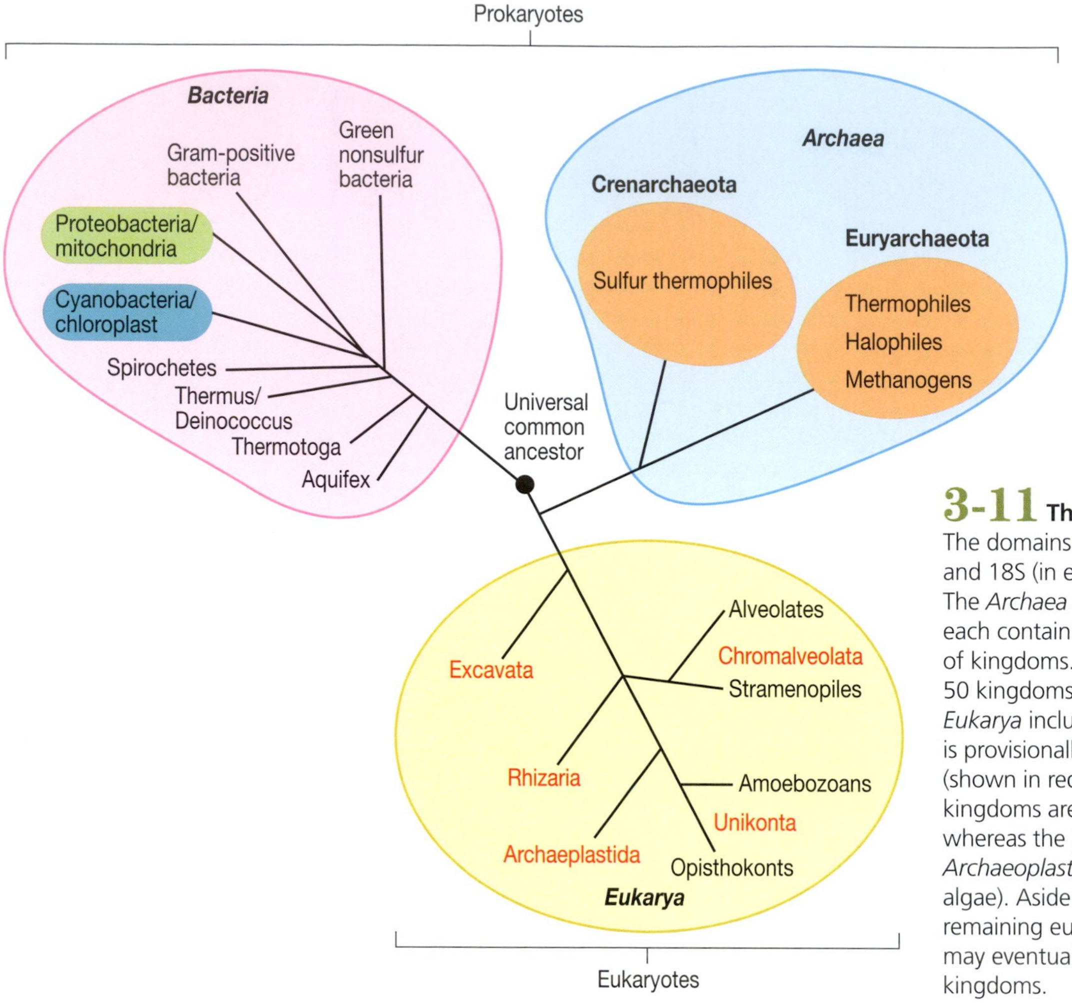
Prokaryotes
Bacteria
Green nonsulfur bacteria
Gram-positive bacteria
Proteobacteria/ mitochondria
Cyanobacteria/ chloroplast
Spirochetes
Thermus/ Deinococcus
Thermotoga
Aquifex
Universal common ancestor
Archaea
Crenarchaeota
Euryarchaeota
Sulfur thermophiles
Thermophiles
Halophiles
Methanogens
Alveolates
Chromalveolata
Stramenopiles
Excavata
Rhizaria
Amoebozoans
Unikonta
Archaeplastida
Opisthokonts
Eukarya
Eukaryotes

3-11 The Three Domains of Life ◆ The domains are based on 16S (in prokaryotes) and 18S (in eukaryotes) rRNA sequencing results. The *Archaea* and *Bacteria* are prokaryotic domains, each containing an as yet undetermined number of kingdoms. (*Bacteria* may contain as many as 50 kingdoms, *Archaea* perhaps only 3.) Kingdom *Eukarya* includes all the eukaryotic organisms and is provisionally divided into five "supergroups" (shown in red). The familiar animal and fungus kingdoms are in the Unikonta supergroup, whereas the plant kingdom aligns with the *Archaeoplastida* (which includes green and red algae). Aside from these three kingdoms, the remaining eukaryotes (formerly known as protists) may eventually be divided into five or more kingdoms.

resting stage called a **cyst**. Four subgroups are recognized. These are, **parabasalids**, **diplomonads**, **kinetoplastids**, and **euglenozoans**.

- ◆ The parabasalid *Trichomonas vaginalis* (Figure 3-13A) is the causative agent of vulvovaginitis (trichomoniasis) in human females (males can also be infected). It has four anterior flagella and an **undulating membrane** made from a fin-like extension of the cytoplasmic membrane. Hydrogenosomes, organelles that make ATP anaerobically with H_2 as an end product, are present. There is evidence that in some cases these are degenerated mitochondria. Only *Trichomonas* trophozoites are known.
- ◆ Diplomonads are unicellular flagellates with two large nuclei (visible in the *Giardia* specimen shown in Figure 3-13B). At one time diplomonads were thought to lack mitochondria, but recent evidence shows they have degenerated into mitosomes that lack the electron transport chain and are thus nonfunctional in generating ATP for the cell. Instead, diplomonads use cytoplasmic biochemical pathways, such as glycolysis. *Giardia* is the causative agent of giardiasis, a diarrheal disease.
- ◆ Euglenozoans are unicellular, green, photosynthetic autotrophs when light is available, but can live heterotrophically when it is not (making them **mixotrophic**). They usually have two flagella, or more rarely only one. The red "eyespot" is a distinctive characteristic that is easily seen with the light microscope. A euglenid is shown in Figure 3-13C.
- ◆ The presence of a DNA mass within their single mitochondrion, called a kinetoplast, is a unifying feature of kinetoplastids. Many are free living, but *Trypanosoma* (Figure 3-13D) is a parasite that causes sleeping sickness in humans.

Supergroup Archaeplastida

There is strong evidence to support the conclusion that members of this group are all descended from an ancestor that engulfed a cyanobacterium (by **primary endosymbiosis**), which ultimately evolved into the chloroplasts of green algae, red algae, and true plants. Thus, all members are autotrophic. Most also possess walls made of cellulose. Two groups are covered here: **chlorophytes** and **charophytes**.

- ◆ The chlorophytes are commonly called green algae. Most are freshwater, unicellular, filamentous, or colonial

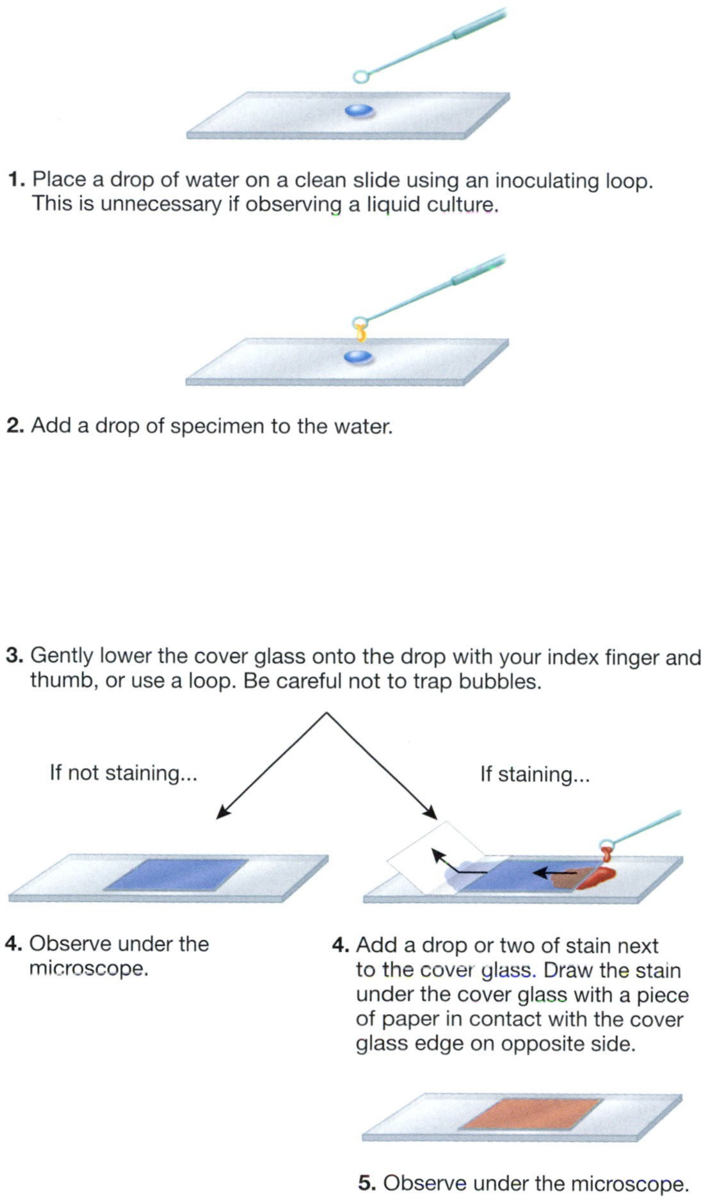
1. Place a drop of water on a clean slide using an inoculating loop. This is unnecessary if observing a liquid culture.
2. Add a drop of specimen to the water.
3. Gently lower the cover glass onto the drop with your index finger and thumb, or use a loop. Be careful not to trap bubbles.
If not staining...
If staining...
4. Observe under the microscope.
4. Add a drop or two of stain next to the cover glass. Draw the stain under the cover glass with a piece of paper in contact with the cover glass edge on opposite side.
5. Observe under the microscope.

3-12 Procedural Diagram: Wet Mount ◆ Use these instructions to make wet mounts of living specimens. The preparation may be stained or not, but staining will eventually kill the organisms. Of course, so will drying. You can expect a wet mount to last only 15–20 minutes.

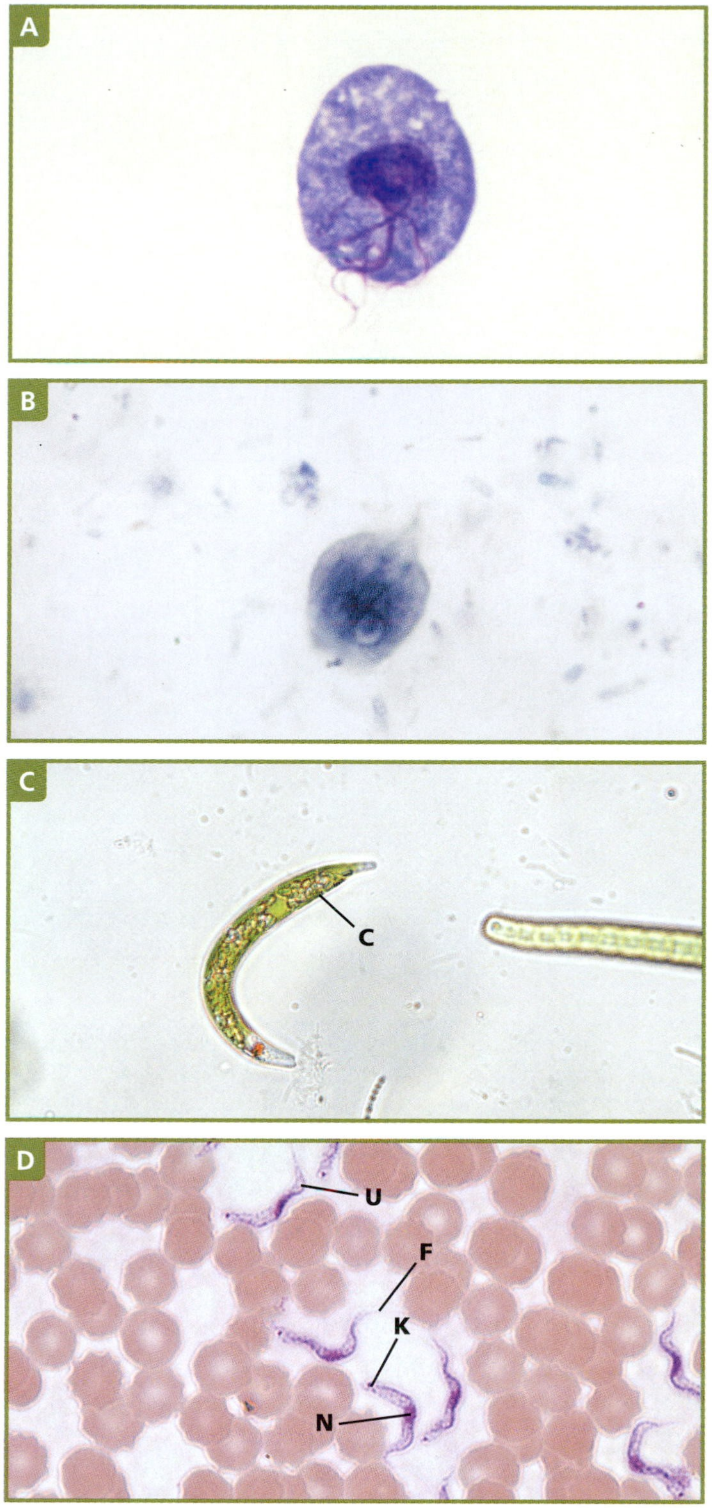
A
B
C
C
D
U
F
K
N

3-13 Supergroup Excavata ◆ **A** This *Trichomonas vaginalis* cell shows several flagella. Cells range in size from 5-15 μm wide by 7-23 μm long. **B** This *Giardia lamblia* trophozoite illustrates the characteristic long, tapering posterior end and the two nuclei, but the flagella are not visible in this preparation. Trophozoites range in size from 9-21 μm by 5-15 μm. **C** *Euglena* is a large genus of mixotrophic flagellates. Most species have chloroplasts (**C**), which are discoid in this specimen. A red "eyespot" is located in the colorless anterior of the cell. The single flagellum also emerges from the anterior. **D** Note the following features: nucleus (**N**), undulating membrane (**U**) kinetoplast (**K**), and flagellum (**F**). Cells range in size from 1.5–3.5 μm wide by 14–33 μm long.

species. Flagella are commonly used for motility, as in the haploid, unicellular *Chlamydomonas* (Figure 3-14A). *Volvox* (Figure 3-14B) is an example of a colonial species made of *Chlamydomonas*-like cells. *Volvox* colonies can undergo sexual reproduction; there are male colonies that produce sperm bundles and female colonies that produce eggs. They also can undergo asexual reproduction by forming daughter colonies within, which are ultimately released as new individuals.

◆ Closely related to the chlorophytes, but even more closely related to true plants (because of certain aspects of cell ultrastructure and physiology), are the charophytes, which include *Spirogyra* (Figure 3-15).

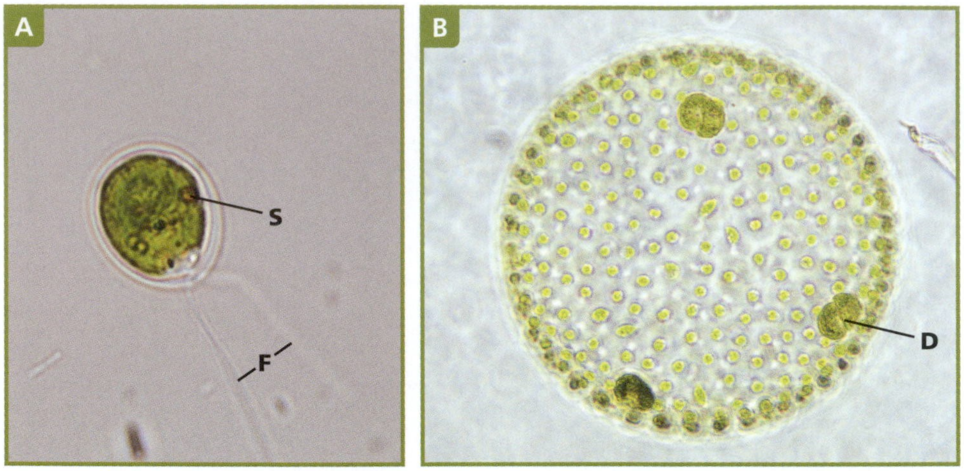

3-14 **Supergroup Archaeplastida (Chlorophyte Subgroup)** ◆ **A** *Chlamydomonas* is a unicellular green alga with two flagella and a single, cup-shaped chloroplast. A red-orange stigma is found with in the chloroplast and is involved in phototaxis. Cells are approximately 30μm long and 20μm wide. **B** *Volvox* is a colonial green alga made of cells similar in shape to *Chlamydomonas*. Daughter colonies form asexually from special cells in the parent colony. Release of the daughter colony involves its eversion as it exits through an enzymatically produced pore. Mature colonies can reach a size visible to the naked eye.

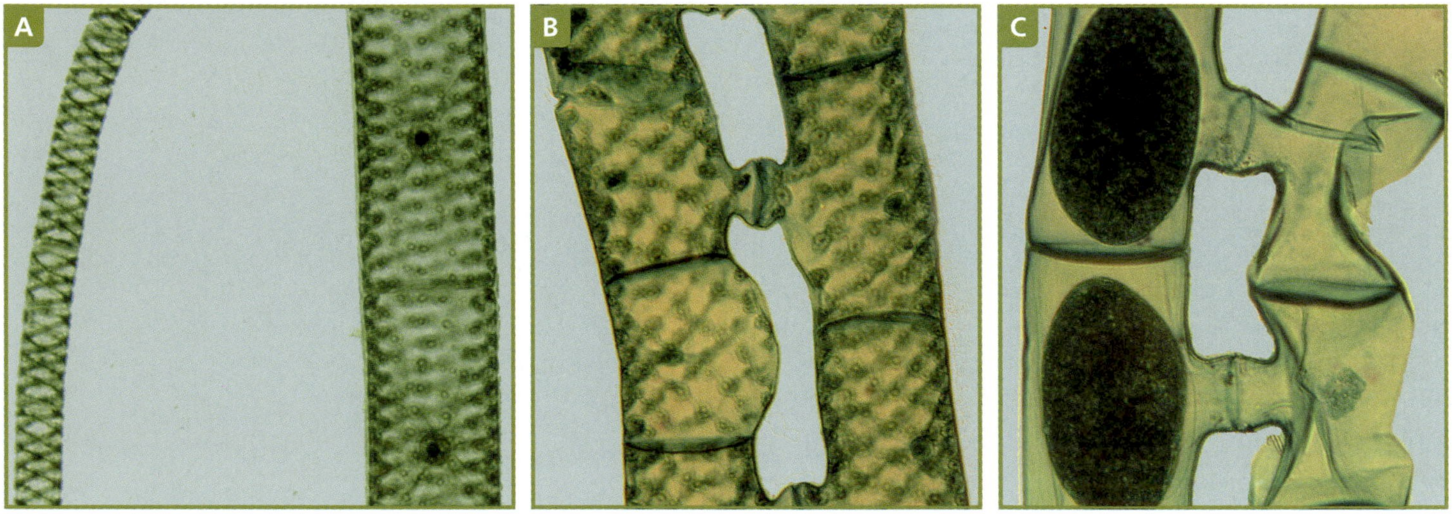

3-15 **Supergroup Archaeplastida (Charophyte Subgroup)** ◆ *Spirogyra* is a filamentous alga with distinctive spiral chloroplasts. The cells are haploid, that is, they have only one complete set of chromosomes. **A** Two vegetative filaments are shown here. Also note the nuclei in the cells on the right. **B** These two filaments have begun **scalariform conjugation**, a sexual process that occurs between different filaments. The protrusions are called **conjugation tubes** and serve as a connection through which the cytoplasm of one cell (acting as a gamete) can be moved and join with the cytoplasm of a cell in the other filament. **C** The cells and their two haploid nuclei fuse to form a diploid **zygospore** and conjugation is complete. When conditions are favorable, the zygospore germinates by undergoing meiosis. Only one haploid nucleus survives and the zygospore cell develops into a new vegetative filament. Note the absence of spiral chloroplasts in the empty cells and in the zygospores.

Supergroup Chromalveolata

The main unifying feature (though not demonstrated in all groups) is evidence of being descendants of an ancestor that engulfed a red algal cell that evolved into a plastid with red algal characteristics. (Because the red alga already had a chloroplast from primary endosymbiosis, the plastids derived from it are said to have been formed by **secondary endosymbiosis.**) In some modern species, the plastid is still present; in others, only red algal plastid DNA remains in the genome. More work needs to be done on this group before we have a clear picture. There are two subgroups within Chromalveolata: the **alveolates** and the **stramenopiles.**

- Alveolates may be either heterotrophic or autotrophic, but they all have small, membranous sacs beneath their cytoplasmic membrane. They are represented here by three ciliates—*Paramecium* (Figure 3-16A), *Stentor* (Figure 3-16B), and *Balantidium coli* (Figure 3-16C)—and one apicomplexan, *Plasmodium* (Figure 3-16D).

Ciliates are characterized by cilia covering their outer surface that provide motility when they sweep back-and-forth. Cilia also line the **oral groove** and sweep food particles inward toward the **cytostome** where it is engulfed. *B. coli* is a human pathogen that can cause ulceration of the colon's mucosa. Bloody, mucoid stool results in acute infections, whereas chronic infections may produce alternating diarrhea and constipation. Most infections are probably asymptomatic.

Apicomplexans have a complex life cycle, often involving more than one host and multiple tissues. For instance, *Plasmodium* spp., the causative agents of malaria, infect both the liver and red blood cells of humans, and the gut and salivary glands of the insect vector, the female *Anopheles* mosquito.

- The stramenopiles are characterized by a unique flagellum with three-parted lateral hairs emerging from it as well as some other physiological features. Even though diatoms (Figure 3-17) lack flagella, they are so successful and are so frequently seen in aquatic samples that we have chosen them to represent the group. Diatoms are single-celled or colonial autotrophs. They have a distinctive golden brown color due to the abundance of the pigment **fucoxanthin**. Their highly ornamented cell wall is made of silica embedded in an organic matrix and consists of two halves, with one half overlapping the other in the same way the lid of a Petri dish overlaps its base. **Centric diatoms** have a round shape, whereas **pennate diatoms** are bilaterally symmetrical. Oil droplets are frequently visible.

Supergroup Unikonta

This group is composed of heterotrophs and includes amoebas (of all sorts), animals, fungi, and others.

- Amoebas belong to two main groups. "Classical" amoebas belong to a subgroup called **Gymnamoebas** (Figure 3-18A). These are free-living in marine, freshwater, and soil environments. The other group comprises the **entamoebas** (Figures 3-18B and 3-18C), which are parasitic. All amoebas move by forming cytoplasmic extensions called **pseudopods**. They also engulf their food with them. Division is by binary fission and they do not have a cell wall.
- Fungi are nonmotile eukaryotes. Their cell wall is usually made of the polysaccharide **chitin**, not cellulose as in plants. Unlike animals and many other unikonts (that ingest then digest their food), fungi are **absorptive**

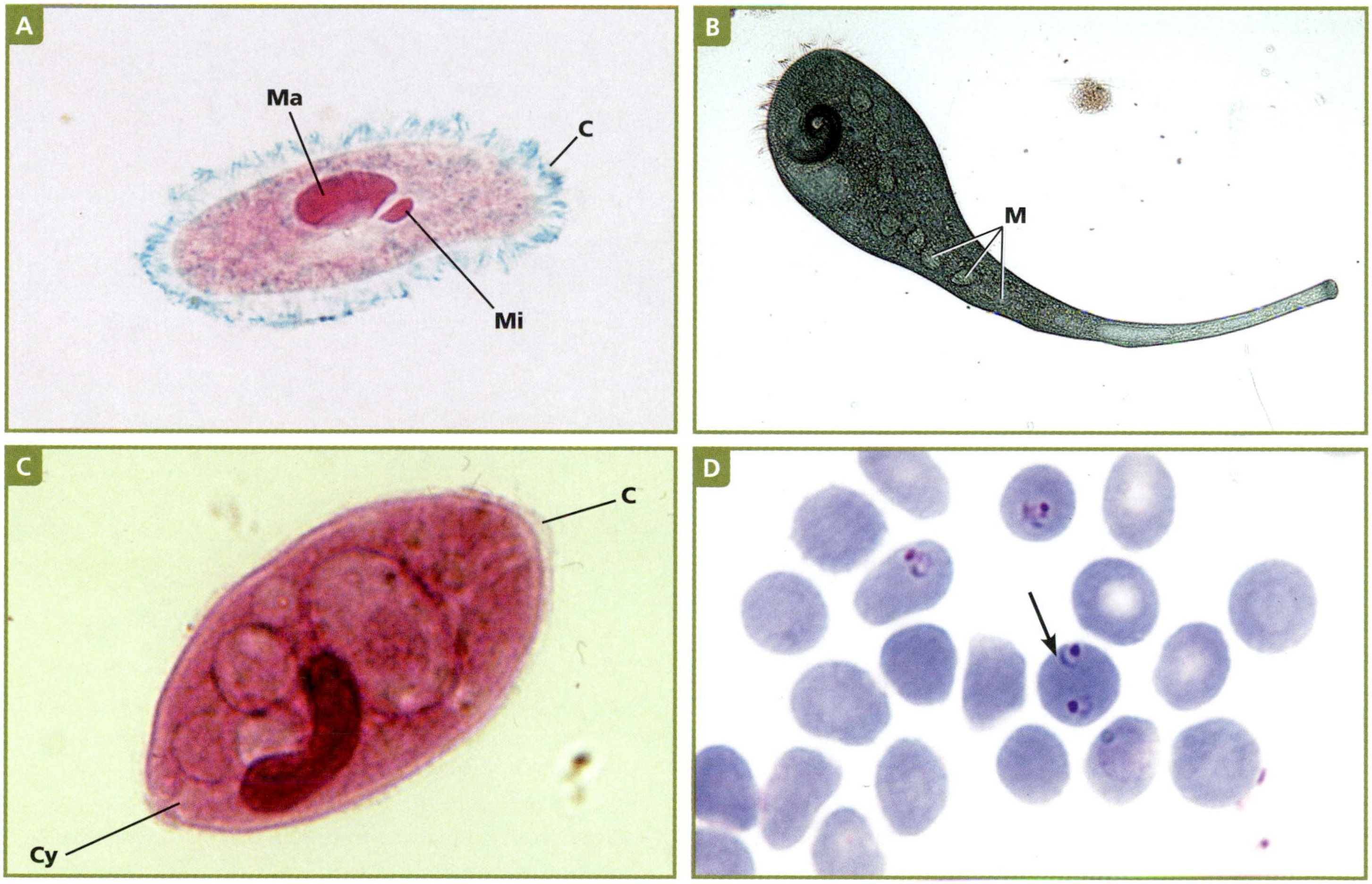

3-16 Supergroup Chromalveolata (Alveolate Subgroup) ◆ **A** Shown is the ciliate *Paramecium bursaria*. Note the cilia (**C**) around the cell's edge and the prominent macronucleus (**Ma**) and micronucleus (**Mi**). Micronuclei are involved in genetic transfer between compatible mating types, whereas the macronucleus is responsible for governing cellular function. At the stage shown, the micronucleus is diploid and the macronucleus has multiple copies of the cell's genome. Cilia serve to move the cell as well as to sweep food into its cytostome ("mouth"). **B** Its trumpet shape, size (up to 2 mm), and beaded macronucleus (**M**) make *Stentor* an easy ciliate to identify. The white spot inside the dark coiled region is the cytostome. This species is naturally green; this is not a stained specimen. **C** *Balantidium coli* trophozoites are oval in shape and have dimensions of 50–100 µm long by 40–70 µm wide. Cilia (**C**) cover the cell surface. Internally, the macronucleus is prominent; the adjacent micronucleus is not. An anterior cytostome (**Cy**) is usually visible. **D** Shown are red blood cells infected with young *Plasmodium falciparum* trophozoites at the ring stage of development. Division of the parasite will eventually cause the red cell to burst, releasing the parasites, which will infect more red blood cells. Note the chromatin dots in the nucleus and the cell (**arrow**) that has been infected by two plasmodia.

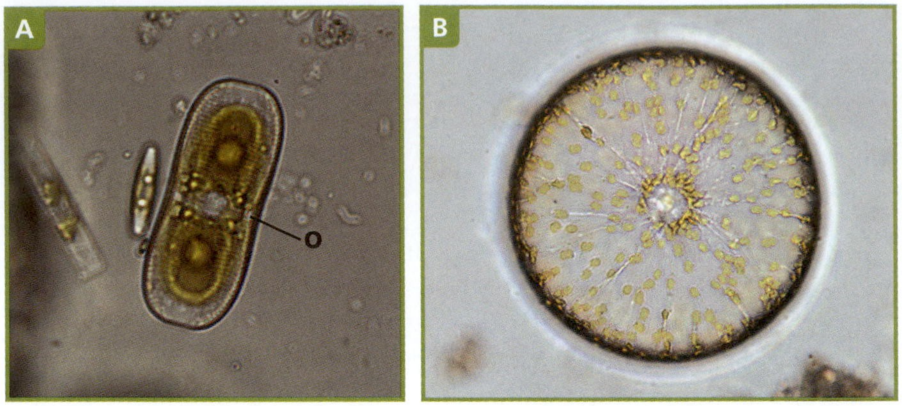

3-17 Supergroup Chromalveolata (Stramenopile Subgroup) ◆ While diatoms lack the flagella characteristic of most stramenopiles, they are so ubiquitous we have chosen them to represent the group. Their cell wall is divided into two halves, one of which fits into the other like the lid and base of a Petri dish, and are composed of silica and organic material. Diatoms are divided into two groups based on their symmetry. Those with bilateral symmetry (**A**) are said to be pennate and those with radial symmetry (**B**) are said to be centric. Note the distinctive golden chloroplasts, the cell wall ornamentation, and the oil droplets (**O**).

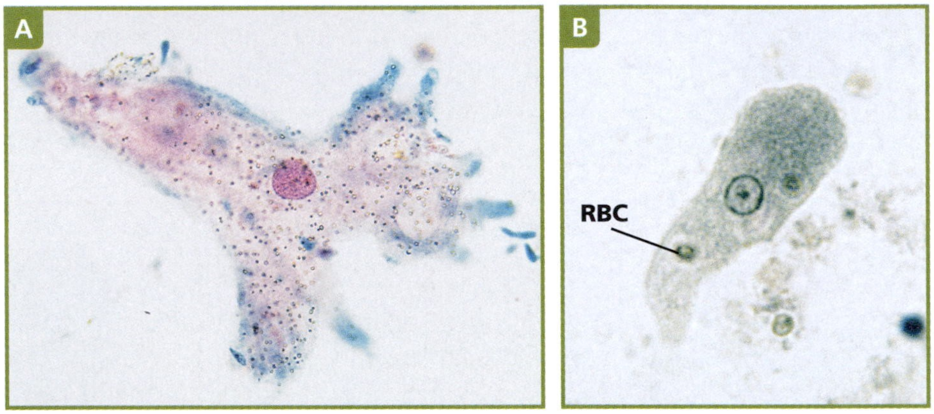

3-18 Supergroup Unikonta (Gymnamoeba and Entamoeba Subgroups) ◆ **A** This *Amoeba* has numerous pseudopods, which are used for movement as well as engulfing prey. **B**. *Entamoeba coli* trophozoites range in size from 12–60 µm. Notice the small, central karyosome, the beaded chromatin at the nucleus' margin, the ingested red blood cells (RBC), and the finely granular cytoplasm. It is the causative agent of amoebic dysentery.

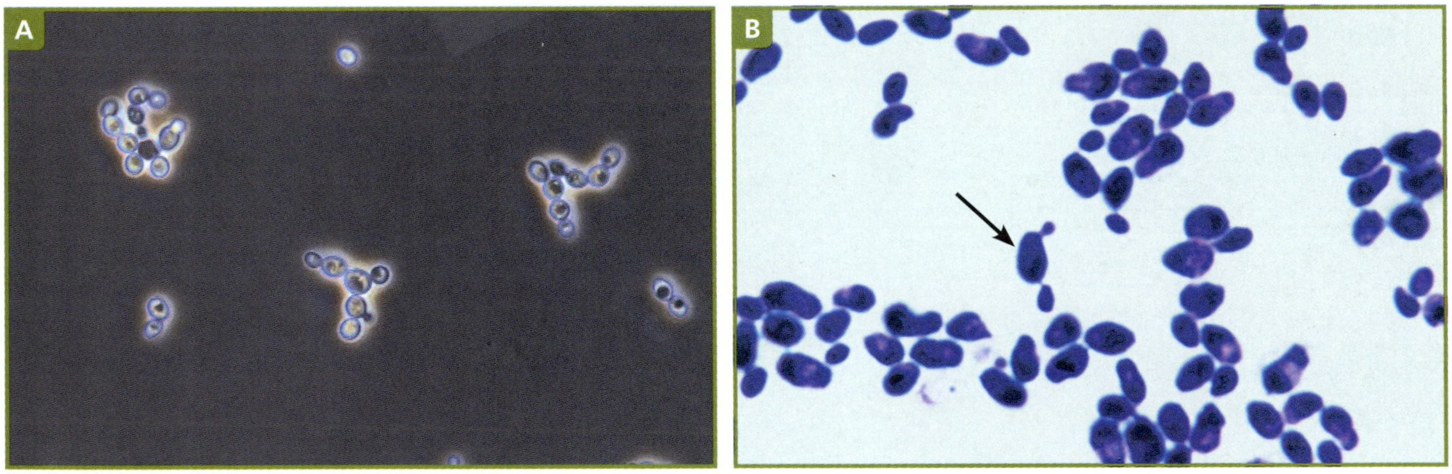

3-19 Supergroup Unikonta (Fungus Subgroup—Yeasts) ◆ **A** The "brewer's yeast" *Saccharomyces cerevisiae* is shown in a wet mount preparation using phase contrast microscopy. The cells are oval with dimensions of 3–8 µm by 5-10 µm. Short chains of cells (pseudohyphae) are visible in this field. **B** Note the oval shape, nuclei, and budding cells (**arrow**) of these *Canida albicans* cells.

heterotrophs. That is, they secrete **exoenzymes** into the environment, and then absorb the digested nutrients. Most are saprophytes that decompose dead organic matter, but some are parasites of plants, animals, or humans. Fungi are informally divided into **unicellular yeasts** and **filamentous molds** based on their overall appearance.

Examples of yeasts are *Saccharomyces cerevisiae* (Figure 3-19A) and *Candida albicans* (Figure 3-19B). Asexual reproduction occurs by uneven cell division called **budding.** Sexual reproduction involves meiosis of the yeast cell nucleus to produce haploid spores that can combine to form new diploid cells. *S. cerevisiae* is used in production of bread, wine, and beer, but is not an important human pathogen. *Candida albicans* is part of the normal respiratory, gastrointestinal, and female urogenital tract microbiota. Under the proper circumstances, it may flourish and produce pathological conditions, such as thrush in the oral cavity, vulvovaginitis of the female genitals, and cutaneous candidiasis of the skin. Systemic candidiasis may follow infection of the lungs, bronchi, or kidneys. Entry into the blood may result in endocarditis.

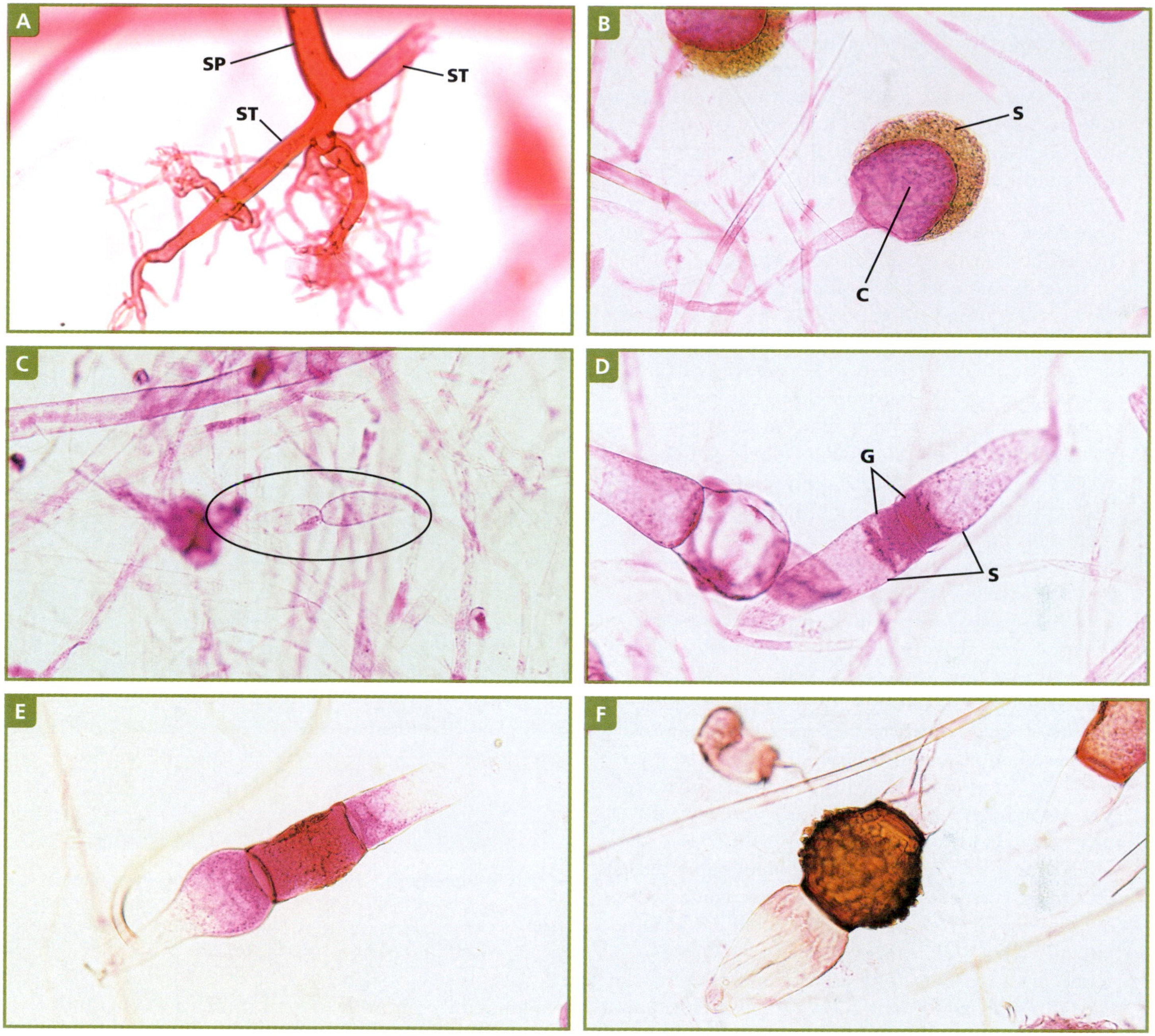

3-20 Supergroup Unikonta (Fungus Subgroup—Molds) ◆ *Rhizopus* is a common mold that is frequently studied in entry-level microbiology courses. **A** Most hyphae run horizontally across the surface. Anchoring rhizoids form at the junction of each sporangiophore (**SP**) and the stolon (**ST**). **B** The sporangium is found at the end of a long, unbranched, sporangiophore. The haploid asexual sporangiospores (**S**) cover the surface of the columella (**C**), which has a flattened base. **C** Progametangia from different hyphae are shown in the center of the field. Contact between the progametangia results in each forming a gamete. **D** Gametangia (**G**) and suspensors (**S**) are shown. Gametangia contain haploid nuclei from each mating type. **E** The zygospore forms when the cytoplasm from the two mating strains fuse (plasmogamy). **F** Haploid nuclei from each strain fuse within the zygospore (karyogamy) to produce many diploid nuclei. Meiosis produces numerous haploid spores.

Individuals most susceptible to *Candida* infections are diabetics, those with immunodeficiency (e.g., AIDS), catheterized patients, and individuals taking antimicrobial medications.

- Molds grow as filaments of cells. These filaments are called **hyphae** and collectively form a **mycelium.** Molds produce a variety of reproductive spores, both sexual and asexual. Three molds are covered here. *Rhizopus*, *Penicillium*, and *Aspergillus*.

Rhizopus species are fast growing molds that produce white or grayish cottony mycelia. The hyphae (called **stolons**) grow more or less horizontally and are anchored to the substrate by rhizoids (Figure 3-20A) The *Rhizopus* life cycle includes sexual and asexual phases. Asexual **sporangiospores** are produced by large, circular sporangia (Figures 3-20B) borne at the ends of long, nonseptate, elevated **sporangiophores.** A hemispherical **columella** supports the sporangium. The spores develop into hyphae identical to those that produced them. On occasion, sexual reproduction occurs when hyphae of different mating types (designated + and − strains) make contact. Initially, **progametangia** (Figure 3-20C) extend from each hypha. Upon contact, a septum separates the end of each progametangium into a gamete (Figure 3-20D). The walls between the two gametangia dissolve and a thick-walled **zygospore** develops (Figures 3-20E and 3-20F). Fusion of nuclei occurs within the zygospore and produces one or more diploid nuclei, or zygotes. After a dormant period, meiosis of the zygotes occurs. The zygospore then germinates and produces a sporangium similar to asexual sporangia. Haploid spores are released and develop into new hyphae, completing the life cycle.

The genus *Aspergillus* is characterized by green to yellow, brown, or black granular colonies with a white edge. The *Aspergillus* fruiting body (Figure 3-21A) is distinctive, with chains of spores (called **conidia**) arising from one or two rows of cells attached to a swollen vesicle at the end of an unbranched stalk. *A. fumigatus* and other species are opportunistic pathogens that cause aspergillosis, an umbrella term covering many diseases. One form of pulmonary aspergillosis (referred to as "fungus ball") involves colonization of the bronchial tree or tissues damaged by tuberculosis. Some species of *Aspergillus* are of commercial importance. Fermentation of soybeans by *A. oryzae* produces soy paste. Soy sauce is produced by fermenting soybeans with a mixture of *A. oryzae* and *A. soyae*. *Aspergillus* is also used in commercial production of citric acid.

Members of the genus *Penicillium* produce distinctive green, powdery, radially furrowed colonies with a white edge. Its **fruiting body** (the spore-producing structure) looks like a brush, and it is the source of its name: in Latin, *penicillus* means paint brush (Figure 3-21B). *Penicillium* is best known for its production of the antibiotic penicillin, but it is also a common contaminant. Some *Penicillium* species are of commercial importance for fermentations used in cheese production. Examples include *P. roquefortii* (Roquefort cheese) and *P. camembertii* (Camembert and Brie cheeses).

3-21 Supergroup Unikonta (Fungus Subgroup—Molds) ◆
A Shown is a section of an *Aspergillus niger* fruiting body bearing asexual spores (conidia). **B** *Penicillium* species produce a characteristic brush-shaped "fruiting body" bearing conidia.

❖ Application

Familiarity with eukaryotic cells not only rounds out your microbiological experience, it is important to be able to differentiate eukaryotic from prokaryotic cells when examining environmental or clinical specimens.

❖ In This Exercise

Most of this lab manual is devoted to *Bacteria*, but in this exercise you will be given the opportunity to examine various eukaryotic microorganisms. This will not only serve to familiarize you with simple eukaryotes, but also give you practice using the microscope, measuring specimens, and making wet-mount preparations.

❖ Materials

Per Class

✓ Living cultures (as available) of eukaryotic microbes
- *Amoeba*
- *Paramecium*
- *Euglena*
- *Spirogyra*
- *Volvox*
- *Saccharomyces*
- *Stentor*

✓ Prepared slides (as available) of
- Any of the above
- *Entamoeba histolytica* trophozoites
- *Balantidium coli* trophozoites
- *Giardia lamblia* trophozoites
- *Plasmodium falciparum* thick smear
- Miscellaneous diatoms
- *Candida albicans*
- *Rhizopus* sporangia and gametangia
- *Aspergillus* conidiophore
- *Penicillium* conidiophore
- *Trichomonas vaginalis*
- *Trypanosoma* spp.

Per Student

✓ clean glass slides and cover glasses
✓ compound microscope
✓ cytological stains (e.g., methylene blue, I_2KI)
✓ methyl cellulose
✓ immersion oil
✓ cotton swabs
✓ lens paper
✓ lens cleaning solution

Procedure

General Instructions for Prepared Slides

1. Obtain a microscope and place it on the table or workspace. Check to be sure the stage is all the way down and the scanning objective is in place.
2. Begin with prepared slides. Select one and clean it with a tissue if it is dirty, then place it on the microscope stage. Center the specimen under the scanning objective.
3. Follow the instructions given in Exercise 3-1 to bring the specimen into focus at the highest magnification that allows you to see the degree of detail you desire.
4. Practice scanning with the mechanical stage until you are satisfied that you have seen everything that is interesting to see. Sketch what you see at optimum magnification in the table provided on the Data Sheet.
5. Measure cellular dimensions, and record these in the table provided on the Data Sheet.
6. Repeat with as many slides as available and as time allows.

Prepared Slides

Trichomonas vaginalis
- Identify nucleus and flagella.

Giardia lamblia trophozoite
- Identify oval shape, flagella (four pairs), nuclei (two), and median bodies (two).

Trypanosoma spp.
- Identify kinetoplast, nucleus, undulating membrane, and flagellum.

Balantidium coli trophozoite
- Identify elongated shape, cilia, macronucleus, and micronucleus (maybe).

Plasmodium falciparum
- Identify ring stage.

Diatoms
- Differentiate centric and pennate forms.

Entamoeba histolytica trophozoite
- Identify pseudopods, beaded nucleus with small, central karyosome, and ingested erythrocytes.

Candida albicans
- Identify vegetative cells and budding cells.

Rhizopus
- Sporangia: Identify sporangiophores, sporangia, and spores.
- Gametangia: Identify progametangia, gametangia, young zygosporangia, and mature zygosporangia.

Aspergillus
- Identify hyphae, conidiophores, and conidia.

Penicillium
- Identify hyphae, conidiophores, and chains of conidia.

General Instructions for Wet-Mount Slides

1. Prepare wet mounts of available specimens by following the Procedural Diagram in Figure 3-12. A drop of methylcellulose may be added to the wet mount if you have fast swimmers.
2. Measure cellular dimensions, and record these in the table provided on the Data Sheet.
3. Repeat with as many slides as available and as time allows.

General Instructions for Clean up

1. When you are finished observing the specimens, blot any oil from the oil immersion lens (if used) with a lens paper. Also, check the high dry lens for oil, and clean it if necessary.
2. Return all lenses and adjustments to their storage positions before putting away the microscope.

3 Return prepared slides to their proper storage box. Dispose of wet-mount slides as directed by your lab instructor.

Wet-Mount Slides

Euglena

- Identify chloroplast and flagellum. Do not stain.

Chlamydomonas **spp.**

- Identify chloroplast, flagella, stigma. Do not stain.

Volvox

- Identify colony and daughter colonies. Do not stain.

Spirogyra

- Identify nuclei, spiral chloroplasts, and cell wall. Do not stain.

Paramecium

- Identify macronucleus, cilia, oral groove, and contractile vacuole.

Stentor **spp.**

- Identify cilia, beaded macronucleus, cytostome.

Amoeba

- Identify nucleus and pseudopods.

Saccharomyces cerevisiae

- Identify vegetative cells, nuclei, and budding cells. Stain with methylene blue.

References

Ash, Lawrence R., and Thomas C. Orihel. 1991. Chapter 14 in *Parasites: A Guide to Laboratory Procedures and Identification.* American Society for Clinical Pathology (ASCP) Press, Chicago, IL.

Collins, C. H., Patricia M. Lyne, and J.M. Grange. 1995. Chapter 51 in *Collins and Lyne's Microbiological Methods*, 7th ed. Butterworth-Heineman, Oxford.

Fisher, Fran, and Norma B. Cook. 1998. Chapter 2 in *Fundamentals of Diagnostic Mycology.* W. B. Saunders Company, Philadelphia, PA.

Freeman, Scott. 2005. Chapter 27 in *Biological Science*, 2nd ed. Pearson Education/Prentice Hall, Upper Saddle River, NJ.

Koneman, Elmer W., Stephen D. Allen, William M. Janda, Paul C. Schreck-enberger, and Washington C. Winn, Jr. 1997. Chapters 19 and 20 in *Color Atlas and Textbook of Diagnostic Microbiology,* 5th ed. J.B. Lippincott Company, Philadelphia, PA.

Lee, John J., Seymour H. Hutner, and Eugene C. Bovee. 1985. *Illustrated Guide to the Protozoa.* Society of Protozoologists, Lawrence, KS.

Madigan, Michael T., and John M. Martinko. 2006. Chapter 11 in Brock's *Biology of Microorganisms*, 11th ed., Pearson Education/Prentice Hall, Upper Saddle River, NJ.

Reece, Jane B., Lisa A. Urry, Michael L. Cain, Steven A. Wasserman, Peter V. Minorsky, and Robert B. Jackson. 2011. Chapters 26, 27, and 28 in *Biology,* 9th ed. Pearson Education/Benjamin Cummings Publishing Co., San Francisco.

Name ____________________ Date ____________________

Lab Section ____________________ I was present and performed this exercise (initials) ________

Microscopic Examination of Eukaryotic Microbes

OBSERVATIONS AND INTERPRETATIONS

Fill in the chart for each eukaryotic microbe you observe.

Organism (include wet mount or prepared slide)	Sketch (include magnification and stain, if any)	Dimensions	Identifying Characteristics (List only those you observed)

Organism (include wet mount or prepared slide)	Sketch (include magnification and stain, if any)	Dimensions	Identifying Characteristics (List only those you observed)

Questions

1 What features did the cells you observed have in common? How were they different?

2 If you observed the same organism on a prepared slide and a wet mount, how did the images compare?

Bacterial Structure and Simple Stains

Importance of Contrast

In Exercise 3-1 you were introduced to two of the three important features of a microscope and microscopy: magnification and resolution. A third feature is **contrast**. To be visible, the specimen must contrast with the background of the microscope field. Because cytoplasm is essentially transparent, viewing cells with the standard light microscope is difficult without stains to provide that contrast. In the upcoming exercises, you will learn how to correctly prepare a bacterial smear for staining and how to perform simple and negative stains. Details of cell structure (morphology, size, and arrangement) then may be seen. In a medical laboratory these usually are determined with a Gram stain (Exercise 3-6), but you will be using simple stains as an introduction to the staining process, as well as an introduction to these cellular characteristics.

Bacterial Cell Morphologies

Bacterial cells are much smaller than eukaryotic cells (Figure 3-10) and come in a variety of **morphologies** (shapes) and **arrangements**. Determining cell morphology is an important first step in identifying a bacterial species. Cells may be spheres (**cocci**, singular **coccus**), rods (**bacilli**, singular **bacillus**) or spirals (**spirilla**, singular **spirillum**). Variations of these shapes include slightly curved rods (**vibrios**), short rods (**coccobacilli**) and flexible spirals (**spirochetes**). Examples of cell shapes are shown in Figures 3-22 through 3-28. In Figure 3-29, *Corynebacterium xerosis* illustrates **pleomorphism**, where a variety of cell shapes—slender, ellipsoidal, or ovoid rods—may be seen in a given sample.

Bacterial Cell Arrangements

Cell arrangement, determined by the number of planes in which division occurs and whether the cells separate after division, is also useful in identifying bacteria. Spirilla rarely are seen as anything other than single cells, but cocci and bacilli do form multicellular associations. Because cocci exhibit the most variety in arrangements, they are used for illustration in Figure 3-30. If the two daughter cells remain attached after a coccus divides, a **diplococcus** is formed. The same process happens in bacilli that produce **diplobacilli**. If the cells continue to divide in the same plane and remain attached to form a chain, they exhibit a **streptococcus** or **streptobacillus** arrangement.

If a second division occurs in a plane perpendicular to the first, a **tetrad** is formed. A third division plane perpendicular to the other two produces a cube-shaped arrangement of eight cells called a **sarcina**. Tetrads and sarcinae are seen only in cocci. If the division planes of a coccus are irregular, a cluster of cells is produced to form a **staphylococcus**. Figures 3-31 through 3-36 illustrate common cell arrangements.

Arrangement and morphology often are easier to see when the organisms are grown in a broth rather than on a solid medium, or are observed from a direct smear. If you have difficulty identifying cell morphology or arrangement, consider transferring the organism to a broth culture, incubating it, and trying again.

One last bit of advice: Don't expect nature to conform perfectly to our categories of morphology and cell arrangement. These are convenient descriptive categories that will not be applied easily in all cases. When examining a slide, look for the most common morphology and most complex arrangement. Do not be afraid to report what you see. For instance, it's okay to say, "Cocci in singles, pairs, and chains."

Surface Area and Volume of Cells

Cells are three-dimensional objects with a surface that contacts the environment and a volume made up of cytoplasm. The ability to transport nutrients into the cell and wastes out of the cell is proportional to the amount of surface area doing the transport. The demand for nutrients and production of wastes is proportional to a cell's volume. It's a mathematical fact of life that as an object gets bigger, its volume increases more rapidly than its surface area. Therefore, a cell can achieve a size at which its surface area is not adequate to supply the nutrient needs of its cytoplasm. That is, its **surface area-to-volume ratio** is too small. At this point, a cell usually divides its volume to increase its surface area. This phenomenon is a major factor in limiting cell size and determining a cell's habitat.

Bacilli, cocci, and spirilla with the same volume have different amounts of surface area. A sphere has a lower surface-to-volume ratio than a bacillus or spirillum of the same volume. A streptococcus, however, would have approximately the same surface-to-volume ratio as a bacillus of the same volume.

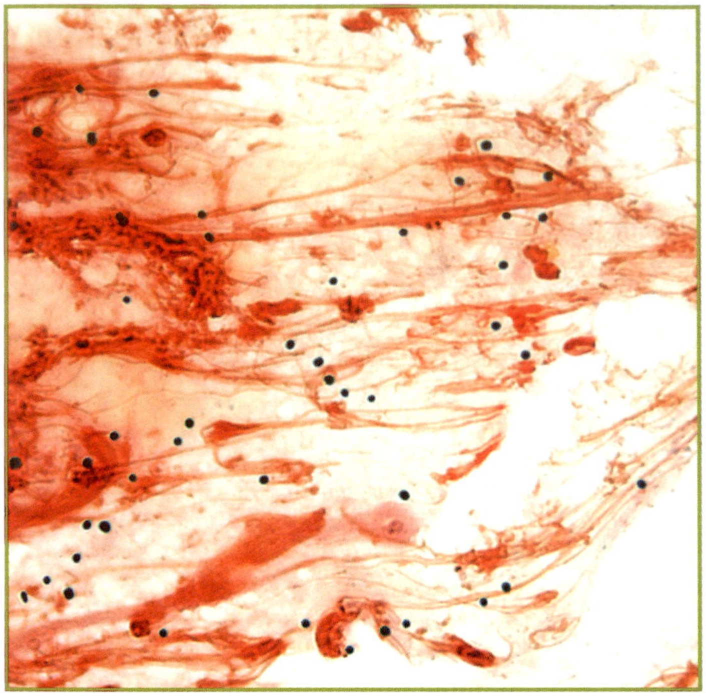

3-22 Single Cocci from a Nasal Swab (Gram Stain) ◆ This direct smear of a nasal swab illustrates unidentified cocci (dark circles) stained with crystal violet. Each cell is about 1 μm in diameter. The red background material is mostly mucus.

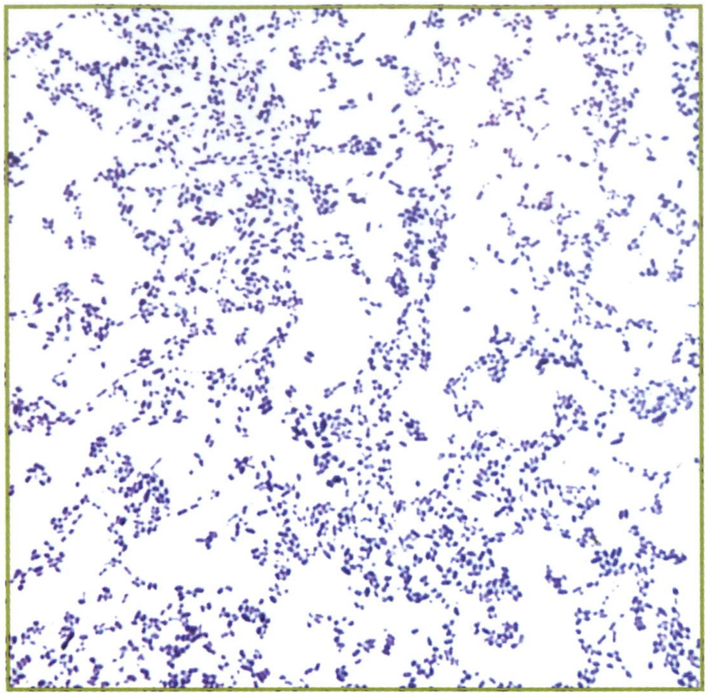

3-23 Ovoid Coccus (Gram Stain) ◆ *Lactococcus lactis* is an elongated coccus that a beginning microbiologist might confuse with a rod. Notice the slight elongation of the cells, and also that most cells are not more than twice as long as they are wide. *L. lactis* is found naturally in raw milk and milk products, but these cells were grown in culture. Cell dimensions are about 1 μm wide by 1–2 μm long.

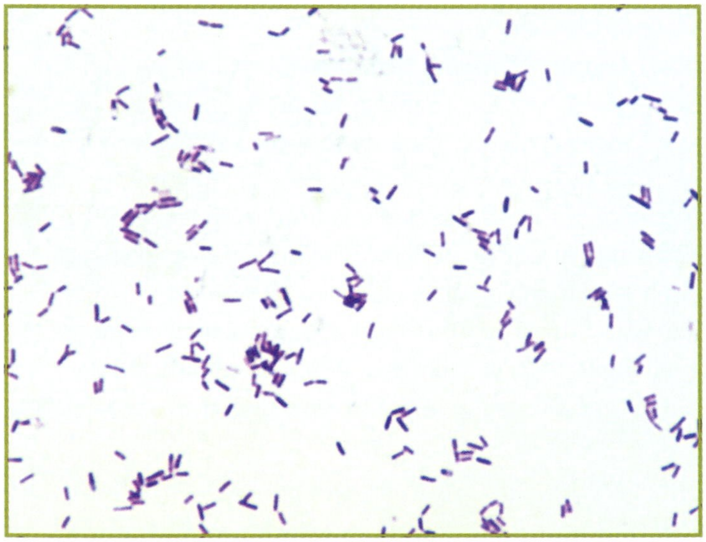

3-24 "Typical" Bacillus (Crystal Violet Stain) ◆ Notice the variability in rod length (because of different ages of the cells) in this stain of the soil organism *Bacillus subtilis* grown in culture. Cell dimensions are 0.7 μm wide by 2.0–3.0 μm long.

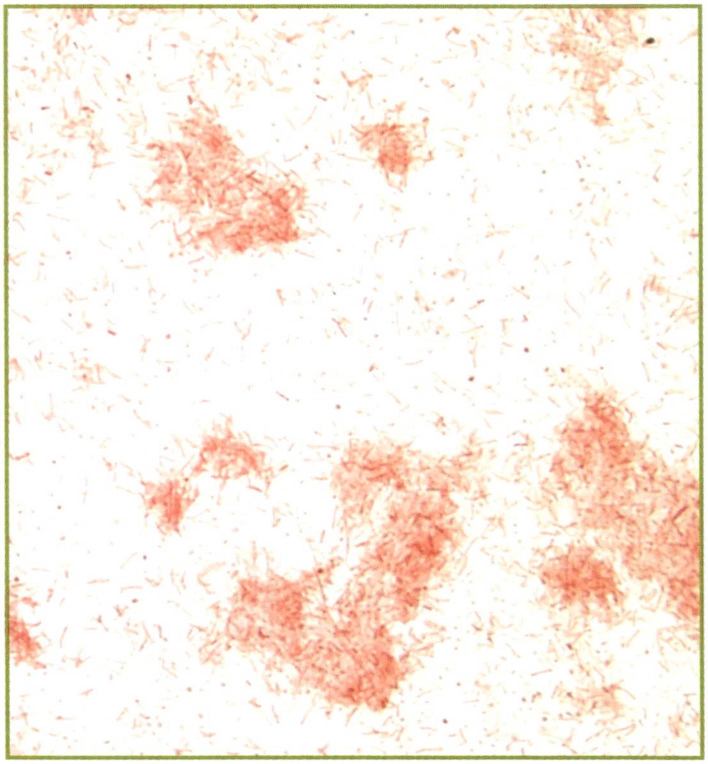

3-25 Long, Thin Bacillus (Gram Stain) ◆ The cells of *Aeromonas sobria*, a freshwater organism, are considerably longer than they are wide. Notice that a cell can be a bacillus without being in the genus *Bacillus*. Cell dimensions are about 0.5 μm wide by 1–3.5 μm long. These cells were grown in culture.

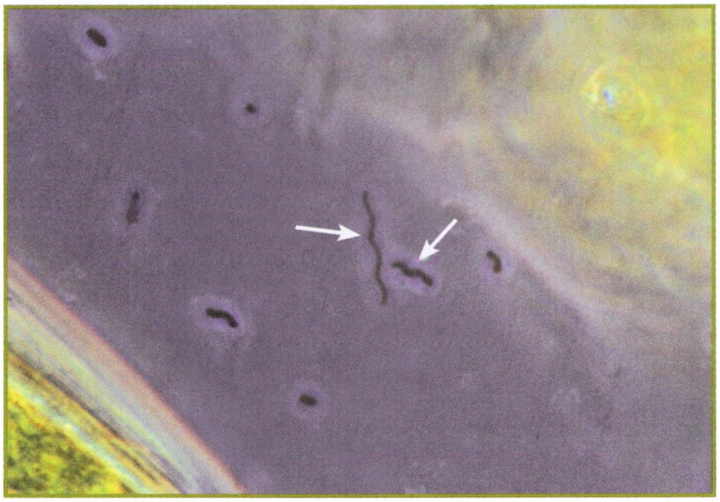

3-26 **Two Different Spirilla (Phase Contrast Wet Mount)** ◆ The two different spirilla are undoubtedly different species based on their different morphologies: one is long and slender with loose spirals, the other is shorter and fatter with tighter coils. Cell dimensions are less than 1 μm wide and up to 35 μm long.

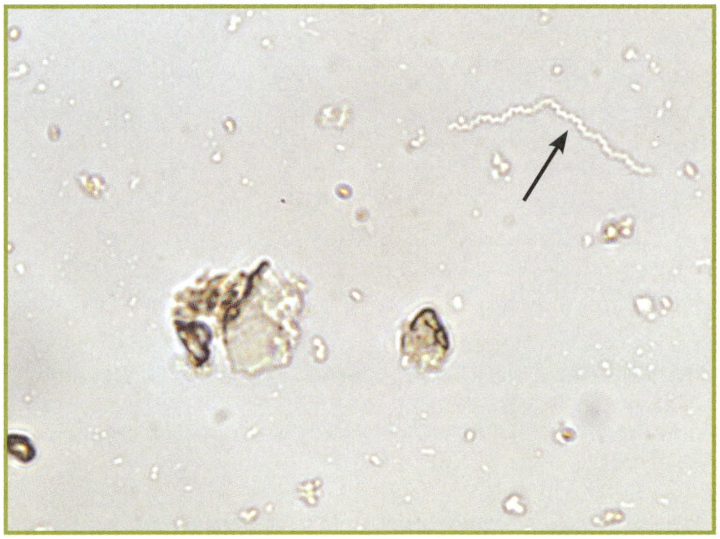

3-27 **A Spirochaete** ◆ Spirochaetes are tightly coiled, flexible rods. This bright-field micrograph shows a marine species of *Spirochaeta* (**arrow**). Notice the bend in the center of the cell. Cell dimensions of the genus are less than 1 μm wide by 5–500 μm long.

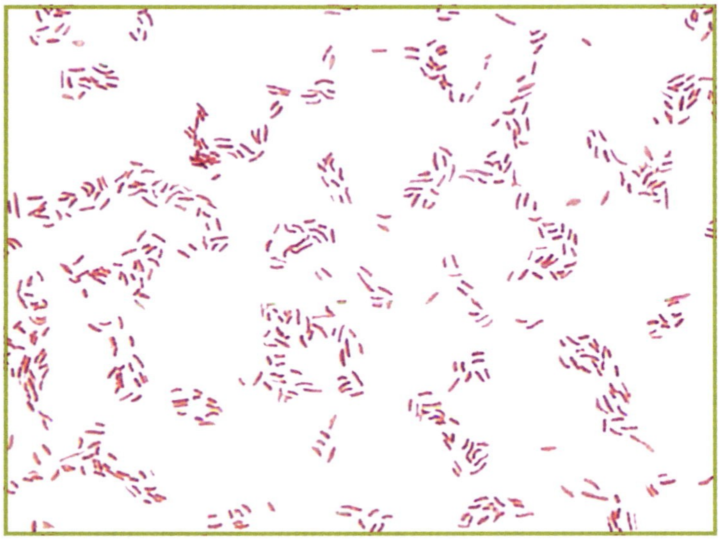

3-28 **Marine *Vibrio* (Gram Stain)** ◆ Slightly curved rods are demonstrated by this unidentified species isolated from a Southern California estuary. Notice that not all rods are curved. Cells are approximately 2–3 μm by 0.5 μm.

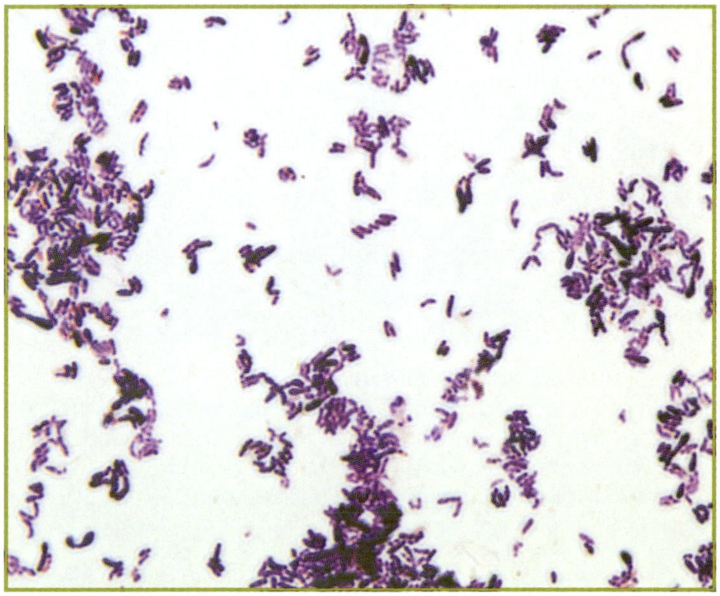

3-29 **Bacterial Pleomorphism (Gram stain)** ◆ Some organisms grow in a variety of shapes and are said to be **pleomorphic**. Notice that the rods of *Corynebacterium xerosis* range from almost spherical to many times longer than wide. This organism normally inhabits skin and mucous membranes and may be an opportunistic pathogen in compromised patients.

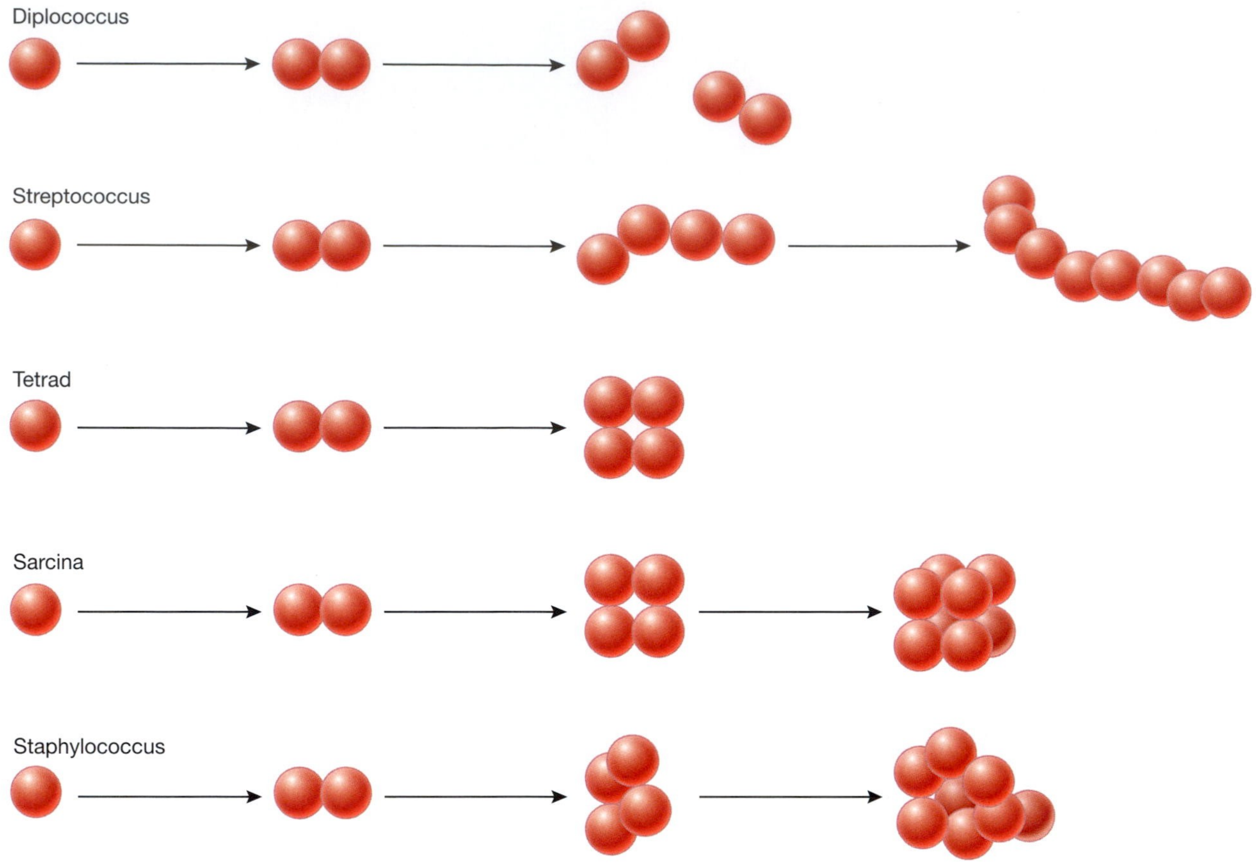
Diplococcus
Streptococcus
Tetrad
Sarcina
Staphylococcus

3-30 Division Patterns Among Cocci ◆ Diplococci have a single division plane and the cells generally occur in pairs. Streptococci also have a single division plane, but the cells remain attached to form chains of variable length. If there are two perpendicular division planes, the cells form tetrads. Sarcinae have divided in three perpendicular planes to produce a regular cuboidal arrangement of cells. Staphylococci have divided in more than three planes to produce a characteristic grapelike cluster of cells. (Note: Rarely will a sample be composed of just one arrangement. Report what you see, and emphasize the most complex arrangement.)

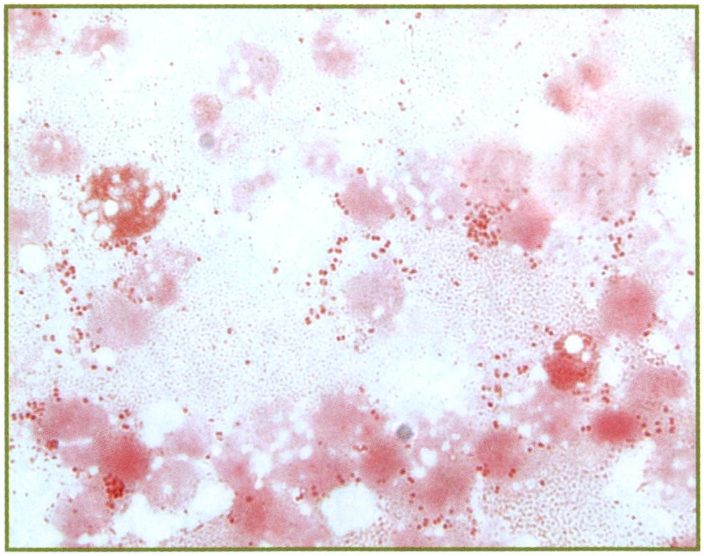

3-31 Diplococcus Arrangement (Gram Stain) ◆ *Neisseria gonorrhoeae*, a diplococcus, causes gonorrhea in humans. Members of this genus produce diplococci with flattened adjacent sides. Each cell is less than 1 μm in diameter.

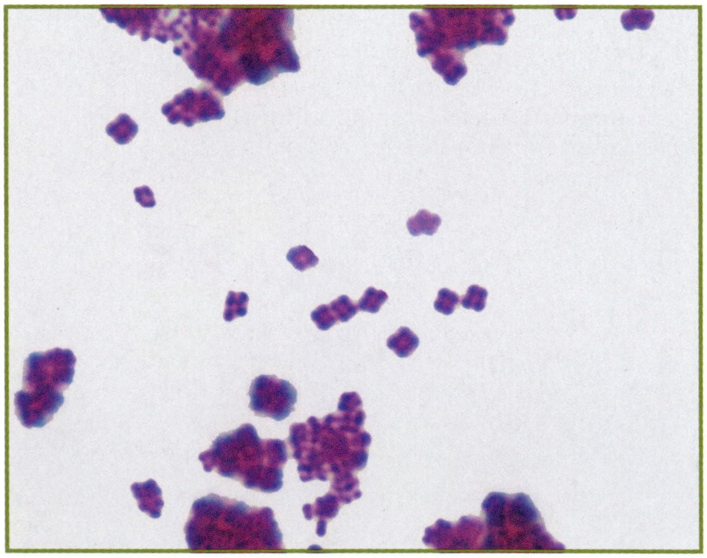

3-32 Tetrad Arrangement (Gram Stain) ◆ *Micrococcus roseus* grows in squared packets of cells, evident even when they are bunched together. The normal habitat for *Micrococcus* species is the skin, but the ones here were obtained from culture. Each cell is approximately 1 μm in diameter.

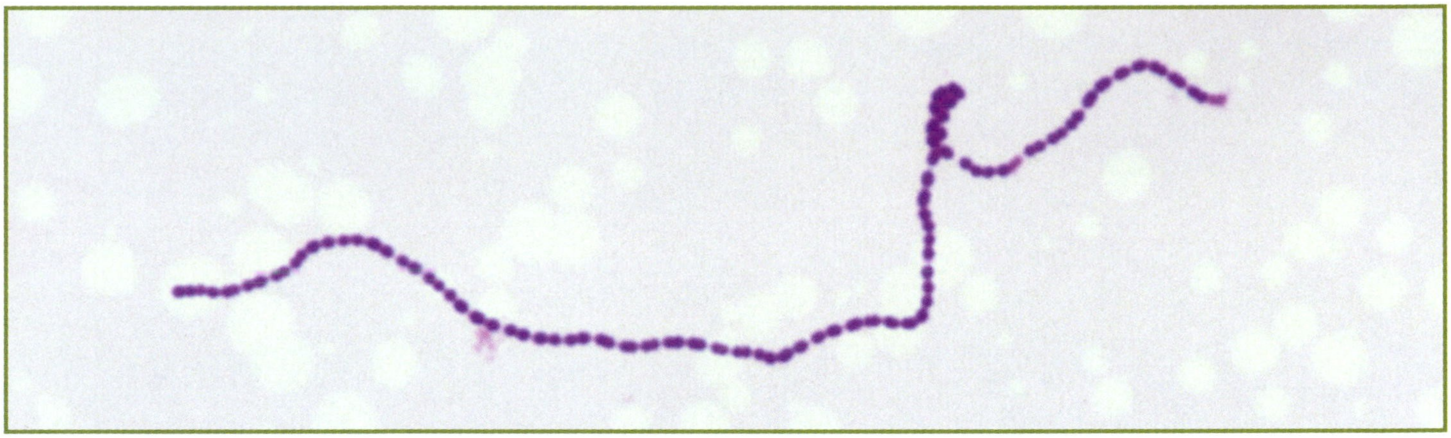

3-33 Streptococcus Arrangement (Gram Stain) ◆ *Enterococcus faecium* is a streptococcus that inhabits the digestive tract of mammals. This specimen is from a broth culture (which enables the cells to form long chains) and was stained with crystal violet. Notice the slight elongation of these cells along the axis of the chain. Cells are up to 2 μm wide by 2.5 μm long.

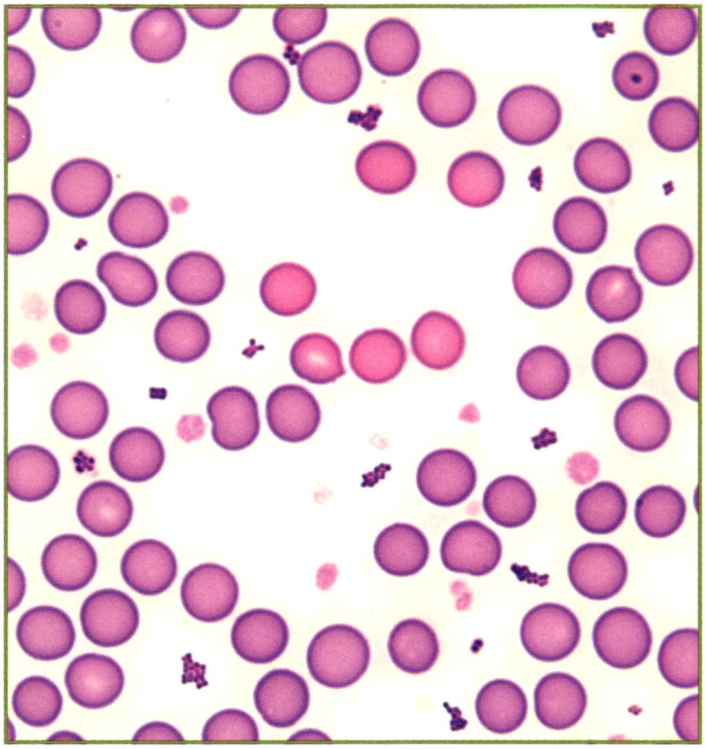

3-34 Staphylococcus Arrangement ◆ *Staphylococcus aureus* is shown in a blood smear. Note the staphylococci interspersed between the erythrocytes. *S. aureus* is a common opportunistic pathogen of humans and finding it in the blood (sepsis) carries with it a 20–40% chance of patient death. Cells are approximately 1 μm in diameter.

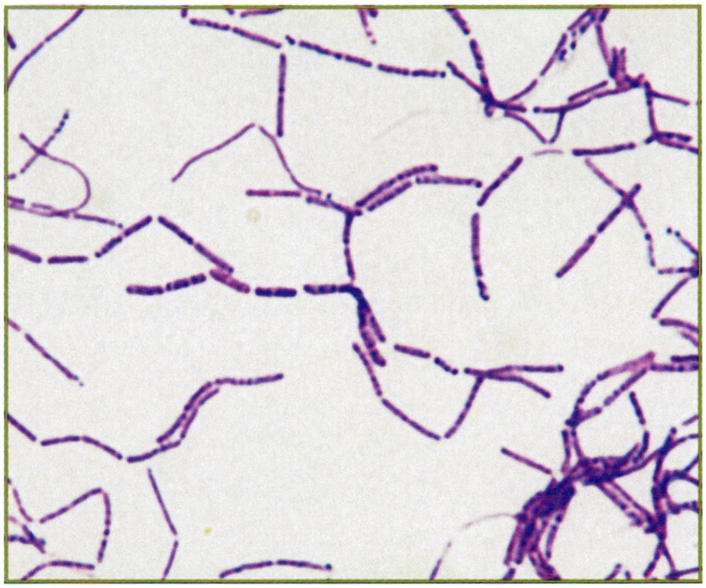

3-35 Streptobacillus Arrangement (Crystal Violet Stain) ◆ *Bacillus megaterium* is a streptobacillus. These cells were obtained from culture and are 1.2–1.5 μm wide by 2–5 μm long.

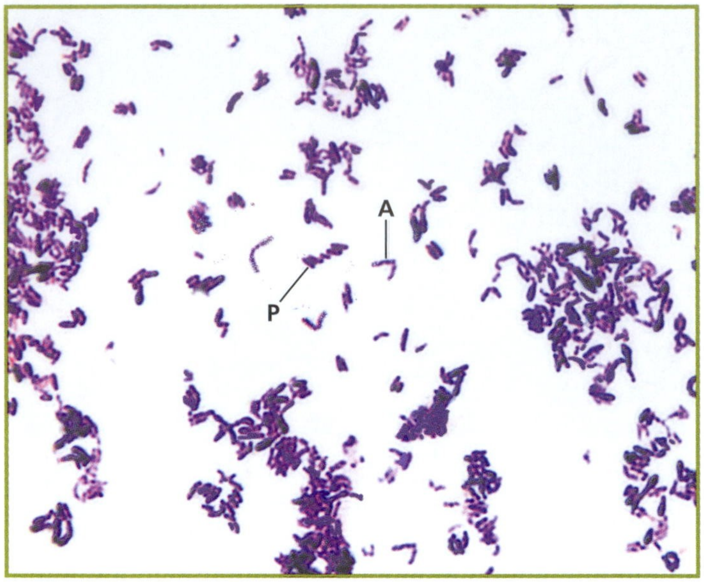

3-36 Palisades and Angular Arrangements of Arthrobacter (Gram Stain) ◆ Stacking of rod-shaped cells side-by-side is the palisades arrangement (**P**). An angular arrangement (**A**) is when a pair of rods is bent where the cells join. Both arrangements are due to "snapping division" in which the cells divide lengthwise and remain more or less attached. *Corynebacterium* also shows these arrangements.

EXERCISE 3-4

Simple Stains

Theory

Stains are solutions consisting of a solvent (usually water or ethanol) and a colored molecule (often a benzene derivative), the **chromogen**. The portion of the chromogen that gives it its color is the **chromophore**. A chromogen may have multiple chromophores, with each intensifying the color. The **auxochrome** is the charged portion of a chromogen and allows it to act as a dye through ionic or covalent bonds between the chromogen and the cell. **Basic stains**[1] (where the auxochrome becomes positively charged as a result of picking up a hydrogen ion or losing a hydroxide ion) are attracted to the negative charges on the surface of most bacterial cells. Thus, the cell becomes colored (Figure 3-37). Common basic stains include methylene blue, crystal violet, and safranin. Examples of basic stains may be seen in Figures 3-24, 3-35, and 3-38.

Basic stains are applied to bacterial smears that have been **heat-fixed.** Heat-fixing kills the bacteria, makes them adhere to the slide, and coagulates cytoplasmic proteins to make them more visible. It also distorts the cells to some extent.

Application

Because cytoplasm is transparent, cells usually are stained with a colored dye to make them more visible under the microscope. Then cell morphology, size, and arrangement can be determined. In a medical laboratory, these are usually determined with a Gram stain (Exercise 3-6), but you will be using simple stains as an introduction to these.

In This Exercise

Today you will learn how to prepare a bacterial smear (emulsion) and perform simple stains. Several different organisms will be supplied so you can begin to see the variety of cell morphologies and arrangements in the bacterial world. We suggest that you perform all the stains on one or two organisms (to get practice) and look at your lab partners' stains to see the variety of cell types. Be sure that you view all the available organisms.

[1] Notice that the term "basic" means "alkaline," not "elementary"; however, coincidentally, *basic* stains can be used for *simple* staining procedures.

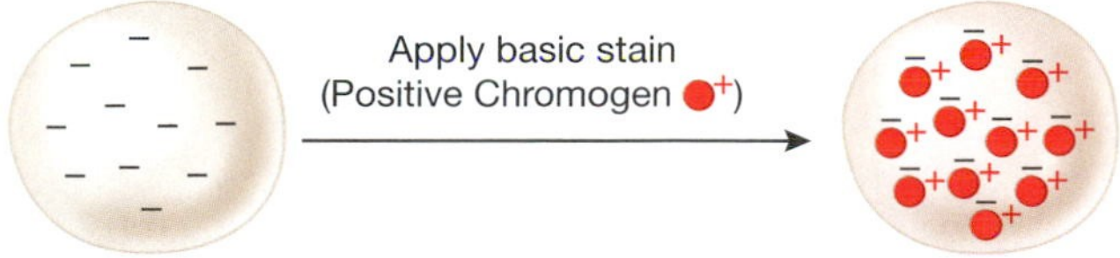
Apply basic stain
(Positive Chromogen ●+)
Negatively charged cell
Cell is stained

3-37 Chemistry of Basic Stains ◆ Basic stains have a positively charged chromogen (●+), which forms an ionic bond with the negatively charged bacterial cell, thus colorizing the cell.

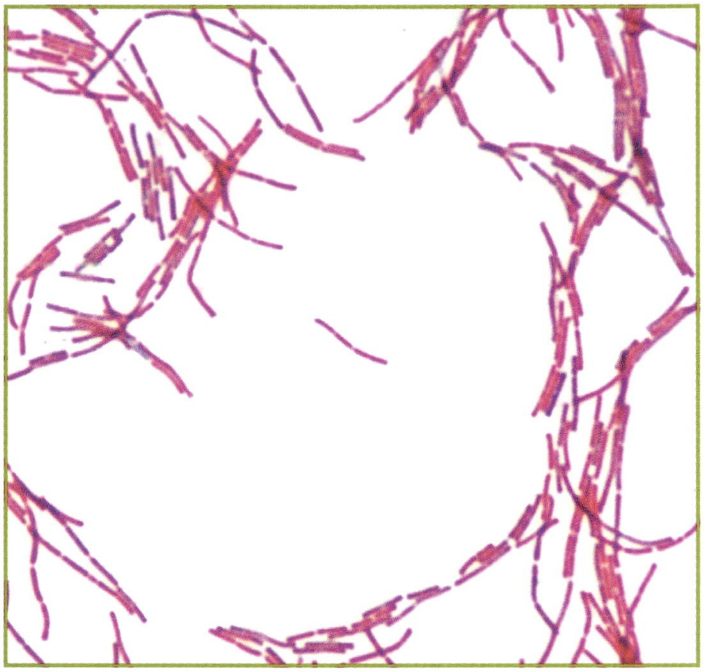

3-38 Safranin Dye in a Simple Stain ◆ This is a simple stain using safranin, a basic stain. Notice that the stain is associated with the cells and not the background. The organism is *Bacillus subtilis*, the type species for the genus *Bacillus*, grown in culture.

Materials

Per Group

- ✓ clean glass microscope slides
- ✓ methylene blue stain
- ✓ safranin stain
- ✓ crystal violet stain
- ✓ chemical eye protection (recommended)
- ✓ disposable gloves
- ✓ lab coat
- ✓ squirt bottle with water
- ✓ staining tray
- ✓ staining screen
- ✓ bibulous paper (or paper towels)
- ✓ slide holder
- ✓ compound microscope with oil objective lens and ocular micrometer
- ✓ immersion oil
- ✓ lens paper

✓ recommended organisms:

- *Bacillus cereus*
- *Micrococcus luteus*
- *Moraxella catarrhalis*
- *Rhodospirillum rubrum*
- *Staphylococcus epidermidis*
- *Vibrio harveyi*

Procedure

1 A bacterial smear (emulsion) is made prior to most staining procedures. Follow the Procedural Diagram in Figure 3-39 to prepare bacterial smears of each organism. With care, emulsions of three different organisms can be put on a single slide. We recommend making three slides with the first three organisms and three slides with the other three for a total of six slides. Be sure to write the names of each organism so they correspond to the correct emulsion! Prepare all the slides at once so they can be air drying at the same time.

2 Heat-fix each slide as described in Figure 3-39.

3 Following the basic staining procedure illustrated in the Procedural Diagram in Figure 3-40, stain one of each slide (that is, those with different organisms) with each stain using the following times:

crystal violet:	stain for 30 to 60 seconds
safranin:	stain for up to 1 minute
methylene blue:	stain for 30 to 60 seconds

Be sure to wear gloves when staining. Some labs require eye protection. Your instructor will advise you on this. Record your actual staining times in the table provided in the Data Sheet so you can adjust for over staining or under staining. You may be asked to remove your gloves prior to using the microscope to prevent getting stain on it.

4 Using the oil immersion lens, observe each slide. Record your observations of cell morphology, arrangement, and size in the chart provided on the Data Sheet.

5 Dispose of the slides and used stain according to your laboratory's policy.

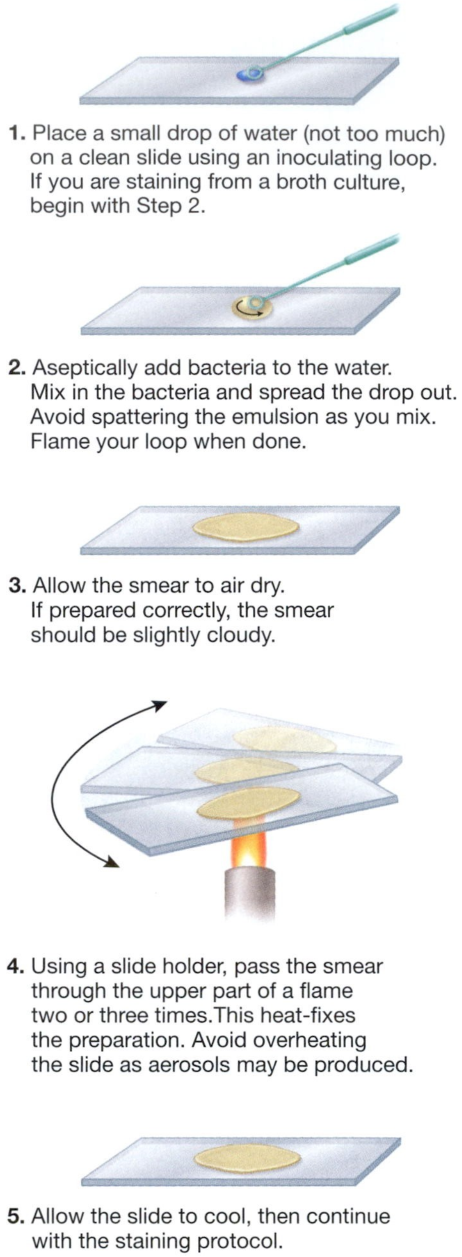

1. Place a small drop of water (not too much) on a clean slide using an inoculating loop. If you are staining from a broth culture, begin with Step 2.
2. Aseptically add bacteria to the water. Mix in the bacteria and spread the drop out. Avoid spattering the emulsion as you mix. Flame your loop when done.
3. Allow the smear to air dry. If prepared correctly, the smear should be slightly cloudy.
4. Using a slide holder, pass the smear through the upper part of a flame two or three times.This heat-fixes the preparation. Avoid overheating the slide as aerosols may be produced.
5. Allow the slide to cool, then continue with the staining protocol.

3-39 Procedural Diagram: Making a Bacterial Smear (Emulsion) ◆ Preparation of uniform bacterial smears will make consistent staining results easier to obtain. Heat-fixing the smear kills the bacteria, makes them adhere to the slide, and coagulates their protein for better staining. Caution: Avoid producing aerosols. Do not spatter the smear as you mix it, do not blow on or wave the slide to speed-up air drying, and do not overheat when heat-fixing.

1. Begin with a heat-fixed emulsion (see Figure 3-39). More than one organism can be put on a slide. In this exercise, we recommend putting three organisms on each slide.

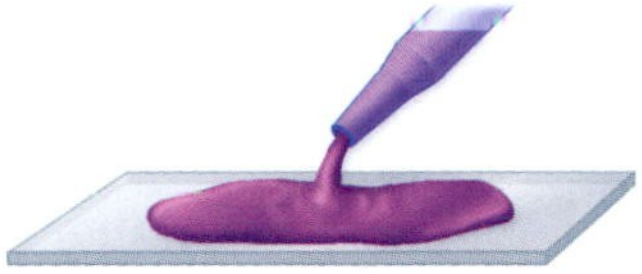

2. Cover the smear with stain for the designated time. Use a staining tray to catch excess stain. Be sure to wear gloves.

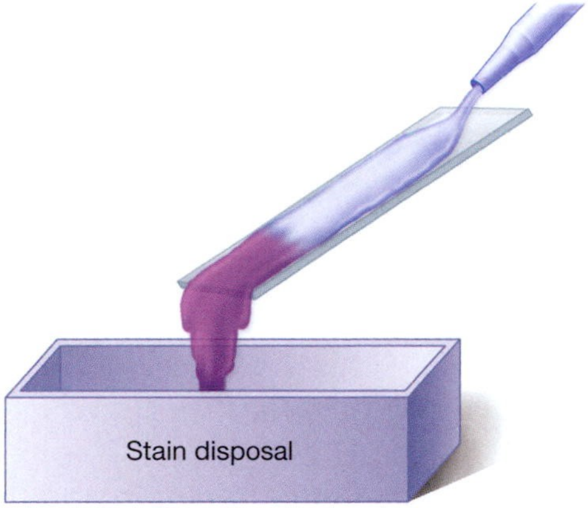
Stain disposal

3. Grasp the slide with a slide holder. Rinse the slide with water. Dispose of the excess stain according to your lab's practices.

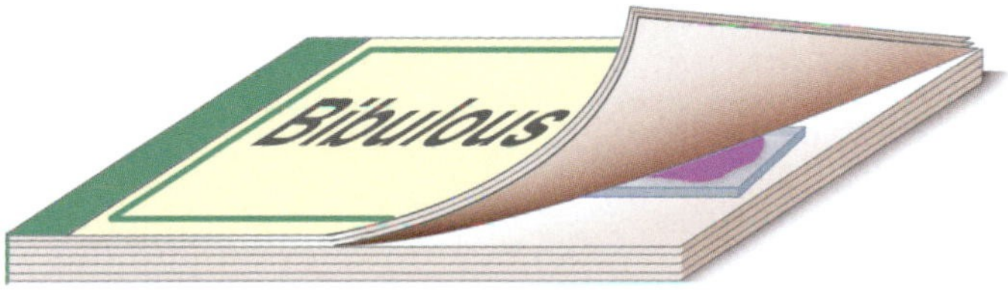
Bibulous

4. Gently blot dry in a tablet of bibulous paper or paper towels. (Alternatively, a page from the tablet can be removed and used for blotting.) Do not rub. When dry, observe under oil immersion.

3-40 Procedural Diagram: Simple Stain ◆ Staining times differ for each stain, but cell density of your smear also affects staining time. Strive for consistency in making your smears. Caution: Be sure to flame your loop after cell transfer and properly dispose of the slide when you are finished observing it.

References

Chapin, Kimberle. 1995. Chapter 4 in *Manual of Clinical Microbiology,* 6th ed., edited by Patrick R. Murray, Ellen Jo Baron, Michael A. Pfaller, Fred C. Tenover, and Robert H. Yolken. American Society for Microbiology, Washington, DC.

Chapin, Kimberle C., and Patrick R. Murray. 2003. Pages 257–259 in *Manual of Clinical Microbiology,* 8th ed., edited by Patrick R. Murray, Ellen Jo Baron, James H. Jorgensen, Michael A. Pfaller, and Robert H. Yolken. American Society for Microbiology, Washington, DC.

Forbes, Betty A., Daniel F. Sahm, and Alice. S. Weissfeld. 2002. Chapter 9 in *Bailey & Scott's Diagnostic Microbiology*, 11th ed. Mosby-Year Book, Inc. St. Louis.

Murray, R. G. E., Raymond N. Doetsch, and C. F. Robinow. 1994. Page 27 in *Methods for General and Molecular Bacteriology*, edited by Philipp Gerhardt, R. G. E. Murray, Willis A. Wood, and Noel R. Krieg. American Society for Microbiology, Washington, DC.

Norris, J. R., and Helen Swain. 1971. Chapter II in *Methods in Microbiology,* Vol. 5A, edited by J. R. Norris and D. W. Ribbons. Academic Press, Ltd., London.

Power, David A., and Peggy J. McCuen. 1988. Page 4 in *Manual of BBL™ Products and Laboratory Procedures*, 6th ed. Becton Dickinson Microbiology Systems, Cockeysville, MD.

DATA SHEET 3-4

NAME ______________________ DATE ______________________

LAB SECTION ______________________ I WAS PRESENT AND PERFORMED THIS EXERCISE (INITIALS) ________

Simple Stains

OBSERVATIONS AND INTERPRETATIONS

Record your observations in the chart below.

Organism	Stain and Duration	Cellular Morphology and Arrangement (Include a detailed sketch of a few representative cells from one of your preparations of each organism)	Cell Dimensions

Questions

1 What is the consequence of leaving a stain on the bacterial smear too long (over staining)?

2 What is the consequence of not leaving a stain on the smear long enough (under staining)?

3 Choose a coccus and a bacillus from the organisms you observed, and calculate their surface-to-volume ratios. Consider the coccus to be a perfect sphere and the bacillus to be a cylinder. Use the equations supplied.

Cell Morphology	Surface Area	Volume
Coccus	$SA = 4\pi r^2$	$V = \frac{4}{3}\pi r^3$
Bacillus	$SA = 2(H \times W) + 4(L \times H)$	$V = \pi r^2 h$

r = radius, H = height, L = length, W = width, $\pi = 3.14$

Surface-to-Volume Ratio of Sample Cells

Organism	Cell Morphology	Surface Area (μm^2)	Volume (μm^3)	Surface-to-Volume Ratio

4 Consider a coccus and a rod of equal volume.

a. Which is more likely to survive in a dry environment? Explain your answer.

b. Which would be better adapted to a moist environment? Explain your answer.

EXERCISE 3-5

Negative Stains

Theory

The negative staining technique uses a dye solution in which the chromogen is acidic and carries a negative charge. (An acidic chromogen gives up a hydrogen ion, which leaves it with a negative charge.) The negative charge on the bacterial surface repels the negatively charged chromogen, so the cell remains unstained against a colored background (Figure 3-41). A specimen stained with the acidic stain nigrosin is shown in Figure 3-42.

Application

The negative staining technique is used to determine morphology and cellular arrangement in bacteria that are too delicate to withstand heat-fixing. A primary example is the spirochete *Treponema*, which is distorted by the heat-fixing of other staining techniques. Also, where determining the accurate size is crucial, a negative stain can be used because it produces minimal cell shrinkage.

In This Exercise

Today you will perform negative stains on three different organisms. You also will have the opportunity to compare the size of *M. luteus* measured with the negative stain to its size as determined using a simple stain.

Materials

Per Student Group

- ✓ nigrosin stain or eosin stain
- ✓ clean glass microscope slides
- ✓ chemical eye protection (recommended)
- ✓ disposable gloves
- ✓ lab coat
- ✓ compound microscope with oil objective lens and ocular micrometer
- ✓ immersion oil
- ✓ lens paper
- ✓ bibulous paper or paper towel
- ✓ recommended organisms:
  - ♦ *Micrococcus luteus*
  - ♦ *Bacillus cereus*
  - ♦ *Rhodospirillum rubrum*

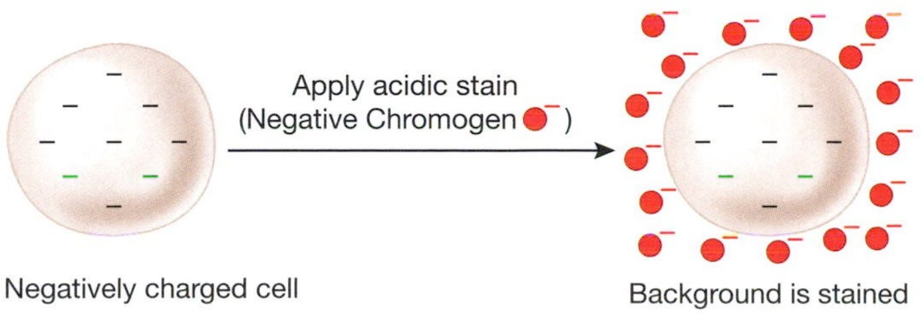

3-41 Chemistry of Acidic Stains ♦ Acidic stains have a negatively charged chromogen (●⁻) that is repelled by negatively charged cells. Thus, the background is colored and the cell remains transparent.

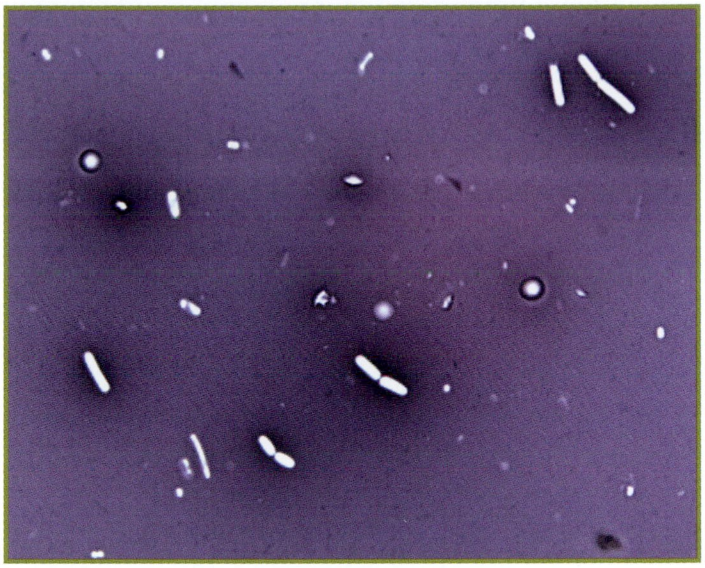

3-42 A Nigrosin Negative Stain ♦ Notice that the *Bacillus megaterium* cells are unstained against a dark background. The circular objects are bubbles.

Procedure

1. Follow the Procedural Diagram in Figure 3-43 to prepare a negative stain of each organism. Be sure to wear gloves and eye protection (if required). Label each slide with the organism's name.
2. Dispose of the spreader slide in a disinfectant jar or sharps container immediately after use.
3. If required to do so, remove your gloves. Then, observe using the oil immersion lens. Record your observations in the chart on the Data Sheet.
4. Dispose of the specimen slide in a disinfectant jar or sharps container after use.

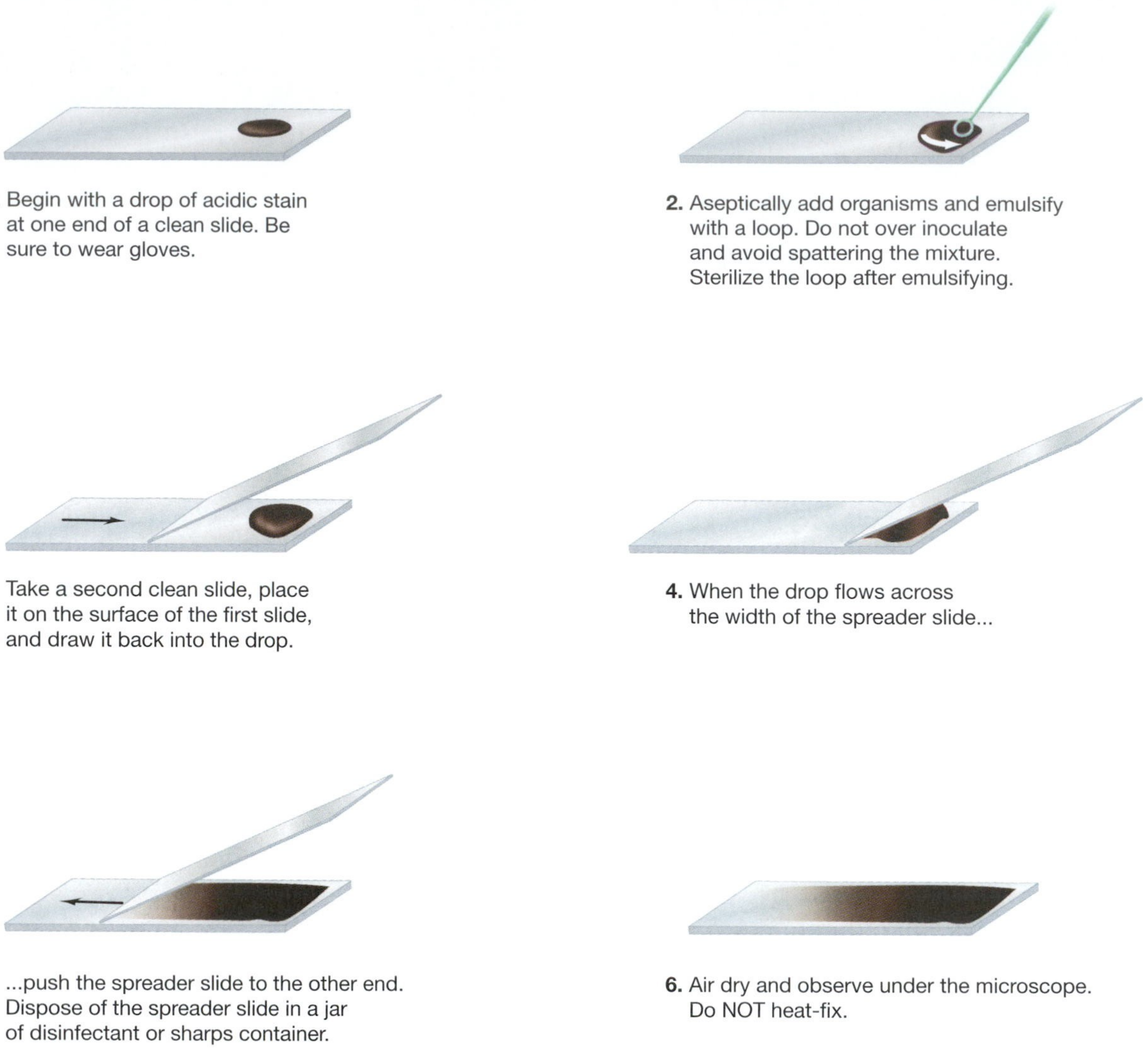
1. Begin with a drop of acidic stain at one end of a clean slide. Be sure to wear gloves.
2. Aseptically add organisms and emulsify with a loop. Do not over inoculate and avoid spattering the mixture. Sterilize the loop after emulsifying.
3. Take a second clean slide, place it on the surface of the first slide, and draw it back into the drop.
4. When the drop flows across the width of the spreader slide...
5. ...push the spreader slide to the other end. Dispose of the spreader slide in a jar of disinfectant or sharps container.
6. Air dry and observe under the microscope. Do NOT heat-fix.

3-43 Procedural Diagram: Negative Stain ◆ Be sure to sterilize your loop after transfer, and to appropriately dispose of the spreader and specimen slides.

References

Claus, G. William. 1989. Chapter 5 in *Understanding Microbes—A Laboratory Textbook for Microbiology*. W. H. Freeman and Co., New York.

Murray, R. G. E., Raymond N. Doetsch, and C. F. Robinow. 1994. Page 27 in *Methods for General and Molecular Bacteriology,* edited by Philipp Gerhardt, R. G. E. Murray, Willis A. Wood, and Noel R. Krieg. American Society for Microbiology, Washington, DC.

NAME ______________________ DATE ______________________

LAB SECTION ______________________ I WAS PRESENT AND PERFORMED THIS EXERCISE (INITIALS) ______

DATA SHEET 3-5

Negative Stains

OBSERVATIONS AND INTERPRETATIONS

Record your observations in the chart below.

Organism	Stain	Cellular Morphology and Arrangement (Include a detailed sketch of a few representative cells)	Cell Dimensions

Questions

1 Why doesn't a negative stain colorize the cells in the smear?

2 Eosin is a red stain and methylene blue is blue. What should be the result of staining a bacterial smear with a mixture of eosin and methylene blue?

3 Compare the diameter of *M. luteus* cells as measured using a basic stain (Exercise 3-4) and an acidic stain in this exercise. What might account for any difference?

Differential and Structural Stains

Differential stains allow a microbiologist to detect differences between organisms or differences between parts of the same organism. In practice, these are used much more frequently than simple stains because they not only allow determination of cell size, morphology, and arrangement (as with a simple stain) but information about other features as well.

The Gram stain is the most commonly used differential stain in bacteriology. Other differential stains are used for organisms not distinguishable by the Gram stain and for those that have other important cellular attributes, such as acid-fastness, a capsule, spores, or flagella. With the exception of the acid-fast stain, these other stains sometimes are referred to as **structural stains**.

EXERCISE 3-6

Gram Stain

Theory

The Gram stain is a differential stain in which a **decolorization** step occurs between the application of two basic stains. The Gram stain has many variations, but they all work in basically the same way (Figure 3-44). The **primary stain** is crystal violet. Iodine is added as a **mordant** to enhance crystal violet staining by forming a **crystal violet–iodine complex**. Decolorization with an alcohol solution (that sometimes contains acetone) follows and is the most critical step in the procedure. Gram-negative cells are decolorized by the solution (that is, they lose the crystal violet stain) whereas Gram-positive cells are not. Gram-negative cells can thus be colorized by the **counterstain** safranin. Upon successful completion of a Gram stain, Gram-positive cells appear purple and Gram-negative cells appear reddish-pink (Figure 3-45).

Electron microscopy and other evidence indicate that the ability to resist decolorization or not is based on the different wall constructions of Gram-positive and Gram-negative cells. Gram-negative cell walls have a higher lipid content (because of the outer membrane) and a thinner peptidoglycan layer than Gram-positive cell walls (Figure 3-46). The alcohol/acetone in the decolorizer extracts the lipid, making the Gram-negative wall more porous and incapable of retaining the crystal violet–iodine complex,

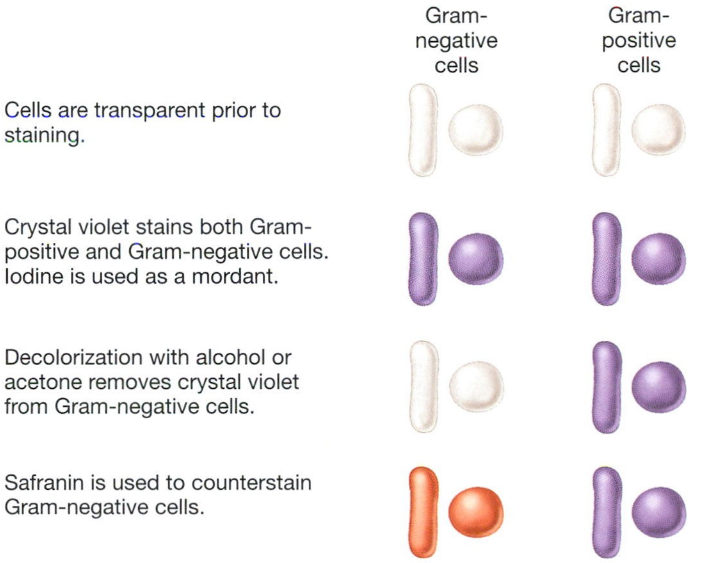

Gram-negative cells
Gram-positive cells
Cells are transparent prior to staining.
Crystal violet stains both Gram-positive and Gram-negative cells. Iodine is used as a mordant.
Decolorization with alcohol or acetone removes crystal violet from Gram-negative cells.
Safranin is used to counterstain Gram-negative cells.

3-44 Gram Stain ◆ After application of the primary stain (crystal violet), decolorization, and counterstaining with safranin, Gram-positive cells stain violet and Gram-negative cells stain pink/red. Notice that crystal violet and safranin are both basic stains, and that the decolorization step is what makes the Gram stain differential.

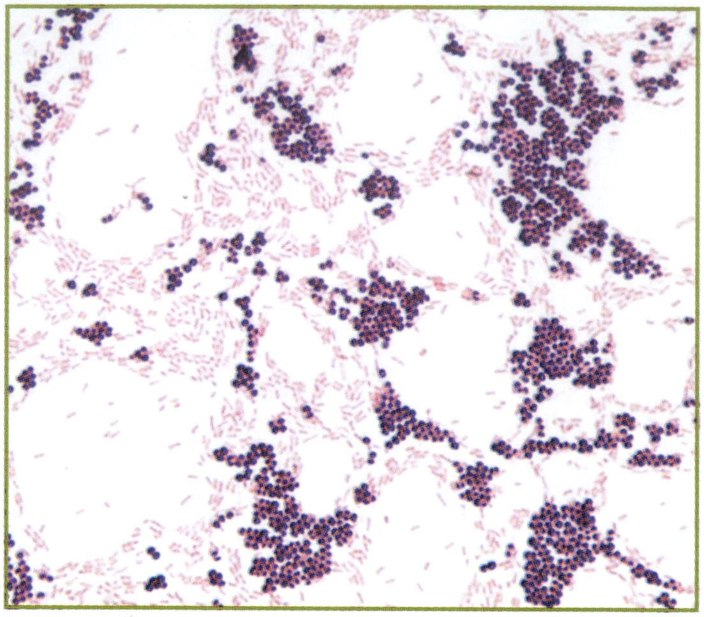

3-45 Gram Stain of *Staphylococcus* (+) and *Proteus* (–) ◆ *Staphylococcus* has a staphylococcal arrangement, whereas *Proteus* is a bacillus.

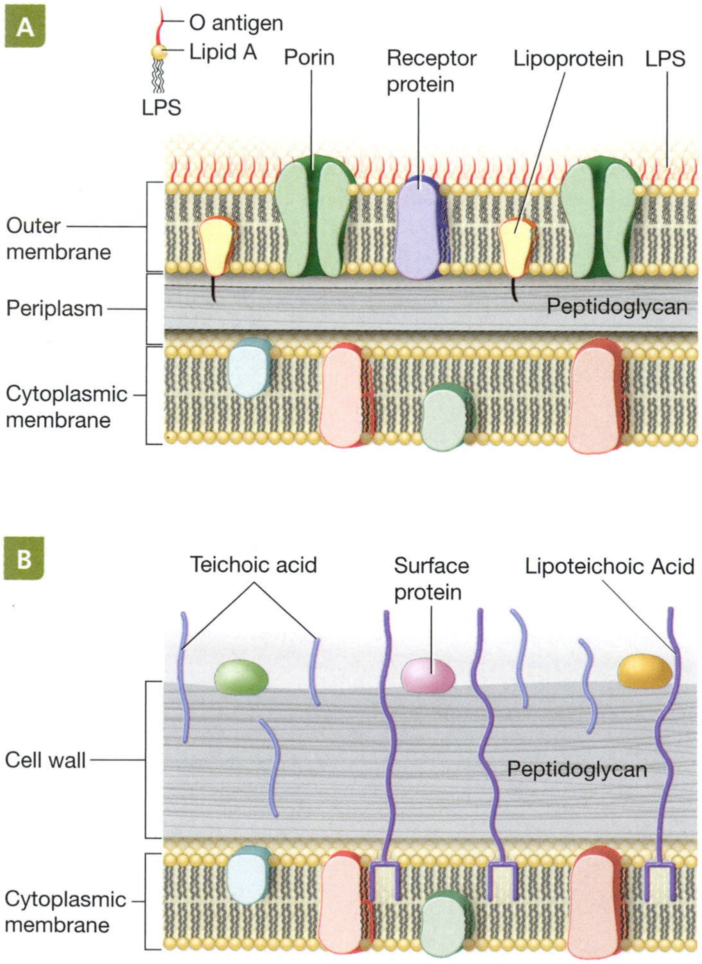
A
O antigen
Lipid A
LPS
Porin
Receptor protein
Lipoprotein
LPS
Outer membrane
Periplasm
Peptidoglycan
Cytoplasmic membrane
B
Teichoic acid
Surface protein
Lipoteichoic Acid
Cell wall
Peptidoglycan
Cytoplasmic membrane

3-46 Bacterial Cell Walls ◆ **A** The Gram-negative wall is composed of less peptidoglycan (as little as a single layer) and more lipid (due to the outer membrane) than the Gram-positive wall **B**.

thereby decolorizing it. The thicker peptidoglycan and greater degree of cross-linking (because of teichoic acids) trap the crystal violet–iodine complex more effectively, making the Gram-positive wall less susceptible to decolorization.

Although some organisms give Gram-variable results, most variable results are a consequence of poor technique. The decolorization step is the most crucial and most likely source of Gram stain inconsistency. It is possible to **over decolorize** by leaving the alcohol on too long and get reddish Gram-*positive* cells. It also is possible to **under decolorize** and produce purple Gram-*negative* cells. Neither of these situations changes the actual Gram reaction for the organism being stained. Rather, these are false results because of poor technique.

A second source of poor Gram stains is inconsistency in preparation of the emulsion. Remember, a good emulsion dries to a faint haze on the slide.

A third source of inconsistent Gram stains is the organisms themselves. Some Gram-positives (notably *Bacillus* species) lose their ability to retain the crystal violet in as little as 24 hours. Figure 3-47 illustrates a *Bacillus* culture stained at 24 hours of growth. Always plan on doing your Gram stains on cultures no older than 24 hours for best results.

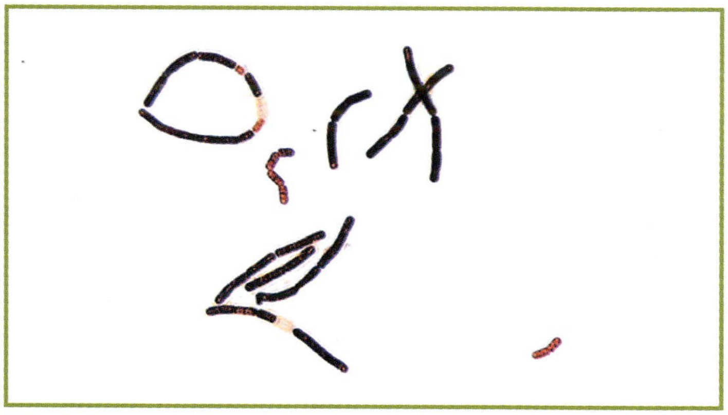

3-47 24-Hour *Bacillus* Illustrating Loss of Gram-positive Reaction (Gram Stain) ◆ This *Bacillus* was isolated from aerobic estuarine mud and stained within 24 hours of making the pure culture. Notice some cells are pink and have lost their ability to resist decolorization.

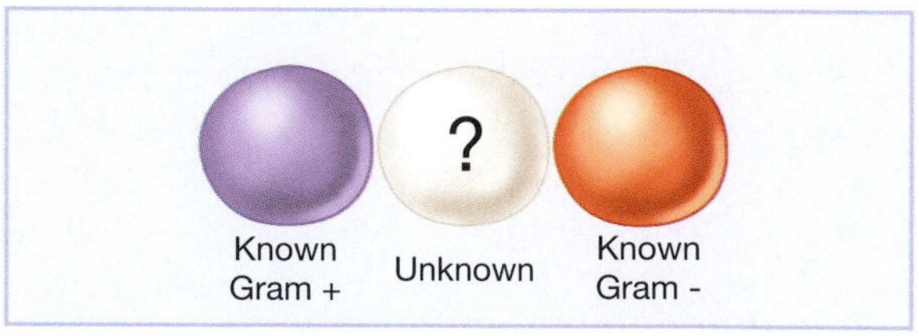
?
Known Gram +
Unknown
Known Gram -

3-48 Positive Controls To Check Your Technique ◆ Staining known Gram-positive and Gram-negative organisms on either side of your unknown organism act as positive controls for your technique. Try to make the emulsions as close to one another as possible. Spreading them out across the slide makes it difficult to stain and decolorize them equally.

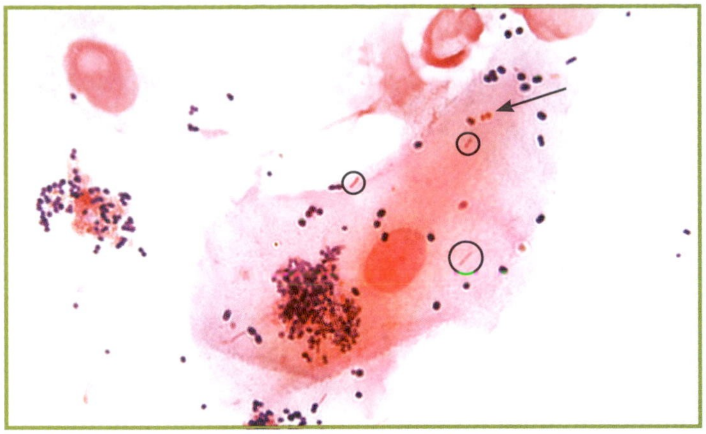

3-49 Direct Smear Positive Control (Gram Stain) ◆ A direct smear made from the gumline may also be used as a Gram stain control. Expect numerous Gram-positive bacteria (especially cocci) and some Gram-negative cells, including your own epithelial cells. In this slide, Gram-positive cocci predominate, but a few Gram-negative cells are visible, including Gram-negative rods (**circled**) and a Gram-negative diplococcus (**arrow**) on the surface of the epithelial cell.

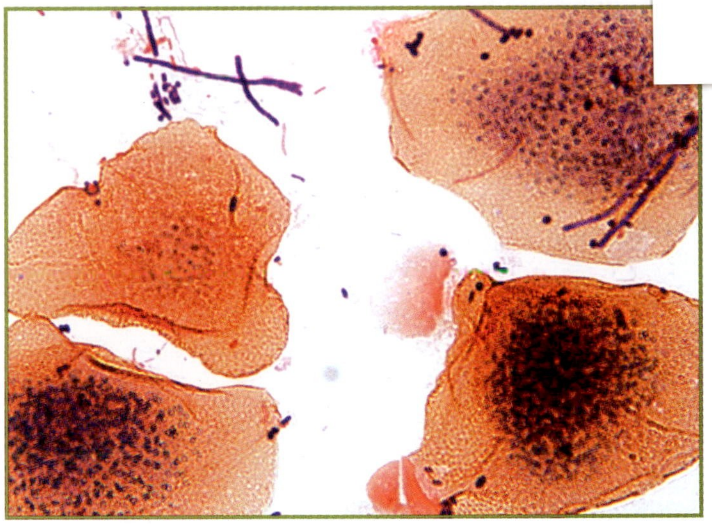

3-50 Under-Decolorized Gram Stain ◆ This is a direct smear from the gumline. Notice the purple patches of stain on the epithelial cells. Also notice the variable quality of this stain—the epithelial cell to the left of center is stained better than the others.

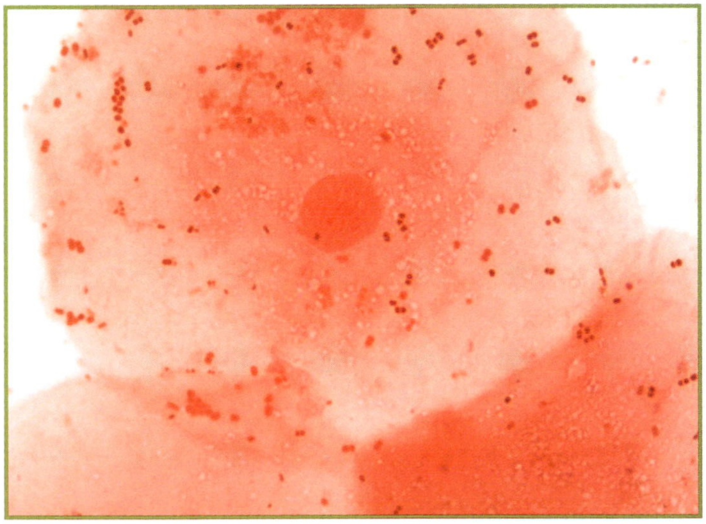

3-51 Over-Decolorized Gram Stain ◆ This also is a direct smear from the gumline. Notice the virtual absence of any purple cells, a certain indication of over decolorization.

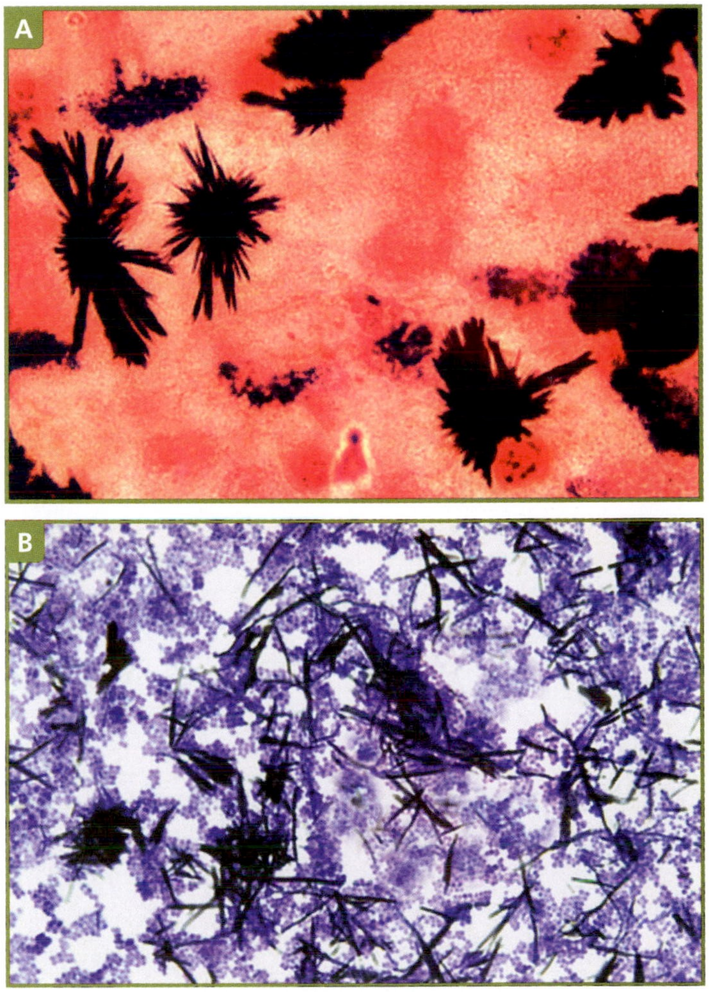

3-52 Crystal Violet Crystals (Gram Stain) ◆ If the staining solution is not adequately filtered or is old, crystal violet crystals may appear. Although they are pleasing to the eye, they obstruct your view of the specimen. Crystals from two different Gram stains are shown here: **A** a gumline direct smear; **B** *Micrococcus roseus* grown in culture.

Until correct results are obtained consistently, it is recommended that control smears of Gram-positive and Gram-negative organisms be stained along with the organism in question (Figure 3-48). As an alternative control, a direct smear made from the gumline may be Gram stained (Figure 3-49) with the expectation that both Gram-positive and Gram-negative organisms will be seen. Under-decolorized and over-decolorized gumline direct smears are shown for comparison (Figures 3-50 and 3-51). Positive controls also should be run when using new reagent batches.

Interpretation of Gram stains can be complicated by nonbacterial elements. For instance, stain crystals from an old or improperly made stain solution can disrupt the field (Figure 3-52) or stain precipitate may be mistakenly identified as bacteria (Figure 3-53).

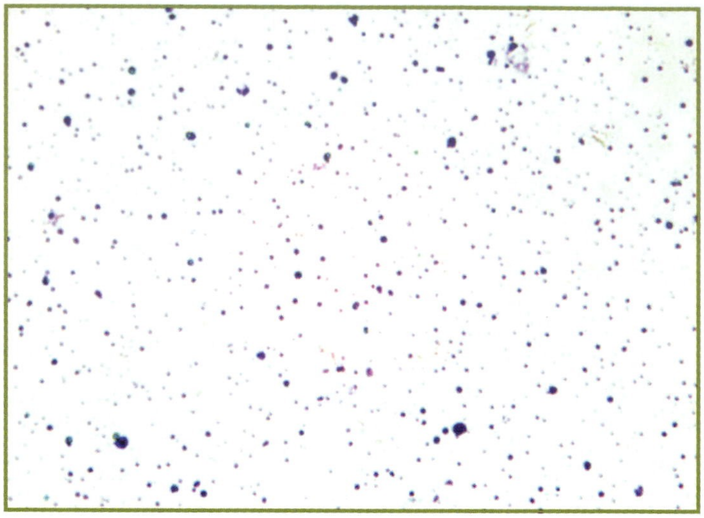

3-53 Stain Precipitate (Gram Stain) ◆ If the slide is not rinsed thoroughly or the stain is allowed to dry on the slide, spots of stain precipitate may form and may be confused with bacterial cells. Their variability in size is a clue that they are not bacteria.

❖ Application

The Gram stain, used to distinguish between Gram-positive and Gram-negative cells, is the most important and widely used microbiological differential stain. In addition to Gram reaction, this stain allows determination of cell morphology, size, and arrangement. It typically is the first differential test run on a specimen brought into the laboratory for identification. In some cases, a rapid, presumptive identification of the organism or elimination of a particular organism is possible.

❖ In This Exercise

The Gram stain is the single most important differential stain in bacteriology. Therefore, you will have to practice it and practice it some more to become proficient in obtaining consistent results. The organisms should be used in the combinations given so you will have one Gram-positive and one Gram-negative on each slide.

❖ Materials

Per Student Group

- ✓ compound microscope with oil objective lens and ocular micrometer
- ✓ clean glass microscope slides
- ✓ sterile toothpick
- ✓ Gram stain solutions (commercial kits are available)
- ✓ Gram crystal violet
- ✓ Gram iodine
- ✓ 95% ethanol (or ethanol/acetone solution)
- ✓ Gram safranin
- ✓ squirt bottle with water
- ✓ bibulous paper or paper towels
- ✓ chemical eye protection (recommended)
- ✓ disposable gloves
- ✓ lab coat
- ✓ staining tray
- ✓ staining screen
- ✓ slide holder
- ✓ recommended organisms (overnight cultures grown on agar slants or turbid broth cultures):
  - ◆ *Staphylococcus epidermidis*
  - ◆ *Escherichia coli*
  - ◆ *Moraxella catarrhalis*
  - ◆ *Corynebacterium xerosis*

Procedure

1. Because Gram stains require much practice, we recommend that you prepare several slides of each combination and let them air dry simultaneously. Then they'll be ready when you need them.
2. Follow the procedure illustrated in Figure 3-39 to prepare and heat-fix smears of *Staphylococcus epidermidis* and *Escherichia coli* immediately next to one another on two or three different clean glass slides. (If you make the emulsions at opposite ends of the slide, you may find it difficult to stain and decolorize each equally.) Strive to prepare smears of uniform thickness, as thick smears risk being under decolorized. Continue with Step 3 as you allow these slides to air dry. When dry, heat-fix them.
3. Repeat step 2 for *Moraxella catarrhalis* and *Corynebacterium xerosis* on two or three clean glass slides, air dry, and heat-fix them.
4. Wash your hands, then use the sterile toothpick to obtain a sample from your teeth at the gumline. (Do not draw blood! What you want is easily removed from your gingival pockets.) Transfer the sample to a drop of water on a clean glass slide, air dry, and heat-fix.
5. Follow the basic staining procedure illustrated in Figure 3-54. We recommend staining the pure cultures first. **Important: Stain one slide at a time and look at it before you stain the next one so you can correct any errors in technique you may make.** After your technique is consistent, stain the gumline sample. Be sure to wear gloves for all stains. You may be advised to remove them before using the microscope to prevent stain from getting on the microscope.

6 If required to do so, remove your gloves and observe using the oil immersion lens. Record your observations of cell morphology and arrangement, dimensions, and Gram reactions in the chart provided on the Data Sheet. For the gumline, choose a region with epithelial cells and bacterial cells to observe. Record your observations of the multiple cell types.

7 Dispose of the specimen slides in a jar of disinfectant or a sharps container after use.

8 Put your microscope away properly, making sure there is no oil on the lenses or stage.

References

Chapin, Kimberle C., and Patrick R. Murray. 2003. Pages 258–260 in *Manual of Clinical Microbiology*, 8th ed., edited by Patrick R. Murray, Ellen Jo Baron, James H. Jorgensen, Michael A. Pfaller, and Robert H. Yolken. American Society for Microbiology, Washington, DC.

Forbes, Betty A., Daniel F. Sahm, and Alice. S. Weissfeld. 2002. Chapter 9 in *Bailey & Scott's Diagnostic Microbiology*, 11th ed. Mosby-Year Book, St. Louis.

Koneman, Elmer W., Stephen D. Allen, William M. Janda, Paul C. Schreckenberger, and Washington C. Winn, Jr. 1997. Chapter 14 in *Color Atlas and Textbook of Diagnostic Microbiology*, 5th ed. J. B. Lippincott Co., Philadelphia.

Murray, R. G. E., Raymond N. Doetsch, and C. F. Robinow. 1994. Pages 31 and 32 in *Methods for General and Molecular Bacteriology*, edited by Philipp Gerhardt, R. G. E. Murray, Willis A. Wood, and Noel R. Krieg. American Society for Microbiology, Washington, DC.

Norris, J. R., and Helen Swain. 1971. Chapter II in *Methods in Microbiology*, Vol 5A, edited by J. R. Norris and D. W. Ribbons. Academic Press, Ltd., London.

Power, David A., and Peggy J. McCuen. 1988. Page 261 in *Manual of BBL™ Products and Laboratory Procedures*, 6th ed. Becton Dickinson Microbiology Systems, Cockeysville, MD.

1. Begin with a heat-fixed emulsion. (See Figure 3-39.)

2. Cover the smear with crystal violet stain for 1 minute. Use a staining tray to catch excess stain. Be sure to wear gloves and eye protection if it is required in your laboratory.

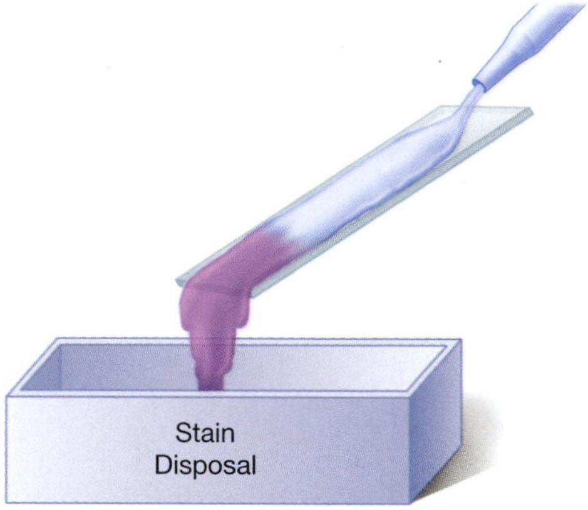

3. Grasp the slide with a slide holder.

Gently rinse the slide with distilled water. Alternatively, you can tap the edge of the slide to remove the excess crystal violet stain and eliminate the wash step.

4. Cover the smear with Gram's stain for 1 minute. Use a staining tray to catch excess stain.

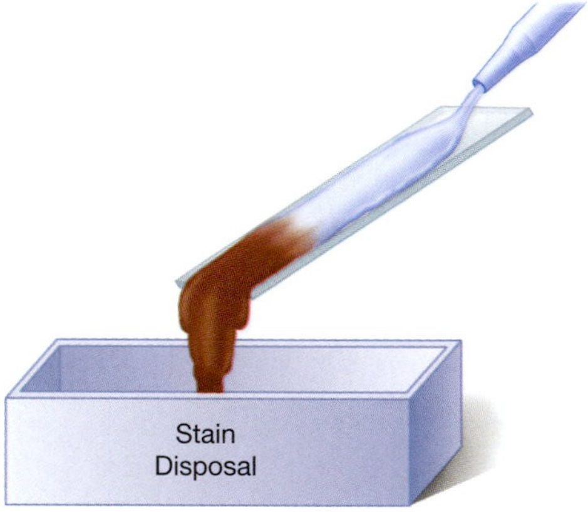

5. Grasp the slide with a slide holder. Gently rinse the slide with distilled water.

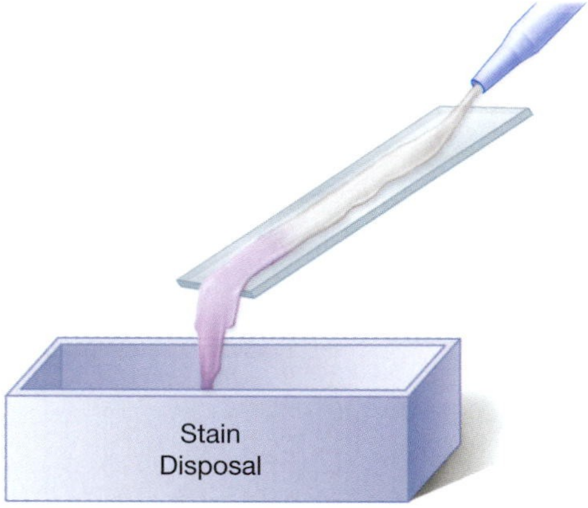

6. Decolorize with 95% ethanol or ethanol/acetone by allowing it to trickle down the slide until the run-off is clear. Generally, 10 seconds is the maximum time needed for this.

Gently rinse the slide with distilled water. Remember, though, that the decolorizer is working while you are changing from the decolorizer bottle to the water bottle. Make this transition quickly, but not carelessly.

7. Counterstain with safranin stain for 1 minute. Rinse with distilled water.

8. Gently blot dry in a tablet of bibulous paper or paper towels. (Alternatively, a page from the tablet can be removed and used for blotting.) Do not rub. When dry, observe under oil immersion.

3-54 Procedural Diagram: Gram Stain ◆ Pay careful attention to the staining times. If your preparations do not give "correct" results, the most likely source of error is in the decolorization step. Adjust its timing accordingly on subsequent stains.

Name ______________________ Date ______________________

Lab Section ______________________ I was present and performed this exercise (initials) ________

DATA SHEET 3-6

Gram Stain

OBSERVATIONS AND INTERPRETATIONS

Record your observations in the chart below.

Organism or Source	Cellular Morphology and Arrangement (Include a detailed sketch of a few representative cells)	Cell Dimensions	Color	Gram Reaction (+/−)

Questions

1 Predict the effect of the following "mistakes" made when performing a Gram stain. Your answer needs to address the predicted effect on both Gram-positive and Gram-negative cells. Consider each mistake independently.

a. Failure to add the iodine.

b. Failure to apply the decolorizer.

c. Failure to apply the safranin.

d. Reversal of crystal violet and safranin stains.

2 Both crystal violet and safranin are basic stains and may be used to do simple stains on Gram-positive and Gram-negative cells. This being the case, explain how they stain different cell types in the Gram stain.

3 If you saw large, eukaryotic cells in the preparation made from your gumline, they were most likely your own epithelial cells. Are you Gram-positive or Gram-negative? What is it about your cell structure that would allow you to predict this result?

4 One of your lab partners has followed the recommended procdure of running Gram-positive and Gram-negative control organisms on her Gram stain of an unknown species. Her choices of controls were *Escherichia coli* and *Bacillus subtilis*. She tries several times and each time concludes she is decolorizing too long because both controls have pink cells (one more than the other). What might you suggest she try and why?

EXERCISE 3-7

Acid-Fast Stains

❖ Theory

The presence of mycolic acids in the cell walls of acid-fast organisms is the cytological basis for the acid-fast differential stain. Mycolic acid is a waxy substance that gives acid-fast cells a higher affinity for the primary stain and resistance to decolorization by an acid alcohol solution. A variety of acid-fast staining procedures are employed, two of which are the Ziehl-Neelsen (ZN) method and the Kinyoun (K) method. These differ primarily in that the ZN method uses heat as part of the staining process, whereas the K method is a "cold" stain. In both protocols the bacterial smear may be prepared in a drop of serum to help the "slippery" acid-fast cells adhere to the slide. The two methods provide comparable results.

The waxy wall of acid-fast cells repels typical aqueous stains. (As a result, most acid-fast positive organisms are only weakly Gram-positive.) In the ZN method (Figure 3-56), the phenolic compound carbolfuchsin is used as the primary stain because it is lipid-soluble and penetrates the waxy cell wall. It also stains acid-fast negative cells. Staining of acid-fast cells by carbolfuchsin is further enhanced by steam-heating the preparation to melt the wax and allow the stain to move into the cell. Acid alcohol is used to decolorize nonacid-fast cells; acid-fast cells resist this decolorization. A counterstain, such as methylene blue, then is applied. Upon completion, acid-fast cells are reddish-purple; nonacid-fast cells are the color of the counterstain (Figure 3-56).

The Kinyoun (K) method uses a slightly more lipid-soluble and concentrated carbolfuchsin as the primary stain. These properties allow the stain to penetrate the acid-fast walls without the use of heat but make this method slightly less sensitive than the ZN method. Decolorization with acid alcohol is followed by a contrasting counterstain, such as brilliant green (Figure 3-57) or methylene blue.

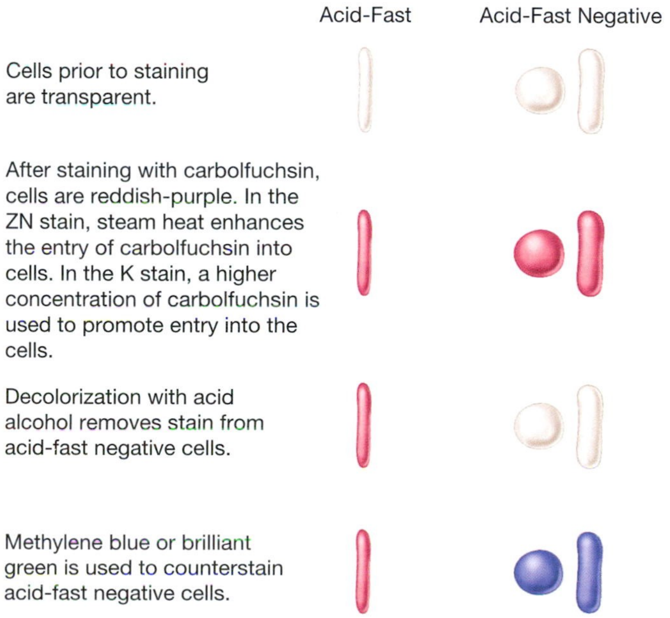
Acid-Fast
Acid-Fast Negative
Cells prior to staining are transparent.
After staining with carbolfuchsin, cells are reddish-purple. In the ZN stain, steam heat enhances the entry of carbolfuchsin into cells. In the K stain, a higher concentration of carbolfuchsin is used to promote entry into the cells.
Decolorization with acid alcohol removes stain from acid-fast negative cells.
Methylene blue or brilliant green is used to counterstain acid-fast negative cells.

3-55 Acid-Fast Stain Basics ◆ Acid-fast cells stain reddish-purple; acid-fast negative cells stain blue or the color of the counterstain if a different one is used.

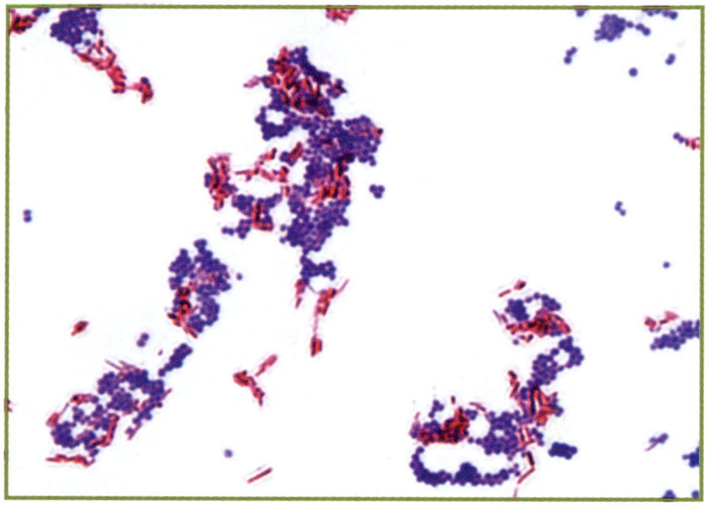

3-56 An Acid-Fast Stain Using the Ziehl-Neelson (ZN) Method ◆ Notice how most of the *Mycobacterium phlei* (AF+) cells are in clumps, an unusual state for most rods. They do this because their waxy cell walls make them sticky. A few individual cells are visible, however, and they clearly are rods that measure approximately 0.5 μm by 2–3 μm. The *Staphylococcus epidermidis* cells (AF−) are also in clumps, but that is because they grow as grape-like clusters. Each cell's diameter is approximately 1μm. Compare this micrograph with Figure 3-57.

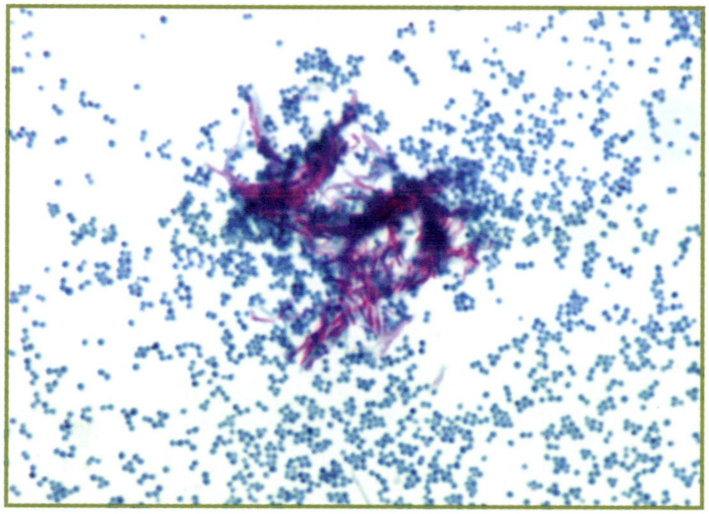

3-57 An Acid-Fast Stain Using the Kinyoun (K) Method ◆ This is an acid-fast stain of *Mycobacterium smegmatis* (AF+) and *Staphylococcus epidermidis* (AF−) from cultures. Again, notice the clumping of the acid-fast organisms. Brilliant green dye or methylene blue can be used as a counterstain. Brilliant green commonly stains more of a gray color, but it still contrasts with the carbolfuchsin of acid-fast positive cells. Compare this micrograph with the one in Figure 3-56.

❖ Application

The acid-fast stain is a differential stain used to detect cells capable of retaining a primary stain when treated with an acid alcohol. It is an important differential stain used to identify bacteria in the actinomycete genus *Mycobacterium,* some of which are pathogens (e.g., *M. leprae* and *M. tuberculosis*, causative agents of leprosy and tuberculosis, respectively). Members of the actinomycete genus *Nocardia* (*N. brasiliensis* and *N. asteroides* are opportunistic pathogens) are partially acid-fast. Oocysts of coccidian parasites, such as *Cryptosporidium* and *Isospora,* are also acid-fast. Because so few organisms are acid-fast, the acid-fast stain is run only when infection by an acid-fast organism is suspected.

Acid-fast stains are useful in identifying **acid-fast bacilli** (**AFB**) and in rapid, preliminary diagnosis of tuberculosis (with greater than 90% predictive value from sputum samples). It also can be performed on patient samples to track the progress of antibiotic therapy and determine the degree of contagiousness. A prescribed number of microscopic fields is examined and the number of AFB is determined and reported using a standard scoring system.

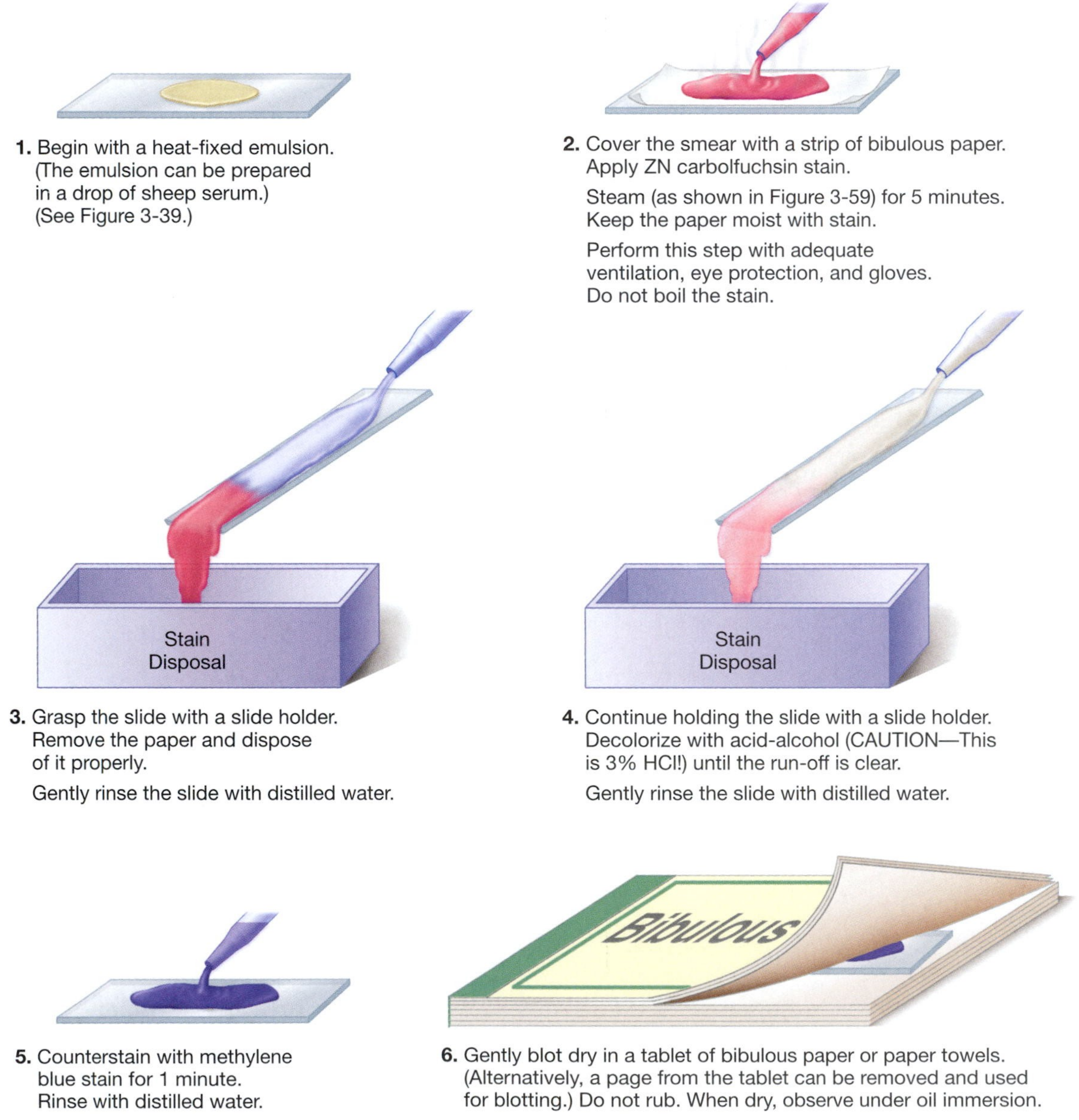

1. Begin with a heat-fixed emulsion. (The emulsion can be prepared in a drop of sheep serum.) (See Figure 3-39.)
2. Cover the smear with a strip of bibulous paper. Apply ZN carbolfuchsin stain.
Steam (as shown in Figure 3-59) for 5 minutes. Keep the paper moist with stain.
Perform this step with adequate ventilation, eye protection, and gloves. Do not boil the stain.
Stain Disposal
3. Grasp the slide with a slide holder. Remove the paper and dispose of it properly.
Gently rinse the slide with distilled water.
Stain Disposal
4. Continue holding the slide with a slide holder. Decolorize with acid-alcohol (CAUTION—This is 3% HCl!) until the run-off is clear.
Gently rinse the slide with distilled water.
Bibulous
5. Counterstain with methylene blue stain for 1 minute. Rinse with distilled water.
6. Gently blot dry in a tablet of bibulous paper or paper towels. (Alternatively, a page from the tablet can be removed and used for blotting.) Do not rub. When dry, observe under oil immersion.

3-58 Procedural Diagram: ZN Acid-Fast Stain ◆ Be sure to perform this stain in a fume hood or well-ventilated area. Carry the slide to and from the steaming apparatus in a covered Petri dish.

❖ In This Exercise

Today you will perform an acid-fast stain designed primarily to identify members of the genus *Mycobacterium*. Because you know ahead of time which organisms should give a positive result and which should give a negative result, it is okay to mix them into a single emulsion.

❖ Materials

Per Student Group

- ✓ compound microscope with oil objective lens and ocular micrometer
- ✓ clean glass microscope slides
- ✓ staining tray
- ✓ staining screen
- ✓ bibulous paper or paper towel
- ✓ slide holder
- ✓ Ziehl-Neelsen Stains (complete kits are commercially available)
  - methylene blue stain
  - Ziehl's carbolfuchsin stain
  - acid alcohol (95% ethanol + 3% HCl)
- ✓ Kinyoun Stains (complete kits are commercially available)
  - Kinyoun carbolfuchsin
  - acid alcohol (95% ethanol + 3% HCl)
  - brilliant green stain
- ✓ sheep serum
- ✓ squirt bottle with water
- ✓ heating apparatus (steam or hotplate)—for ZN only
- ✓ nonsterile Petri dish for transporting slides
- ✓ disposable gloves
- ✓ lab coat
- ✓ chemical eye protection for ZN (otherwise, recommended)
- ✓ recommended organisms:
  - *Mycobacterium smegmatis*
  - *Staphylococcus epidermidis*

Procedure: Ziehl-Neelsen (ZN) Method

1. Prepare a smear of each organism on a clean glass slide as illustrated in Figure 3-39, substituting a drop of sheep serum for the drop of water. Air dry and then heat-fix the smears. Note: You may make two separate smears right next to one another on the slide or mix the two organisms in one smear because you already know what results to expect in this exercise. Mixing wouldn't be done working with an unknown.
2. Follow the staining protocol shown in the Procedural Diagram (Figure 3-58). Be sure to wear gloves and eye protection. Use a steaming apparatus (such as the one in Figure 3-59) to heat the slide. If the slide must be carried to and from the steaming apparatus, put it in a covered Petri dish. Be sure to have adequate ventilation while staining. A fume hood or biological safety cabinet is recommended.
3. If required to do so, remove your gloves and observe using the oil immersion lens. Record your observations of cell morphology and arrangement, dimensions, and acid-fast reaction in the chart on the Data Sheet.
4. When finished, dispose of slides in a disinfectant jar or sharps container.

Procedure: Kinyoun (K) Method

1. Prepare a smear of each organism on a clean glass slide as illustrated in Figure 3-39, substituting a drop of sheep serum for the drop of water. Air dry and then heat-fix the smears. Note: You may make two separate smears right next to one another on the slide or mix the two organisms in one smear because you already know what results to expect in this exercise. Mixing wouldn't be done working with an unknown.
2. Follow the staining protocol shown in the Procedural Diagram (Figure 3-60). Be sure to wear gloves and perform the stain with adequate ventilation.

3-59 Steaming the Slide During the Ziehl-Neelsen Procedure ❖ Carefully steam the slide to melt the waxy wall so the carbolfuchsin can get into acid-fast cells. Do not boil the slide or let it dry out. Keep it moist with stain for the entire 5 minutes of steaming. Caution: This should be performed in a well-ventilated area with hand, clothing, and eye protection.

3 If required to do so, remove your gloves and observe using the oil immersion lens. Record your observations of cell morphology and arrangement, dimensions, and acid-fast reaction on the Data Sheet.

4 When finished, dispose of slides in a disinfectant jar or a sharps container.

References

Chapin, Kimberle C., and Patrick R. Murray. 2003. Pages 259–261 in *Manual of Clinical Microbiology,* 8th ed., edited by Patrick R. Murray, Ellen Jo Baron, James H. Jorgensen, Michael A. Pfaller, and Robert H. Yolken. American Society for Microbiology, Washington, DC.

Doetsch, Raymond N., and C. F. Robinow. 1994. Page 32 in *Methods for General and Molecular Bacteriology*, edited by Philipp Gerhardt, R. G. E. Murray, Willis A. Wood, and Noel R. Krieg. American Society for Microbiology, Washington, DC.

Forbes, Betty A., Daniel F. Sahm, and Alice. S. Weissfeld. 2002. Chapter 9 in *Bailey & Scott's Diagnostic Microbiology*, 11th ed. Mosby-Year Book, St. Louis.

Norris, J. R., and Helen Swain. 1971. Chapter II in *Methods in Microbiology,* Vol. 5A, edited by J. R. Norris and D. W. Ribbons. Academic Press, Ltd., London.

Power, David A., and Peggy J. McCuen. 1988. Page 5 in *Manual of BBL™ Products and Laboratory Procedures,* 6th ed. Becton Dickinson Microbiology Systems, Cockeysville, MD.

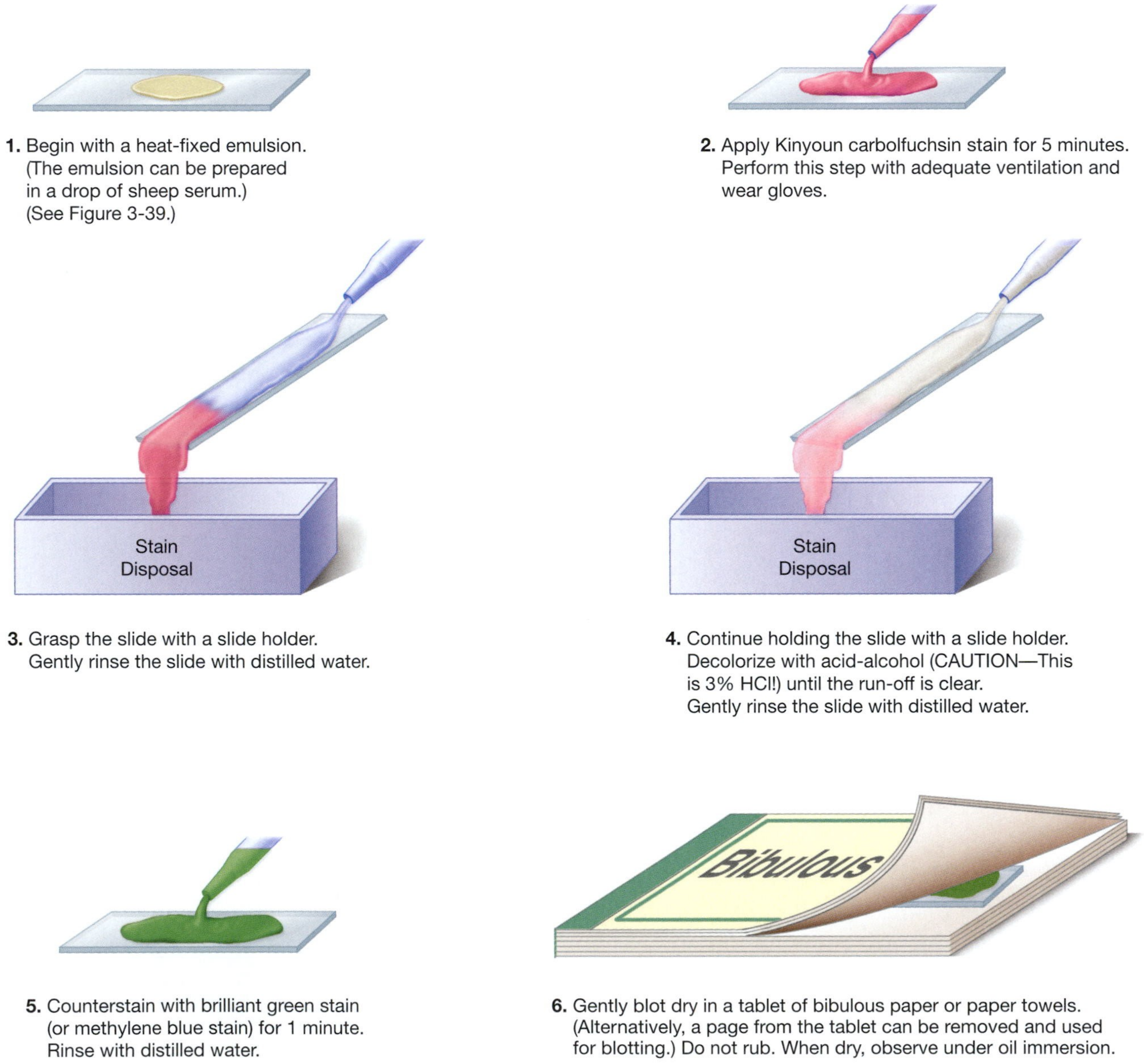

1. Begin with a heat-fixed emulsion. (The emulsion can be prepared in a drop of sheep serum.) (See Figure 3-39.)
2. Apply Kinyoun carbolfuchsin stain for 5 minutes. Perform this step with adequate ventilation and wear gloves.
Stain Disposal
3. Grasp the slide with a slide holder. Gently rinse the slide with distilled water.
Stain Disposal
4. Continue holding the slide with a slide holder. Decolorize with acid-alcohol (CAUTION—This is 3% HCl!) until the run-off is clear. Gently rinse the slide with distilled water.
Bibulous
5. Counterstain with brilliant green stain (or methylene blue stain) for 1 minute. Rinse with distilled water.
6. Gently blot dry in a tablet of bibulous paper or paper towels. (Alternatively, a page from the tablet can be removed and used for blotting.) Do not rub. When dry, observe under oil immersion.

3-60 Procedural Diagram: Kinyoun Acid-Fast Stain ◆

Name ______________________ Date ______________________

Lab Section ______________________ I was present and performed this exercise (initials) ________

Acid-Fast Stains

OBSERVATIONS AND INTERPRETATIONS

Record your observations in the chart below.

Organism	Staining Method (ZN or K)	Cellular Morphology and Arrangement (Include a detailed sketch of a few representative cells)	Cell Dimensions	Color	Acid-Fast Reaction (+/−)

DATA SHEET 3-7

Questions

1 How does heating the bacterial smear during a ZN stain promote entry of carbolfuchsin into the acid-fast cell wall?

2 Are acid-fast negative cells stained by carbolfuchsin? If so, how can this be a differential stain?

3 Why do you suppose the acid-fast stain is not as widely used as the Gram stain? When is it more useful than the Gram stain?

EXERCISE 3-8

Capsule Stain

Theory

Capsules are composed of mucoid polysaccharides or polypeptides that repel most stains. The capsule stain technique takes advantage of this characteristic by staining *around* the cells. Typically an acidic stain such as Congo red or nigrosin, which stains the background, and a basic stain that colorizes the cell proper, are used. The capsule remains unstained and appears as a white halo between the cells and the colored background (Figure 3-61).

This technique begins in the negative stain; cells are spread in a film across the slide with an acidic stain and are not heat-fixed. Heat-fixing causes the cells to shrink, leaving an artifactual white halo around them that might be interpreted as a capsule. In place of heat-fixing, cells may be emulsified in a drop of serum to promote their adhering to the glass slide, but this is not essential.

Application

The capsule stain is a differential stain used to detect cells capable of producing an extracellular capsule. Capsule production increases virulence in some microbes (such as the anthrax bacillus *Bacillus anthracis* and the pneumococcus *Streptococcus pneumoniae*) by making them less vulnerable to phagocytosis.

In This Exercise

The capsule stain allows you to visualize an extracellular capsule, if present. Be careful to distinguish between a tiny white halo (as a result of cell shrinkage) and a true capsule. A true capsule will be a substantial clearing of stain around at least some of the cells.

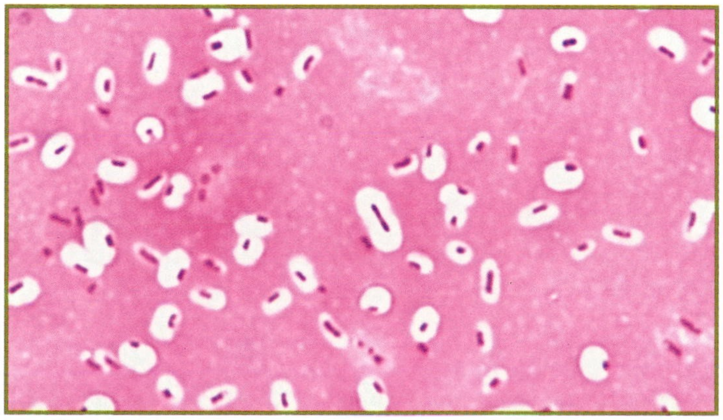

3-61 Capsule Stain ◆ The acidic stain colorizes the background while the basic stain colorizes the cell, leaving the capsules as unstained, white clearings around the cells. Notice the lack of uniform capsule size, and even the absence of a capsule in some cells.

Materials

Per Student

- ✓ clean glass slides
- ✓ sheep serum (optional)
- ✓ Maneval's stain
- ✓ Congo red stain
- ✓ squirt bottle with water
- ✓ staining tray
- ✓ staining screen
- ✓ bibulous paper or paper towel
- ✓ slide holder
- ✓ chemical eye protection (recommended)
- ✓ disposable gloves
- ✓ lab coat
- ✓ sterile toothpicks
- ✓ compound microscope with oil objective lens and ocular micrometer
- ✓ immersion oil
- ✓ lens paper
- ✓ recommended organisms (18 to 24 hour skim milk or Trypticase Soy Agar slant pure cultures)
  - ◆ *Klebsiella pneumoniae* (BSL-2)
  - ◆ *Staphylococcus epidermidis*

Procedure

1. Follow the protocol in the Procedural Diagram (Figure 3-62) to make stains of the organisms supplied. Use a separate slide for each specimen. These specimens are not heat-fixed. Be sure to wear gloves.
2. Using a sterile toothpick, obtain a sample from below the gumline in your mouth. (Do not draw blood!) Mix the sample into a drop of Congo red stain on a slide, then perform a capsule stain on it.
3. If required to do so, remove your gloves, then observe, using the oil immersion lens. Record your observations of cell morphology and arrangement, cell dimensions, and presence or absence of a capsule in the chart on the Data Sheet.
4. Dispose of the specimen slides in a jar of disinfectant or sharps container after use.

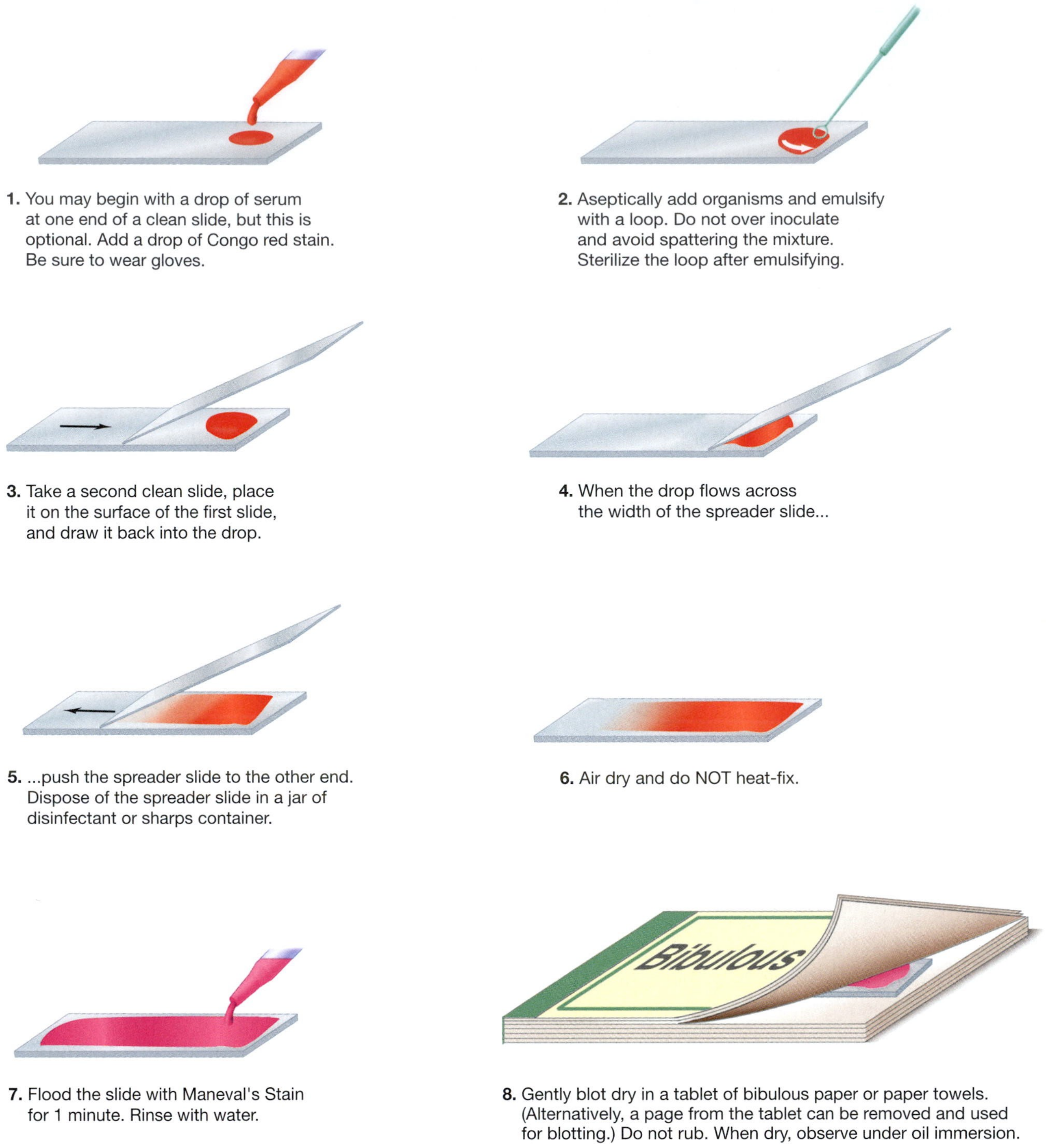
1. You may begin with a drop of serum at one end of a clean slide, but this is optional. Add a drop of Congo red stain. Be sure to wear gloves.
2. Aseptically add organisms and emulsify with a loop. Do not over inoculate and avoid spattering the mixture. Sterilize the loop after emulsifying.
3. Take a second clean slide, place it on the surface of the first slide, and draw it back into the drop.
4. When the drop flows across the width of the spreader slide...
5. ...push the spreader slide to the other end. Dispose of the spreader slide in a jar of disinfectant or sharps container.
6. Air dry and do NOT heat-fix.
7. Flood the slide with Maneval's Stain for 1 minute. Rinse with water.
Bibulous
8. Gently blot dry in a tablet of bibulous paper or paper towels. (Alternatively, a page from the tablet can be removed and used for blotting.) Do not rub. When dry, observe under oil immersion.

3-62 **Procedural Diagram: Capsule Stain** ◆

References

Murray, R. G. E., Raymond N. Doetsch, and C. F. Robinow. 1994. Page 35 in *Methods for General and Molecular Bacteriology*, edited by Philipp Gerhardt, R. G. E. Murray, Willis A. Wood, and Noel R. Krieg. American Society for Microbiology, Washington, DC.

Norris, J. R., and Helen Swain. 1971. Chapter II in *Methods in Microbiology*, Vol 5A, edited by J. R. Norris and D. W. Ribbons. Academic Press, Ltd, London.

NAME ______________________ DATE ______________________

LAB SECTION ______________________ I WAS PRESENT AND PERFORMED THIS EXERCISE (INITIALS) ________

Capsule Stain

OBSERVATIONS AND INTERPRETATIONS

Record your observations in the chart below.

Organism	Cellular Morphology and Arrangement (Include a detailed sketch of a few representative cells)	Cell Dimensions	Capsule (+/−)	Width of Capsule (If present)

Questions

1 Capsules are neutrally charged. This being the case, what is the purpose of emulsifying the sample in serum in this staining procedure?

2 Some oral bacteria produce an extracellular "capsule." Of what benefit is a capsule to these cells?

3 Sketch any cells from your mouth sample that display an unusual morphology or arrangement.

EXERCISE 3-9

Endospore Stain

Theory

An **endospore** is a dormant form of the bacterium that allows it to survive poor environmental conditions. Spores are resistant to heat and chemicals because of a tough outer covering made of the protein **keratin**. The keratin also resists staining, so extreme measures must be taken to stain the spore. In the Schaeffer-Fulton method (Figure 3-63), a primary stain of malachite green is forced into the spore by steaming the bacterial emulsion. Alternatively, malachite green can be left on the slide for 15 minutes or more to stain the spores. Malachite green is water-soluble and has a low affinity for cellular material, so **vegetative cells** and **spore mother cells** can be decolorized with water (while the spores retain it) and counterstained with safranin (Figure 3-64).

Spores may be located in the middle of the cell (**central**), at the end of the cell (**terminal**), or between the end and middle of the cell (**subterminal**). Spores also may be differentiated based on shape—either **spherical** or **elliptical** (**oval**)—and size relative to the cell (i.e., whether they cause the cell to look swollen or not). These structural features are shown in Figures 3-65 through Figure 3-67.

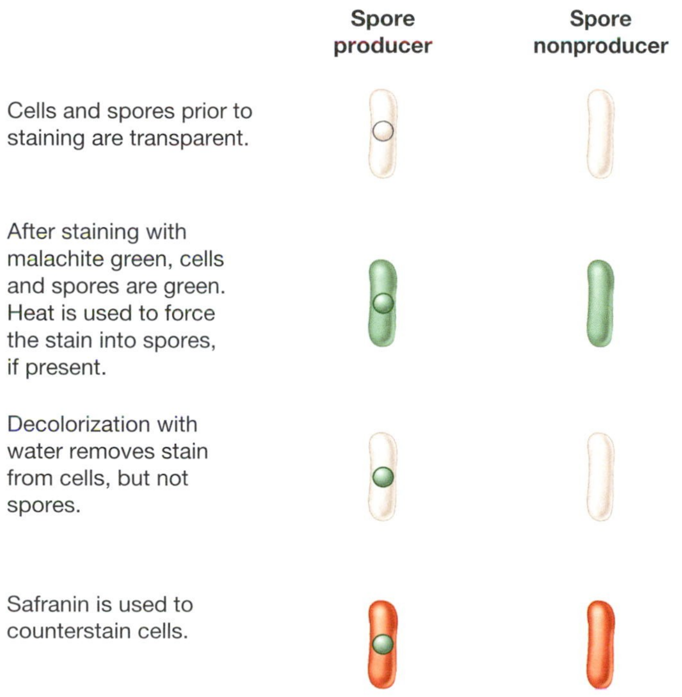
Spore producer
Spore nonproducer
Cells and spores prior to staining are transparent.
After staining with malachite green, cells and spores are green. Heat is used to force the stain into spores, if present.
Decolorization with water removes stain from cells, but not spores.
Safranin is used to counterstain cells.

3-63 The Schaeffer-Fulton Spore Stain ◆ Upon completion, spores are green, and vegetative and spore mother cells are red.

Application

The spore stain is a differential stain used to detect the presence and location of spores in bacterial cells. Only a few genera produce spores. Among them are the genera *Bacillus* and *Clostridium*. Most members of *Bacillus* are soil, freshwater, or marine **saprophytes**, but a few are pathogens, such as *B. anthracis*, the causal agent of anthrax. Most members of *Clostridium* are soil or aquatic saprophytes or inhabitants of human intestines, but four pathogens are fairly well known: *C. tetani*, *C. botulinum*, *C. perfringens*, and *C. difficile,* which produce tetanus, botulism, gas gangrene, and pseudomembranous colitis, respectively.

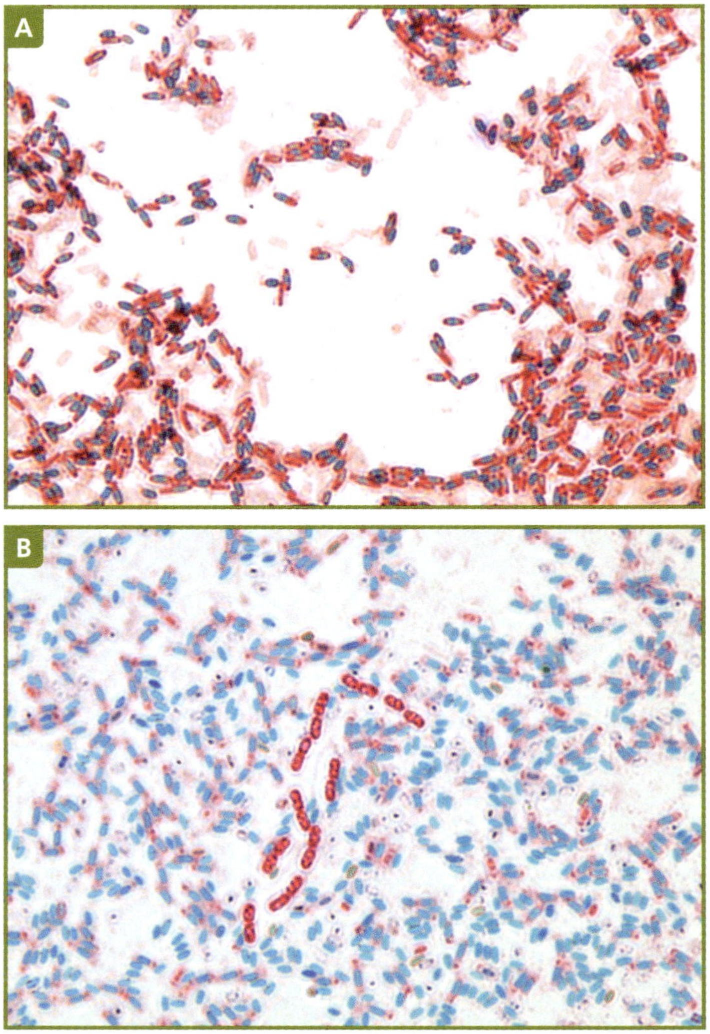
A
B

3-64 Culture Age Can Affect Sporulation ◆ Bacteria capable of producing spores don't do so uniformly during their culture's growth or in the wild. Sporulation is done in response to nutrient depletion, and so is characteristic of older cultures. These two cultures illustrate different degrees of sporulation. **A** Most cells in this specimen contain spores; very few have been released. The lighter cells in this preparation are most likely disintegrating spore mother cells. **B** This specimen consists mostly of released spores. Note the young vegetative cells in the middle of the field. The white patches are storage granules. (How would you know they aren't spores?)

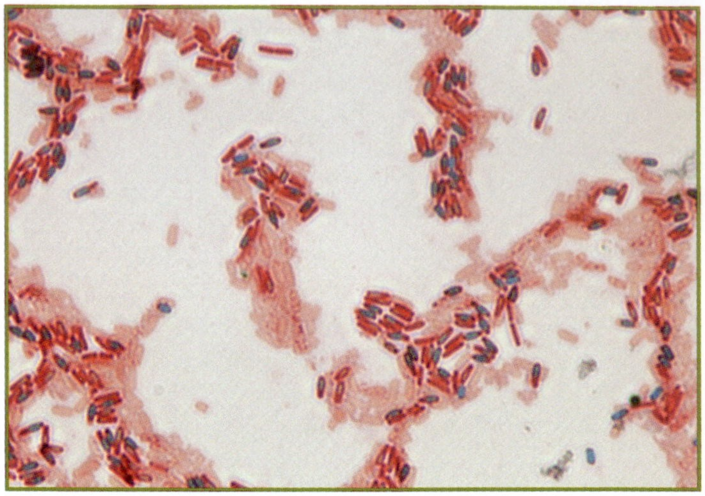

3-65 Central Elliptical Endospores ◆ Most of these *Bacillus megaterium* spores are centrally positioned, but some are subterminal. All are elliptical. The spores do not distend the mother cell.

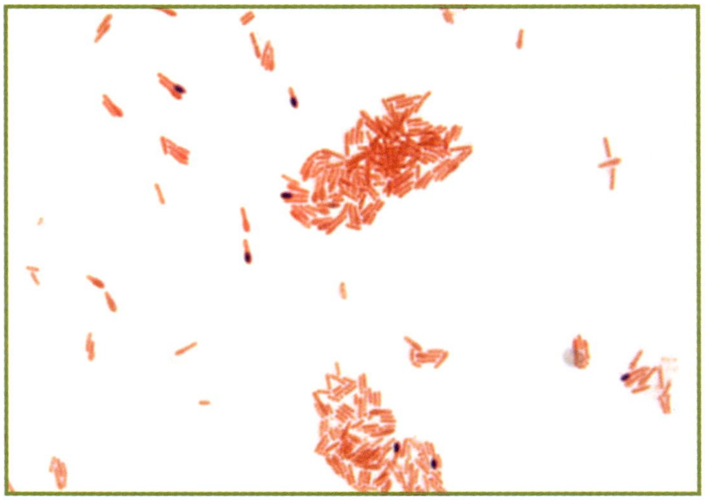

3-66 Subterminal Spores ◆ This is a stained preparation of *Clostridium botulinum* using an alternative procedure. The black, elliptical spores slightly distend the cell.

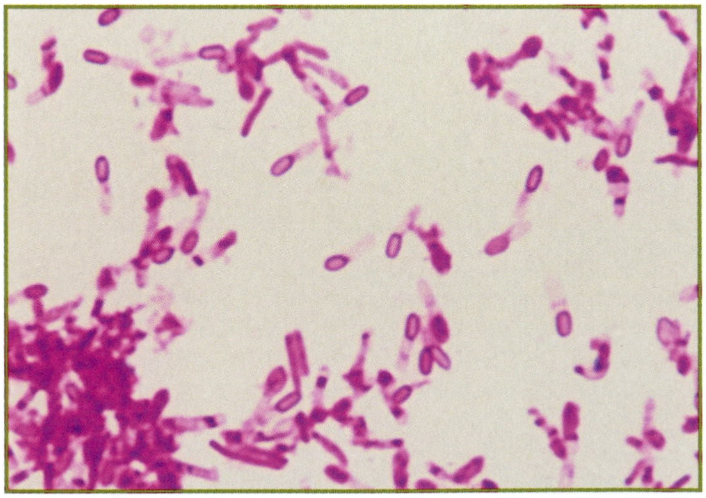

3-67 Elliptical Terminal Endospores ◆ This is *Clostridium tetani* stained by a different spore stain protocol using carbolfuchsin. Notice how the spores have caused the ends of the cells to swell.

In This Exercise

In the spore stain you will examine spores from different-aged *Bacillus* cultures. You will not have to stain each species. Divide the work within your lab group, but be sure to look at each others' slides. Also be sure to compare spore shape and location for each species.

Materials

Per Student Group

- ✓ clean glass microscope slides
- ✓ malachite green stain
- ✓ safranin stain
- ✓ squirt bottle with water
- ✓ heating apparatus (steam apparatus or hot plate)
- ✓ bibulous paper or paper towel
- ✓ staining tray
- ✓ staining screen
- ✓ slide holder
- ✓ disposable gloves
- ✓ lab coat
- ✓ chemical eye protection
- ✓ compound microscope with oil objective lens and ocular micrometer
- ✓ immersion oil
- ✓ lens paper
- ✓ nonsterile Petri dish for transporting slides
- ✓ recommended organisms (per student group):
    - 48 hour and 5 day Sporulating Agar slant pure cultures of *Bacillus cereus*
    - 48 hour and 5 day Sporulating Agar slant pure cultures of *Bacillus coagulans*
    - 48 hour and 5 day Sporulating Agar slant pure cultures of *Bacillus megaterium*
    - 48 hour and 5 day Sporulating Agar slant pure cultures of *Bacillus subtilis*

Note: *Trypticase Soy Agar can be substituted for Sporulating Agar)*

Medium Recipe

Sporulating Agar

◆ Pancreatic digest of gelatin	6.0 g
◆ Pancreatic digest of casein	4.0 g
◆ Yeast extract	3.0 g
◆ Beef extract	1.5 g
◆ Dextrose	1.0 g

- Agar 15.0 g
- Manganous sulfate 0.3 g
- Distilled or deionized water 1.0 L

Final ph 6.6 ± 0.2 at 25°C

Procedure

1 Prepare and heat-fix as many as four small smears of different cultures on the same slide as illustrated in Figure 3-39. Divide the work within your lab group. Minimally, each student should prepare a smear of the 48-hour and 5-day cultures of one species.

2 Follow the instructions in the Procedural Diagram in Figure 3-68. Use a steaming apparatus like the one shown in Figure 3-69. Be sure to have adequate ventilation, eye protection, and gloves. If you must carry the slide to and from the steaming apparatus, put it in a covered Petri dish.

3 Observe using the oil immersion lens. Record your observations of cell morphology and arrangement, cell dimensions, and spore presence, position, and shape in the chart provided on the Data Sheet.

4 Dispose of specimen slides in a disinfectant jar or a sharps container.

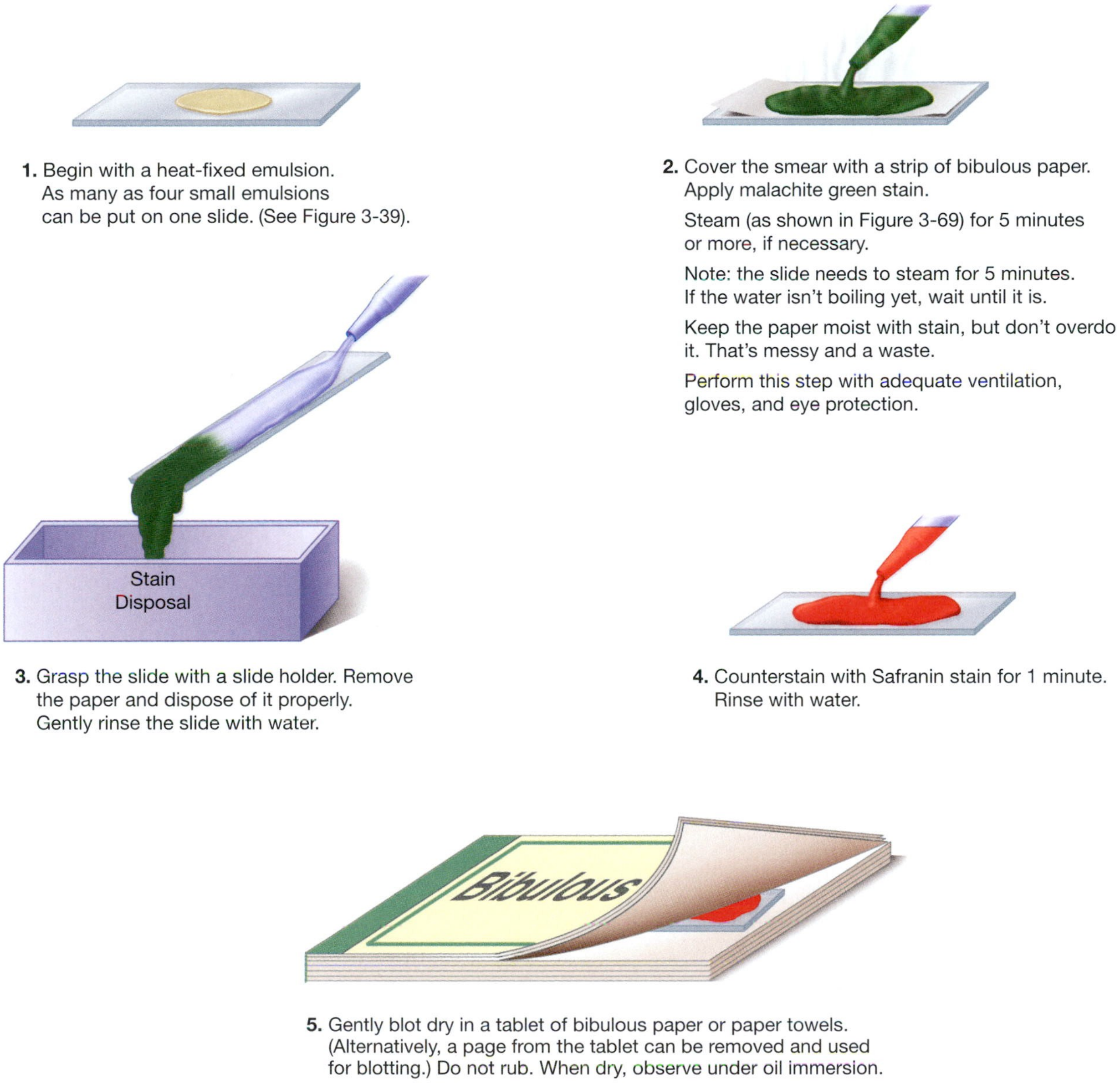
1. Begin with a heat-fixed emulsion. As many as four small emulsions can be put on one slide. (See Figure 3-39).
2. Cover the smear with a strip of bibulous paper. Apply malachite green stain.
Steam (as shown in Figure 3-69) for 5 minutes or more, if necessary.
Note: the slide needs to steam for 5 minutes. If the water isn't boiling yet, wait until it is.
Keep the paper moist with stain, but don't overdo it. That's messy and a waste.
Perform this step with adequate ventilation, gloves, and eye protection.
Stain Disposal
3. Grasp the slide with a slide holder. Remove the paper and dispose of it properly. Gently rinse the slide with water.
4. Counterstain with Safranin stain for 1 minute. Rinse with water.
Bibulous
5. Gently blot dry in a tablet of bibulous paper or paper towels. (Alternatively, a page from the tablet can be removed and used for blotting.) Do not rub. When dry, observe under oil immersion.

3-68 Procedural Diagram: Schaeffer-Fulton Spore Stain ◆ Steam the staining preparation; do not boil it. Be sure to perform this procedure with adequate ventilation (preferably a fume hood), eye protection, a lab coat, and gloves.

3-69 Steaming the Slide During the Endospore Stain ◆
Carefully steam the slide to force the malachite green into the endospores. Do not boil the slide or let it dry out. Keep it moist with stain as it is steaming for 5 minutes (or more, if necessary). ***Caution:*** This should be performed in a well-ventilated area (preferably a fume hood) with hand, clothing, and eye protection.

References

Claus, G. William. 1989. Chapter 9 in *Understanding Microbes—A Laboratory Textbook for Microbiology*. W. H. Freeman and Co., New York.

Murray, R. G. E., Raymond N. Doetsch, and C. F. Robinow. 1994. Page 34 in *Methods for General and Molecular Bacteriology*, edited by Philipp Gerhardt, R. G. E. Murray, Willis A. Wood, and Noel R. Krieg. American Society for Microbiology, Washington, DC.

Name ____________________ Date ____________________

Lab Section ____________________ I was present and performed this exercise (initials) ________

DATA SHEET 3-9

Endospore Stain

OBSERVATIONS AND INTERPRETATIONS

Record your observations in the chart below.

Organism (Include culture age)	**Cellular Morphology and Arrangement** (Include a detailed sketch of a few representative cells and spores)	**Cell Dimensions**	**Spores** (Present or absent)	**Spore Shape and Position** (If present)

Questions

1 Why does this exercise call for an older (5-day) culture of *Bacillus*?

2 What does a positive result for the spore stain indicate about the organism? What does a negative result for the spore stain indicate about the organism?

3 Why is it not necessary to include a negative control for this stain procedure?

4 Spores do not stain easily. Perhaps you have seen them as unstained white objects inside *Bacillus* species in other staining procedures. If they are visible as unstained objects in other stains, of what use is the endospore stain? (**Hint:** See Figure 3-64.)

EXERCISE 3-10

Bacterial Motility: Wet Mount and Hanging Drop Preparations

Theory

A wet mount preparation is made by placing the specimen in a drop of water on a microscope slide and covering it with a cover glass. Because no stain is used and most cells are transparent, viewing is best done by closing the iris diaphragm to minimize the amount of light entering the objective. Your instructor can help you find the proper adjustment (Figure 3-70). Motility often can be observed at low or high dry magnification, but viewing must be done quickly because of drying of the preparation. As the water recedes, bacteria will appear to be herded across the field. This is not motility. You should look for independent darting of the cells.

A hanging drop preparation allows longer observation of the specimen because it doesn't dry out as quickly. A thin ring of petroleum jelly is applied around the well of a depression slide. A drop of water then is placed in the center of the cover glass and living microbes are gently transferred into it. The depression slide is carefully placed over the cover glass in such a way that the drop is received into the depression and is undisturbed. The petroleum jelly causes the cover glass to stick to the slide.

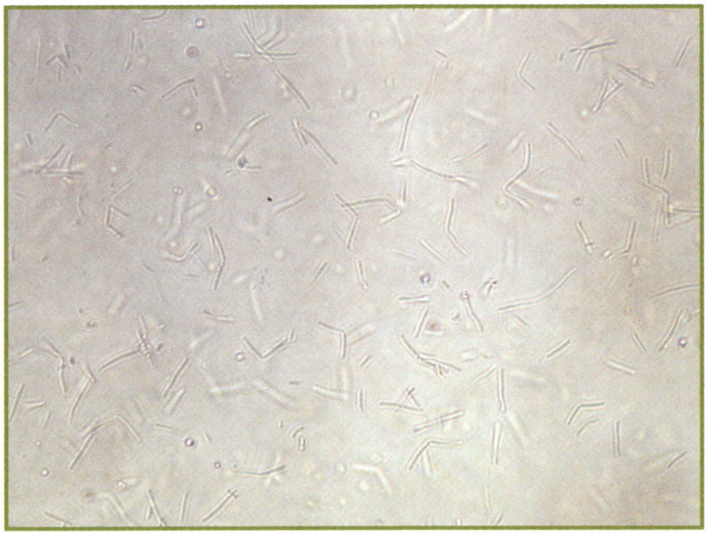

3-70 Wet Mount ◆ Shown is an unstained wet-mount preparation of a motile Gram-negative rod. Because of the thickness of the water in the wet mount, cells show up in many different focal planes and are mostly out of focus. To get the best possible image, adjust the condenser height and reduce the light intensity with the iris diaphragm.

The preparation then may be picked up, inverted so the cover glass is on top, and placed under the microscope for examination. As with the wet mount, viewing is best done by closing the iris diaphragm to minimize the amount of light entering the objective. Your instructor can help you find the proper adjustment. The petroleum jelly forms an airtight seal that slows drying of the drop, allowing a long period for observation of cell size, shape, binary fission, and motility.

If these techniques are done to determine motility, the observer must be careful to distinguish between true motility and the **Brownian motion** created by collisions with water molecules. In the latter, cells will appear to vibrate in place. With true motility, cells will exhibit independent movement over greater distances.

Application

Most bacterial microscopic preparations result in death of the microorganisms as a result of heat-fixing and staining. Simple **wet mounts** and the **hanging drop technique** allow observation of living cells to determine motility. They also are used to see natural cell size (because they haven't been heat-fixed), arrangement, and shape. All of these characteristics may be useful in identification of a microbe.

In This Exercise

Today you will have the opportunity to view living bacterial cells swimming. The hanging drop preparation allows longer viewing, whereas the simple wet mount can be used to view motility, and it is the starting point for some flagella stain protocols.

Materials

Per Student

- ✓ compound microscope with oil lens and ocular micrometer
- ✓ depression slide and cover glass
- ✓ clean microscope slides and cover glasses
- ✓ petroleum jelly
- ✓ toothpick
- ✓ immersion oil
- ✓ lens paper
- ✓ disposable gloves
- ✓ recommended organisms (overnight cultures grown in broth media):
  - ♦ *Aeromonas sobria*
  - ♦ *Staphylococcus epidermidis*
  - ♦ pond water (optional)

Procedure: Wet Mount Preparation

1 Place a loopful of water on a clean glass slide.

2 Carefully add bacteria from the culture to the drop. Don't over inoculate and don't mix vigorously. Flame the loop after transfer. If you are using pond water, just transfer a drop to the slide.

3 Gently lower a cover glass with your loop supporting one side over the drop of water. Avoid trapping air bubbles.

4 Observe under high dry or oil immersion and record your results in the chart on the Data Sheet. Avoid hitting the cover slip with the oil objective. (Depending on the thickness of your preparation and cover glass, there may not be room for the oil lens to be used. If that is the case, simply observe with high dry.)

5 Dispose of the slide and cover glass in a disinfectant jar or a sharps container when finished.

Procedure: Hanging Drop Preparation

1 Follow the procedure illustrated in Figure 3-71 for each specimen.

2 Observe under high dry or oil immersion (if possible without hitting the cover slip with the oil objective lens) and record your results in the chart on the Data Sheet.

3 When finished, remove the cover glass from the depression slide with an inoculating loop and soak both in a disinfectant jar for at least 15 minutes. Flame the loop. After soaking, rinse the slide with 95% ethanol to remove the petroleum jelly, then with water to remove the alcohol. Dry the slide for reuse.

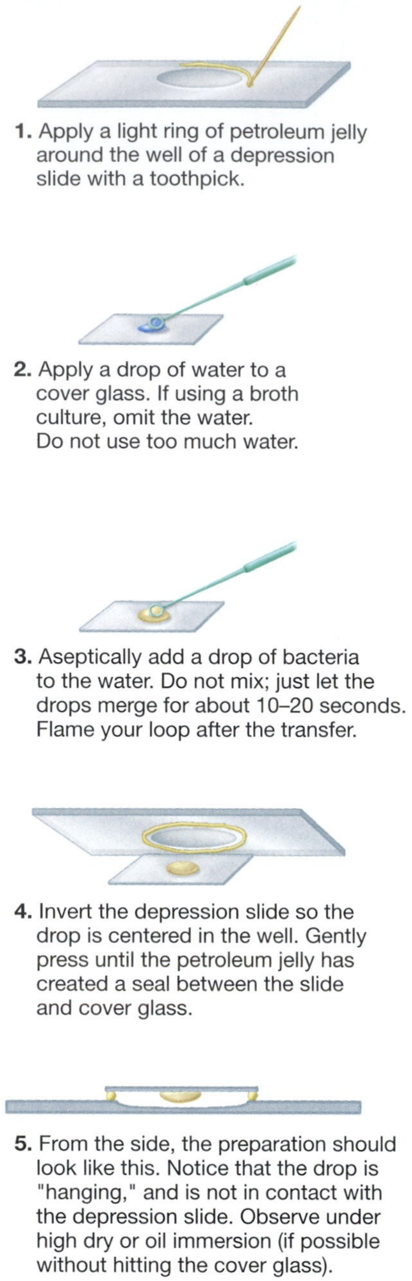
1. Apply a light ring of petroleum jelly around the well of a depression slide with a toothpick.
2. Apply a drop of water to a cover glass. If using a broth culture, omit the water. Do not use too much water.
3. Aseptically add a drop of bacteria to the water. Do not mix; just let the drops merge for about 10–20 seconds. Flame your loop after the transfer.
4. Invert the depression slide so the drop is centered in the well. Gently press until the petroleum jelly has created a seal between the slide and cover glass.
5. From the side, the preparation should look like this. Notice that the drop is "hanging," and is not in contact with the depression slide. Observe under high dry or oil immersion (if possible without hitting the cover glass).

3-71 Procedural Diagram: Hanging Drop Preparation ◆
The hanging drop method is used for long-term observation of a living specimen.

References

Iino, Tetsuo, and Masatoshi Enomoto. 1969. Chapter IV in *Methods in Microbiology*, Vol 1, edited by J. R. Norris and D. W. Ribbins. Academic Press, Ltd., London.

Murray, R. G. E., Raymond N. Doetsch, and C. F. Robinow. 1994. Page 26 in *Methods for General and Molecular Bacteriology*, edited by Philipp Gerhardt, R. G. E. Murray, Willis A. Wood, and Noel R. Krieg. American Society for Microbiology, Washington, DC.

Quesnel, Louis B. 1969. Chapter X in *Methods in Microbiology*, Vol 1, edited by J. R. Norris and D. W. Ribbins. Academic Press, Ltd., London.

NAME ____________________ DATE ____________________

LAB SECTION ____________________ I WAS PRESENT AND PERFORMED THIS EXERCISE (INITIALS) ________

Bacterial Motility: Wet Mount and Hanging Drop Preparations

OBSERVATIONS AND INTERPRETATIONS

Record your observations in the chart below.

Organism	Procedure (Wet Mount or Hanging Drop)	Cellular Morphology and Arrangement (Include a detailed sketch of a few representative cells)	Cell Dimensions	Motility (+/−)

Questions

1 In a wet mount, each of the following complications *could* lead to a false interpretation of motility. For each, write "false positive," "false negative," or "no effect," depending on how it could interfere with your reading of a motile organism and a nonmotile organism.

Complication	Motile Organism	Nonmotile Organism
Over inoculation of the slide with organisms		
Cells attaching to the glass slide or cover glass		
Receding water line		
Using an old culture		

2 You are told that viewing is best done with the iris diaphragm closed down. Why will transparent cells be easier to view with less light?

3 Would you expect Brownian motion to increase the longer you observe a hanging drop or wet-mount preparation? Why?

4 If you examined pond water, did you notice any bacteria swimming in "runs and tumbles"? Did you see any swimming in a more-or-less straight line?

EXERCISE 3-11

Bacterial Motility: Flagella

Theory

Bacterial flagella typically are too thin to be observed with the light microscope and ordinary stains. Various special flagella stains have been developed that use a **mordant** to assist in coating flagella with stain to a visible thickness. Most require experience and advanced techniques, and typically are not performed in beginning microbiology classes. So, you will be examining prepared slides.

The number and arrangement of flagella may be observed with a flagella stain. A single flagellum is said to be **polar** and the cell has a **monotrichous** arrangement (Figure 3-72). Other arrangements (shown in Figures 3-73 through 3-75) include **amphitrichous**, with flagella at both ends of the cell; **lophotrichous**, with tufts of flagella at the end of the cell; and **peritrichous**, with flagella emerging from the entire cell surface.

Application

The flagella stain allows direct observation of flagella. The presence and arrangement of flagella may be useful in identifying bacterial species.

In This Exercise

You will examine prepared slides illustrating various flagellar arrangements.

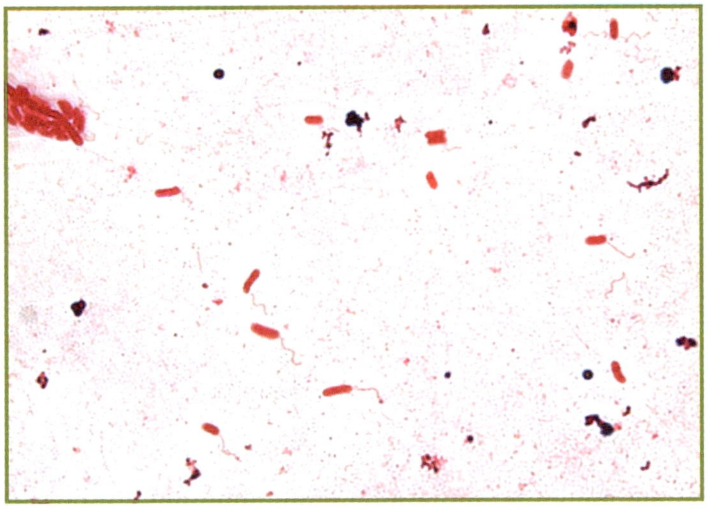

3-72 Polar Flagella ◆ *Pseudomonas aeruginosa* is often suggested as a positive control for flagella stains. Notice the single flagellum emerging from the ends of many (but not all) cells. This is a result of the fragile nature of flagella, which can be broken from the cells during slide preparation.

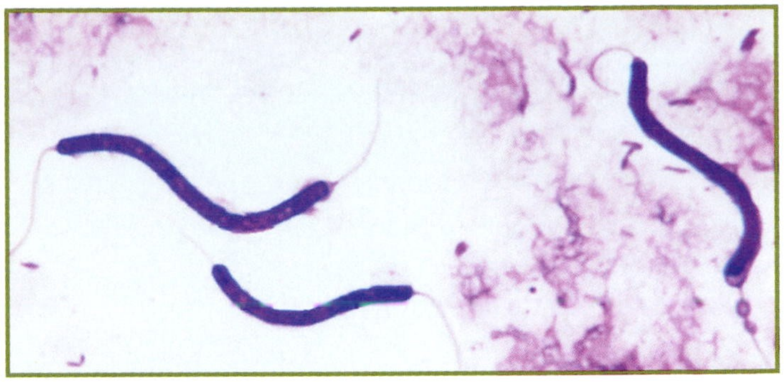

3-73 Amphitrichous Flagella ◆ *Spirillum volutans* cells have a flagellum at each end.

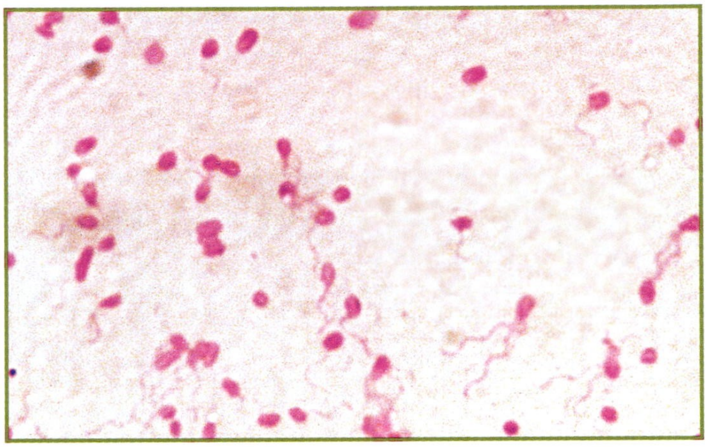

3-74 Lophotrichous Flagella ◆ Several flagella emerge from one end of this *Pseudomonas* species. Not all cells have flagella because they were too delicate to stay intact during the staining procedure.

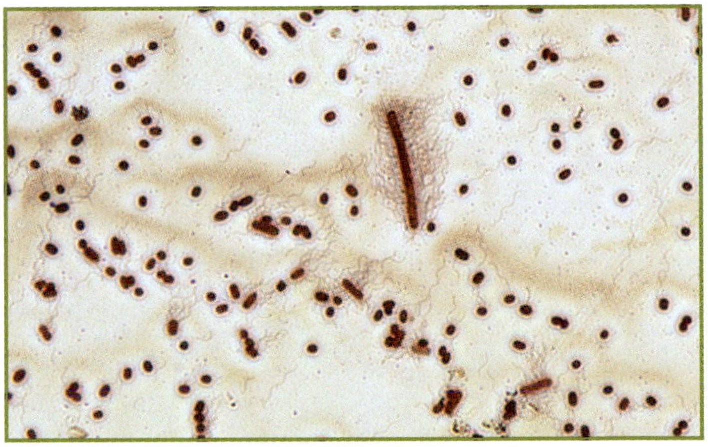

3-75 Peritrichous Flagella of *Proteus vulgaris* ◆ Notice the flagella emerging from the entire cell surface. *P. vulgaris* is capable of swarming motility (Figure 2-24), in which the cells spread across the agar surface at specific intervals, and then remain in place until the next swarm. The smaller cells are called "swimmers" and are the form seen when grown in a liquid medium. When transferred to a solid medium or under certain environmental conditions, swimmers differentiate into "swarmers." A swarmer is seen in the center of this micrograph. Swarmers are larger, contain multiple nucleoids (the site of DNA), and produce an extracellular slime or capsule that assists in swarming. After swarming, they break up into swimmer cells. Swarmers are the more virulent form.

Materials

- ✓ compound microscope with ocular micrometer
- ✓ immersion oil
- ✓ lens paper
- ✓ commercially prepared slides of motile organisms showing flagella.

Procedure

1 Examine the available prepared microscope slides of bacteria with flagella. You will need to search carefully, because not all cells will have flagella and they are extremely thin even in the best preparations. Using the oil immersion lens is essential.

2 Record your observations of flagellar length, arrangement and number on the Data Sheet. Also record cell morphology, arrangement, and dimensions.

References

Heimbrook, Margaret E., Wen Lan L. Wang, and Gail Campbell (1989). "Staining Bacterial Flagella Easily". *J. Clin. Microbiol.* 27: 2612–2615.

Iino, Tetsuo and Masatoshi Enomoto. 1971. Chapter IV in *Methods in Microbiology,* Vol 5A, edited by J. R. Norris and D. W. Ribbins. Academic Press, Ltd., London.

Murray, R. G. E., Raymond N. Doetsch and C. F. Robinow. 1994. Page 35 in *Methods for General and Molecular Bacteriology*, edited by Philipp Gerhardt, R. G. E. Murray, Willis A. Wood, and Noel R. Krieg. American Society for Microbiology, Washington, DC.

NAME ______________________ DATE ______________________

LAB SECTION ______________________ I WAS PRESENT AND PERFORMED THIS EXERCISE (INITIALS) ________

DATA SHEET 3-11

Bacterial Motility: Flagella

OBSERVATIONS AND INTERPRETATIONS

Record your observations in the chart below.

Organism	Cellular Morphology and Arrangement (Include a detailed sketch of a few representative organisms with their flagella)	Cell Dimensions	Flagellar Length

Questions

1 Why can't flagella be observed in action?

2 Flagella have a diameter of about 1 nm. A hypothetical question: To resolve flagella, what is the maximum wavelength of the electromagnetic spectrum that could be used to create the image (given numerical apertures of 1.25 for both the condenser and the oil lens)? Refer to Exercise 3-1 for help with this question.

3 Based on the samples of motile bacteria you have observed in Exercise 3-10 and this exercise, what conclusion can you draw about cell morphology and the possession of flagella?

EXERCISE 3-12

Morphological Unknown

❖ Theory

In this exercise you will be given one pure bacterial culture selected from the organisms listed in Table 3-4. Your job will be to identify it using only the staining techniques covered in Section Three.

Your first task is to convert the information organized in Table 3-4 into flowchart form. A flowchart is simply a visual tool to illustrate the process of elimination that is the foundation of unknown identification. We have started it for you on the Data Sheet by giving you a few of the branches and listing appropriate organisms. You must complete it by adding necessary branches until you have shown a path to identify each of the organisms. The Procedure below contains a detailed explanation of the process.

Once you have designed your flowchart, you will run one stain at a time on your organism. As you match your staining results with those in the flowchart, you will follow a path to its identification.

A final differential stain will be performed as a **confirmatory test.** It will serve as further evidence that you have identified your unknown correctly.

❖ Application

Schemes employing differential tests are the main strategy for microbial identification.

❖ In This Exercise

You will practice the stains you have learned to this point, as well as familiarize yourself with the standard process of elimination used in bacterial identification.

❖ Materials

Per Student Group

- ✓ compound microscope with oil immersion lens
- ✓ immersion oil
- ✓ clean microscope slides
- ✓ clean cover glasses
- ✓ lens paper
- ✓ Gram stain kit
- ✓ acid-fast stain kit
- ✓ capsule stain kit
- ✓ disposable gloves
- ✓ lab coat
- ✓ chemical eye protection
- ✓ spore stain kit
- ✓ Bunsen burner
- ✓ striker
- ✓ inoculating loop
- ✓ unknown organisms in numbered tubes (one per student). Fresh slant or broth cultures of[1]
  - *Bacillus subtilis*
  - *Corynebacterium xerosis*
  - *Lactococcus lactis*
  - *Micrococcus luteus*
  - *Mycobacterium smegmatis*
  - *Neisseria sicca*
  - *Pseudomonas putida*
  - *Rhizobium leguminosarum*
  - *Rhodospirillum rubrum*
  - *Shigella flexneri* (BSL-2)
  - *Staphylococcus epidermidis*

  (Note: Your instructor will choose organisms appropriate to your specific microbiology course and facilities.)
- ✓ Gram-positive and Gram-negative control organisms (one set per student group)—18–24 hour Trypticase Soy Agar slant cultures in tubes labeled with Gram reaction.
- ✓ A set of organisms (listed above) to be used for positive controls (per class)
- ✓ sterile Trypticase Soy Broth tubes (one per student)
- ✓ sterile Trypticase Soy Agar slants (one per student)
- ✓ sterile Sporulating Agar slants (one per student—optional)

Procedure

1 Using the information contained in Table 3-4, complete the flowchart on the Data Sheet. Each of the 11 organisms should occupy a solitary position at the end of a branch. Do not include stains in a branch that do not differentiate any organisms. Also, be sure the names and numbers of organisms on any level are equal to the names and numbers of organisms leading into that branch. Have your instructor check your flowchart.

2 Obtain one unknown slant culture. Record its number in the space labeled "Unknown Number" on the Data Sheet.

[1] Cultures should have abundant growth.

TABLE 3-4 Table of Results for Organisms Used in the Exercise ◆ These results are typical for the species listed. Your specific strains may vary because of their genetics, their age, or the environment in which they are grown. It is important that you record your results in case strain variability leads to misidentification. Also, pay attention to the footnotes!

TABLE OF RESULTS

Organism	Gram Stain	Cell Morphology	Cell Arrangement[1]	Acid-Fast Stain	Motility (Wet Mount)[2]	Capsule[3]	Spore Stain[4]
Bacillus subtilis	+	rod	usually single cells	–	+	±	+
Corynebacterium xerosis	+	irregular rods	single cells or multiples in angular or palisade arrangement	–	–	–	–
Lactococcus lactis	+	ovoid cocci (appearing stretched in the direction of the chain or pair)	(pairs or) chains	–	–	+	–
Micrococcus luteus	+	cocci	(pairs or) tetrads	–	–	–	–
Mycobacterium smegmatis	weak +	thin rods	single cells or branched; sometimes in dense clusters	+	–	–	–
Neisseria sicca	–	cocci	pairs with adjacent sides flattened	–	–	+	–
Pseudomonas putida	–	straight to slightly curved rods	single cells	–	+	–	–
Rhizobium leguminosarum	–	short rods	usually single rods	–	–	+	–
Rhodospirillum rubrum	–	thick spirilla or bent rods	single	–	+	–	–
Shigella flexneri (BSL-2)	–	rods	single	–	–	–	–
Staphylococcus epidermidis	+	cocci	(singles, pairs, tetrads, or) clusters	–	–	–	–

[1] It may be necessary to grow your unknown in broth culture to see the arrangement clearly. Because the purpose of this lab is to practice the stains, use cell arrangement in the identification flow chart only as a last resort.

[2] You might have better luck looking for motility if you transfer some culture to a small amount of sterile broth (in a microtube, for instance) and incubate it for an hour or so at 35°C. Then, make your wet mount from that culture.

[3] For purposes of this exercise, to be considered a capsule, the white halo needs to be at least the thickness of the cell.

[4] If spores are not readily visible, continue incubating your culture for another 24–48 hours and check again. You may also inoculate sporulation agar and incubate for 24–48 hours.

3 Perform a Gram stain of the organism. The stain should be run with known Gram-positive and Gram-negative controls to verify your technique (see Figure 3-48). In addition to providing information on Gram reaction, the Gram stain will allow observation of cell size, shape, and arrangement. Enter all these and the date in the appropriate boxes of the chart provided on the Data Sheet. (Note: Cell size for the unknown candidates is not given and should not be used to differentiate the species. To give you an idea of typical cell sizes, the cocci should be about 1 μm in diameter, and the rods 1 to 5 μm in length and about 1 μm in width.)

4 If you have cocci, or are unclear about cell arrangement, aseptically transfer your unknown to sterile Trypticase Soy Broth and examine it again in 24–48 hours using a crystal violet or carbolfuchsin simple stain. Cell arrangement often is easier to interpret using cells grown in broth.

5 Based on the Gram stain and cell morphology results, follow the appropriate branches in the flowchart until you reach the list of remaining possible organisms that matches the results of your unknown.

6 Perform the next stain and determine to which new branch of the flowchart your unknown belongs. Record the result and date of this stain in the chart provided on the Data Sheet.

7 Repeat Step 6 until you eliminate all but one organism in the flowchart—this is your unknown! If you are not doing all the stains on the day the unknowns were handed out, be sure to inoculate and incubate appropriate media so you will have fresh cultures to work with. If you need to do a spore stain, incubate your original culture for another 24–48 hours, or inoculate Sporulation Agar and incubate it for 24–48 hours. If checking for motility, inoculate a broth and incubate it for 24–48 hours.

8 After you identify your unknown, perform one more stain to confirm your result. This **confirmatory test** must not be one you used in the identification and should be added to your flowchart. The result of this test should agree with the predicted result (as given in Table 3-4). If your confirmatory test result doesn't match the expected result, repeat any suspect stains and find the source of error. (Note: The source of error may be your technique or bacterial strain variability. The results in Table 3-4 are typical for each organism, but some strains may vary. By rerunning suspect stain(s) *with controls*, you probably will be able to identify the source of error.)

9 Use a colored marker to highlight the path on the flowchart that leads to your unknown. Have your instructor check your work.

Name ____________________ Date ____________________

Lab Section ____________________ I was present and performed this exercise (initials) ________

Morphological Unknown

OBSERVATIONS AND INTERPRETATIONS

1 Complete the flowchart using the information in Table 3-4. Each path should end with a single organism.

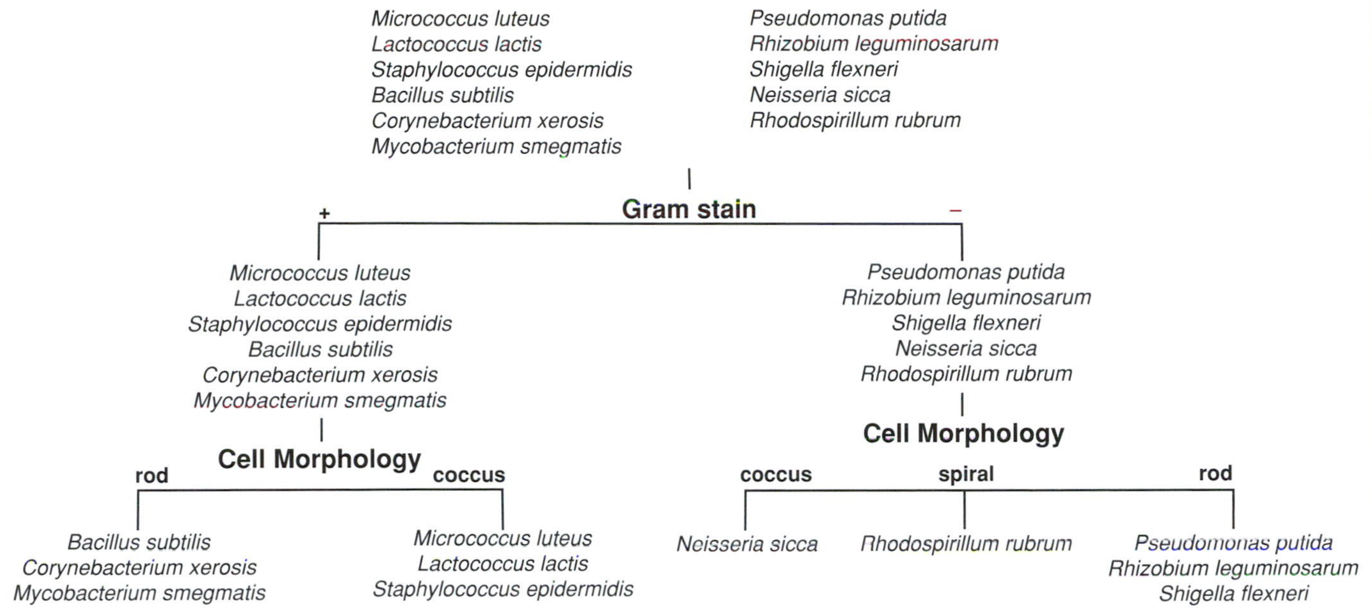

2 Record your stain results and the date each was run in the chart below. Include sketches (as appropriate) along with your written descriptions. Sketch the cells carefully—don't scribble a bunch of dots or squiggles.

Unknown number: ________________

	Gram Reaction Cell Morphology and Arrangement Cell Dimensions	Acid-Fast Stain	Motility (wet mount)	Capsule Stain	Spore Stain (shape and location)
Date(s) Run					
Results	Gram Reaction: Cell Morphology and Arrangement Cell Dimensions: Sketch:			Sketch:	Sketch:

3 Match your results with those in the flowchart. Highlight the path on the flowchart that leads to identification of your unknown.

4 Highlight your confirmatory test in the chart above.

5 Write the identity of your unknown organism in the space below.

My Unknown is:

SECTION 4

Selective Media

Individual microbial species in a mixed culture must be isolated and cultivated as **pure cultures** before they can be tested and properly identified. The most common means of isolating an organism from a mixed culture is to streak for isolation. Refer to Exercise 1-4 for a description of the streak-plate method of isolation.

The **media** introduced in this section are designed to enhance the isolation procedure by inhibiting growth of some organisms while encouraging the growth of others. Thus, they are referred to as **selective media**. Many types of selective media contain indicators to expose differences between organisms and are called **differential media.** The media illustrated in this section are used specifically to isolate pathogenic Gram-negative bacilli and Gram-positive cocci from human or environmental samples containing a mixture of microorganisms.

Clinical microbiologists often choose selective media to isolate suspected pathogens from normal flora typically found in patient samples. Environmental microbiologists often choose selective and differential media to detect **coliform bacteria.** Coliforms are a subgroup of *Enterobacteriaceae* (*Escherichia coli* being the most prominent member) that produce gas from lactose fermentation. Fermentation will be discussed more fully in Section 5. Most coliforms are normal inhabitants of the human intestinal tract; therefore, their presence in the environment may be evidence of fecal contamination.

In the exercises that follow, you will examine some commonly used selective media for isolating Gram-positive cocci and Gram-negative rods. Most of the exercises contain two examples of growth—spot inoculations with pure cultures to magnify growth characteristics and streak-plates of mixed samples to illustrate colony morphology.

All of the exercises contain a Table of Results, identifying the various reactions produced on a specific medium. These include color results, interpretations of results, symbols used to quickly identify the various reactions, and presumptive identification of typical organisms encountered with these media. Presumptive identification is not final identification. It is an "educated guess" based on evidence provided by the selective/differential medium coupled with information about the origin of a sample.

A Word About Selective Media

All of the selective media chosen for this section share four fundamental constituents—nutritional components, inhibitors, differentiating substrates, and indicators. As you read the Theories and lists of ingredients, watch for these components; it will help you to understand how each medium functions to target and isolate specific organisms or metabolically similar groups.

1. Nutritional Components

Nutritional components are selected to obtain the optimum growth of the organisms being tested for and suspected of being in the sample. Although many ingredients are suitable for many different types of microorganisms, typically, slight variations are made to tailor the medium for a specified group. Yeast extract can be added for fastidious organisms requiring specific vitamins, or sodium chloride (although not really a nutritional component) can be added for organisms where osmotic equilibrium is critical. As you will see, the inhibitor is the major selective component added to the medium, but nutritional components can enhance the overall selectivity.

2. Inhibitors

Inhibitors make the medium selective. They are designed to exploit weaknesses in specific groups of organisms and prevent or inhibit their growth. Some inhibitors function by interrupting DNA synthesis or expression of a gene. Other inhibitors function at the enzymatic level, interfering with a critical reaction. Still other inhibitors interfere with membrane permeability, thus upsetting homeostasis and starting a cascade of catastrophic changes inside the cell.

3. Substrates

Substrates make the medium differential. Differentiation and identification of organisms frequently relies on their differing abilities to perform specific chemical reactions and to do them in a way that can be observed. This can all be controlled, in the artificial environment of differential media, by including the basic components for growth and at least one substrate that only one organism or group of organisms can utilize. Following incubation of a mixed culture on the medium, organisms that might otherwise be indistinguishable may become (with the help of indicators—see #4) easily differentiated. For example, if a fecal sample is being tested for the presence of a pathogen known not to ferment lactose, then lactose is an important substrate needed to distinguish it from *E. coli* and other resident fecal coliforms, all of which ferment lactose.

4. Indicators

Indicators make a desired or expected reaction visible. An indicator is virtually always included in differential media because bacterial growth, by itself, is usually not enough to reliably differentiate between microbial groups. The medium or growth must change fairly dramatically for us to notice it. Subtle changes can be easily missed and lead to false readings. One very common type of indicator is a dye that changes color as pH changes. All dyes have a specific functional range and are chosen based on the starting pH of the medium, which in turn is based on the optimum pH range of the organisms being sought.

Indicators can also be chemicals that produce a color change by reacting with substrate breakdown products. For example, if the organism being tested reduces the medium substrate, an indicator that is oxidized (ferric sulfate in solution releases Fe^{3+}) can produce a pretty spectacular change. Oxidized iron is a common reactant used to form a visible precipitate.

Selective Media for Isolation of Gram-Positive Cocci

Gram-negative organisms frequently inhibit growth of Gram-positive organisms when they are cultivated together. Therefore, when looking for staphylococci or streptococci in a clinical sample, it may be necessary to begin by streaking the unknown mixture onto a selective medium that inhibits Gram-negative growth. The three selective media introduced in this unit—Phenylethyl Alcohol Agar, Columbia CNA Agar, and Mannitol Salt Agar—are all used to isolate streptococci, enterococci, or staphylococci in human samples.

EXERCISE 4-1

Phenylethyl Alcohol Agar

Theory

Phenylethyl Alcohol Agar (PEA) is an undefined, selective medium that encourages growth of Gram-positive organisms and inhibits growth of most Gram-negative organisms (Figure 4-1). It is not a differential medium because it does not distinguish between different organisms; it merely encourages or discourages growth. Digests of casein and soybean meal provide nutrition while sodium chloride provides a stable osmotic environment suitable for the addition of sheep blood if desired. Phenylethyl alcohol is the selective agent that inhibits Gram-negative organisms by breaking down their membrane permeability barrier, thus allowing influx of substances ordinarily blocked and leakage of large amounts of cellular potassium. This ultimately disrupts or halts DNA synthesis.

Application

PEA is used to isolate staphylococci and streptococci (including enterococci and lactococci) from specimens containing mixtures of bacterial flora. Typically, it is used to screen out the common contaminants *Escherichia coli* and *Proteus* species. When prepared with 5% sheep blood, it is used for cultivation of Gram-positive anaerobes.

In This Exercise

You will spot-inoculate one PEA plate and one Nutrient Agar (NA) plate with three test organisms. The NA plate will serve as a comparison for growth quality on the PEA plate.

Materials

Per Student Group

- ✓ one PEA plate
- ✓ one NA plate
- ✓ fresh broth cultures of:
  - *Escherichia coli*
  - *Staphylococcus epidermidis*

Medium Recipes

Phenylethyl Alcohol Agar

Pancreatic digest of casein	15.0 g
Papaic digest of soybean meal	5.0 g

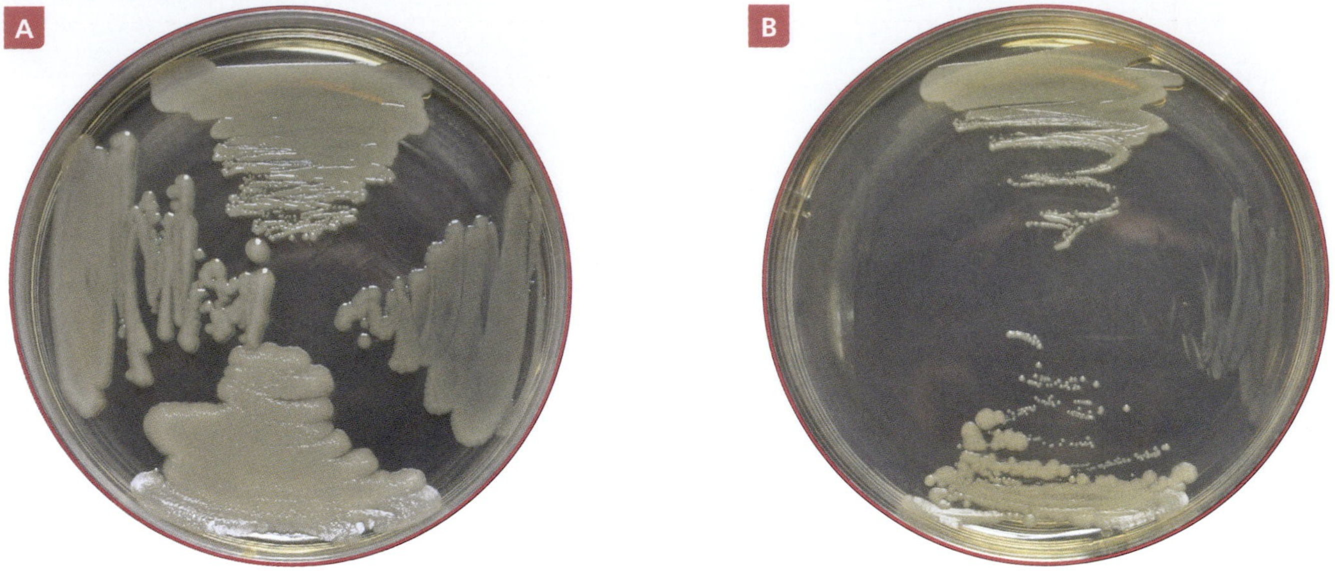

4-1 Brain–Heart Infusion Agar vs. Phenylethyl Alcohol ◆ **A** Two Gram-positive organisms (inoculated top and bottom) and two Gram-negative organisms (left and right) demonstrating good growth on BHIA. **B** The same organisms inoculated in the same positions on a PEA plate. Notice that the two Gram-positive organisms grow well on both plates. The two Gram-negative organisms are, to differing degrees, inhibited by PEA. Note the faint growth on the right, suggesting some resistance to the selective agent.

TABLE 4-1 PEA Results and Interpretations

TABLE OF RESULTS		
Result	**Interpretation**	**Presumptive ID**
Poor growth or no growth (P)	Organism is inhibited by phenylethyl alcohol	Probable Gram-negative organism
Good growth (G)	Organism is not inhibited by phenylethyl alcohol	Probable *Staphylococcus*, *Streptococcus*, *Enterococcus*, or *Lactococcus*

- Sodium chloride 5.0 g
- β-Phenylethyl alcohol 2.5 g
- Agar 15.0 g
- Distilled or deionized water 1.0 L

pH 7.1–7.5 at 25°C

Nutrient Agar

- Beef extract 3.0 g
- Peptone 5.0 g
- Agar 15.0 g
- Distilled or deionized water 1.0 L

pH 6.6 – 7.0 at 25°C

Procedure

Lab One

1 Using a permanent marker, divide the bottom of each plate into three sectors.

2 Label the plates with the organisms' names, your name, and the date.

3 Mix each culture well.

4 Spot-inoculate the sectors on the PEA plate with the test organisms. Refer to Appendix B if necessary.

5 Repeat Step 4 with the Nutrient Agar plate.

6 Invert and incubate the plates at 35 ± 2°C for 24 to 48 hours.

Lab Two

1 Examine and compare the plates for color and quality of growth.

2 Record your results on the Data Sheet. Refer to Table 4-1 as needed.

References

Forbes, Betty A., Daniel F. Sahm, and Alice S. Weissfeld. 2002. Chapter 10 in *Bailey & Scott's Diagnostic Microbiology*, 11th ed. Mosby, St. Louis.

Zimbro, Mary Jo, and David A. Power. 2003. Page 443 in *Difco™ & BBL™ Manual—Manual of Microbiological Culture Media*. Becton, Dickinson and Co., Sparks, MD.

Name ______________________ Date ______________________

Lab Section ______________________ I was present and performed this exercise (initials) ________

DATA SHEET 4-1

Phenylethyl Alcohol Agar

OBSERVATIONS AND INTERPRETATIONS

Refer to Table 4-1 when recording your results and interpretations in the chart below.

Organism	Growth (P/G)		Interpretation
	PEA	NA	

Questions

1 You were instructed to compare the growth on the NA and PEA plates. Because the two media used in this exercise were completely different and likely to produce differing amounts of growth, why not simply compare the organisms to each other on the PEA plate? In your explanation, include the information provided by the growth on the NA plate.

2 PEA contains only 0.25% phenylethyl alcohol because high concentrations inhibit both Gram-negative organisms and Gram-positive organisms. List some possible reasons why this is true.

3 If you observed growth of Gram-negative organisms on your PEA plate, does this negate the usefulness of PEA as a selective medium? Why or why not? (Hint: Compare the NA and PEA plates.)

4 Is PEA a defined medium or an undefined medium? Why is this formulation desirable?

5 Which ingredient(s) in PEA supply(ies)

a. Carbon?

b. Nitrogen?

EXERCISE 4-2

Columbia CNA With 5% Sheep Blood Agar

Theory

Columbia CNA with 5% Sheep Blood Agar is an undefined, differential, and selective medium that allows growth of Gram-positive organisms (especially staphylococci, streptococci, and enterococci) and stops or inhibits growth of most Gram-negative organisms (Figure 4-2). Casein, digest of animal tissue, beef extract, yeast extract, corn starch, and sheep blood provide a range of carbon and energy sources to support a wide variety of organisms. In addition, sheep blood supplies the X factor (heme) and yeast extract provides B-vitamins. The antibiotics colistin and nalidixic acid (CNA) act as selective agents against Gram-negative organisms by affecting membrane integrity and interfering with DNA replication, respectively. They are particularly effective against *Klebsiella, Proteus,* and *Pseudomonas* species. Further, sheep blood makes possible differentiation of Gram-positive organisms based on hemolytic reaction.

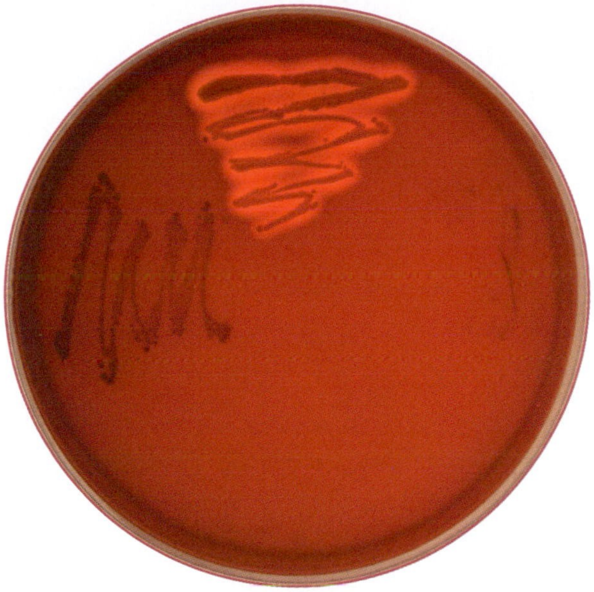

4-2 **Columbia CNA with 5% Sheep Blood Agar** ◆ The plate was inoculated with four organisms—two Gram-positive cocci, and two Gram-negative rods. Only the Gram-positive organisms (left and top quadrants) grow well on the Columbia CNA agar. The two Gram-negatives either didn't grow (bottom) or grew poorly (right). Further, the top Gram-positive is β-hemolytic, whereas the one on the left is nonhemolytic.

Application

Columbia CNA with 5% Sheep Blood Agar is used to isolate staphylococci, streptococci, and enterococci, primarily from clinical specimens.

In This Exercise

You will spot-inoculate one Columbia CNA with 5% Sheep Blood Agar plate and one Nutrient Agar (NA) plate with four test organisms. The NA plate will serve as a comparison for growth quality on the Columbia CNA with 5% Sheep Blood Agar plate.

Materials

Per Student Group

- ✓ one Columbia CNA with 5% Sheep Blood Agar plate
- ✓ one Nutrient Agar plate
- ✓ fresh broth cultures of:
  - *Escherichia coli* (or other Gram-negative, as available)
  - *Streptococcus* spp.
  - *Enterococcus* spp.
  - *Staphylococcus* spp.

Medium Recipes

Columbia CNA with 5% Sheep Blood Agar

◆ Pancreatic digest of casein	12.0 g
◆ Peptic digest of animal tissue	5.0 g
◆ Yeast extract	3.0 g
◆ Beef extract	3.0 g
◆ Corn starch	1.0 g
◆ Sodium chloride	5.0 g
◆ Colistin	10.0 mg
◆ Nalidixic acid	10.0 mg
◆ Agar	13.5 g
◆ Distilled or deionized water *pH 7.1–7.5 at 25°C*	1.0 L
◆ Sheep blood (defibrinated)	50 mL added to 950 mL of above

Nutrient Agar

◆ Beef extract	3.0 g
◆ Peptone	5.0 g
◆ Agar	15.0 g
◆ Distilled or deionized water *pH 6.6–7.0 at 25°C*	1.0 L

Procedure

Lab One

1. Using a permanent marker, divide the bottom of each plate into four sectors.
2. Label the plates with the organisms' names, your name, and the date.
3. Mix each culture well.
4. Spot-inoculate (Appendix B) the sectors on the CNA plate with the test organisms.
5. Repeat step 4 with the Nutrient Agar plate. Use the same position for each specimen.
6. Invert and incubate the plates at 35±2°C for 24 to 48 hours. If possible, incubate the plates in a candle jar or anaerobic jar with 3–5% CO_2.

Lab Two

1. Referring to Table 4-2, examine and compare the plates for quality of growth and hemolytic reaction. For more information on hemolysis, see Exercise 5-25.
2. Record your results in the space provided on the Data Sheet.

References

Chapin, Kimberle C., and Tsai-Ling Lauderdale. 2007. Chapter 21 in *Manual of Clinical Microbiology*, 8th Ed., edited by Patrick R. Murray, Ellen Jo Baron, James. H. Jorgensen, Marie Louise Landry, and Michael A. Pfaller. ASM Press, Washington, DC.

Forbes, Betty A., Daniel F. Sahm, and Alice S. Weissfeld. 2002. Pages 136 and 138 in *Bailey & Scott's Diagnostic Microbiology*, 11th Ed. Mosby, Inc., St. Louis.

Zimbro, Mary Jo, and David A. Power. 2003. Page 156 in *DIFCO & BBL Manual—Manual of Microbiological Culture Media*. Becton, Dickinson and Company, Sparks, MD.

TABLE 4-2 Columbia CNA with 5% Sheep Blood Agar Results and Interpretations

TABLE OF RESULTS

Result	Interpretation	Presumptive ID
Poor growth or no growth (P)	Organism is inhibited by colistin and nalidixic acid	Probable Gram-negative organism
Good growth (G) with clearing (β-hemolysis)	Organism is not inhibited by colistin and nalidixic acid and completely hemolyzes RBCs	Probable β-hemolytic *Staphylococcus*, *Streptococcus*, or *Enterococcus* (Note: Artifactual greening may occur with some β-hemolytic organisms)
Good growth (G) with greening of medium (α-hemolysis)	Organism is not inhibited by colistin and nalidixic acid and partially hemolyzes RBCs	Probable α-hemolytic *Staphylococcus*, *Streptococcus*, or *Enterococcus*
Good growth (G) with no change of medium's color (γ-hemolysis)	Organism is not inhibited by colistin and nalidixic acid and does not hemolyze RBCs	Probable γ-hemolytic *Staphylococcus*, *Streptococcus*, or *Enterococcus*

NAME ______________________________ DATE ______________________________

LAB SECTION ______________________________ I WAS PRESENT AND PERFORMED THIS EXERCISE (INITIALS) ________

Columbia CNA With 5% Sheep Blood Agar

OBSERVATIONS AND INTERPRETATIONS

Refer to Table 4-2 when recording your results and interpretations in the chart below.

Organism	Growth (P/G)		Hemolysis (α, β, γ)	Interpretation
	Columbia CNA Agar	NA		

Questions

1 You were instructed to compare the growth on the NA and Columbia CNA Agar plates. Because the two media used in this exercise were completely different and likely to produce differing amounts of growth, would it have been better to compare the organisms to each other on the CNA plate? Include what information is provided by the NA plate in your explanation.

2 Earlier formulations of Columbia CNA Agar were made using a slightly higher concentration of colistin and nalidixic acid, but this was reduced to improve recovery of Gram-positive organisms. List some possible reasons why the higher concentration was less effective at allowing growth of Gram-positive organisms.

EXERCISE 4-3

Mannitol Salt Agar

❖ Theory

Mannitol Salt Agar (MSA) contains the carbohydrate mannitol, 7.5% sodium chloride (NaCl), and the pH indicator phenol red. Phenol red is yellow below pH 6.8, red at pH 7.4 to 8.4, and pink at pH 8.4 and above. Mannitol provides the substrate for fermentation and makes the medium differential. Sodium chloride makes the medium selective because its concentration is high enough to dehydrate and kill most bacteria. Staphylococci thrive in the medium, largely because of their adaptation to salty habitats such as human skin. Phenol red indicates whether fermentation has taken place by changing color as the pH changes. (See Section 5, A Word About Biochemical Tests and Acid-Base Reactions.)

Most staphylococci are able to grow on MSA, but do not ferment the mannitol, so the growth appears pink or red and the medium remains unchanged. *Staphylococcus aureus* ferments the mannitol, which produces acids and lowers the pH of the medium (Figure 4-3). The result is formation of bright yellow colonies usually surrounded by a yellow halo (Figures 4-4 and 4-5). For more information on fermentation, refer to Appendix A.

❖ Application

Mannitol Salt Agar is used for isolation and differentiation of *Staphylococcus aureus*.

❖ In This Exercise

Today you will spot-inoculate one MSA plate and one Nutrient Agar (NA) plate with three test organisms. The NA will serve as a comparison for growth quality on the MSA plate.

❖ Materials

Per Student Group

- ✓ one MSA plate
- ✓ one NA plate
- ✓ fresh broth cultures of:
  - *Staphylococcus aureus* (BSL-2)
  - *Staphylococcus epidermidis*
  - *Escherichia coli*

❖ Medium Recipes

Mannitol Salt Agar

♦ Beef extract	1.0 g
♦ Peptone	10.0 g
♦ Sodium chloride	75.0 g
♦ D-Mannitol	10.0 g
♦ Phenol red	0.025 g
♦ Agar	15.0 g
♦ Distilled or deionized water	1.0 L

pH 7.2–7.6 at 25°C

Nutrient Agar

♦ Beef extract	3.0 g
♦ Peptone	5.0 g
♦ Agar	15.0 g
♦ Distilled or deionized water	1.0 L

pH 6.6–7.0 at 25°C

Procedure

Lab One

1. Using a permanent marker, divide the bottom of each plate into three sectors.
2. Label the plates with the organisms' names, your name, and the date.

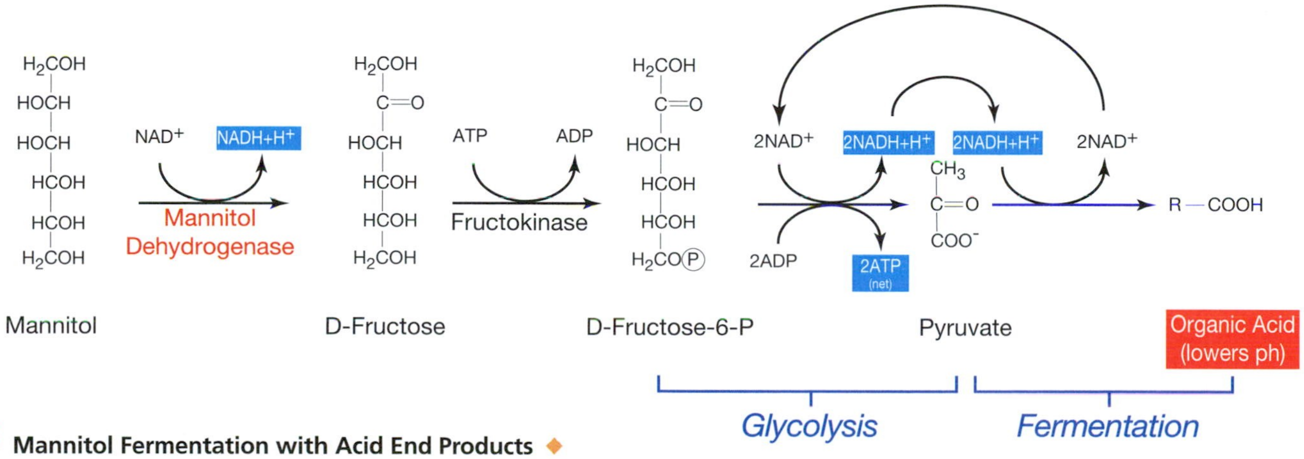

4-3 Mannitol Fermentation with Acid End Products ♦

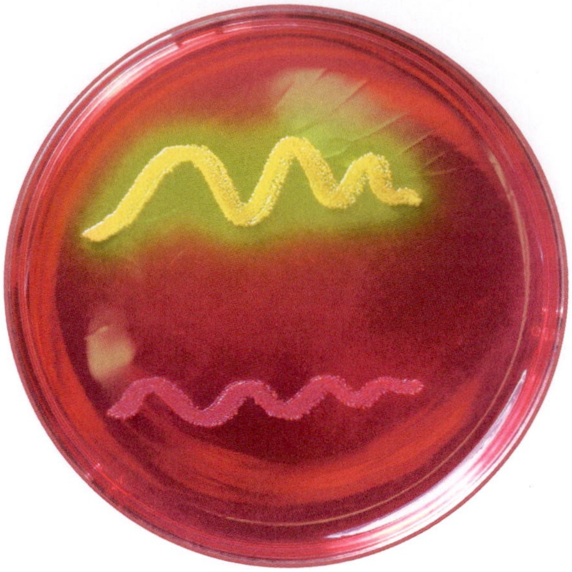

4-4 Mannitol Salt Agar ◆ This MSA was inoculated with two of the organisms used in today's lab. Both grew well, but only the top one fermented mannitol and produced acid end products. This is evidenced by the yellow growth and halo surrounding it. Compare this plate with the one in Figure 4-5 and see if you can identify the organisms.

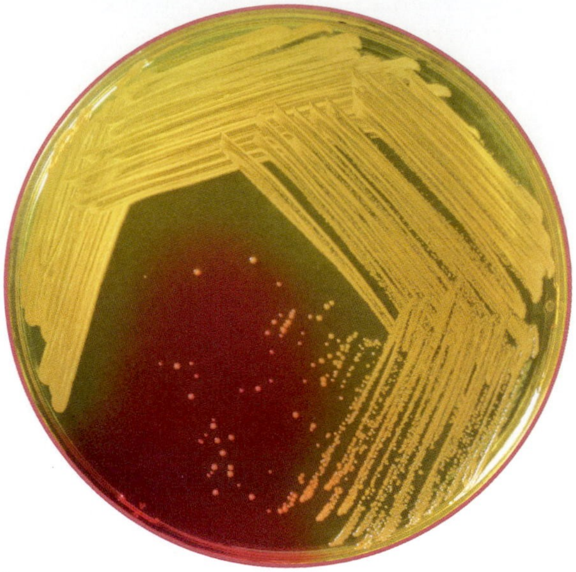

4-5 Mannitol Salt Agar Streaked for Isolation ◆ This MSA was inoculated with the same two organisms as in Figure 4-4. The solid growth in the first three quadrants is still a mixture of the two organisms, so disregard the color in this region. Note the small red colonies and larger yellow colonies in the fourth streak.

TABLE 4-3 Mannitol Salt Agar Results and Interpretations ◆

TABLE OF RESULTS		
Result	**Interpretation**	**Presumptive ID**
Poor growth or no growth (P)	Organism is inhibited by NaCl	Not *Staphylococcus*
Good growth (G)	Organism is not inhibited by NaCl	*Staphylococcus*
Yellow growth or halo (Y)	Organism produces acid from mannitol fermentation	Possible pathogenic *Staphylococcus aureus*
Red growth (no halo) (R)	Organism does not ferment mannitol. No reaction	*Staphylococcus* other than *S. aureus*

3 Mix each culture well.

4 Spot-inoculate the sectors on the Mannitol Salt Agar plate with the test organisms. Refer to Appendix B if necessary.

5 Repeat Step 4 with the Nutrient Agar plate.

6 Invert and incubate the plates at 35 ± 2°C for 24 to 48 hours.

Lab Two

1 Examine and compare the plates for color and quality of growth.

2 Record your results on the Data Sheet.

References

Delost, Maria Dannessa. 1997. Page 112 in *Introduction to Diagnostic Microbiology, a Text and Workbook*. Mosby, St. Louis.

Forbes, Betty A., Daniel F. Sahm, and Alice S. Weissfeld. 2002. Chapter 19 in *Bailey & Scott's Diagnostic Microbiology*, 11th ed. Mosby, St. Louis.

Zimbro, Mary Jo, and David A. Power. 2003. Page 349 in *Difco™ & BBL™ Manual—Manual of Microbiological Culture Media*. Becton, Dickinson and Co., Sparks, MD.

Name ______________________ Date ______________________

Lab Section ______________________ I was present and performed this exercise (initials) ________

Mannitol Salt Agar

OBSERVATIONS AND INTERPRETATIONS

Refer to Table 4-3 when recording your results and interpretations in the chart below. Use abbreviations or symbols as needed.

Organism	Growth (P/G)		MSA Growth Color (Y/R)	Interpretation
	MSA	NA		

Questions

1 What purpose does the Nutrient Agar plate serve? In what way does it increase the validity of the test result?

DATA SHEET 4-3

2 What would be the likely consequences of omitting the NaCl in Mannitol Salt Agar? Why?

3 Would omitting the NaCl alter the medium's specificity or sensitivity? Explain. (See "A Word About Experimental Design" on page 6.)

4 Which ingredient(s) supply(ies)

a. Carbon?

b. Nitrogen?

5 With the diversity of microorganisms in the world, how can a single test such as MSA be used to confidently identify *Staphylococcus aureus*?

Selective Media for Isolation of Gram-negative Rods

The healthy human body is residence to billions of beneficial bacteria, especially members of the family *Enterobacteriaceae*, also called enteric or "gut" bacteria. Some of these bacteria are relatively benign; others are potentially serious pathogens. But virtually all are potentially harmful when introduced to an area of the body outside their normal habitat. As you will see, the media in this unit are designed to isolate and differentiate these organisms from each other and from non-enterics in a variety of ways.

The three examples selected for this unit are MacConkey Agar, Eosin Methylene Blue (EMB) Agar, and Hektoen Enteric (HE) Agar. MacConkey Agar and EMB Agar are selective for Gram-negative organisms and both contain indicators to differentiate lactose fermenters from lactose nonfermenters. EMB Agar is commonly used to test for the presence of coliforms in environmental samples. As mentioned in the introduction to this section, the presence of coliforms in the environment suggests fecal contamination and the possible presence of dangerous pathogens. HE Agar differentiates *Salmonella* and *Shigella* from each other and from other enterics based on their ability to overcome the inhibitory effects of bile, reduce sulfur to H_2S, and ferment lactose, sucrose, or salicin. (For more information about fermentation reactions, refer to the introductory material in Section 5 and to Appendix A.)

EXERCISE 4-4

MacConkey Agar

❖ Theory

MacConkey Agar is a selective and differential medium containing lactose, bile salts, neutral red, and crystal violet. Bile salts and crystal violet inhibit growth of Gram-positive bacteria. Neutral red dye is a pH indicator that is colorless above a pH of 6.8 and red at a pH less than 6.8. Acid accumulating from lactose fermentation turns the dye red. Lactose fermenters turn a shade of red on MacConkey Agar, whereas lactose nonfermenters retain their normal color or the color of the medium (Figures 4-6 and 4-7). Formulations without crystal violet allow growth of *Enterococcus* and some species of *Staphylococcus*, which ferment the lactose and appear pink on the medium.

❖ Application

MacConkey Agar is used to isolate and differentiate members of the *Enterobacteriaceae* based on the ability to ferment lactose. Variations on the standard medium include MacConkey Agar w/o CV (without crystal violet) to allow growth of Gram-positive cocci, or MacConkey Agar CS to control swarming bacteria (*Proteus*) that interfere with other results.

❖ In This Exercise

You will spot-inoculate one MacConkey Agar plate and one Nutrient Agar (NA) plate with three test organisms. The NA plate will serve as a comparison for growth quality on the MacConkey Agar plate.

❖ Materials

Per Student Group

- ✓ one MacConkey Agar plate
- ✓ one Nutrient Agar plate
- ✓ fresh broth cultures of:
  - *Staphylococcus epidermidis*
  - *Escherichia coli*
  - *Providencia stuartii*

❖ Medium Recipes

MacConkey Agar

◆ Pancreatic digest of gelatin	17.0 g
◆ Pancreatic digest of casein	1.5 g
◆ Peptic digest of animal tissue	1.5 g

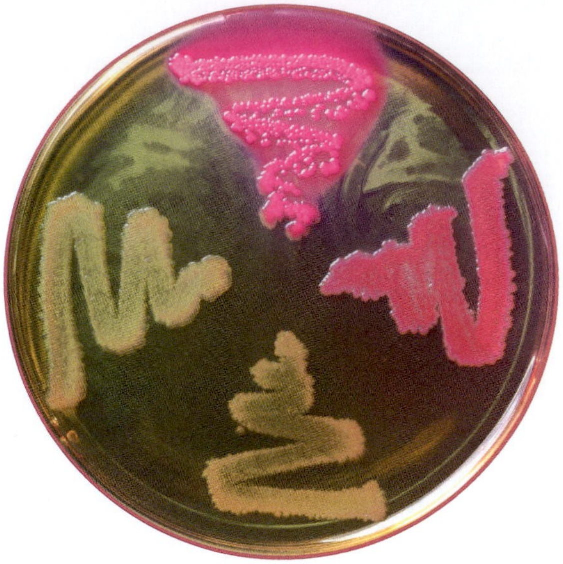

4-6 MacConkey Agar ◆ This MacConkey Agar was inoculated with four members of *Enterobacteriaceae*, all of which show abundant growth. The top organism and the one on the right are lactose fermenters, as evidenced by the pink color. Note the bile precipitate around the top organism. The organisms on the bottom and on the left produced no color, so they do not appear to be lactose fermenters.

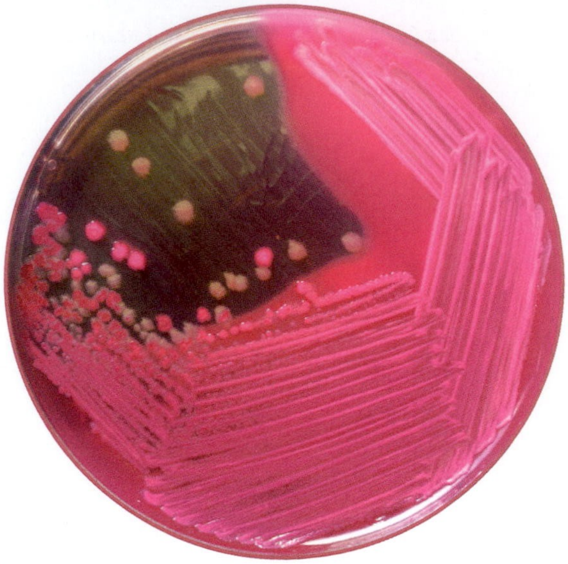

4-7 MacConkey Agar Streaked for Isolation ◆ This MacConkey Agar was inoculated with two enteric organisms similar to those selected for today's exercise. Both grew well, but only one fermented the lactose. The solid growth in the first three quadrants is a mixture of the two organisms, so the color produced in that region is not determinative. Note the colors of the individual colonies.

◆ Lactose	10.0 g
◆ Bile salts	1.5 g
◆ Sodium chloride	5.0 g
◆ Neutral red	0.03 g
◆ Crystal violet	0.001 g
◆ Agar	13.5 g
◆ Distilled or deionized water	1.0 L
pH 6.9–7.3 at 25°C	

Nutrient Agar

◆ Beef extract	3.0 g
◆ Peptone	5.0 g
◆ Agar	15.0 g
◆ Distilled or deionized water	1.0 L
pH 6.6–7.0 at 25°C	

Procedure

Lab One

1 Using a permanent marker, divide the bottom of each plate into three sectors.

2 Label the plates with the organisms' names, your name, and the date.

3 Mix each culture well.

4 Spot-inoculate the sectors on the MacConkey Agar plate with the test organisms. Refer to Appendix B if necessary.

5 Repeat Step 4 with the Nutrient Agar plate.

6 Invert and incubate the plates at 35 ± 2°C for 24 to 48 hours.

Lab Two

1 Examine and compare the plates for color and for quality of growth.

2 Refer to Table 4-4 when recording your results on the Data Sheet.

References

Forbes, Betty A., Daniel F. Sahm, and Alice S. Weissfeld. 2002. Chapter 10 in *Bailey & Scott's Diagnostic Microbiology*, 11th ed. Mosby-Yearbook, St. Louis.

Winn, Washington, C., et al. 2006. *Koneman's Color Atlas and Textbook of Diagnostic Microbiology*, 6th ed. Lippincott Williams & Wilkins, Baltimore.

Zimbro, Mary Jo, and David A. Power. 2003. Page 334 in *Difco™ & BBL™ Manual—Manual of Microbiological Culture Media*. Becton, Dickinson and Co., Sparks, MD.

TABLE 4-4 MacConkey Agar Results and Interpretations

TABLE OF RESULTS		
Result	**Interpretation**	**Presumptive ID**
Poor growth or no growth (P)	Organism is inhibited by crystal violet and/or bile	Gram-positive
Good growth (G)	Organism is not inhibited by crystal violet or bile	Gram-negative
Pink to red growth with or without bile precipitate (R)	Organism produces acid from lactose fermentation	Probable coliform
Growth is "colorless" (not red or pink) (C)	Organism does not ferment lactose. No reaction	Noncoliform

Name ______________________ Date ______________________

Lab Section ______________________ I was present and performed this exercise (initials) ________

MacConkey Agar

OBSERVATIONS AND INTERPRETATIONS

Refer to Table 4-4 when recording your results and interpretations in the chart below. Use abbreviations or symbols as needed.

Organism	Growth (P/G)		MSA Growth Color (R/C)	Interpretation
	MAC	NA		

Questions

1 What purpose does the Nutrient Agar plate serve? In what way does it increase the validity of the test results?

2 With respect to the MacConkey Agar, what would be the possible consequence of:

a. replacing the lactose with glucose?

b. replacing the neutral red with phenol red (yellow when acidic; red or pink when alkaline)?

3 How would removing crystal violet from MacConkey Agar alter the sensitivity and specificity of the medium? (See "A Word About Experimental Design" on page 6 for assistance.)

4 Compare the recipes of Nutrient Agar and MacConkey Agar. If an organism can grow on both media, on which would you expect it to grow better? Why?

5 MacConkey Agar is a selective medium. Is it also a defined or an undefined medium? Why is that formulation desirable?

EXERCISE 4-5

Eosin Methylene Blue Agar

❖ Theory

Eosin Methylene Blue (EMB) Agar is a complex (chemically undefined), selective, and differential medium. It contains peptone, lactose, sucrose, and the dyes eosin Y and methylene blue. The peptone provides a complex mixture of carbon, nitrogen, and other nutritional components. The sugars are included to encourage growth of enteric bacteria and to differentiate them based on color reactions created when combined with the dyes. Lactose supports coliforms such as *Escherichia coli* and sucrose supports pathogens such as *Proteus* or *Salmonella* species (Figure 4-8).

The purpose of the dyes is twofold: 1) they inhibit the growth of Gram-positive organisms and 2) they react with vigorous lactose fermenters and (in the acidic environment) turn the growth dark purple or black. This dark growth is typical of *Escherichia coli* and is usually accompanied by a green metallic sheen (Figures 4-9 and 4-10). Other less aggressive lactose fermenters such as *Enterobacter* or *Klebsiella* species produce colonies that can range from pink to dark purple on the medium. Nonfermenters and sucrose fermenters typically retain their normal color or take on the coloration of the medium.

❖ Application

EMB Agar is used for the isolation of fecal coliforms. It can be streaked for isolation or used in the Membrane Filter Technique as discussed in Exercise 8-12.

❖ In This Exercise

You will spot-inoculate one EMB Agar plate and one Nutrient Agar (NA) plate with four test organisms. The NA plate will serve as a comparison for growth quality on the EMB Agar plate.

❖ Materials

Per Student Group

- ✓ one EMB plate
- ✓ one NA plate
- ✓ fresh broth cultures of:
  - *Enterobacter aerogenes*
  - *Staphylococcus epidermidis*
  - *Escherichia coli*
  - *Providencia stuartii*

❖ Medium Recipes

Eosin Methylene Blue Agar

♦ Peptone	10.0 g
♦ Lactose	5.0 g
♦ Sucrose	5.0 g
♦ Dipotassium phosphate	2.0 g
♦ Agar	13.5 g
♦ Eosin Y	0.4 g
♦ Methylene blue	0.065 g
♦ Distilled or deionized water	1.0 L

pH 6.9–7.3 at 25°C

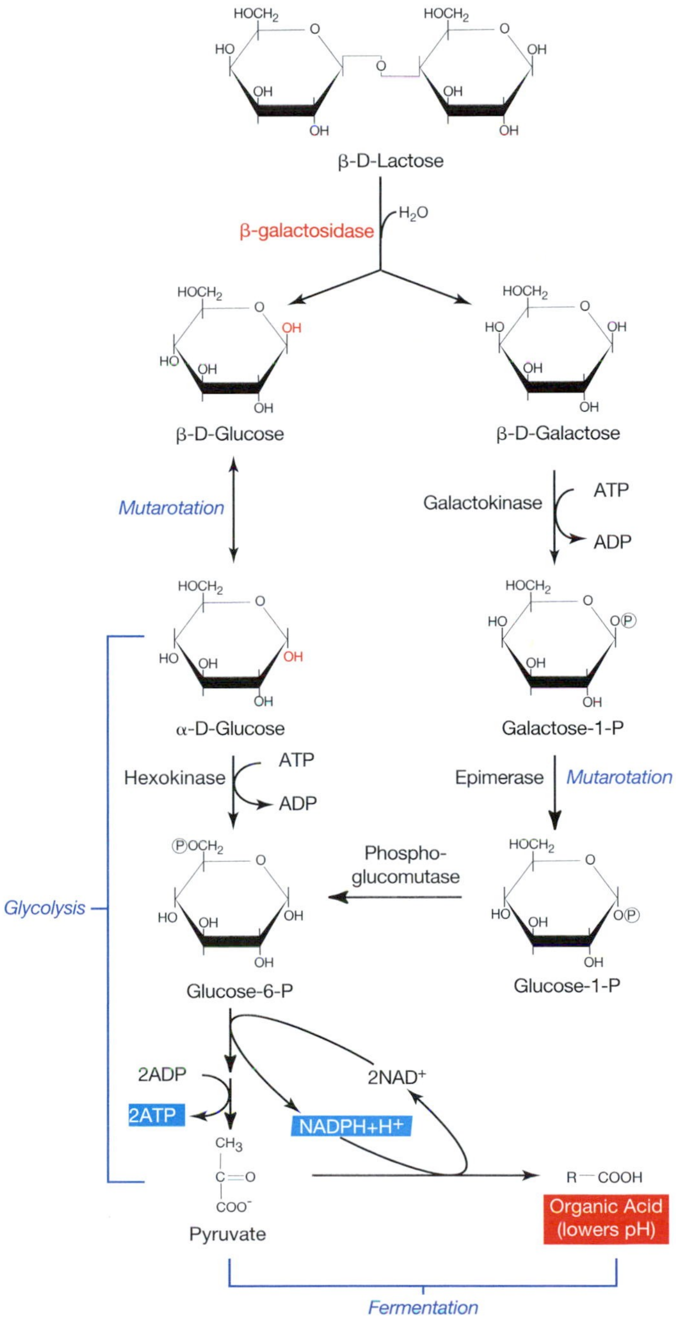

4-8 Lactose Fermentation with Acid End Products ♦

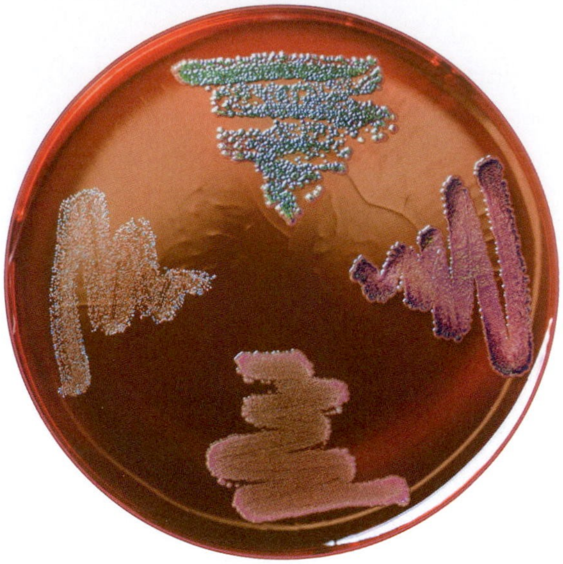

4-9 Eosin Methylene Blue Agar ◆ This EMB Agar was inoculated with (clockwise from the top) two coliforms, a Gram-negative noncoliform, and a Gram-positive organism. Note the characteristic green metallic sheen of the coliform at the top and the pink coloration of the one at the right. Both organisms are lactose fermenters; the difference in color results from the degree of acid production. The organism on the bottom is a nonfermenter, as indicated by the lack of color. Growth of the Gram-positive organism on the left was inhibited by eosin Y and methylene blue.

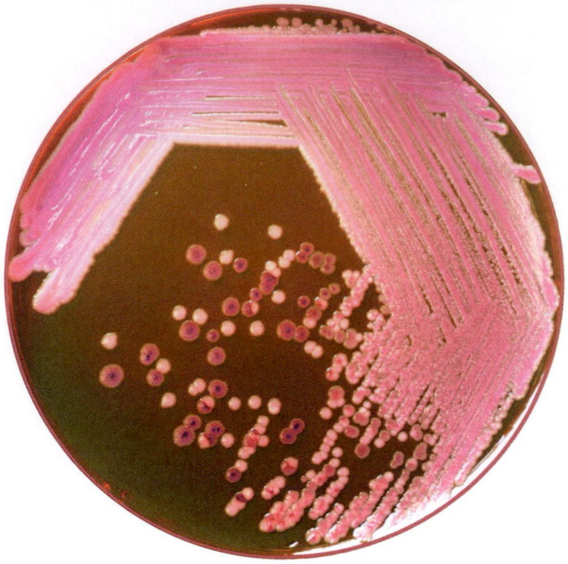

4-10 Eosin Methylene Blue Agar Streaked for Isolation ◆ This EMB agar was inoculated with two of the Gram-negative rods shown in Figure 4-9. One is a coliform; the other is not. Color produced by mixtures of bacteria is not determinative, so disregard the first three quadrants and pay attention only to the individual colonies. The coliform that produced a green metallic sheen on the plate shown in Figure 4-9 produced pink colonies with dark centers in this photo. The noncoliform colonies are beige.

Nutrient Agar

◆ Beef extract	3.0 g
◆ Peptone	5.0 g
◆ Agar	15.0 g
◆ Distilled or deionized water	1.0 L

pH 6.6–7.0 at 25°C

Procedure

Lab One

1. Using a permanent marker, divide the bottom of each plate into four sectors.
2. Label the plates with the organisms' names, your name, and the date.
3. Mix each culture well.
4. Spot-inoculate the four sectors on the EMB plate with the test organisms. Refer to Appendix B if necessary.
5. Repeat step 4 with the Nutrient Agar plate.
6. Invert and incubate the plates at 35 ± 2°C for 24 to 48 hours.

Lab Two

1. Examine and compare the plates for color and quality of growth.
2. Record your results on the Data Sheet.

References

Forbes, Betty A., Daniel F. Sahm, and Alice S. Weissfeld. 2002. Chapter 10 in *Bailey & Scott's Diagnostic Microbiology*, 11th ed. Mosby, St. Louis.

Winn, Washington, C., et al. 2006. *Koneman's Color Atlas and Textbook of Diagnostic Microbiology*, 6th ed. Lippincott Williams & Wilkins, Baltimore.

Zimbro, Mary Jo, and David A. Power. 2003. Page 218 in *Difco™ & BBL™ Manual—Manual of Microbiological Culture Media*. Becton, Dickinson and Co., Sparks, MD.

TABLE 4-5 EMB Results and Interpretations ◆

TABLE OF RESULTS		
Result	**Interpretation**	**Presumptive ID**
Poor growth or no growth (P)	Organism is inhibited by eosin and methylene blue	Gram-positive
Good growth (G)	Organism is not inhibited by eosin and methylene blue	Gram-negative
Growth is pink and mucoid (Pi)	Organism ferments lactose with little acid production	Possible coliform
Growth is "dark" (purple to black, with or without green metallic sheen) (D)	Organism ferments lactose and/or sucrose with acid production	Probable coliform
Growth is "colorless" (no pink, purple, or metallic sheen) (C)	Organism does not ferment lactose or sucrose. No reaction	Noncoliform

Name ________________ Date ________________

Lab Section ________________ I was present and performed this exercise (initials) ________

Eosin Methylene Blue Agar

OBSERVATIONS AND INTERPRETATIONS

Refer to Table 4-5 when recording your results and interpretations in the chart below. Use abbreviations or symbols as needed.

Organism	Growth (P/G)		EMB Growth Color (Pi/D/C)	Interpretation
	EMB	NA		

Questions

1 What purpose does the Nutrient Agar plate serve? In what way does it increase the validity of the test result?

2 Dipotassium phosphate is a buffer added to EMB that adjusts the pH to the proper starting level. What would be a possible consequence of adding buffers to raise the starting pH to 7.8?

DATA SHEET 4-5

3 Would the change in starting pH suggested in question #2 alter the medium's sensitivity or specificity? Explain.

4 You are becoming aware of the diversity and abundance of bacteria, and that we always have some uncertainty about the identity of an isolate. With this in mind, how can we be certain that growth with a green metallic sheen is truly a coliform bacterium?

5 Compare the recipes of Nutrient Agar and EMB Agar. If an organism can grow on both media, on which would you expect it to grow better? Why?

6 EMB is a selective medium. Is it also a defined or an undefined medium? Why is that formulation desirable?

7 Which ingredient(s) in EMB supply(ies)

a. Carbon?

b. Nitrogen?

EXERCISE 4-6

Hektoen Enteric Agar

Theory

Hektoen Enteric (HE) Agar is a complex (chemically undefined), moderately selective, and differential medium designed to isolate *Salmonella* and *Shigella* species from other enterics. The test is based on the ability to ferment lactose, sucrose, or salicin, and to reduce sulfur to hydrogen sulfide gas (H_2S). Sodium thiosulfate is included as the source of oxidized sulfur. Ferric ammonium citrate is included as a source of oxidized iron to react with any sulfur that becomes reduced (H_2S), thus forming the black precipitate ferrous sulfide (FeS). Bile salts are included to prevent or inhibit growth of Gram-positive organisms. The bile salts also have a moderate inhibitory effect on enterics, so relatively high concentrations of animal tissue and yeast extract are included to offset this situation. Bromthymol blue and acid fuchsin dyes are added to indicate pH changes.

Differentiation is possible as a result of the various colors produced in the colonies and in the agar. Enterics that produce acid from fermentation will produce yellow to salmon-pink colonies. Neither *Salmonella* nor *Shigella* species ferment any of the sugars; instead they break down the animal tissue, which raises the pH of the medium slightly and gives the colonies a blue-green color. Additionally, *Salmonella* species reduce sulfur to H_2S, so the colonies formed also contain FeS, which makes them partially or completely black. Refer to Figures 4-11 and 4-12.

Application

HE Agar is used to isolate and differentiate *Salmonella* and *Shigella* species from other Gram-negative enteric organisms.

In This Exercise

You will streak-inoculate one Hektoen Enteric Agar plate and one Nutrient Agar plate with four test organisms. The Nutrient Agar will serve as a comparison for growth quality on the Hektoen Enteric Agar plate.

Materials

Per Student Group

- ✓ one HE Agar plate
- ✓ one NA plate
- ✓ fresh broth cultures of:
  - *Enterococcus faecalis*
  - *Escherichia coli*
  - *Salmonella typhimurium* (BSL-2)
  - *Shigella flexneri* surrogate (*Providencia stuartii*)

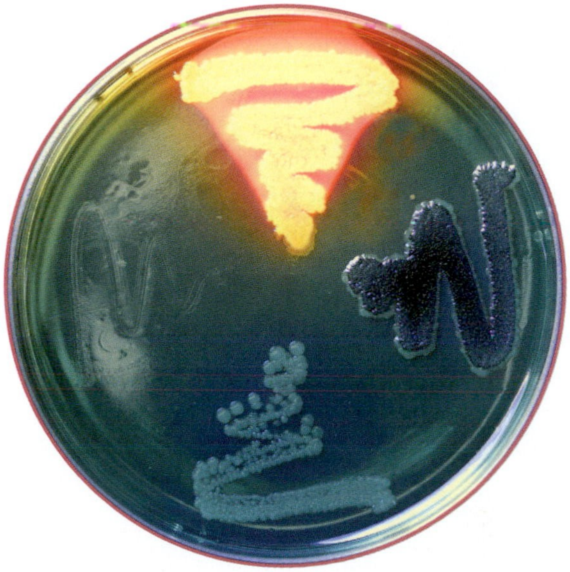

4-11 Hektoen Enteric Agar ◆ This HE Agar was inoculated with (clockwise from top), a Gram-negative lactose fermenter (indicated by the yellow growth), two lactose nonfermenters, and a Gram-positive organism. Note the blue-green growth of the lactose nonfermenters. Note also the black precipitate in the growth on the right, indicative of a reaction between ferric ammonium citrate and H_2S produced from sulfur reduction. The Gram-positive organism on the left was severely inhibited by the bile salts.

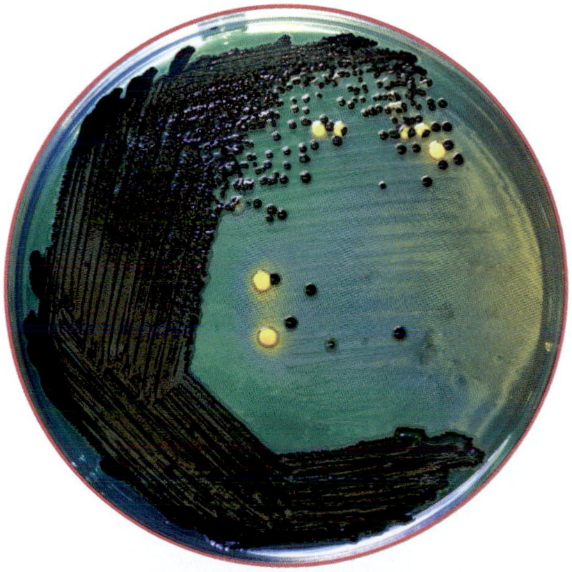

4-12 Hektoen Enteric Agar Streaked for Isolation ◆ This HE Agar was streaked with two of the organisms chosen for today's exercise. One is a lactose fermenter, evidenced by the yellow growth; the other is a sulfur reducer, as indicated by the black precipitate in the colonies. Can you identify these two organisms?

Medium Recipes

Hektoen Enteric Agar

- Yeast extract 3.0 g
- Peptic digest of animal tissue 12.0 g
- Lactose 12.0 g
- Sucrose 12.0 g
- Salicin 2.0 g
- Bile salts 9.0 g
- Sodium chloride 5.0 g
- Sodium thiosulfate 5.0 g
- Ferric ammonium citrate 1.5 g
- Bromthymol blue 0.064 g
- Acid fuchsin 0.1 g
- Agar 13.5 g
- Distilled or deionized water 1.0 L

    pH 7.4–7.8 at 25°C

Nutrient Agar

- Beef extract 3.0 g
- Peptone 5.0 g
- Agar 15.0 g
- Distilled or deionized water 1.0 L

    pH 6.6–7.0 at 25°C

Procedure

Lab One

1 Using a permanent marker, divide the bottom of each plate into four sectors.

2 Label the plates with the organisms' names, your name, and the date.

3 Mix each culture well.

4 Spot-inoculate the four sectors on the HE plate with the test organisms. Refer to Appendix B if necessary.

5 Repeat Step 4 with the Nutrient Agar plate.

6 Invert and incubate the plates aerobically at 35 ± 2°C for 48 hours.

Lab Two

1 Examine and compare the plates for color and quality of growth.

2 Record your results on the Data Sheet.

References

Forbes, Betty A., Daniel F. Sahm, and Alice S. Weissfeld. 2002. Chapter 10 in *Bailey & Scott's Diagnostic Microbiology,* 11th ed. Mosby-Yearbook, St. Louis.

Winn, Washington, C., et al. 2006. *Koneman's Color Atlas and Textbook of Diagnostic Microbiology,* 6th ed. Lippincott Williams & Wilkins, Baltimore.

Zimbro, Mary Jo, and David A. Power. 2003. Page 265 in *Difco™ & BBL™ Manual—Manual of Microbiological Culture Media.* Becton, Dickinson and Co., Sparks, MD.

TABLE 4-6 Hektoen Enteric Agar Results and Interpretations

TABLE OF RESULTS

Result	Interpretation	Presumptive ID
Poor growth or no growth (P)	Organism is inhibited by bile and/or one of the dyes included	Gram-positive
Good growth (G)	Organism is not inhibited by bile or any of the dyes included	Gram-negative
Pink to orange growth (Pi)	Organism produces acid from lactose fermentation	Not *Shigella* or *Salmonella*
Blue-green growth with black precipitate (Bppt)	Organism does not ferment lactose, but reduces sulfur to hydrogen sulfide (H_2S)	Possible *Salmonella*
Blue-green growth without black precipitate (B)	Organism does not ferment lactose or reduce sulfur. No reaction	Possible *Shigella* or *Salmonella*

NAME ______________________ DATE ______________________

LAB SECTION ______________________ I WAS PRESENT AND PERFORMED THIS EXERCISE (INITIALS) ________

Hektoen Enteric Agar

OBSERVATIONS AND INTERPRETATIONS

Refer to Table 4-6 when recording your results and interpretations in the chart below. Use abbreviations or symbols as needed.

Organism	Growth (P/G)		HE Growth Color (Pi/Bppt/B)	Interpretation
	HE	NA		

Questions

1 What purpose does the Nutrient Agar plate serve? In what way does it increase the validity of the test result?

2 Which ingredient(s) in this medium supply(ies)

a. Carbon?

b. Nitrogen?

3 Compare the recipes of Nutrient Agar and HE Agar. If an organism can grow on both media, on which would you expect it to grow better? Why?

4 HE Agar is a selective medium. Is it also a defined or an undefined medium? Why is that formulation desirable?

5 This medium was designed to differentiate *Salmonella* and *Shigella* from other enterics. *Salmonella* species sometimes produce a black precipitate in their growth; *Shigella* species do not. If you were designing a medium to differentiate these two genera, which ingredients from this medium would you include? Explain.

6 All enterics ferment glucose. What would be some consequences of replacing the sugars in this medium with glucose? What color combinations would you expect to see?

7 List all the things you know about a viable organism that produces green colonies with black centers on the medium in question #6.

SECTION

5

Differential Tests

In the last century, bacteria have been differentiated based on their enormous biochemical diversity. The diagnostic systems that microbiologists use for this purpose are called **differential tests** or **biochemical tests** because they differentiate and sometimes identify microorganisms based on specific biochemical characteristics. As you will see repeatedly in this section, microbial biochemical characteristics are brought about by enzymatic reactions that are part of specific and well-understood metabolic pathways. The tests used in this section were selected to illustrate those pathways and the means to identify them.

The first half of Section Five will introduce you to the tests individually so you can familiarize yourself with the metabolic pathways and the mechanisms by which specific tests expose them. The second half offers the opportunity for you to use some multiple test media designed for rapid identification.

The categories of differential tests in this section are:

- Energy metabolism, including fermentation of carbohydrates and aerobic and anaerobic respiration
- Utilization of a specified medium component
- Decarboxylation and deamination of amino acids
- Hydrolytic reactions requiring intracellular or extracellular enzymes
- Multiple reactions performed in a single combination medium
- Miscellaneous differential tests

A Word About Biochemical Tests and Acid-Base Reactions

It might interest you to know that although biochemical tests differentiate bacteria based on a variety of physiological characteristics many of these tests do so by making a simple qualitative evaluation of shifts in pH. Differential test media provide specific growth conditions (usually including a single key substrate such as glucose or lactose in the case of fermentation media) whereby an organism can carry out a specified metabolic reaction or series of reactions. But however complex the chemistry may be, the evaluation of the test is often a simple check to see if acid or base has been formed. An indicator dye is usually included in the test medium (phenol red, bromcresol purple, or bromthymol green) to change color when the pH changes. The reason for the different dyes is that each one operates best in a specific pH range and allows media to be tailored to the type of organism being tested. Oral bacteria, for example, prefer an acidic environment (around pH 4) whereas enteric bacteria prefer a more neutral environment (around pH 7). Obviously there are more metabolic reactions than can be discussed in this manual, but it is helpful to know a little bit about why certain metabolic products are formed and how they affect their environment.

pH change is an excellent indicator of what products (or kinds of products) are formed during metabolic activity. As such, it is an extremely useful tool in diagnostic microbiology. The principle is simple—if the pH changes (as evidenced by a color change) in one direction, the organism is said to have done one thing; if the pH (color) changes in the other direction, the organism is said to have done something else. If there is no change in color, it is determined that there was no reaction or *that the organism did not grow*.

Biochemical media allow us to observe a variety of microbial metabolic pathways but, as you will see as you progress through this section, the one recurring theme is fermentation. Microbial fermentation produces three principal products: gas, alcohol, and acid. Gas (frequently in the form of CO_2) can be an easy indicator of fermentation because it is easy to see. Typically, this is accomplished by the addition of a small inverted test tube, called a Durham tube, to the broth media. When gas-producing fermentation occurs, bubbles form inside the Durham tube. Figure A-5 on page 530 illustrates several reactions where carbon compounds are split and CO_2 is released into the medium. Figure 5-4 on page 247 illustrates various reactions in PR Glucose Broth. Alcohols produced by fermentation are liquid and colorless, so they are not seen with the naked eye, and because they do not ionize in water they have minimal effect if any on the pH of the medium. The weak acids (e.g., acetic acid, propionic acid, formic acid, succinic acid, and butyric acid) have a dramatic effect on the pH of any medium and can easily be detected by addition of one of the dyes mentioned above.

Acids, as defined by the Brønsted/Lowry theory of acids and bases are substances that donate hydrogen ions (H^+) while bases are substances that accept hydrogen ions. All of the above-mentioned acids are called **carboxylic acids** because they contain the carboxyl group–COOH. Because carboxylic acids can exist as either R–COOH or $R\text{–}COO^-$, they are said to be **conjugate pairs**. R–COOH is the conjugate acid and $R\text{–}COO^-$ is the conjugate base. A typical reaction would look like this: $R\text{–}COOH + H_2O \rightleftharpoons R\text{–}COO^- + H_2O + H^+$. As you can see, when the reaction goes to the right, hydrogen ions are released into the medium and this lowers the pH (Exercise 2-10).

Conversely, the removal of hydrogen ions or the addition of hydroxide ions (OH^-) to the water raises the pH. A typical way that an organism raises the pH in fermentation and other growth media is by breaking down proteins. Breaking down proteins raises the pH of media because the reaction releases ammonia, which further reacts with the water to form ammonium and hydroxide ions. The overall reaction looks like this: $NH_3 + H_2O \rightarrow NH_4 + OH^-$.

It is important to note that a differential medium prepared in the laboratory is not intended to duplicate an organism's natural environment. Differential media are artificial mixtures of organic and inorganic chemicals intended to exploit different microbial abilities, encouraging them to perform in ways that might never actually be expressed in nature. So even though an organism may never radically lower the pH of its natural environment (because of low substrate availability or competing protein catabolism) we can manipulate it to do so in the lab by cultivating it in medium spiked with a specific sugar or other substrate. And when two or more species, otherwise indistinguishable by way of staining, microscopy, or colony characteristics, are thus compared, the differences frequently become clear.

Introduction to Energy Metabolism Tests

Heterotrophic bacteria obtain their energy by means of either **respiration** or **fermentation**. Both catabolic systems convert the chemical energy of organic molecules to high-energy bonds in **adenosine triphosphate (ATP)**. Energy production in respiration is accomplished by means of three well-understood pathways, **glycolysis**, the **tricarboxylic acid cycle (Krebs cycle)**, and **oxidative phosphorylation** (sometimes called the **electron transport chain**, or **ETC**). Energy production in fermentation is accomplished by a variety of pathways, all referred to as **substrate level phosphorylation**.

Glycolysis is the splitting of the six-carbon glucose molecule into two three-carbon **pyruvate** molecules while reducing **coenzymes** and producing ATP. Coenzymes are the part of the enzyme-coenzyme-substrate complex that directly takes part in the transfer of electrons with the substrate. Because they can exist in either the oxidized or reduced state, coenzymes are responsible for both oxidations and reductions. As you will see, all of the major respiratory reactions directly involved in the conversion of glucose to carbon dioxide—glycolysis, the conversion of pyruvate to acetyl-CoA, and the Krebs cycle—oxidize organic substrates and reduce coenzymes. Refer to Appendix A to see some important reactions involving the coenzymes, NAD (nicotinamide adenine dinucleotide) and FAD (flavin adenine dinucleotide).

Following glycolysis is a reaction in which pyruvate is oxidized, producing the two-carbon molecule acetyl-CoA and one molecule of CO_2. The acetyl-CoA enters the Krebs cycle where, in a complex series of reactions, it too is converted to CO_2. These reactions release energy used to produce more ATP and in the process reduce many coenzymes.

If respiration were to end at this point, the cell would lose its ability to oxidize glucose because of a relative shortage of oxidized coenzymes. However, as mentioned above, cells perform another critical set of reactions called the electron transport chain. The ETC is a system whereby reduced coenzymes donate electrons to molecules that pass them down a chain of other molecules in a series of **oxidation-reduction** reactions to an inorganic molecule called the **final** (or **terminal) electron acceptor**. When the final electron acceptor is oxygen, the respiration is **aerobic**. When the final electron acceptor is an inorganic molecule other than oxygen (e.g., sulfate or nitrate) the respiration is **anaerobic**. Of all the respiratory reactions, the ETC releases the greatest amount of energy and produces the most ATP.

In contrast to respiration, fermentation is the metabolic process whereby the electrons stripped from glucose ultimately are accepted by one or more of its organic products. Reduced carbon compounds in the form of acids and organic solvents, as well as CO_2, are the typical end products of fermentation. This **substrate-level phosphorylation** releases much less energy and, consequently, produces less ATP. See "A Word About Reduction Potential" on page 261.

EXERCISE 5-1

Oxidation–Fermentation Test

Theory

The Oxidation–Fermentation (O–F) Test is designed to differentiate bacteria on the basis of fermentative or oxidative metabolism of carbohydrates. In this medium, oxidative organisms oxidize the carbohydrate to CO_2, H_2O, and energy by way of glycolysis, the oxidation of pyruvate, the Krebs cycle, and finally the electron transport chain with oxygen (the final electron acceptor) being reduced to H_2O. Similarly, fermentative organisms convert the carbohydrate to pyruvate, but because oxygen is not available as an electron acceptor, the pyruvate and other organic intermediates are reduced to organic acids, gas, or alcohol. Consequently, fermenters acidify O–F medium to a much greater extent than do oxidizers.

Hugh and Leifson's O–F medium includes a high sugar-to-peptone ratio to reduce the possibility that alkaline products from peptone utilization will neutralize weak acids produced by oxidation of the carbohydrate. Bromthymol blue dye, which is yellow at pH 6.0, green at pH 7.1, and blue at pH 7.6, is included as the indicator. A low agar concentration makes it a semisolid medium that allows determination of motility.

The medium is prepared with glucose, lactose, sucrose, maltose, mannitol, or xylose and is not slanted. Two tubes of the specific sugar medium are stab-inoculated several times with the test organism. After inoculation, one tube is sealed with a layer of sterile mineral oil to promote anaerobic growth and fermentation (Figure 5-1). The mineral oil creates an environment unsuitable for oxidation because it prevents the diffusion of oxygen from the air into the medium. The other tube is left unsealed to allow aerobic growth and oxidation. (Note: Tubes of O–F medium are heated in boiling water and then cooled prior to inoculation. This removes free oxygen from the medium and ensures an anaerobic environment in all tubes. The tubes covered with oil will remain anaerobic, whereas the uncovered medium will become aerobic as oxygen diffuses back in.)

Organisms that are able to ferment the carbohydrate or ferment *and* oxidize the carbohydrate will turn the sealed and unsealed media yellow throughout. Organisms that are able to oxidize only will turn the unsealed medium yellow (or partially yellow) and leave the sealed medium green or blue. Slow or weak fermenters will turn both tubes slightly yellow at the top. Organisms that are not able to metabolize the sugar will either produce no color change or turn the medium blue because of alkaline products from amino acid degradation. The results are summarized in Table 5-1 and shown in Figure 5-2.

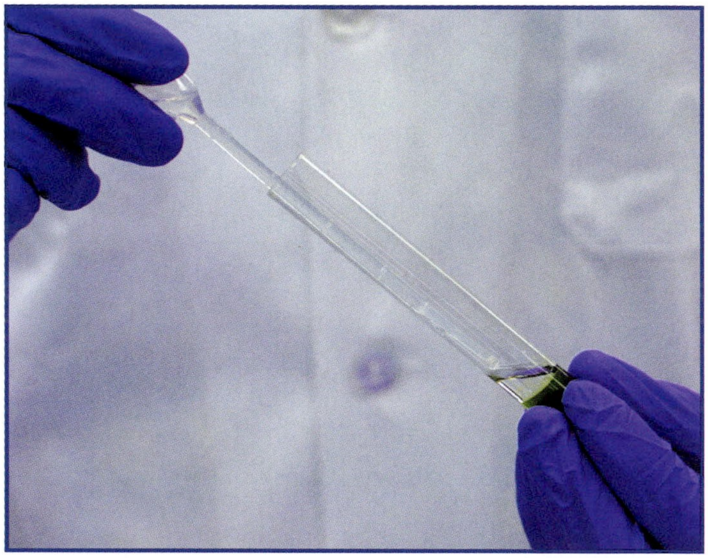

5-1 Adding the Mineral Oil Layer ◆ Tip the tube slightly to one side, and gently add 3–4 mm of sterile mineral oil. Be sure to use a sterile pipette for each tube.

TABLE 5-1 O–F Medium Results and Interpretations ◆

TABLE OF RESULTS

Sealed	Unsealed	Interpretation	Symbol
Green or blue	Any amount of yellow	Oxidation	O
Yellow throughout	Yellow throughout	Oxidation and fermentation or fermentation only	O–F or F[1]
Slightly yellow at the top	Slightly yellow at the top	Oxidation and slow fermentation or slow fermentation only	O–F or F[1]
Green or blue	Green or blue	No sugar metabolism; organism is nonsaccharolytic	N

[1] The results of a fermentative organism and one capable of both fermentative and oxidative metabolism (of the appropriate carbohydrate) look the same in this medium. Therefore, when both tubes are yellow, the organism is recorded as (O–F) or (F).

5-2 **Oxidation–Fermentation Test** ◆ These pairs of tubes illustrate three possible results in the O-F test. Reading from left to right, pair #1 was uninoculated for color comparison. Pair #2 was inoculated with an organism capable of both oxidative and fermentative utilization of glucose (O-F). Pair #3 was inoculated with an organism capable only of oxidative metabolism (O). Pair #4 was inoculated with an organism incapable of utilizing glucose. Note the blue color in the upper portion of the unsealed tube suggesting that the organism is both nonsaccharolytic (N) and strictly aerobic.

Application

The O–F Test is used to differentiate bacteria based on their ability to oxidize or ferment specific sugars. It allows presumptive separation of the fermentative *Enterobacteriaceae* from the oxidative *Pseudomonas* and *Bordetella*, and the nonreactive *Alcaligenes* and *Moraxella*.

In This Exercise

Eight O–F media tubes will be used for this exercise—six for inoculation (two for each organism) and two for controls. One of each pair will receive a mineral oil overlay to maintain an anaerobic environment and force fermentation if the organism is capable of doing so. You will not do a motility determination, as motility is covered in Exercise 5-24. For best results, use a heavy inoculum and several stabs for each tube. Use only sterile pipettes to add the mineral oil.

Materials

Per Student Group

- ✓ eight O–F glucose tubes
- ✓ sterile mineral oil
- ✓ sterile transfer pipettes
- ✓ fresh agar slants of:
  - *Escherichia coli*
  - *Kocuria rosea*
  - *Alcaligenes faecalis*

Medium Recipe

Hugh and Leifson's O–F Medium with Glucose

◆ Pancreatic digest of casein	2.0 g
◆ Sodium chloride	5.0 g
◆ Dipotassium phosphate	0.3 g
◆ Agar	2.5 g
◆ Bromthymol blue	0.08 g
◆ Glucose	10.0 g
◆ Distilled or deionized water	1.0 L

pH = 6.6–7.0 at 25°C

Procedure

Lab One

1. Obtain eight O–F tubes. Label six (in three pairs) with the names of the organisms, your name, and the date. Label the last pair "control."
2. Stab-inoculate each pair with the appropriate test organism. Stab several times to a depth of about 1 cm from the bottom of the agar. Do not inoculate the controls.
3. Overlay one of each pair of tubes (including the controls) with 3–4 mm of sterile mineral oil (see Figure 5-1).
4. Incubate all tubes at 35 ± 2°C for 48 hours.

Lab Two

1 Examine the tubes for color changes. Be sure to compare them to the controls before making a determination.

2 Record your results on the Data Sheet.

References

Collins, C. H., Patricia M. Lyne, and J. M. Grange. 1995. Page 112 in *Collins and Lyne's Microbiological Methods*, 7th ed. Butterworth-Heinemann, United Kingdom.

Delost, Maria Dannessa. 1997. Pages 218–219 in *Introduction to Diagnostic Microbiology.* Mosby, St. Louis.

Forbes, Betty A., Daniel F. Sahm, and Alice S. Weissfeld. 2002. Pages 154–155 in *Bailey & Scott's Diagnostic Microbiology,* 11th ed. Mosby, St. Louis.

MacFaddin, Jean F. 2000. Page 379 in *Biochemical Tests for Identification of Medical Bacteria*, 3rd ed. Lippincott Williams & Wilkins, Philadelphia.

Smibert, Robert M., and Noel R. Krieg. 1994. Page 625 in *Methods for General and Molecular Bacteriology*, edited by Philipp Gerhardt, R. G. E. Murray, Willis A. Wood, and Noel R. Krieg, American Society for Microbiology, Washington, DC.

Zimbro, Mary Jo, and David A. Power, eds. 2003. Page 410 in *Difco™ and BBL™ Manual—Manual of Microbiological Culture Media.* Becton Dickinson and Co., Sparks, MD.

Name ______________________________ Date ______________________________

Lab Section ______________________________ I was present and performed this exercise (initials) ________

Oxidation–Fermentation Medium

OBSERVATIONS AND INTERPRETATIONS

Refer to Table 5-1 when recording your results and interpretations in the chart below.

Organism	Color Results		Symbol	Interpretation
	Sealed	Unsealed		
Control				

Questions

1 What is the purpose of the control tubes used in this test? Why is it necessary to use two controls rather than one? What are the specific purposes of each? Be sure to list all functions that you can think of.

DATA SHEET 5-1

2 Some microbiologists recommend inoculating a pair of O–F basal media (without carbohydrate) along with the carbohydrate media. Why do you think this is done?

3 All enterics are facultative anaerobes; that is, they have both respiratory and fermentative enzymes. What color results would you expect for organisms in O–F glucose media inoculated with an enteric? Remember to describe both sealed and unsealed tubes.

4 Suppose that when you examined your tubes (in this exercise) after incubating them, you noticed that the unsealed control contained slight yellowing at the top. Suppose further that pair #1 showed complete yellowing of both tubes and pairs #2 and #3 showed slight yellowing of the unsealed tube. Assuming all other tubes were green, what conclusions could you safely make?

Which results, if any, are reliable? Why?

Which results, if any, are not reliable? Why not?

Fermentation Tests

As defined at the beginning of this section, carbohydrate fermentation is the metabolic process by which an organic molecule acts as an electron donor (becoming oxidized in the process) and one or more of its organic products act as the final (or terminal) electron acceptor. In actuality, the term "carbohydrate fermentation" is used rather broadly to include hydrolysis of disaccharides prior to the fermentation reaction. Thus, a "lactose fermenter" is an organism that splits the disaccharide lactose into the monosaccharides glucose and galactose and then ferments the monosaccharides. In this section you will see the term "fermenter" frequently. Unless it is expressly used otherwise, this term should be assumed to include the initial hydrolysis and/or conversion reactions.

Fermentation of glucose begins with the production of pyruvate. Although some organisms use alternative pathways, most bacteria accomplish this by glycolysis. The end products of pyruvate fermentation include a variety of organic acids, alcohols, and hydrogen or carbon dioxide gas. The specific end products depend on the specific organism and the substrate fermented. Refer to Figure 5-3 and Appendix A.

In this unit you will perform tests using two differential fermentation media—Phenol Red (PR) Broth and MR-VP Medium. Phenol Red Broth is a general-purpose fermentation medium. Typically, it contains any one of several carbohydrates (e.g., glucose, lactose, sucrose) and a pH indicator to detect acid formation. MR-VP broth is a dual-purpose medium that tests an organism's ability to follow either (or both) of two specific fermentation pathways. The methyl red (MR) test detects what is called a **mixed acid fermentation**. The Voges-Proskauer (VP) test identifies bacteria that are able to produce acetoin as part of a **2,3-butanediol fermentation**. See Appendix A for more information about fermentation.

EXERCISE 5-2

Phenol Red Broth

Theory

Phenol Red (PR) Broth is a differential test medium prepared as a base to which a carbohydrate is added. Included in the base medium are peptone and the pH indicator phenol red. Phenol red is yellow below pH 6.8, pink to magenta above pH 7.4, and red in between. During preparation the pH is adjusted to approximately 7.3 so it appears red. Finally, an inverted Durham tube is added to each tube to trap a portion of any gas produced.

Acid production from fermentation of the carbohydrate lowers the pH below the neutral range of the indicator and turns the medium yellow (Figure 5-4). Deamination of peptone amino acids produces ammonia (NH_3), which raises the pH and turns the broth pink. Gas production is indicated by a bubble or pocket in the Durham tube where the broth has been displaced.

Application

PR broth is used to differentiate members of *Enterobacteriaceae* and to distinguish them from other Gram-negative rods.

In This Exercise

Today you will inoculate PR Glucose Broths with three organisms to determine their fermentation characteristics. Use Table 5-2 as a guide when interpreting and recording your results. Be sure to use an uninoculated control for color comparison.

Materials

Per Student Group

✓ four PR Glucose Broths with Durham tubes

✓ fresh cultures of:

- *Escherichia coli*
- *Alcaligenes faecalis*
- *Providencia stuartii*

Medium Recipe

PR Glucose Broth

Pancreatic digest of casein	10.0 g
Sodium chloride	5.0 g
Carbohydrate (glucose, lactose, sucrose)	5.0 g
Phenol red	0.018 g
Distilled or deionized water	1.0 L

pH = 7.1–7.5 at 25°C

Procedure

Lab One

1 Obtain four PR Glucose Broths. Label each with the name of the organism, your name, the medium name, and the date. Label the fourth tube "control."

2 Inoculate each broth with a test organism. Do not inoculate the control.

3 Incubate all of the tubes at 35 ± 2°C for 48 hours.

Lab Two

Using the uninoculated control for color comparison and Table 5-2 as a guide, examine all of the tubes and enter your results on the Data Sheet. When indicating the various reactions, use the standard symbols as shown, with the acid reading first, followed by a slash and then the gas reading. K indicates alkalinity and –/– symbolizes no reaction. Note: Do not try to read these results without comparing them to the control.

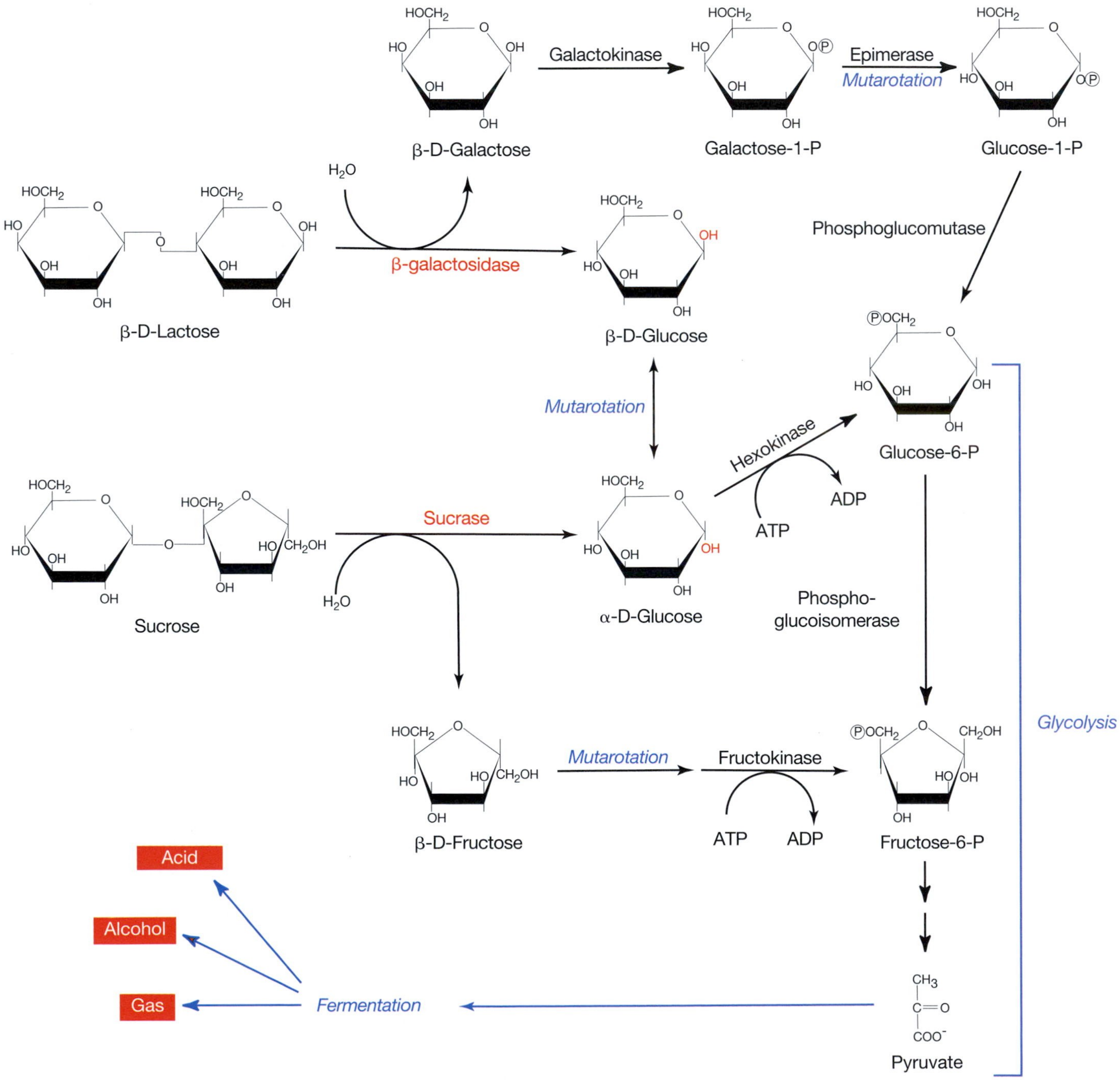

5-3 Fermentation of the Disaccharides Lactose and Sucrose ◆ Notice that fermentation of the disaccharides lactose and sucrose relies on enzymes to hydrolyze them into two monosaccharides. Once the monosaccharides are formed, they are fermented via glycolysis, just as glucose is.

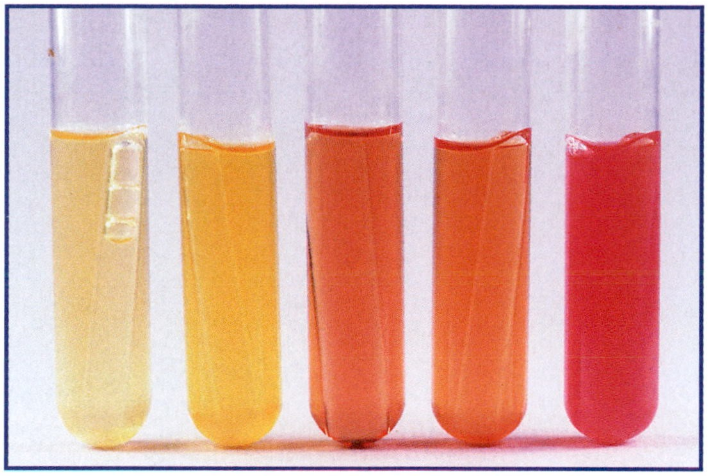

5-4 PR Glucose Broth Results ◆ From left to right, these five PR Glucose Broths demonstrate acid and gas production (A/G), acid production without gas (A/–), uninoculated control tube for comparison, no reaction (–/–), and alkaline condition as a result of peptone degradation (K).

References

Lányi, B. 1987. Page 44 in *Methods in Microbiology*, Vol. 19, edited by R. R. Colwell and R. Grigorova. Academic Press, New York.

MacFaddin, Jean F. 2000. Page 57 in *Biochemical Tests for Identification of Medical Bacteria*, 3rd ed. Lippincott Williams & Wilkins, Philadelphia.

Zimbro, Mary Jo, and David A. Power, eds. 2003. Page 440 in *Difco™ and BBL™ Manual—Manual of Microbiological Culture Media*. Becton Dickinson and Co., Sparks, MD.

TABLE 5-2 PR Broth Results and Interpretations ◆

TABLE OF RESULTS		
Result	**Interpretation**	**Symbol**
Yellow broth, bubble in tube	Fermentation with acid and gas end products	A/G
Yellow broth, no bubble in tube	Fermentation with acid end products; no gas produced	A/–
Red broth, no bubble in tube	No fermentation	–/–
Pink broth, no bubble in tube	Degradation of peptone; alkaline end products	K

Name ______________________ Date ______________________

Lab Section ______________________ I was present and performed this exercise (initials) ________

Phenol Red Broth

OBSERVATIONS AND INTERPRETATIONS

Enter your results in the chart below, using the symbols shown in Table 5-2.

Organism	Results	Interpretation
Control		

Questions

1 When conducting this test, some microbiologists inoculate two sets of broth, one with carbohydrate and one without carbohydrate (PR Base Broth). What function would PR Base Broth serve? *Hint:* It is another type of control.

2 Assuming, as in Question 1, you tested an organism using PR Glucose and PR Base broths, which of the combinations of results in the following table would be reliable. Interpret each of the combinations and explain how the results might be obtained.

PR Glucose	PR Base	Interpretation	Reliable Y/N
A/G	A/G		
A/G	–/–		
A/G	K		
A/G	A/–		
K	K		
K	–/–		
A/–	K		
A/–	A/G		

3 Suppose you inoculate a PR broth with a slow-growing fermenter. After 48 hours, you see slight turbidity but score it as (–/–). Is this result a false positive or a false negative? Is this a result of poor specificity or poor sensitivity of the test system?

4 Early formulations of this medium used a smaller amount of carbohydrate and occasionally produced false alkaline (pink) results after 48 hours. Why do you think this happened? List at least two steps, as a microbiologist, that you could take to prevent the problem.

EXERCISE 5-3

Methyl Red and Voges-Proskauer Tests

Theory

Methyl Red and Voges-Proskauer (MR-VP) Broth is a combination medium used for both Methyl Red (MR) *and* Voges-Proskauer (VP) tests. It is a simple solution containing only peptone, glucose, and a phosphate buffer. The peptone and glucose provide protein and fermentable carbohydrate, respectively, and the potassium phosphate resists pH changes in the medium.

The MR test is designed to detect organisms capable of performing a **mixed acid fermentation**, which overcomes the phosphate buffer in the medium and lowers the pH (Figures 5-5 and 5-6). The acids produced by these organisms tend to be stable, whereas acids produced by other organisms tend to be unstable and subsequently are converted to more neutral products.

Mixed acid fermentation is verified by the addition of methyl red indicator dye following incubation. Methyl red is red at pH 4.4 and yellow at pH 6.2. Between these two pH values, it is various shades of orange. Red color is the only true indication of a positive result. Orange is negative or inconclusive. Yellow is negative (Figure 5-7).

The Voges-Proskauer test was designed for organisms that are able to ferment glucose, but quickly convert their acid products to acetoin and 2,3-butanediol (Figures 5-6 and 5-8). Adding VP reagents to the medium oxidizes the **acetoin** to **diacetyl**, which in turn reacts with **guanidine nuclei** from peptone to produce a red color (Figure 5-9). A positive VP result, therefore, is red. No color change (or development of copper color) after the addition of reagents is negative. The copper color is a result of interactions between the reagents and should not be confused with the true red color of a positive result (Figure 5-10). Use of positive and negative controls for comparison is usually recommended.

After incubation, two 1 mL volumes are transferred from the MR-VP broth to separate test tubes. Methyl red indicator reagent is added to one tube, and VP reagents are added to the other. Any resulting color changes are observed and documented. For help with this procedure, refer to the diagram in Figure 5-11.

Application

The Methyl Red and Voges-Proskauer tests are components of the *IMViC* battery of tests (*I*ndole, *M*ethyl red, *V*oges-Proskauer, and *C*itrate) used to distinguish between members of the family *Enterobacteriaceae* and differentiate them from other Gram-negative rods. For more information, refer to Appendix A.

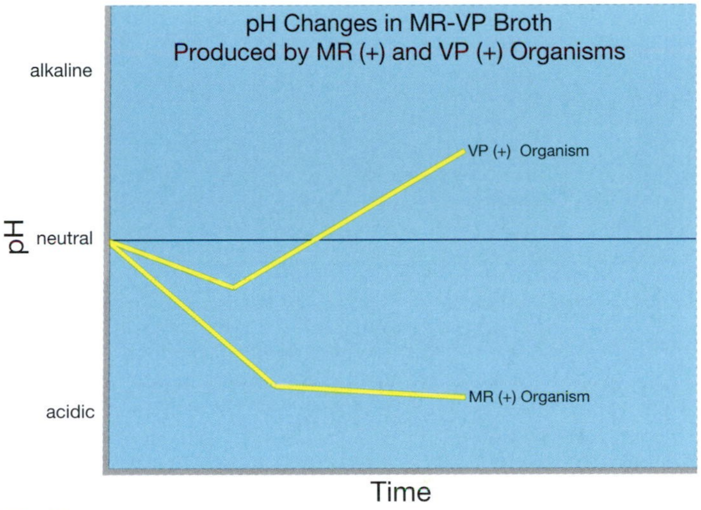

5-6 pH Changes in MR-VP Broth ◆ The MR test identifies organisms that perform a mixed acid fermentation and produce stable acid end products. MR (+) organisms lower the broth's pH permanently. The VP test is used to identify organisms that perform a 2,3-butanediol fermentation. VP (+) organisms initially may produce acid and temporarily lower the pH, but because the 2,3-butanediol fermentation end products are neutral, the pH at completion of the test is near neutral.

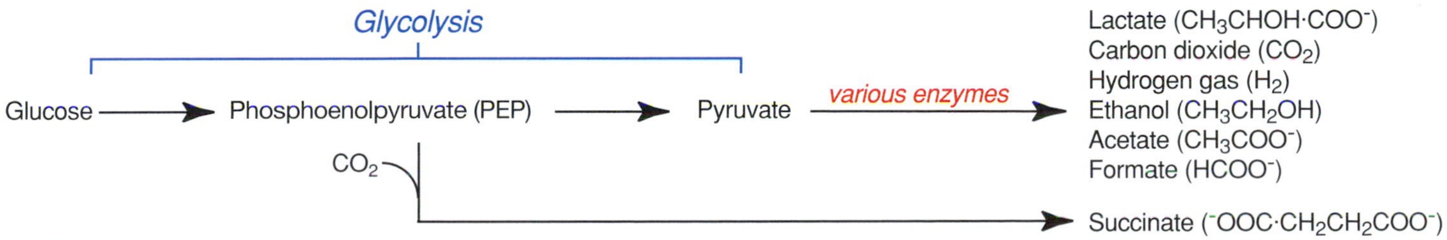

5-5 Mixed Acid Fermentation of *E. coli* ◆ *E. coli* is a representative Methyl Red positive organism and is recommended as a positive control for the test. Its mixed acid fermentation produces the end products listed in order of abundance. Most of the formate is converted to H_2 and CO_2 gases. **Note:** The amount of succinate falls between acetate and formate but is derived from PEP, not pyruvate. *Salmonella* and *Shigella* also are Methyl Red-positive.

❖ In This Exercise

Today you will perform Methyl Red and Voges-Proskauer tests on *Escherichia coli* and *Enterobacter aerogenes*. After a 48-hour incubation period, you will observe the results of a mixed-acid fermentation and a 2,3-butanediol fermentation. The exercise involves transfer of a portion of broth containing living organisms and the measured addition of reagents to complete the observable reactions. Please use care in your transfers and in your measurements. For help with the procedure, refer to Figure 5-11. For help interpreting your results, refer to Figures 5-7 and 5-10 and Tables 5-3 and 5-4.

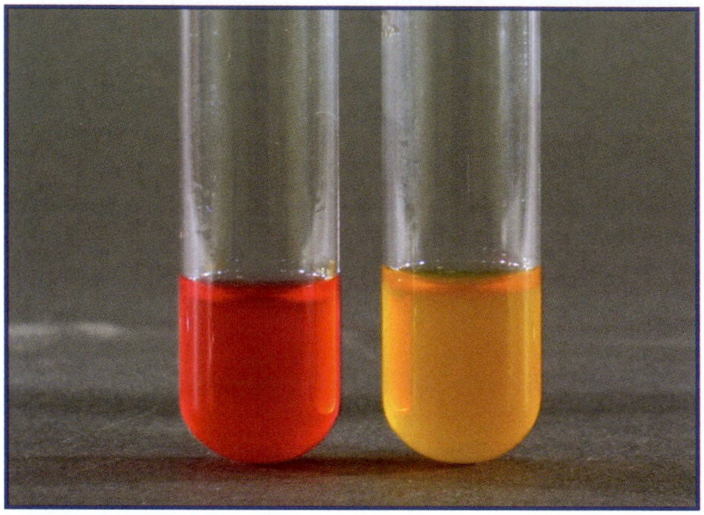

5-7 Methyl Red Test ◆ The tube on the left is MR-positive. The tube on the right is MR-negative.

❖ Materials

Per Student Group

- ✓ three MR-VP broths
- ✓ Methyl Red reagent
- ✓ VP reagents A and B
- ✓ six nonsterile test tubes
- ✓ three nonsterile 1 mL pipettes
- ✓ fresh cultures of:
  - ◆ *Escherichia coli*
  - ◆ *Enterobacter aerogenes*

❖ Medium and Reagent Recipes

MR-VP Broth

◆ Buffered peptone	7.0 g
◆ Dipotassium phosphate	5.0 g
◆ Dextrose (glucose)	5.0 g
◆ Distilled or deionized water	1.0 L

pH 6.7–7.1 at 25°C

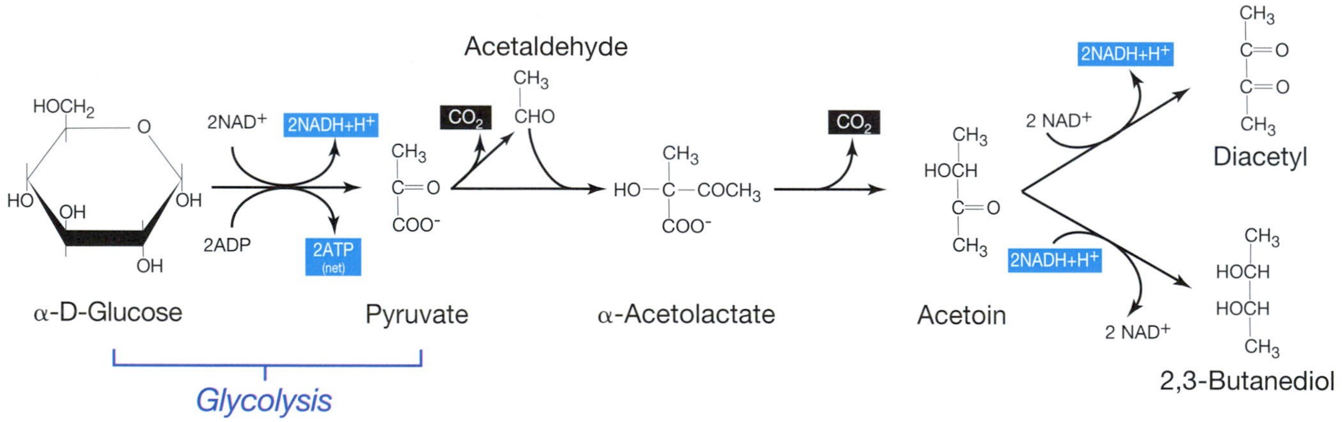

5-8 2,3-Butanediol Fermentation ◆ Acetoin is an intermediate in this fermentation. Reduction of acetoin by NADH leads to the end product 2,3-butanediol. Oxidation of acetoin produces diacetyl, as in the indicator reaction for the VP test (see Figure 5-9).

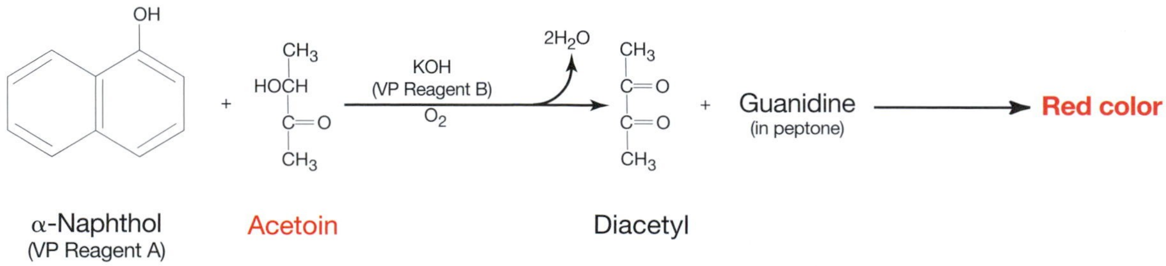

5-9 Indicator Reaction of Voges-Proskauer Test ◆ Reagents A and B are added to VP broth after 48 hours' incubation. These reagents react with acetoin and oxidize it to diacetyl, which in turn reacts with guanidine (from the peptone in the medium) to produce a red color.

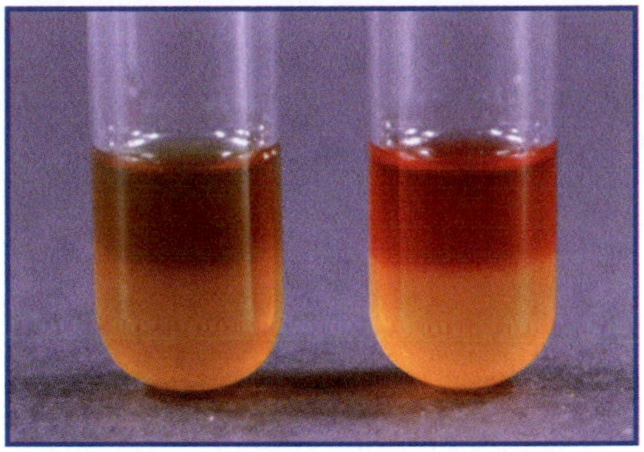

5-10 The Voges-Proskauer Test ◆ The tube on the left is VP-negative. The tube on the right is VP-positive. The copper color at the top of the VP-negative tube is the result of the reaction of KOH and α-naphthol and should not be confused with a positive result.

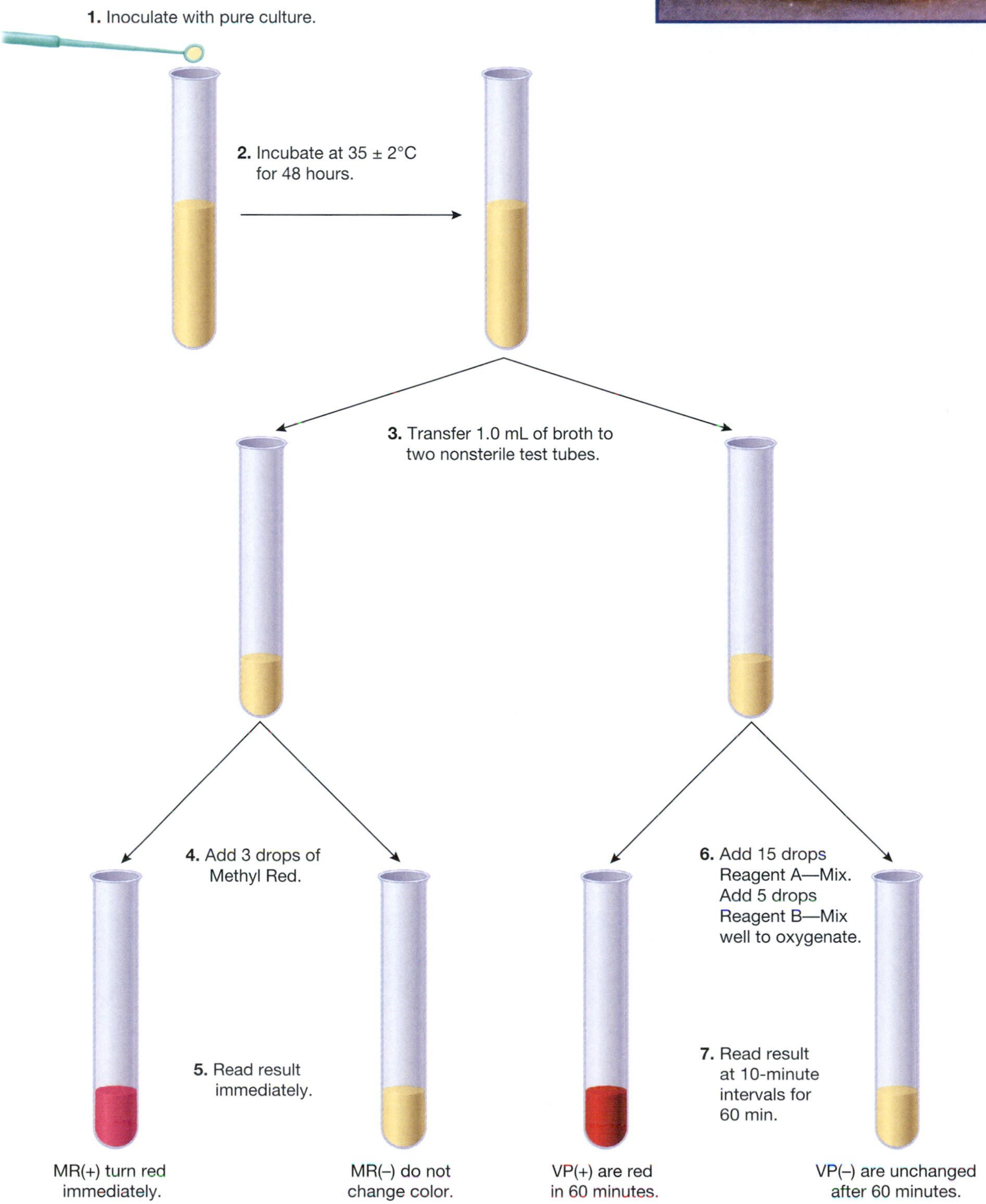

1. Inoculate with pure culture.
2. Incubate at 35 ± 2°C for 48 hours.
3. Transfer 1.0 mL of broth to two nonsterile test tubes.
4. Add 3 drops of Methyl Red.
5. Read result immediately.
6. Add 15 drops Reagent A—Mix. Add 5 drops Reagent B—Mix well to oxygenate.
7. Read result at 10-minute intervals for 60 min.
MR(+) turn red immediately.
MR(–) do not change color.
VP(+) are red in 60 minutes.
VP(–) are unchanged after 60 minutes.

5-11 Procedural Diagram for the Methyl Red and Voges-Proskauer Tests ◆

Methyl Red Reagent

- Methyl Red dye — 0.1 g
- Ethanol — 300.0 mL
- Distilled water to bring volume to — 500.0 mL

VP Reagent A

- α-naphthol — 5.0 g
- Absolute alcohol to bring volume to — 100.0 mL

VP Reagent B

- Potassium hydroxide — 40.0 g
- Distilled water to bring volume to — 100.0 mL

Procedure

Lab One

1 Obtain three MR-VP broths. Label two of them with the names of the organisms, your name, and the date. Label the third tube "control."

2 Inoculate two broths with the test cultures. Do not inoculate the control.

3 Incubate all tubes at 35 ± 2°C for 48 hours.

Lab Two

1 Mix each broth well. Transfer 1.0 mL from each broth to each of two nonsterile test tubes. The pipettes should be sterile, but the tubes need not be. Refer to the procedural diagram in Figure 5-11.

2 For the three pairs of test tubes, add the reagents as follows.

Tube #1: Methyl Red Test

a. Add three drops of Methyl Red reagent. Observe for red color formation immediately.

b. Using Table 5-3 as a guide, record your results on the Data Sheet.

Tube #2: VP Test

a. Add 15 drops (0.6 mL) of VP reagent A. Mix well to oxygenate the medium.

b. Add 5 drops (0.2 mL) of VP reagent B. Mix well again to oxygenate the medium.

c. Place the tubes in a test tube rack and observe for red color formation at regular intervals starting at 10 minutes for up to 1 hour. Mix again periodically to oxygenate the medium.

d. Using Table 5-4 as a guide, record your results on the Data Sheet. Sometimes it is difficult to differentiate weak positive reactions from negative reactions. VP (−) reactions often produce a copper color. Weak VP (+) reactions produce a pink color. The strongest color change should be at the surface.

TABLE 5-3 Methyl Red Test Results and Interpretations

TABLE OF RESULTS		
Result	**Interpretation**	**Symbol**
Red	Mixed acid fermentation	+
No color change	No mixed acid fermentation	−

TABLE 5-4 Voges-Proskauer Test Results and Interpretations

TABLE OF RESULTS		
Result	**Interpretation**	**Symbol**
Red	2,3-butanediol fermentation (acetoin produced)	+
No color change	No 2,3-butanediol fermentation (acetoin is not produced)	−

References

Chapin, Kimberle C., and Tsai-Ling Lauderdale. 2003. Page 361 in *Manual of Clinical Microbiology,* 8th ed. Edited by Patrick R. Murray, Ellen Jo Baron, James H. Jorgensen, Michael A. Pfaller, and Robert H. Yolken. ASM Press, American Society for Microbiology, Washington, DC.

Delost, Maria Dannessa. 1997. Pages 187–188 in *Introduction to Diagnostic Microbiology.* Mosby, St. Louis.

Forbes, Betty A., Daniel F. Sahm, and Alice S. Weissfeld. 2002. Page 275 in *Bailey & Scott's Diagnostic Microbiology,* 11th ed. Mosby, St. Louis.

MacFaddin, Jean F. 2000. Pages 321 and 439 in *Biochemical Tests for Identification of Medical Bacteria,* 3rd ed. Williams & Wilkins, Baltimore.

Zimbro, Mary Jo, and David A. Power, eds. 2003. Page 330 in *Difco™ and BBL™ Manual—Manual of Microbiological Culture Media.* Becton Dickinson and Co., Sparks, MD.

Name ______________________ Date ______________________

Lab Section ______________________ I was present and performed this exercise (initials) ________

Methyl Red and Voges-Proskauer Tests

OBSERVATIONS AND INTERPRETATIONS

1 Refer to Tables 5-3 and 5-4 as you record your results and interpretations in the chart below.

Organism	MR Result	VP Result	Interpretation
Control			

Questions

1 Some protocols call for a shorter incubation time for the MR and VP tests. Other protocols allow for up to 10 days incubation with virtually no risk of producing a false positive.

a. Which of the two tests would likely produce more false negatives with a shorter incubation time. Why?

b. Which test would likely benefit most from a longer incubation time? Why?

2 Would a false negative result for the VP test more likely be attributable to poor sensitivity or to poor specificity of the test system?

3 Why were you told to shake the VP tubes after the reagents were added? *Hint:* See Figure 5-9.

4 Why is the Methyl Red Test read immediately and the Voges-Proskauer read after 60 minutes? *Hint:* Consider the functions of the MR and VP reagents.

5 Some microbiologists recommend reincubating organisms producing Methyl Red-negative results for an additional 2 to 3 days. Why do you think this is done?

Tests Identifying Microbial Ability to Respire

In this unit we will examine several techniques designed to differentiate bacteria based on their ability to respire. As mentioned earlier in this section, respiration is the conversion of glucose to energy in the form of ATP by way of glycolysis, the Krebs cycle, and oxidative phosphorylation in the electron transport chain (ETC). In respiration, the final electron acceptor at the end of the ETC is always an inorganic molecule.

Tests that identify an organism as an aerobic or an anaerobic respirer generally are designed to detect specific products of (or constituent enzymes used in) the reduction of the final (or terminal) electron acceptor. Aerobic respirers reduce oxygen to water or other compounds. Anaerobic respirers reduce other inorganic molecules such as nitrate or sulfate. Nitrate is reduced to nitrogen gas (N_2) or other nitrogenous compounds such as nitrite (NO_2), and sulfate is reduced to hydrogen sulfide gas (H_2S). Metabolic pathways are described in detail in Appendix A.

The exercises and organisms chosen for this unit directly or indirectly demonstrate the presence of an ETC. The catalase test detects organismal ability to produce catalase—an enzyme that detoxifies the cell by converting hydrogen peroxide produced in the ETC to water and molecular oxygen. The oxidase test identifies the presence of cytochrome c oxidase in the ETC. The nitrate reduction test examines bacterial ability to transfer electrons to nitrate at the end of the ETC and thus respire anaerobically. For more information on biochemical tests that demonstrate anaerobic respiration, refer to Exercise 5-18, SIM Medium, and Exercise 5-19, Triple Sugar Iron Agar/Kligler Iron Agar.

EXERCISE 5-4

Catalase Test

❖ Theory

The electron transport chains of aerobic and facultatively anaerobic bacteria are composed of molecules capable of accepting and donating electrons as conditions dictate. As such, these molecules alternate between the oxidized and reduced forms, passing electrons down the chain to the final electron acceptor. Energy lost by electrons in this sequential transfer is used to perform oxidative phosphorylation (i.e., produce ATP from ADP).

One carrier molecule in the ETC called **flavoprotein** can bypass the next carrier in the chain and transfer electrons directly to oxygen (Figure 5-12). This alternative pathway produces two highly potent cellular toxins—hydrogen peroxide (H_2O_2) and superoxide radical (O_2^-).

Organisms that produce these toxins also produce enzymes capable of breaking them down. **Superoxide dismutase** catalyzes conversion of superoxide radicals (the more lethal of the two compounds) to hydrogen peroxide. **Catalase** converts hydrogen peroxide into water and gaseous oxygen (Figure 5-13).

Bacteria that produce catalase can be detected easily using typical store-grade hydrogen peroxide. When hydrogen peroxide is added to a catalase-positive culture, oxygen gas bubbles form immediately (Figures 5-14 and 5-15). If no bubbles appear, the organism is catalase-negative. This test can be performed on a microscope slide or by adding hydrogen peroxide directly to the medium containing bacterial growth.

$$\underset{\text{Reduced Flavoprotein}}{FPH_2} + O_2 \longrightarrow \underset{\text{Oxidized Flavoprotein}}{FP} + \underset{\text{Hydrogen Peroxide}}{H_2O_2}$$

$$2H^+ + \underset{\text{Superoxide Radical}}{2\,O_2^-} \xrightarrow{\text{Superoxide dismutase}} \underset{\text{Hydrogen Peroxide}}{H_2O_2} + O_2$$

5-12 Microbial Production of H_2O_2 ◆ Hydrogen peroxide may be formed through the transfer of electrons from reduced flavoprotein to oxygen or from the action of superoxide dismutase.

$$\underset{\text{Hydrogen Peroxide}}{2H_2O_2} \xrightarrow{\text{Catalase}} 2H_2O + O_2\,(g)$$

5-13 Catalase Mediated Conversion of H_2O_2 ◆ Catalase is an enzyme of aerobes, microaerophiles, and facultative anaerobes that converts hydrogen peroxide to water and oxygen gas.

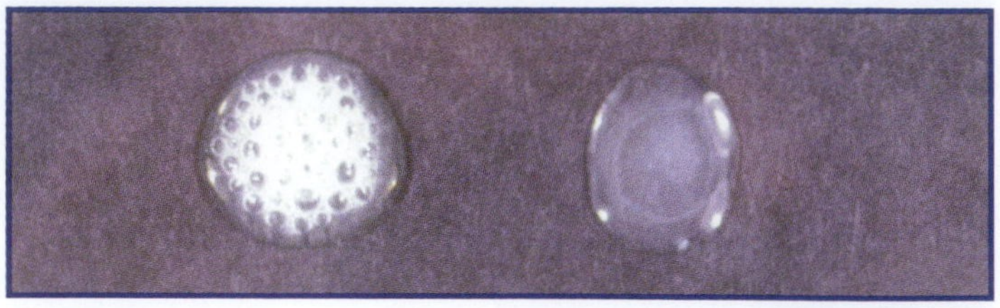

5-14 **Catalase Slide Test** ◆ Visible bubble production indicates a positive result in the catalase slide test. A catalase-positive organism is on the left. A catalase-negative organism is on the right.

5-15 **Catalase Tube Test** ◆ The catalase test also may be performed on an agar slant. A catalase-positive organism is on the left. A catalase-negative organism is on the right.

Application

This test is used to identify organisms that produce the enzyme catalase. It is used most commonly to differentiate members of the catalase-positive *Micrococcaceae* from the catalase-negative *Streptococcaceae*. Variations on this test also may be used in identification of *Mycobacterium* species.

In This Exercise

You will perform the Catalase Test. The materials provided will enable you to conduct both the slide test and the slant test. The slide test has to be done first because the tube test requires the addition of hydrogen peroxide directly to the slant, after which the culture will be unusable.

Materials

Per Student Group

✓ hydrogen peroxide (3% solution)
✓ transfer pipettes
✓ microscope slides
✓ one Nutrient Agar slant
✓ fresh Nutrient Agar cultures of:
- *Staphylococcus epidermidis*
- *Lactococcus lactis*

Procedure

Slide Test

1. Transfer a large amount of growth to a microscope slide. (Be sure to perform this test in the proper order. Placing the metal loop into H_2O_2 could catalyze a false-positive reaction.)
2. Aseptically place one or two drops of hydrogen peroxide directly onto the bacteria and immediately observe for the formation of bubbles. (When running this test on an unknown, a positive control should be run simultaneously to verify the quality of the peroxide.)
3. If you get a negative result, observe the slide, using the scanning objective on your microscope.
4. Record your results on the Data Sheet.

Slant Test

1. Add approximately 1 mL of hydrogen peroxide to each of the three slants (one uninoculated control and two cultures). Do this *one slant at a time*, observing and recording the results as you go.
2. Record your results on the Data Sheet.

References

Collins, C. H., Patricia M. Lyne, and J. M. Grange. 1995. Page 110 in *Collins and Lyne's Microbiological Methods*, 7th ed. Butterworth-Heinemann, Oxford, United Kingdom.

Forbes, Betty A., Daniel F. Sahm, and Alice S. Weissfeld. 2002. Pages 151 and 166 in *Bailey & Scott's Diagnostic Microbiology*, 11th ed. Mosby, St. Louis.

Lányi, B. 1987. Page 20 in *Methods in Microbiology*, Vol. 19, edited by R. R. Colwell and R. Grigorova. Academic Press, New York.

MacFaddin, Jean F. 2000. Page 78 in *Biochemical Tests for Identification of Medical Bacteria*, 3rd ed. Williams & Wilkins, Baltimore.

Smibert, Robert M., and Noel R. Krieg. 1994. Page 614 in *Methods for General and Molecular Bacteriology*, edited by Philipp Gerhardt, R. G. E. Murray, Willis A. Wood, and Noel R. Krieg. American Society for Microbiology, Washington, DC.

TABLE 5-5 **Catalase Test Results and Interpretations** ◆

TABLE OF RESULTS

Result	Interpretation	Symbol
Bubbles	Catalase is present	+
No bubbles	Catalase is absent	–

NAME ______________________ DATE ______________________

LAB SECTION ______________________ I WAS PRESENT AND PERFORMED THIS EXERCISE (INITIALS) ________

Catalase Test

OBSERVATIONS AND INTERPRETATIONS

Using Table 5-5 as a guide, record your results and interpretations in the chart below.

Organism	Bubbles? Y / N	+ / −	Interpretation
Control			

Questions

1 When flavoprotein transfers electrons directly to the final electron acceptor, hydrogen peroxide is produced. What other consequences might result from electron carriers in the ETC being bypassed? (*Hint*: See Section 5 Introduction and Appendix A.)

2 Would a false positive from the reaction between the inoculating loop and hydrogen peroxide be caused by poor specificity or poor sensitivity of the test system? Explain.

3 Why is it advisable to perform this test on a known catalase-positive organism along with the organism you are testing?

4 What is the purpose of adding hydrogen peroxide to the uninoculated tube?

A Word About Reduction Potential

All physical and chemical processes obey the Second Law of Thermodynamics—the law of entropy. Without constant energy input, order evolves to disorder, and high energy dissipates and becomes low energy. In living systems, losses and gains of energy are often cyclic— some of the energy lost in one reaction is regenerated in another.

The energy contained in a cellular molecule, called its energy of formation, comes from the energy released during reactions involved in its assembly from individual components. And again, obeying the law of entropy, *some* of this energy is retained by the molecule and can be released in subsequent reactions. The energy released in metabolic reactions is called **free energy**. Among other things, free energy is used to produce high-energy molecules such as ATP.

As you will remember from chemistry class, a major type of reaction in all metabolic processes is the **oxidation-reduction** reaction, also known as **"redox."** Remember that oxidation is defined as the loss of electrons and reduction is the gain of electrons. A molecule that gives up electrons is oxidized, but is called a reductant because it is reducing another molecule. Similarly, a molecule that accepts electrons is an oxidant.

Molecules either accept or donate electrons based on their relative **reduction potentials**. Reduction potential, simply defined, is the tendency of a molecule to donate electrons. Without the input of energy, electron transfers will always go from a molecule with high reduction potential to a molecule with low reduction potential. The "Electron Tower" in Figure 5-16 graphically illustrates the relationship of many important cellular oxidants and reductants.

Redox Pair	Reduction Potential (E_o')	
CO_2/glucose	−0.43V	($24e^-$)
$2H^+/H_2$	−0.42V	($2e^-$)
$S_2O_3^{2-}/HS^-$	−0.40V	($2e^-$)
NAD^+/NADH	−0.32V	($2e^-$)
S^0/H_2S	−0.28V	($2e^-$)
SO_4^{2-}/H_2S	−0.22V	($8e^-$)
Pyruvate/lactate	−0.19V	($2e^-$)
Fumarate/succinate	+0.03V	($2e^-$)
Cytochrome $b_{ox/red}$	+0.035V	($1e^-$)
Fe^{3+}/Fe^{2+} (pH 7)	+0.20V	($1e^-$)
$Ubiquinone_{ox/red}$	+0.11V	($2e^-$)
Cytochrome $c_{ox/red}$	+0.25V	($1e^-$)
Cytochrome $a_{ox/red}$	+0.39V	($1e^-$)
NO_3^-/NO_2^-	+0.42V	($2e^-$)
NO_3^-/ ½ N_2	+0.74V	($5e^-$)
Fe^{3+}/Fe^{2+} (pH 2)	+0.76V	($1e^-$)
½ O_2/H_2O	+0.82V	($2e^-$)

5-16 Electron Tower ◆ Redox pairs are listed from top to bottom in order of decreasing reduction potential. Positively charged molecules are more likely to accept electrons than negatively charged molecules. Therefore, the more positive the charge is, the lower the reduction potential is. In the redox pairs, the oxidant (oxidized form of the molecule) is on the left; the reductant (reduced form of the molecule) is on the right. Electrons are transferred downward from the reductant of one pair to the oxidant of a pair containing lower reduction potential. The actual reduction potentials of each pair is in red. Also included are the numbers of electrons (in parentheses) transferred in any one reaction. (Not to scale.)

EXERCISE 5-5

Oxidase Test

❖ Theory

In the introduction to this section, we discussed the major metabolic reactions and talked a little bit about coenzymes. Understanding the Oxidase Test requires familiarization with these fundamental energy transformations and the role of coenzymes. See also "A Word About Reduction Potential" on page 261.

Consider the life of a glucose molecule entering an aerobically respiring cell (see Appendix A, page 526). It is first split (oxidized) in glycolysis where it is converted to two molecules of pyruvate and reduces two NAD (coenzyme) molecules to NADH ($+H^+$). Then each of the pyruvate molecules becomes oxidized and converted to a two-carbon molecule called acetyl–CoA and one molecule of CO_2, which reduces another NAD to NADH. Then the Krebs cycle finishes the oxidation by producing two more molecules of CO_2 (per acetyl–CoA) and reduces three more NADs and one FAD to $FADH_2$.

As you can see, by this time the cell is becoming quite full of reduced coenzymes. Therefore, in order to continue oxidizing glucose, these coenzymes must be converted back to the oxidized state. This is the job of the electron transport chain (Figure 5-17).

Many aerobes, microaerophiles, facultative anaerobes (and even some anaerobes) have ETCs. The functions of the ETC are to, 1) transport electrons down a chain of molecules with increasingly positive reduction potentials

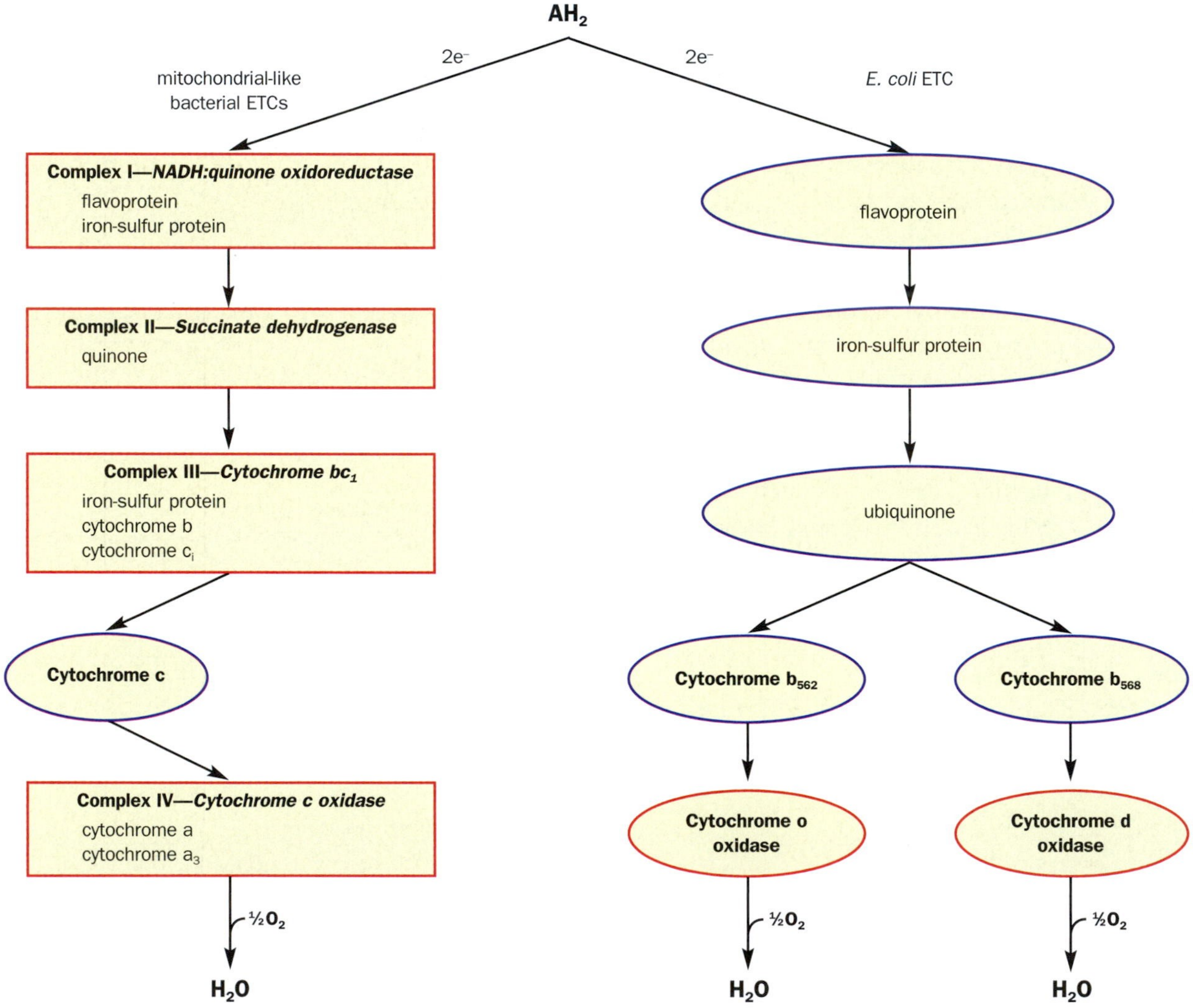

5-17 Aerobic Electron Transport Chains (ETCs) ◆ There are a variety of different bacterial aerobic electron transport chains. All begin with flavoproteins that pick up electrons from coenzymes, such as NADH or $FADH_2$ (represented by AH_2). Bacteria with chains that resemble the mitochondrial ETC in eukaryotes contain cytochrome c oxidase (Complex IV), which transfers electrons from cytochrome c to oxygen. These organisms give a positive result for the oxidase test. Other bacteria, such as members of the *Enterobacteriaceae*, are capable of aerobic respiration, but have a different terminal oxidase system and give a negative result for the oxidase test. The two paths shown on the right in the diagram are both found in *E. coli*. The amount of available O_2 determines which pathway is most active. Enzymatic complexes and oxidases are outlined in red.

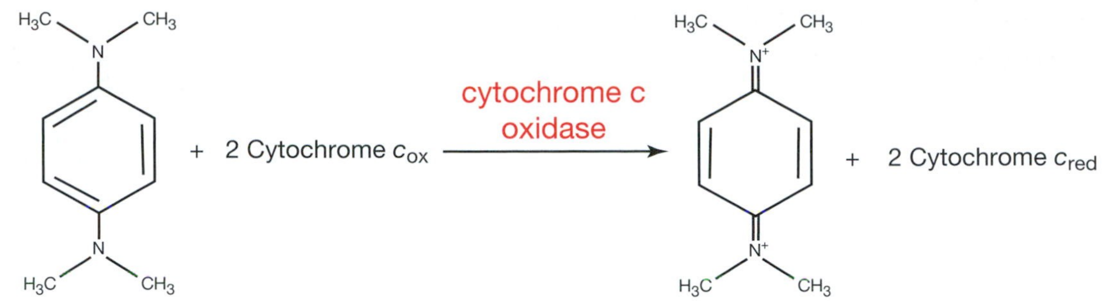

Tetramethyl-*p*-phenylenediamine$_{red}$ (colorless)

Tetramethyl-*p*-phenylenediamine$_{ox}$ (deep purple/blue)

5-18 Chemistry of the Oxidase Reaction ◆ The oxidase enzyme shown is not involved directly in the indicator reaction as shown. Rather, it removes electrons from cytochrome c, making it available to react with the phenylenediamine reagent.

(A Word About Reduction Potential, page 261) to the terminal electron acceptor—½O_2, in the case of an aerobic ETC and, 2) generate a proton motive force by pumping H^+ out of the cell, thus creating an ionic imbalance that will drive the production of ATP by way of membrane ATPases. The protons pumped out of the cell come from the hydrogen atoms whose electrons are being transferred down the chain. Because only alternating ETC molecules are able to carry associated protons along with their electrons, the excess positively charged ions are expelled from the cell. **Flavoproteins, iron-sulfur proteins,** and **cytochromes** are important ETC molecules unable to donate protons.

There are many different types of electron transport chains, but all share the characteristics listed above. Some organisms use more than one type of ETC depending on the availability of oxygen or other preferred terminal electron acceptor. *Escherichia coli*, for example, has two pathways for respiring aerobically and at least one for respiring anaerobically. Many bacteria have ETCs resembling mitochondrial ETCs in eukaryotes. These chains contain a series of four large enzymes broadly named Complexes I, II, III, and IV, each of which contains several molecules jointly able to transfer electrons and use the free energy released in the reactions. The last enzyme in the chain, Complex IV, is called cytochrome c oxidase because it makes the final electron transfer of the chain from cytochrome c, residing in the periplasm, to oxygen inside the cell.

The oxidase test is designed to identify the presence of cytochrome c oxidase. It is able to do this because cytochrome c oxidase has the unique ability to not only oxidize cytochrome c, but to catalyze the *reduction* of cytochrome c by a **chromogenic reducing agent** called tetramethyl-*p*-phenylenediamine. Chromogenic reducing agents are chemicals that develop color as they become oxidized (Figure 5-18).

In the oxidase test, the reducing reagent is added directly to bacterial growth on solid media (Figure 5-19), or (more conveniently) a bacterial colony is transferred to paper saturated with the reagent (Figure 5-20). A dramatic

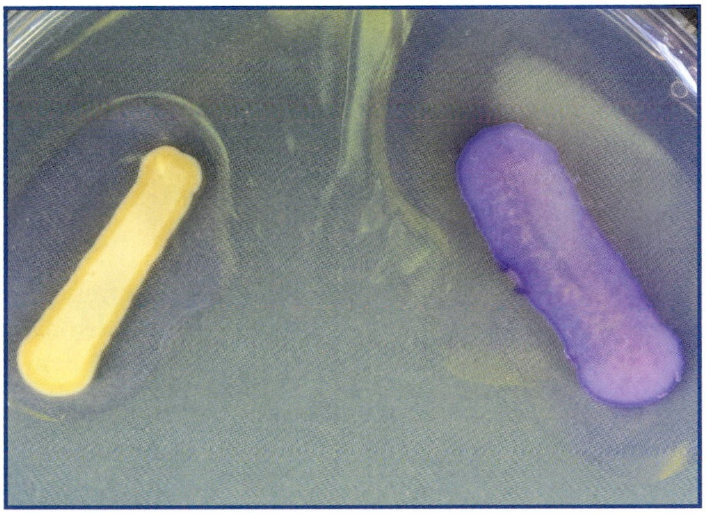

5-19 Oxidase Test on Bacterial Growth ◆ A few drops of reagent on oxidase-positive bacteria will produce a purple-blue color immediately.

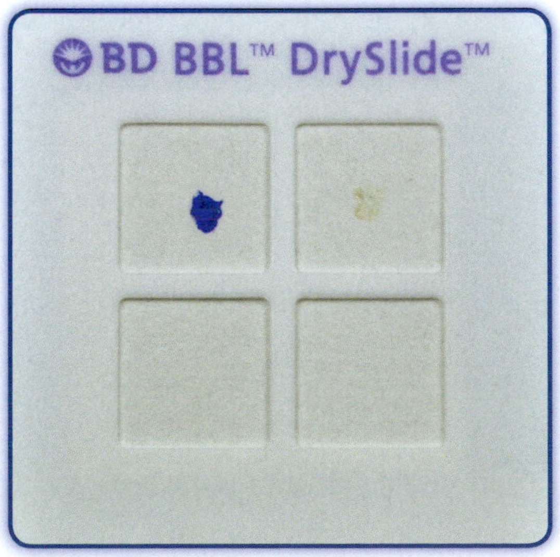

5-20 Oxidase Slide Test ◆ Positive results with this test should appear within 20 seconds. The dark blue inoculum in the upper left is a positive result. No color change (upper right) is a negative result. The bottom two panels were not inoculated. (BBL™ *Dry*Slide™ systems available from Becton Dickinson, Sparks, MD.)

color change occurs within seconds if the reducing agent becomes oxidized, thus indicating that cytochrome c oxidase is present. Lack of color change within the allotted time means that cytochrome c oxidase is not present and signifies a negative result (Table 5-6).

Application

The Oxidase Test is used to identify bacteria containing the respiratory enzyme cytochrome c oxidase. Among its many uses is the presumptive identification of the oxidase-positive *Neisseria*. It also can be useful in differentiating the oxidase-negative *Enterobacteriaceae* from the oxidase-positive *Pseudomonadaceae*.

In This Exercise

You will be performing the oxidase test (likely) using one of several commercially available systems. Your instructor will have chosen the system that is best for your laboratory. Generally, test systems produce a color change from colorless to blue/purple. The system shown in Figure 5-20 is BBL™ *Dry*Slide™, which calls for a transfer of inoculum to the slide and a reading within 20 seconds. Reagents used for this test are unstable and may oxidize independently shortly after they become moist. Therefore, it is important to take your readings within the recommended time.

Materials

Per Student Group

✓ sterile cotton-tipped applicator or plastic loop

✓ sterile water

✓ fresh slant cultures of:

- *Escherichia coli*
- *Moraxella catarrhalis*

Procedure

1. Following your instructor's guidelines, transfer some of the culture to the reagent slide. Get a visible mass of growth on the wooden end of a sterile cotton applicator or plastic loop, and roll it on the reagent slide.
2. Observe for color change within 20 seconds.
3. Repeat steps one and two for the second organism.
4. Enter your results on the Data Sheet.
5. When performing this test on an unknown, run a positive control simultaneously to verify the quality of the test system.

TABLE 5-6 Oxidase Test Results and Interpretations

TABLE OF RESULTS

Result	Interpretation	Symbol
Dark blue within 20 seconds	Cytochrome c oxidase is present	+
No color change to blue within 20 seconds	Cytochrome c oxidase is present	–

References

BBL™ *Dry*Slide™ package insert.

Collins, C. H., Patricia M. Lyne, and J. M. Grange. 1995. Page 116 in *Collins and Lyne's Microbiological Methods,* 7th ed. Butterworth-Heinemann, Oxford, United Kingdom.

Forbes, Betty A., Daniel F. Sahm, and Alice S. Weissfeld. 2002. Pages 151–152 in *Bailey & Scott's Diagnostic Microbiology,* 11th ed. Mosby, St. Louis.

Lányi, B. 1987. Page 18 in *Methods in Microbiology,* Vol. 19, edited by R. R. Colwell and R. Grigorova. Academic Press, New York.

MacFaddin, Jean F. 2000. Page 368 in *Biochemical Tests for Identification of Medical Bacteria,* 3rd ed. Lippincott Williams & Wilkins, Philadelphia.

Smibert, Robert M., and Noel R. Krieg. 1994. Page 625 in *Methods for General and Molecular Bacteriology*, edited by Philipp Gerhardt, R. G. E. Murray, Willis A. Wood, and Noel R. Krieg. American Society for Microbiology, Washington, DC.

White, David, 2000. Chapter 4 in *The Physiology and Biochemistry of Prokaryotes,* 2nd ed. Oxford University Press, New York.

Name ______________________ Date ______________________

Lab Section ______________________ I was present and performed this exercise (initials) ________

Oxidase Test

OBSERVATIONS AND INTERPRETATIONS

Using Table 5-6 as a guide, enter your results in the chart below.

Organism	Color Result	+ / −	Interpretation

Questions

1 Why is it advisable to include an oxidase-positive control organism in any test of an unknown organism?

2 Suppose in a test of an organism that possesses cytochrome c oxidase, the reagent begins to turn blue at 45 seconds. Consider the following possibilities, check all that apply, and defend your choices.

This is an example of (a):	YES	NO
Valid test	☐	☐
Reliable result	☐	☐
False positive	☐	☐
False negative	☐	☐
Poor sensitivity	☐	☐
Poor specificity	☐	☐

3 Provide a possible explanation as to why this test identifies the presence of cytochrome c oxidase and not other oxidases.

EXERCISE 5-6

Nitrate Reduction Test

Theory

Anaerobic respiration involves the reduction of (transfer of electrons to) an inorganic molecule other than oxygen. Nitrate reduction is one such example. Many Gram-negative bacteria (including most *Enterobacteriaceae*) contain the enzyme **nitrate reductase** and perform a single-step reduction of nitrate to nitrite ($NO_3 \rightarrow NO_2$). Other bacteria, in a multistep process known as **denitrification**, are capable of enzymatically converting nitrate to molecular nitrogen (N_2). Some products of nitrate reduction are shown in Figure 5-21.

Nitrate Broth is an undefined medium of beef extract, peptone, and potassium nitrate (KNO_3). An inverted Durham tube is placed in each broth to trap a portion of any gas produced. In contrast to many differential media, no color indicators are included. The color reactions obtained in Nitrate Broth take place as a result of reactions between metabolic products and reagents added after incubation (Figure 5-22).

Before a broth can be tested for nitrate reductase activity (nitrate reduction to nitrite), it must be examined for evidence of denitrification. This is simply a visual inspection for the presence of gas in the Durham tube (Figure 5-23). If the Durham tube contains gas, the test is complete. Denitrification has taken place. Gas produced in a nitrate reduction test by an organism capable of fermenting is not determinative because the source of the gas is unknown. For more information on fermentation, see Exercises 5-1, 5-2 and 5-3.

If there is no visual evidence of denitrification, sulfanilic acid (nitrate reagent A) and naphthylamine (nitrate reagent B) are added to the medium to test for nitrate reduction to nitrite. If present, nitrite will form nitrous acid (HNO_2) in the aqueous medium. Nitrous acid reacts with the added reagents to produce a red, water-soluble compound (Figure 5-24). Therefore, formation of red color after the addition of reagents indicates that the organism reduced nitrate to nitrite. If no color change takes place with the addition of reagents, the nitrate either was not reduced or was reduced

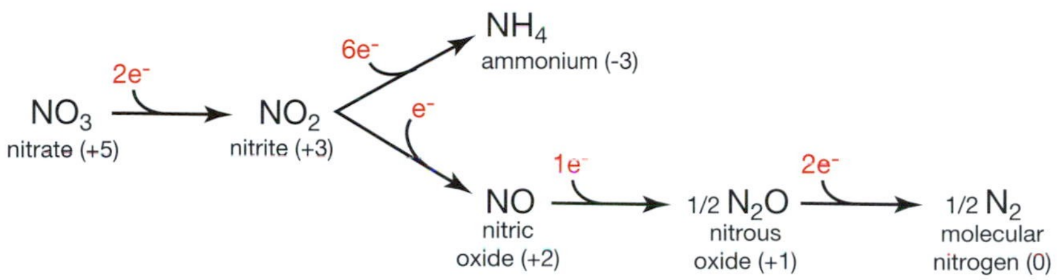

5-21 Possible End Products of Nitrate Reduction ◆ Nitrate reduction is complex. Many different organisms under many different circumstances perform nitrate reduction with many different outcomes. Members of the *Enterobacteriaceae* simply reduce NO_3 to NO_2. Other bacteria, functionally known as "denitrifiers," reduce NO_3 all the way to N_2 via the intermediates shown, and are important ecologically in the nitrogen cycle. Both of these are anaerobic respiration pathways (also known as "nitrate respiration"). Other organisms are capable of assimilatory nitrate reduction, in which NO_3 is reduced to NH_4, which can be used in amino acid synthesis. The oxidation state of nitrogen in each compound is shown in parentheses.

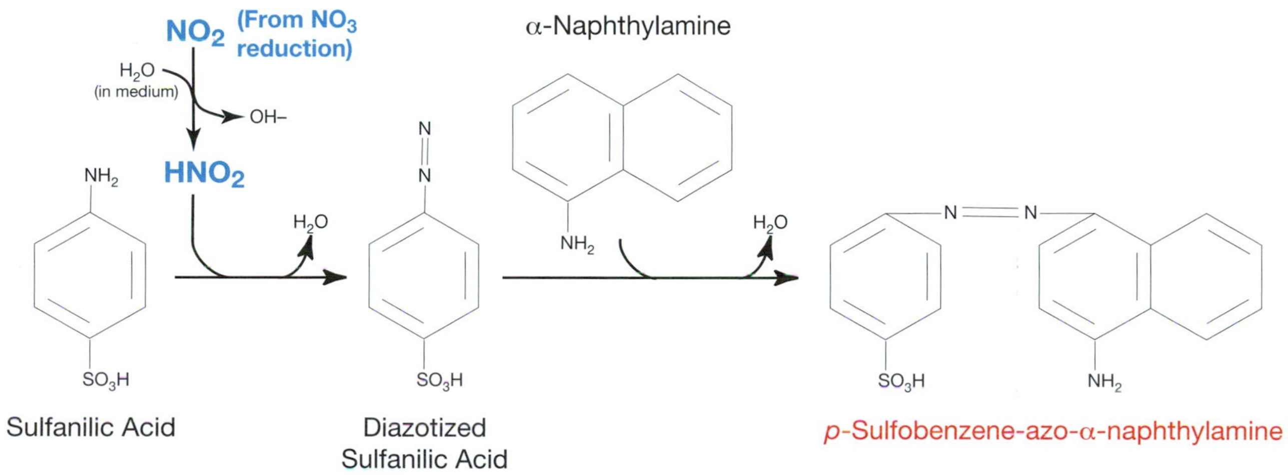

5-22 Indicator Reaction ◆ If nitrate is reduced to nitrite, nitrous acid (HNO_2) will form in the medium. Nitrous acid then reacts with sulfanilic acid to form diazotized sulfanilic acid, which reacts with the α-naphthylamine to form *p*-sulfobenzene-azo-α-naphthylamine, which is red. Thus, a red color indicates the presence of nitrite and is considered a positive result for nitrate reduction to nitrite.

5-23 Incubated Nitrate Broth Before the Addition of Reagents ◆ Numbered 1 through 5 from left to right, these Nitrate Broths have been incubated, but no reagents have been added. Tube 2 is an uninoculated control used for color comparison. Note the gas produced by the organism in tube 5, indicating a positive result. It is a known nonfermenter; therefore, the gas produced is an indication of denitrification. Tube 5 will not receive reagents. Tubes 1 through 4 must now receive reagents. See Figure 5-26.

5-24 Incubated Nitrate Broth After Addition of Reagents ◆ These are the same tubes as in Figure 5-23 after the addition of sulfanilic acid and alpha-naphthylamine to tubes 1–4. Red color is a positive result for reduction of nitrate to nitrite. Tube 1 is positive, tube 2 is the control, tubes 3 and 4 are inconclusive, and tube 5 was already identified as positive. Zinc must now be added to the inconclusive tubes.

to one of the other nitrogenous compounds shown in Figure 5-21. Because it is visually impossible to tell the difference between these two occurrences, another test must be performed.

In this stage of the test, a small amount of powdered zinc is added to the broth to catalyze the reduction of any nitrate (which still may be present as KNO_3) to nitrite. If nitrate is present at the time zinc is added, it will be converted immediately to nitrite, and the above-described reaction between nitrous acid and reagents will follow and turn the medium red. In this instance, the red color indicates that nitrate was *not* reduced by the organism (Figure 5-25). No color change after the addition of zinc indicates that the organism reduced the nitrate to NH_3, NO, N_2O, or some other nongaseous nitrogenous compound.

5-25 Incubated Nitrate Broth After Addition of Reagents and Zinc ◆ These are the same tubes as in Figure 5-24 after the addition of zinc to tubes 2, 3, and 4. The control tube (2) and tube 3 turned red. This color change is a negative result because it indicates, and is verified by the color change in the control, that nitrate is still present in the tube. Tube 4 did not change color, which indicates that the nitrate was reduced by the organism beyond nitrite to some other nitrogenous compound. This is a positive result. After completion of all phases of the test, tubes 1, 4, and 5 are positive. Tube 3 is negative.

❖ In This Exercise

After inoculation and incubation of the Nitrate Broths, you first will inspect the Durham tubes included in the media for bubbles—an indication of gas production (Figure 5-23). One or more of the organisms selected for this exercise are fermenters, and possible producers of gas other than molecular nitrogen; therefore, you will proceed with adding reagents to all tubes. Record any gas observed in the Durham tubes and see if there is a correlation between that and the color results after the reagents and zinc dust have been added. Follow the procedure carefully, and use Table 5-7 as a guide when interpreting the results.

❖ Application

Virtually all members of *Enterobacteriaceae* perform a one-step reduction of nitrate to nitrite. The nitrate test differentiates them from Gram-negative rods that either do not reduce nitrate or reduce it beyond nitrite to N_2 or other compounds.

Materials

Per Student Group

✓ four Nitrate Broths
✓ nitrate test reagents A and B
✓ zinc powder
✓ fresh cultures of:
- *Alcaligenes faecalis)*
- *Escherichia coli*
- *Pseudomonas aeruginosa* (BSL-2)

Medium and Reagent Recipes

Nitrate Broth

♦ Beef extract	3.0 g
♦ Peptone	5.0 g
♦ Potassium nitrate	1.0 g
♦ Distilled or deionized water	1.0 L
pH 6.8–7.2 at 25°C	

Reagent A

♦ *N,N*-Dimethyl-1-naphthylamine	0.6 g
♦ 5N Acetic acid (30%)	100.0 mL

Reagent B

♦ Sulfanilic acid	0.8 g
♦ 5N Acetic acid (30%)	100.0

Procedure

Lab One

1 Obtain four Nitrate Broths. Label three of them with the names of the organisms, your name, and the date. Label the fourth tube "control."

2 Inoculate three broths with the test organisms. Do not inoculate the control.

3 Incubate all tubes at 35 ± 2°C for 24 to 48 hours.

Lab Two

1 Examine each tube for evidence of gas production. Record your results on the Data Sheet. Refer to Table 5-7 when making your interpretations. (Be methodical when recording your results. Red color can have opposite interpretations depending on where you are in the procedure.)

2 Using Figure 5-26 as a guide, proceed to the addition of reagents to *all* tubes as follows:

a. Add eight drops each (approximately 0.5 mL) of reagent A and reagent B to each tube. Mix well, and let the tubes stand undisturbed for 10 minutes. Using the control as a comparison, record your test results in the chart provided in the Data Sheet. Set aside any test broths that are positive. The test is complete for these tubes.

b. Add a pinch of zinc dust to all remaining tubes. (Only a small amount is needed. Dip a wooden applicator into the zinc dust and transfer only the amount that clings to the wood.) Let the tubes stand for 10 minutes. Record your results on the Data Sheet.

TABLE 5-7 Nitrate Test Results and Interpretations

TABLE OF RESULTS

Result	Interpretation	Symbol
Gas (Nonfermenter)	Denitrification—production of nitrogen gas ($NO_3 \rightarrow NO_2 \rightarrow N2$)	+
Gas (Fermenter, or status is unknown)	Source of gas is unknown; requires addition of reagents	
Red color (after addition of reagents A and B)	Nitrate reduction to nitrite ($NO_3 \rightarrow NO_2$)	+
No color (after the addition of reagents)	Incomplete test; requires the addition of zinc dust	
No color change (after addition of zinc)	Nitrate reduction to nongaseous nitrogenous compounds ($NO_3 \rightarrow NO_2 \rightarrow$ nongaseous nitrogenous products)	+
Red color (after addition of zinc dust)	No nitrate reduction	–

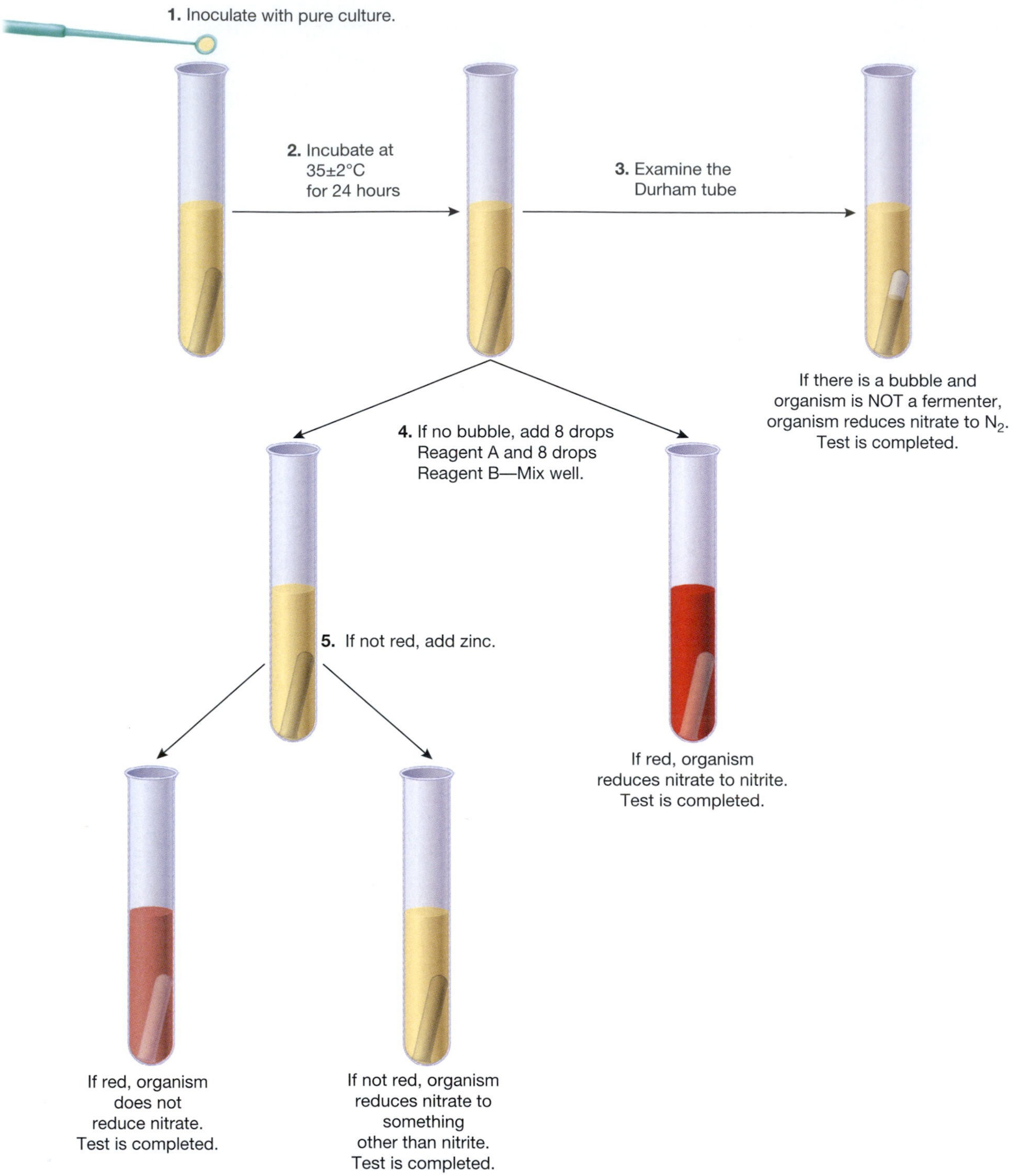
1. Inoculate with pure culture.
2. Incubate at 35±2°C for 24 hours
3. Examine the Durham tube
If there is a bubble and organism is NOT a fermenter, organism reduces nitrate to N_2. Test is completed.
4. If no bubble, add 8 drops Reagent A and 8 drops Reagent B—Mix well.
5. If not red, add zinc.
If red, organism reduces nitrate to nitrite. Test is completed.
If red, organism does not reduce nitrate. Test is completed.
If not red, organism reduces nitrate to something other than nitrite. Test is completed.

5-26 **Procedural Diagram for the Nitrate Reduction Test** ◆

References

Atlas, Ronald M., and Richard Bartha. 1998. Pages 423–425 in *Microbial Ecology—Fundamentals and Applications,* 4th ed. Benjamin/Cummings Science Publishing, Menlo Park, CA.

Forbes, Betty A., Daniel F. Sahm, and Alice S. Weissfeld. 2002. Page 277 in *Bailey & Scott's Diagnostic Microbiology,* 11th ed. Mosby, St. Louis.

Lányi, B. 1987. Page 21 in *Methods in Microbiology,* Vol. 19, edited by R. R. Colwell and R. Grigorova. Academic Press, New York.

MacFaddin, Jean F. 2000. Page 348 in *Biochemical Tests for Identification of Medical Bacteria,* 3rd ed. Williams & Wilkins, Philadelphia.

Moat, Albert G., John W. Foster, and Michael P. Spector. Chapter 14 in *Microbial Physiology,* 4th ed. Wiley-Liss, New York.

Zimbro, Mary Jo, and David A. Power, eds. 2003. Page 400 in *Difco™ and BBL™ Manual—Manual of Microbiological Culture Media.* Becton Dickinson and Co., Sparks, MD.

Name ______________________ Date ______________________

Lab Section ______________________ I was present and performed this exercise (initials) ________

Nitrate Reduction Test

OBSERVATIONS AND INTERPRETATIONS

Using Table 5-7 as a guide, enter your results and interpretations in the chart below.

Organism	Results			Interpretation
	Gas	Color		
	Y/N	After Reagents	After Zinc	
Control				

Questions

1 Why is gas production not recognized as nitrate reduction when the organism is a known fermenter?

2 Suppose you remove your test cultures from the incubator and notice that one of them—a known fermenter—has a gas bubble in the Durham tube. Knowing that fermenters frequently produce gas, you ignore the bubble and proceed to the next step. Adding reagents produces no change, and neither does adding zinc. Is this occurrence consistent with what you have learned about this test? Why?

3 Would you change your answer to #2 above if the control broth did not change color after the addition of reagents? What if the control broth had changed color only after the addition of reagents and zinc?

4 When testing microaerophiles, some microbiologists prefer to use a semisolid nitrate medium that contains a small amount of agar. Why do you think this is done?

Nutrient Utilization Tests

Utilization media are highly defined formulations designed to differentiate organisms based on their ability to grow when an essential nutrient (e.g., carbon or nitrogen) is strictly limited. Simmon's Citrate Medium, the example used in Exercise 5-7, contains sodium citrate as the only carbon-containing compound and ammonium ion as the only nitrogen source. Therefore, only organisms able to produce the enzymes specific to these compounds will be able to survive and grow on the medium.

EXERCISE 5-7

Citrate Test

❖ Theory

The Citrate Test was designed to differentiate members of *Enterobacteriaceae*, all of which are facultative anaerobes. That is, they have the ability to ferment carbohydrates, but preferentially respire using the Citric Acid Cycle (aka Krebs Cycle) given the opportunity and resources to do so. The Citric Acid Cycle is a complex series of redox reactions (as shown in Appendix A and Figure 5-27) whereby acetyl coenzyme A is converted, by means of many intermediates (starting with citrate), to CO_2 and energy.

Simmons Citrate Medium is a defined medium. That is, the amount and source of all ingredients are carefully controlled. Because sodium citrate is the only carbon-containing compound in the medium, it will not support a complex, high energy-yielding respiratory reaction like the Citric Acid Cycle. It does, however, provide the means for a few bacterial species that possess the enzyme **citrate-permease** to transport citrate into the cell and metabolize it by way of the fermentative pathway illustrated in Figure 5-27. As shown in the figure, a variety of products can be formed, depending on the pH of the environment. Bacteria that do not possess citrate-permease will not survive on the medium.

Other important ingredients in Simmons Citrate Medium are ammonium dihydrogen phosphate and bromthymol blue dye. Bromthymol blue dye, which is green at pH 6.9 and blue at pH 7.6, is added as an indicator. Ammonium dihydrogen phosphate provides the sole nitrogen source in the medium. In the process of pulling the nitrogen from the ammonium salt ($NH_4H_2PO_4$), citrate-positive bacteria produce ammonia (NH_3) and ammonium hydroxide (NH_4OH), both of which alkalinize the medium and turn it blue (Figure 5-28). Thus, blue color is a positive citrate test result (Table 5-8).

On very rare occasions a citrate-positive organism will grow on the medium without producing a color change. In most cases, this is because of incomplete incubation or improper starting pH of the medium. Growth on the slant, even in the absence of color change, indicates that citrate is being utilized and is considered a positive test result. To avoid confusion between actual growth and heavy inoculum, which may appear as growth, citrate slants should be inoculated lightly with an inoculating needle rather than a loop.

❖ Application

The citrate utilization test is used to determine the ability of an organism to use citrate as its sole source of carbon. Citrate utilization is one part of a test series referred to as the IMViC (*I*ndole, *M*ethyl Red, *V*oges-Proskauer and

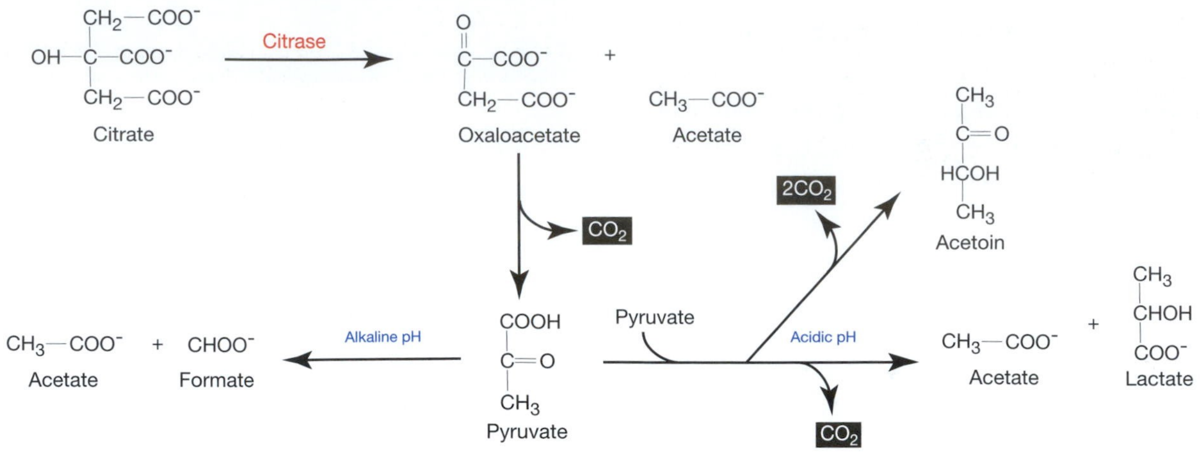

5-27 Citrate Chemistry ◆ In the presence of citrate-permease enzyme, citrate enters the cell and is converted to pyruvate. The pyruvate then is converted to a variety of products depending on the pH of the environment. Because Simmons Citrate Medium starts at or near neutral pH and the alkalinizing effect of the ammonium salt breakdown, positive control organisms for this test typically drive the reaction to the left, producing acetate, formate, and CO_2.

5-28 Citrate Test Results ◆ These Simmons Citrate slants were inoculated with a citrate-positive (+) organism on the left and a citrate-negative (−) organism in the center. The slant on the right is uninoculated.

Citrate tests) that distinguishes between members of the family *Enterobacteriaceae* and differentiate them from other Gram-negative rods.

In This Exercise

You will be inoculating Simmons Citrate slants with a citrate-positive and a citrate-negative organism. To avoid confusing growth on the slant with a heavy inoculum, be sure to inoculate using a needle instead of a loop.

Materials

Per Student Group

✓ three Simmons Citrate slants

✓ fresh slant cultures of:

- *Enterobacter aerogenes*
- *Escherichia coli*

Medium Recipe

Simmons Citrate Medium

◆ Ammonium dihydrogen phosphate	1.0 g
◆ Dipotassium phosphate	1.0 g
◆ Sodium chloride	5.0 g
◆ Sodium citrate	2.0 g
◆ Magnesium sulfate	0.2 g
◆ Agar	15.0 g
◆ Bromthymol blue	0.08 g
◆ Distilled or deionized water	1.0 L

pH 6.7–7.1 at 25°C

Procedure

Lab One

1. Obtain three Simmons Citrate tubes. Label two with the names of the organisms, your name, and the date. Label the third tube "control."
2. Using an inoculating needle and *light inoculum*, streak the slants with the test organisms. Do not inoculate the control.
3. Incubate all tubes at 35 ± 2°C for 48 hours.

Lab Two

1. Observe the tubes for color changes and/or growth.
2. Record your results on the Data Sheet.

TABLE **5-8** **Citrate Test Results and Interpretations**

TABLE OF RESULTS		
Result	**Interpretation**	**Symbol**
Blue (even a small amount)	Citrate is utilized	+
No color change; growth	Citrate is utilized	+
No color change; no growth	Citrate is not utilized	–

References

Collins, C. H., Patricia M. Lyne, and J. M. Grange. 1995. Page 111 in *Collins and Lyne's Microbiological Methods,* 7th ed. Butterworth-Heinemann, Oxford, United Kingdom.

Forbes, Betty A., Daniel F. Sahm, and Alice S. Weissfeld. 2002. Page 266 in *Bailey & Scott's Diagnostic Microbiology,* 11th ed. Mosby, St. Louis.

MacFaddin, Jean F. 2000. Page 98 in *Biochemical Tests for Identification of Medical Bacteria,* 3rd ed. Lippincott Williams & Wilkins, Philadelphia.

Smibert, Robert M., and Noel R. Krieg. 1994. Page 614 in *Methods for General and Molecular Bacteriology*, edited by Philipp Gerhardt, R. G. E. Murray, Willis A. Wood, and Noel R. Krieg, American Society for Microbiology, Washington, DC.

Zimbro, Mary Jo, and David A. Power, eds. 2003. Page 514 in *Difco™ and BBL™ Manual—Manual of Microbiological Culture Media.* Becton Dickinson and Co., Sparks, MD.

Name ________________ Date ________________

Lab Section ________________ I was present and performed this exercise (initials) ________

Citrate Test

OBSERVATIONS AND INTERPRETATIONS

Using Table 5-8 as a guide, enter your results and interpretations in the chart below.

Organism	Color Result	+ / −	Interpretation
Control			

Questions

1 Many bacteria that are able to metabolize citrate (as seen in the Krebs cycle) produce negative results in this test. Why? Be specific. Refer to Appendix A for help.

2 If an organism is able to convert the sodium citrate in Simmons Citrate Agar to pyruvate, list some possible reasons why an organism might ferment it rather than metabolize it oxidatively in the Krebs cycle. Refer to Appendix A for help.

3 Explain how an organism that possesses the citrate enzyme might not test positively on Simmons Citrate Agar. Is this a false negative result? Why or why not?

Decarboxylation and Deamination Tests

Decarboxylation and deamination tests were designed to differentiate members of *Enterobacteriaceae* and to distinguish them from other Gram-negative rods. Most members of *Enterobacteriaceae* produce one or more enzymes that are necessary to break down amino acids. Enzymes that catalyze the removal of an amino acid's carboxyl group (COOH) are called decarboxylases. Enzymes that catalyze the removal of an amino acid's amine group (NH_2) are called deaminases.

Each decarboxylase and deaminase is specific to a particular substrate. Decarboxylases catalyze reactions that produce alkaline products. Thus, we are able to identify the ability of an organism to produce a specific decarboxylase by preparing base medium, adding a known amino acid, and including a pH indicator to mark the shift to alkalinity. Differentiation of an organism in deamination media employs the same principle of substrate exclusivity by including a single known amino acid but requires the addition of a chemical reagent to produce a readable result.

In the next two exercises you will be introduced to the most common of both types of tests. In Exercise 5-8 you will test bacterial ability to decarboxylate the amino acid lysine. In Exercise 5-9 you will test bacterial ability to deaminate the amino acid phenylalanine.

EXERCISE 5-8

Decarboxylation Test

Theory

Møller's Decarboxylase Base Medium contains peptone, glucose, the pH indicator bromcresol purple, and the **coenzyme** pyridoxal phosphate. Bromcresol purple is purple at pH 6.8 and above, and yellow below pH 5.2. Base medium can be used with one of a number of specific amino acid substrates, depending on the decarboxylase to be identified.

After inoculation, an overlay of mineral oil is used to seal the medium from external oxygen and promote fermentation (Figure 5-29). Glucose fermentation in the anaerobic medium initially turns it yellow due to the accumulation of acid end products. The low pH and presence of the specific amino acid induces **decarboxylase-positive** organisms to produce the enzyme. (That is, the specific decarboxylase gene is "switched on.")

Decarboxylation of the amino acid results in accumulation of alkaline end products that turn the medium purple (Figures 5-30 through 5-32). If the organism is a glucose fermenter but does not produce the appropriate decarboxylase, the medium will turn yellow and remain so. If the organism does not ferment glucose, the medium will exhibit no color change. Purple color is the only result that is considered positive; all others are negative (Figure 5-32).

Application

Decarboxylase media can include any one of several amino acids. Typically, these media are used to differentiate organisms in the family *Enterobacteriaceae* and to distinguish them from other Gram-negative rods.

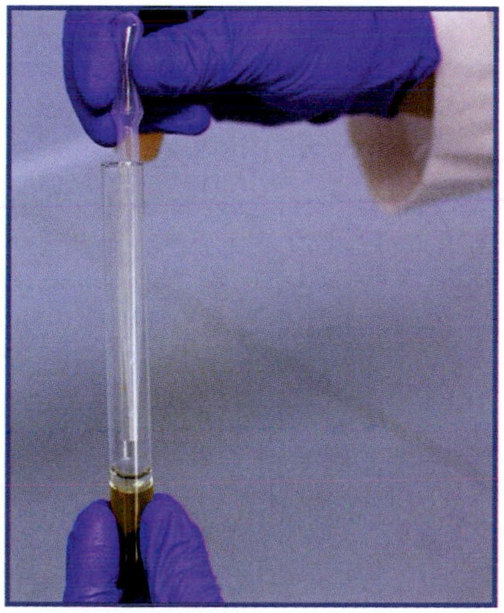

5-29 Adding Mineral Oil to the Tubes

In This Exercise

You will perform decarboxylation tests on two organisms in Møller's Lysine Decarboxylase media. This procedure requires the aseptic addition of mineral oil. Be careful not to introduce contaminants while adding the oil. Refer to Figure 5-32 and Table 5-9 when interpreting your results.

Materials

Per Student Group

- ✓ three Lysine Decarboxylase Broths
- ✓ sterile mineral oil
- ✓ sterile transfer pipettes
- ✓ recommended organisms:
  - *Escherichia coli*
  - *Providencia stuartii*

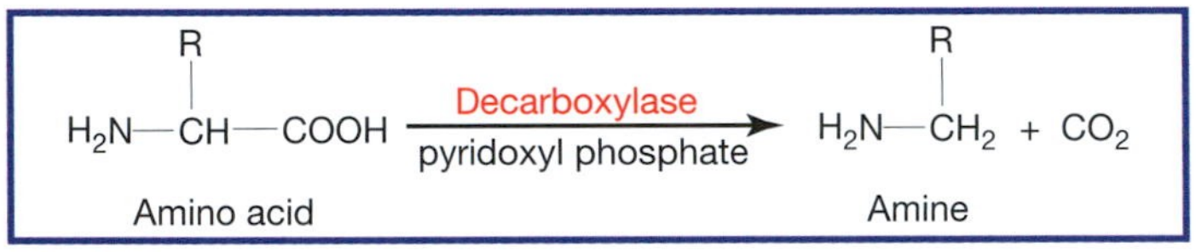

5-30 Amino Acid Decarboxylation ◆ Removal of an amino acid's carboxyl group results in the formation of an amine and carbon dioxide.

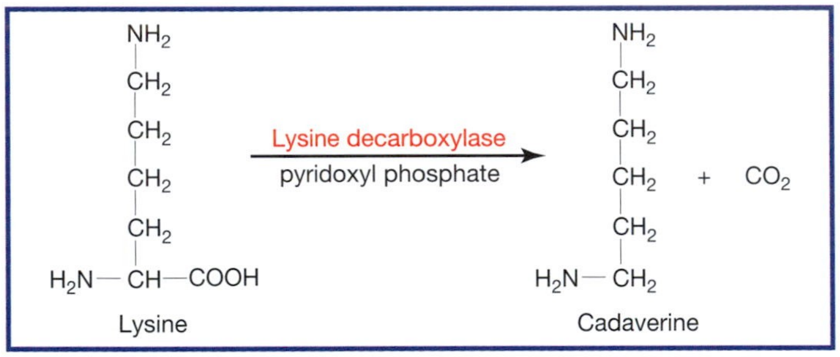

5-31 Lysine Decarboxylation ◆ Decarboxylation of the amino acid lysine produces cadaverine and CO_2.

5-32 Decarboxylation Test Results ◆ A lysine decarboxylase-positive (+) organism is on the left, an uninoculated control is in the center, and a lysine decarboxylase negative (−) is on the right. The color results shown here are representative of arginine and ornithine decarboxylase media as well.

TABLE 5-9 Decarboxylase Test Results and Interpretations ◆

TABLE OF RESULTS		
Result	**Interpretation**	**Symbol**
No color change	No decarboxylation	−
Yellow	Fermentation; no decarboxylation	−
Purple (may be slight)	Decarboxylation; organism produces the specific decarboxylase enzyme	+

Medium Recipe

Lysine Decarboxylase Medium (Møller)

◆ Peptone	5.0 g
◆ Beef extract	5.0 g
◆ Glucose (dextrose)	0.5 g
◆ Bromcresol purple	0.01 g
◆ Cresol red	0.005 g
◆ Pyridoxal	0.005 g
◆ L-Lysine	10.0 g
◆ Distilled or deionized water	1.0 L

pH 5.8–6.2 at 25°C

Procedure

Lab One

1. Obtain three decarboxylase broths. Label two with the name of the organism, your name, and the date. Label the third broth "control."
2. Inoculate the media with the test organisms. Do not inoculate the control.
3. Overlay all tubes with 3 to 4 mm sterile mineral oil (Figure 5-29).
4. Incubate all tubes aerobically at 35 ± 2°C for one week.

Lab Two

1. Remove the tubes from the incubator and examine them for color changes.
2. Record your results on the Data Sheet.

References

Collins, C. H., Patricia M. Lyne, and J. M. Grange. 1995. Page 111 in *Collins and Lyne's Microbiological Methods,* 7th ed. Butterworth-Heinemann, Oxford, United Kingdom.

DIFCO Laboratories. 1984. Page 268 in *DIFCO Manual,* 10th ed. DIFCO Laboratories, Detroit.

Forbes, Betty A., Daniel F. Sahm, and Alice S. Weissfeld. 2002. Page 267 in *Bailey & Scott's Diagnostic Microbiology,* 11th ed. Mosby, St. Louis.

Lányi, B. 1987. Page 29 in *Methods in Microbiology,* Vol. 19, edited by R. R. Colwell and R. Grigorova. Academic Press, New York.

MacFaddin, Jean F. 2000. Page 120 in *Biochemical Tests for Identification of Medical Bacteria,* 3rd ed. Lippincott Williams & Wilkins, Philadelphia.

Name ______________________ Date ______________________

Lab Section ______________________ I was present and performed this exercise (initials) ________

Decarboxylation Test

OBSERVATIONS AND INTERPRETATIONS

Using Table 5-9 as a guide, record your results and interpretations in the chart below.

Organism	Results		Interpretation
	Color	+ / −	
Control			

Questions

1 When conducting this test, some microbiologists inoculate two sets of broth, one with amino acid and one without (Base Broth). What function would Base Broth serve? *Hint:* It is another type of control.

2 How can both *no change* and *yellow color* be considered negative results?

3 Incubation time for this medium is 1 week. Under what circumstances would an early reading be allowable? Not allowable? Explain.

4 How does adding the sterile mineral oil affect the specificity or the sensitivity of the test?

EXERCISE 5-9

Phenylalanine Deaminase Test

Theory

Organisms that produce phenylalanine deaminase can be identified by their ability to remove the amine group (NH_2) from the amino acid phenylalanine. The reaction, as shown in Figure 5-33, requires oxygen and produces ammonia (NH_3) and phenylpyruvic acid. Deaminase activity is evidenced by the presence of phenylpyruvic acid.

Phenylalanine Agar provides a rich source of phenylalanine. A reagent containing ferric chloride ($FeCl_3$) is added to the medium after incubation. The normally colorless phenylpyruvic acid reacts with the ferric chloride and turns a dark green color almost immediately (Figure 5-34). Formation of green color indicates the presence of phenylpyruvic acid and, hence, the presence of phenylalanine deaminase. Yellow is negative (Figure 5-35).

Application

The Phenylalanine Deaminase Test is used to differentiate the genera *Morganella*, *Proteus*, and *Providencia* from other members of the *Enterobacteriaceae*.

In This Exercise

You will inoculate two phenylalanine agar slants with test organisms. After incubation you will add 12% $FeCl_3$ (Phenylalanine Deaminase Test Reagent) and watch for a color change. Refer to Figure 5-35 and Table 5-10 when making your interpretations and recording your results.

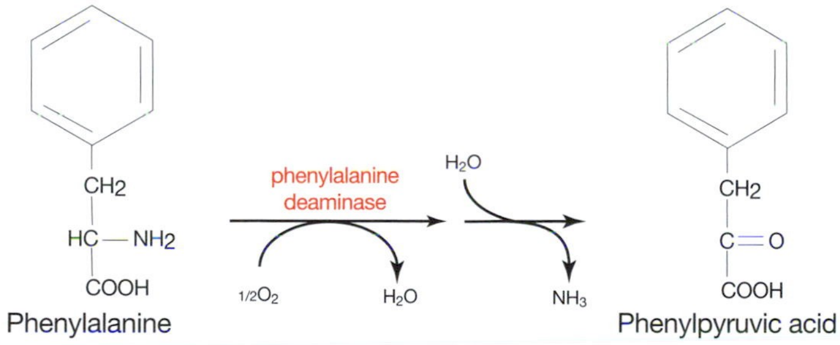

5-33 Deamination of Phenylalanine ◆ Deamination is the removal of an amine (NH_2) from an amino acid.

Phenylpyruvic Acid + $FeCl_3$ ⟶ Green Color

5-34 Indicator Reaction ◆ Phenylpyruvic acid produced by phenylalanine deamination reacts with $FeCl_3$ to produce a green color and indicates a positive phenylalanine test. Be sure to read the result immediately, as the color may fade.

Materials

Per Student Group

- ✓ three Phenylalanine Agar slants
- ✓ 12% ferric chloride solution
- ✓ fresh slant cultures of:
  - *Escherichia coli*
  - *Providencia stuartii*

Medium and Reagent Recipes

Phenylalanine Agar

◆ DL-Phenylalanine	2.0 g
◆ Yeast extract	3.0 g
◆ Sodium chloride	5.0 g
◆ Sodium phosphate	1.0 g
◆ Agar	12.0 g
◆ Distilled or deionized water	1.0 L

pH 7.1–7.5 at 25°C

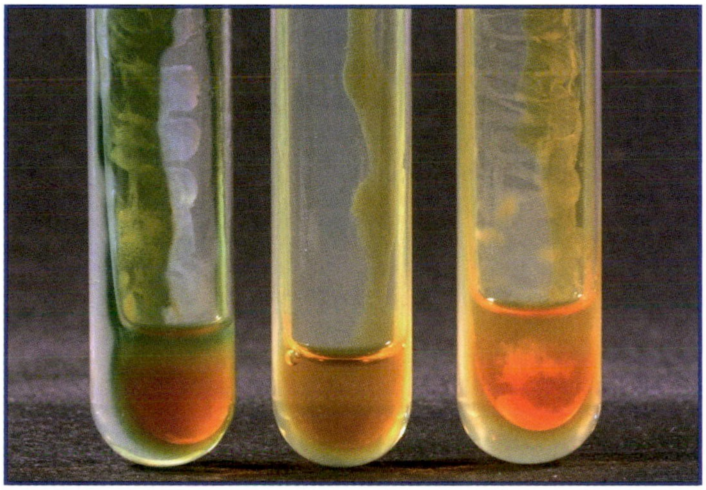

5-35 Phenylalanine Deaminase Test ◆ Note the color produced by the stream of ferric chloride in each tube. A phenylalanine deaminase-positive (+) organism is on the left, an uninoculated control is in the middle, and a phenylalanine deaminase-negative (−) organism is on the right.

TABLE 5-10 Phenylalanine Test Results and Interpretations ◆

TABLE OF RESULTS

Result	Interpretation	Symbol
Green color	Phenylalanine deaminase is present	+
No color change	Phenylalanine deaminase is absent	−

Phenylalanine Deaminase Test Reagent

- Ferric chloride 12.0 g
- Concentrated hydrochloric acid 2.5 mL
- Deionized water (to bring the total to) 100.0 mL

Procedure

Lab One

1. Obtain three Phenylalanine Agar slants. Label two slants with the name of the organism, your name, and the date. Label the third slant "control."
2. Streak two slants with heavy inocula of the test organisms. Do not inoculate the control.
3. Incubate all slants aerobically at 35 ± 2°C for 18 to 24 hours.

Lab Two

1. Add a few drops of 12% ferric chloride solution to each tube and observe for color change. (Note: This color may fade quickly, so read and record your results immediately.)
2. Record your results on the Data Sheet.

References

Forbes, Betty A., Daniel F. Sahm, and Alice S. Weissfeld. 2002. Page 280 in *Bailey & Scott's Diagnostic Microbiology,* 11th ed. Mosby, St. Louis.

Lányi, B. 1987. Page 28 in *Methods in Microbiology,* Vol. 19, edited by R. R. Colwell and R. Grigorova. Academic Press, New York.

MacFaddin, Jean F. 2000. Page 388 in *Biochemical Tests for Identification of Medical Bacteria,* 3rd ed. Lippincott Williams & Wilkins, Philadelphia.

Zimbro, Mary Jo, and David A. Power, eds. 2003. Page 442 in *Difco™ and BBL™ Manual—Manual of Microbiological Culture Media.* Becton Dickinson and Co., Sparks, MD.

Name ______________________ Date ______________________

Lab Section ______________________ I was present and performed this exercise (initials) ________

Phenylalanine Deaminase Test

OBSERVATIONS AND INTERPRETATIONS

Using Table 5-10 as a guide, record your results and interpretations in the chart below.

Organism	Color Result	+ / −	Interpretation
Control			

Questions

1 What is the purpose of the control? What other types of controls would be useful to increase reliability of the test?

DATA SHEET 5-9

2 If you are performing this test on an unknown organism, why is it a good idea to run simultaneous tests on known phenylalanine-positive and phenylalanine-negative organisms?

3 What do deamination and decarboxylation reactions have in common?

4 Phenylalanine medium sometimes is prepared in broth form with a pH indicator similar to that used in Decarboxylase Medium so that increases in pH then can be detected by development of purple color. When testing an organism for the ability to deaminate phenylalanine, would you expect to overlay this medium with mineral oil as is done with the Decarboxylation Test? Why or why not?

Tests Detecting Hydrolytic Enzymes

Reactions that use water to split complex molecules are called **hydrolysis** (or hydrolytic) reactions. The enzymes required for these reactions are called hydrolytic enzymes. When the enzyme catalyzes its reaction inside the cell, it is referred to as **intracellular**. Enzymes secreted from the organism to catalyze reactions outside the cell are called **extracellular** enzymes, or **exoenzymes**. In this unit you will be performing tests that identify the actions of both intracellular and extracellular enzymes.

The hydrolytic enzymes detected by the Bile Esculin and Urea Hydrolysis tests are intracellular and produce identifiable color changes in the medium. Nutrient Gelatin detects gelatinase—an exoenzyme that liquefies the solid medium. DNase Agar, Milk Agar, Tributyrin Agar, and Starch Agar are plated media used to identify extracellular enzymes by the characteristic halo that develops around the bacterial growth following hydrolysis of the specific medium substrate.

EXERCISE **5-10**

Bile Esculin Test

Theory

Bile Esculin Agar is an undefined, selective, and differential medium containing beef extract, digest of gelatin, esculin, oxgall (bile), and ferric citrate. Esculin extracted from the bark of the Horse Chestnut tree, is a glycoside composed of glucose and esculetin. Beef extract and gelatin provide nutrients and energy; bile is the selective agent added to separate the *Streptococcus bovis* group and enterococci from other streptococci. Ferric citrate is added as a source of oxidized iron to indicate a positive test.

Many bacteria can hydrolyze esculin under acidic conditions (Figure 5-36), and many bacteria, especially Gram-negative enterics, demonstrate tolerance to bile. However, among the streptococci, typically only enterococci and members of the *Streptococcus bovis* group (*S. equinus*, *S. gallolyticus*, *S. infantarius*, and *S. alactolyticus*) tolerate bile and hydrolyze esculin.

In this test, when esculin molecules are split, esculetin reacts with the Fe^{3+} from the ferric citrate and forms a dark brown precipitate (Figure 5-37). This precipitate darkens the medium surrounding the growth. An organism that darkens the medium even slightly is Bile Esculin-positive (Figure 5-38). An organism that does not darken the medium is negative.

Application

This test is used most commonly for presumptive identification of enterococci and members of the *Streptococcus bovis* group, all of which are positive.

In This Exercise

Today you will spot inoculate one Bile Esculin Agar plate with two test organisms and incubate it for up to 48 hours. Any blackening of the medium before the end of 48 hours can be recorded as positive. Negative results must incubate the full 48 hours before final determination is made. Use Table 5-11 as a guide.

Materials

Per Student Group

- ✓ one Bile Esculin Agar plate
- ✓ fresh cultures of:
  - ♦ *Lactococcus lactis*
  - ♦ *Enterococcus faecalis* surrogate (*Serratia marcescens*)

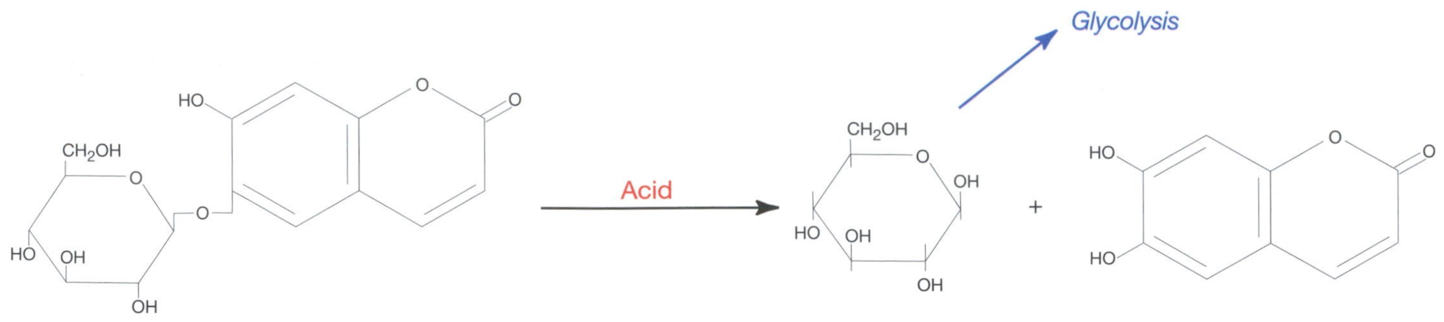

Esculin β-D-Glucose Esculetin

5-36 Acid Hydrolysis of Esculin ◆ Many organisms have the ability to hydrolyze esculin. However, the group D streptococci and enterococci are unique in their ability to do this in the presence of bile salts.

5-37 Bile Esculin Test Indicator Reaction ◆ Esculetin, produced during the hydrolysis of esculin, reacts with $Fe3^{+}$ in the medium to produce a dark brown to black color in the medium.

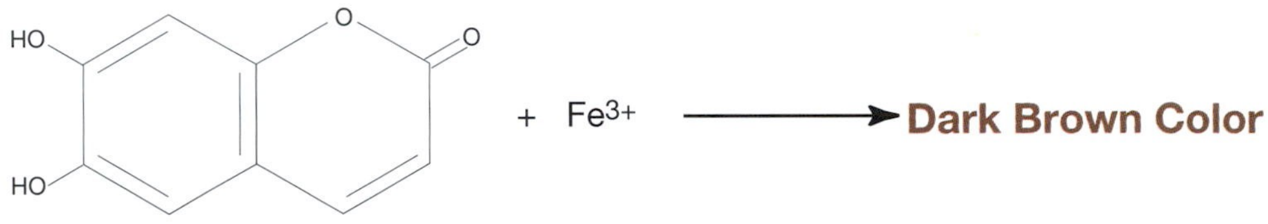

Esculetin

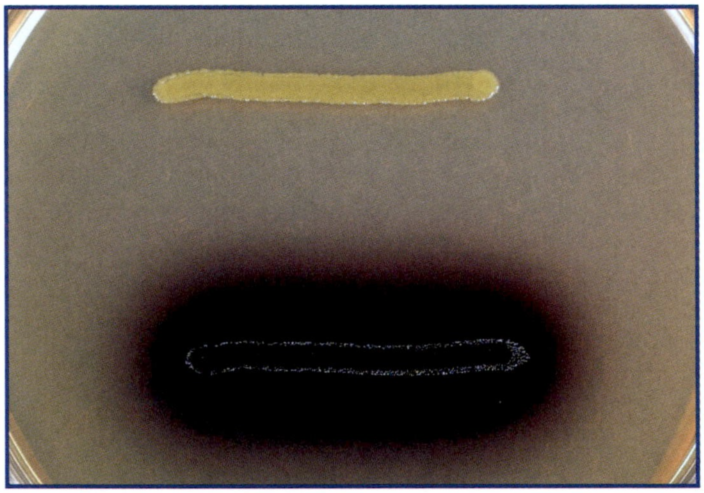

5-38 Bile Esculin Test Results ◆ This plate was inoculated with two Gram-positive cocci. The organism on the bottom is bile esculin-positive. The organism on the top is negative.

Medium Recipe

Bile Esculin Agar

◆ Pancreatic Digest of Gelatin	5.0 g
◆ Beef Extract	3.0 g
◆ Oxgall	20.0 g
◆ Ferric citrate	0.5 g
◆ Esculin	1.0 g
◆ Agar	14.0 g
◆ Distilled or deionized water	1.0 L

pH 6.6–7.0 at 25°C

Procedure

Lab One

1. Obtain a Bile Esculin Agar plate.
2. Using a permanent marker, divide the bottom into two halves.
3. Label the plate with the organisms' names, your name, and the date.
4. Spot-inoculate each half with one test organism. Refer to Appendix B if necessary.
5. Invert and incubate the plate at 35 ± 2°C for 24 to 48 hours.

Lab Two

1. Examine the plate for any darkening of the medium.
2. Record your results on the Data Sheet.

References

Delost, Maria Dannessa. 1997. Page 132 in *Introduction to Diagnostic Microbiology*. Mosby, St. Louis.

Forbes, Betty A., Daniel F. Sahm, and Alice S. Weissfeld. 2002. Page 264 in *Bailey & Scott's Diagnostic Microbiology*, 11th ed. Mosby, St. Louis.

Lányi, B. 1987. Page 56 in *Methods in Microbiology*, Vol. 19, edited by R. R. Colwell and R. Grigorova. Academic Press, New York.

MacFaddin, Jean F. 2000. Page 8 in *Biochemical Tests for Identification of Medical Bacteria*, 3rd ed. Lippincott Williams & Wilkins, Philadelphia.

Zimbro, Mary Jo, and David A. Power, eds. 2003. Page 76 in *Difco™ and BBL™ Manual—Manual of Microbiological Culture Media*. Becton Dickinson and Co., Sparks, MD.

TABLE **5-11** **Bile Esculin Test Results and Interpretations** ◆

TABLE OF RESULTS		
Result	**Interpretation**	**Symbol**
Medium is darkened within 48 hours	Presumptive identification as a member of Group D *Streptococcus* or *Enterococcus*	+
No darkening of the medium after 48 hours	Presumptive determination as not a member of Group D *Streptococcus* or *Enterococcus*	−

NAME ______________________ DATE ______________________

LAB SECTION ______________________ I WAS PRESENT AND PERFORMED THIS EXERCISE (INITIALS) ________

Bile Esculin Test

OBSERVATIONS AND INTERPRETATIONS

Refer to Table 5-11 when recording your results and interpretations in the chart below.

Organism	Color Result	+ / −	Interpretation
Control			

Questions

1 Sometimes Bile Esculin Agar is prepared as slants. With that form of the medium, only organisms able to darken more than half of the agar within 48 hours are considered positive (in contrast to the plated medium where any darkening is positive). Why do you think this is so? *Hint:* Consider the designated purposes of agar slants and plates.

2 In terms of the selectivity and differential capabilities of the medium, how would its utility be affected by not including oxgall? Ferric citrate? Explain.

3 Would it be acceptable to read an early negative result for this test? Why or why not?

EXERCISE 5-11

Starch Hydrolysis

Theory

Starch is a polysaccharide made up of α-D-glucose subunits. It exists in two forms—linear (amylose) and branched (amylopectin)—usually as a mixture with the branched configuration being predominant. The α-D-glucose molecules in both amylose and amylopectin are bonded by 1,4-α-glycosidic (acetal) linkages (Figure 5-39). The two forms differ in that amylopectin contains polysaccharide side chains connected to approximately every 30th glucose in the main chain. These side chains are identical to the main chain except that the number 1 carbon of the first glucose in the side chain is bonded to carbon number 6 of the main chain glucose. The bond, therefore, is a 1,6-α-glycosidic linkage.

Starch is too large to pass through the bacterial cell membrane. Therefore, to be of metabolic value to the bacteria, it first must be split into smaller fragments or individual glucose molecules. Organisms that produce and secrete the extracellular enzymes α-**amylase** and **oligo-1,6-glucosidase** are able to hydrolyze starch by breaking the glycosidic linkages between sugar subunits. As shown in Figure 5-39, the end result of these reactions is the complete hydrolysis of the polysaccharide to its individual α-glucose subunits.

Starch agar is a simple plated medium of beef extract, soluble starch, and agar. When organisms that produce α-amylase and oligo-1,6-glucosidase are cultivated on Starch Agar, they hydrolyze the starch in the area surrounding their growth. Because both starch and its sugar subunits are virtually invisible in the medium, the reagent iodine is used to detect the presence or absence of starch in the vicinity around the bacterial growth. Iodine reacts with starch and produces a blue or dark brown color; therefore, any microbial starch hydrolysis will be revealed as a clear zone surrounding the growth (Figure 5-40 and Table 5-12).

Application

Starch agar originally was designed for cultivating *Neisseria*. It no longer is used for this, but with pH indicators, it is

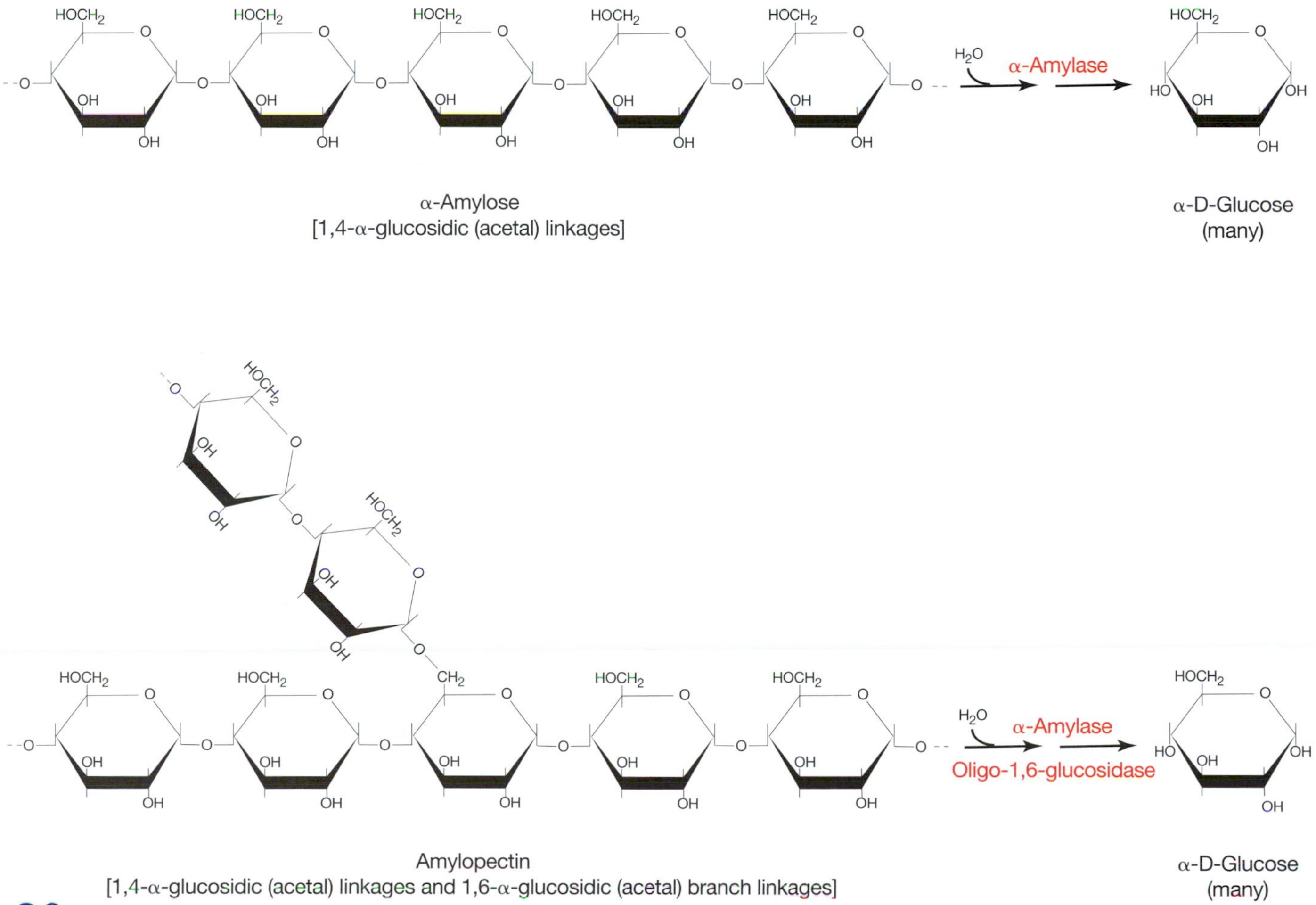

5-39 Starch Hydrolysis by α-Amylase and oligo-1,6-Glucosidase ◆ The enzymes α-amylase and oligo-1,6-glucosidase hydrolyze starch by breaking glycosidic linkages, resulting in release of the component glucose subunits.

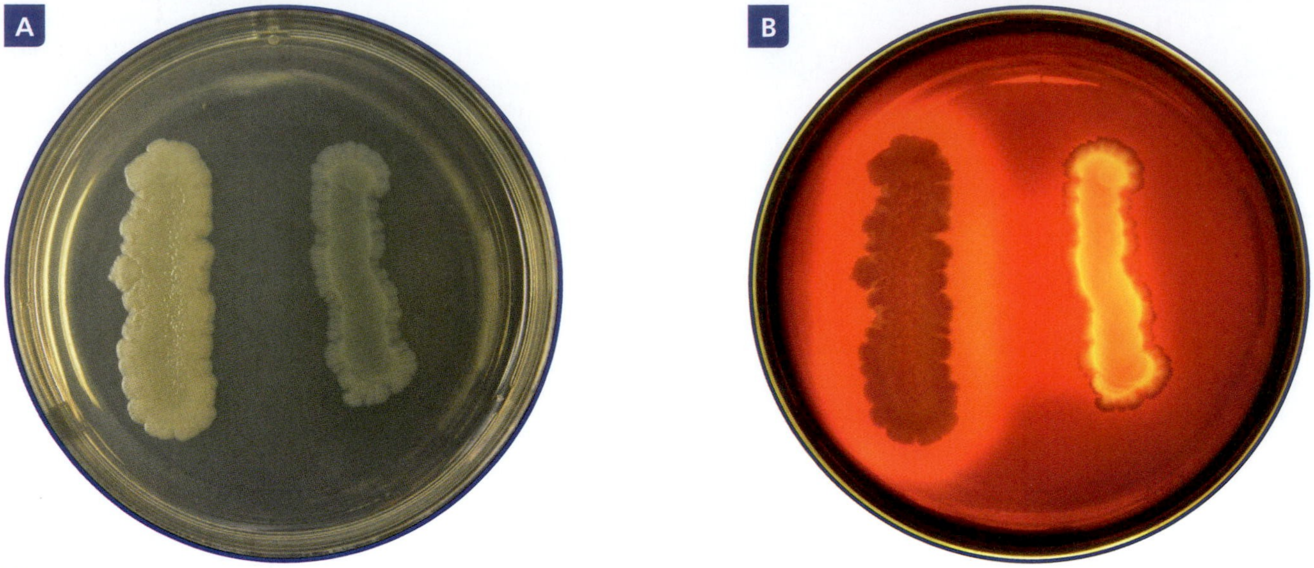

5-40 Starch Hydrolysis Test ◆ **A** This is a Starch Agar plate inoculated with two organisms before iodine has been added. Notice the wavy margins of both organisms. **B** This is the same plate as in photo "A" after iodine has been added. The clearing in the medium around the organism on the left demonstrates a positive result for starch hydrolysis. The organism on the right, with no clearing, is negative. Note that the wavy margin of the negative organism produced a lighter region on the periphery of the growth that might be misinterpreted as clearing. To prevent recording a false positive, it is a good idea to establish the margin of growth for each tested organism prior to adding iodine.

TABLE 5-12 Amylase Test Results and Interpretations ◆

TABLE OF RESULTS		
Result	**Interpretation**	**Symbol**
Clearing around growth	Amylase is present	+
No clearing around growth	No amylase is present	−

used to isolate and presumptively identify *Gardnerella vaginalis*. It aids in differentiating members of the genera *Corynebacterium, Clostridium, Bacillus, Bacteroides, Fusobacterium*, and *Enterococcus*, most of which have positive and negative species.

In This Exercise

You will inoculate a Starch Agar plate with two organisms. After incubation you will add a few drops of iodine to reveal clear areas surrounding the growth. Iodine will not color the agar where the growth is, only the area surrounding the growth. Before you add the iodine, make a note of the growth patterns on the plate so you don't confuse thinning growth at the edges with the halo produced by starch hydrolysis.

Materials

Per Student Group

✓ one Starch Agar plate

✓ Gram iodine (from your Gram stain kit)

✓ recommended organisms:
- *Bacillus cereus*
- *Escherichia coli*

Medium Recipe

Starch Agar

◆ Beef extract	3.0 g
◆ Soluble starch	10.0 g
◆ Agar	12.0 g
◆ Distilled or deionized water	1.0 L

pH 7.3–7.7 at 25°C

Procedure

Lab One

1. Using a marking pen, divide the Starch Agar plate into three equal sectors. Be sure to mark on the bottom of the plate.
2. Label the plate with the names of the organisms, your name, and the date.
3. Spot-inoculate two sectors with the test organisms. Leave the third sector untouched.
4. Invert the plate and incubate it aerobically at 35 ± 2°C for 48 hours.

Lab Two

1. Remove the plate from the incubator and note the location and appearance of the growth.
2. Cover the growth and surrounding areas with Gram iodine. Immediately examine the areas surrounding the growth for clearing. (Usually the growth on the agar prevents contact between the starch and the iodine so no color reaction takes place at that point. Beginning students sometimes look at this lack of color change and judge it incorrectly as a positive result. Therefore, when examining the agar for clearing, look for a halo *around* the growth, not the growth itself.)
3. Record your results on the Data Sheet.

References

Collins, C. H., Patricia M. Lyne, and J. M. Grange. 1995. Page 117 in *Collins and Lyne's Microbiological Methods*, 7th ed. Butterworth-Heinemann, Oxford, United Kingdom.

Lányi, B. 1987. Page 55 in *Methods in Microbiology*, Vol. 19, edited by R. R. Colwell and R. Grigorova. Academic Press, New York.

MacFaddin, Jean F. 2000. Page 412 in *Biochemical Tests for Identification of Medical Bacteria*, 2nd ed. Lippincott Williams & Wilkins, Philadelphia.

Smibert, Robert M., and Noel R. Krieg. 1994. Page 630 in *Methods for General and Molecular Bacteriology*, edited by Philipp Gerhardt, R. G. E. Murray, Willis A. Wood, and Noel R. Krieg. American Society for Microbiology, Washington, DC.

Zimbro, Mary Jo, and David A. Power, eds. 2003. *Difco™ and BBL™ Manual—Manual of Microbiological Culture Media.* Becton Dickinson and Co., Sparks, MD.

NAME ____________________ DATE ____________________

LAB SECTION ____________________ I WAS PRESENT AND PERFORMED THIS EXERCISE (INITIALS) ________

Starch Hydrolysis

OBSERVATIONS AND INTERPRETATIONS

Using Table 5-12 as a guide, enter your results and interpretations in the chart below.

Organism	Result	+ / −	Interpretation
(Control) Sector			

Questions

1 Suppose you had poured iodine on your plate and noticed clearings in the uninoculated area, as well as around both of your transferred cultures. What are some possible explanations for this occurrence? Was integrity of the exercise compromised? What kinds of things might be done to avoid this problem in future exercises?

2 How would you expect the results of this exercise to change if you were to add glucose to the medium?

3 In many tests it is acceptable to read a positive result before the incubation time is completed. Why is this not the case with Starch Agar?

4 Suppose you could selectively prevent production of α-amylase or oligo-1,6-glucosidase in an organism that normally hydrolyzes starch. Which enzyme would the organism miss the most?

EXERCISE 5-12

Urea Hydrolysis

Theory

Urea is a product of decarboxylation of certain amino acids. It can be hydrolyzed to ammonia and carbon dioxide by bacteria containing the enzyme **urease**. Many enteric bacteria (and a few others) possess the ability to metabolize urea (Figure 5-41), but only members of *Proteus*, *Morganella*, and *Providencia* are considered rapid urease-positive organisms.

The only nutrients in Urea Broth are urea and a trace (0.0001%) of yeast extract. It contains buffers that are strong enough to inhibit alkalinization of the medium by all but the rapid urease-positive organisms mentioned above. Phenol red, which is yellow or orange below pH 8.4 and red or pink above, is included to expose any increase in pH. Pink color in the medium in less than 24 hours indicates a rapid urease-positive organism. Orange or yellow is negative (Figure 5-42 and Table 5-13).

Application

The Urea Hydrolysis Test is used to differentiate organisms based on their ability to hydrolyze urea with the enzyme urease. Urinary tract pathogens from the genus *Proteus* may be distinguished from other enteric bacteria by their rapid urease activity.

In This Exercise

You will perform the urea hydrolysis test to determine urease activity of two organisms. This medium usually starts out orange but may turn pale yellow as it incubates. This is an acid reaction and is considered negative. The only positive result is a pink color.

Materials

Per Student Group

✓ three Urea Broths

✓ fresh slant cultures of:

- *Escherichia coli*
- *Proteus vulgaris* surrogate (*Staphylococcus epidermidis*)

Medium Recipe

Rustigian and Stuart's Urea Broth

Yeast extract	0.1 g
Potassium phosphate, monobasic	9.1 g
Potassium phosphate, dibasic	9.5 g
Urea	20.0 g
Phenol red	0.01 g
Distilled or deionized water	1.0 L

pH 6.8 at 25°C

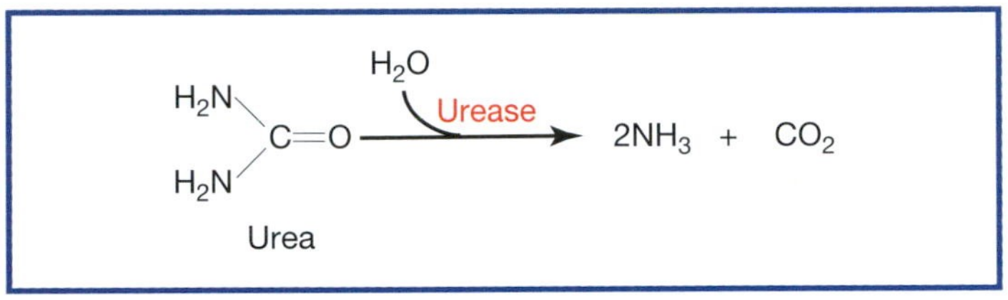

5-41 Urea Hydrolysis ◆ Urea hydrolysis produces ammonia, which raises the pH in the medium and turns the phenol red pink.

5-42 Urea Hydrolysis Test Results ◆ Urease broth tubes after a 24-hour incubation illustrate a urease-positive (+) organism on the left, an uninoculated control in the middle, and urease-negative (−) organism on the right.

TABLE 5-13 Urea Broth Test Results and Interpretations ◆

TABLE OF RESULTS

Result	Interpretation	Symbol
Pink	Rapid urea hydrolysis; strong urease production	+
Orange or yellow	No urea hydrolysis; organism does not produce urease or cannot live in broth	−

Procedure

Lab One

1 Obtain three Urea Broths. Label two with the names of the organisms, your name, and the date. Label the third tube "control."

2 Inoculate two broths with heavy inocula from the test organisms. Do not inoculate the control.

3 Incubate all tubes aerobically at $35 \pm 2°C$ for 24 hours.

Lab Two

1 Remove all tubes from the incubator and examine them for color changes. Record your results on the Data Sheet.

2 Discard all tubes in an appropriate autoclave container.

References

Collins, C. H., Patricia M. Lyne, and J. M. Grange. 1995. Page 117 in *Collins and Lyne's Microbiological Methods,* 7th ed. Butterworth-Heinemann, Oxford, United Kingdom.

Delost, Maria Dannessa. 1997. Page 196 in *Introduction to Diagnostic Microbiology*. Mosby, St. Louis.

DIFCO Laboratories. 1984. Page 1040 in *DIFCO Manual*, 10th, ed. DIFCO Laboratories, Detroit.

Forbes, Betty A., Daniel F. Sahm, and Alice S. Weissfeld. 2002. Page 283 in *Bailey & Scott's Diagnostic Microbiology,* 11th ed. Mosby, St. Louis.

Lányi, B. 1987. Page 24 in *Methods in Microbiology,* Vol. 19, edited by R. R. Colwell and R. Grigorova. Academic Press, New York.

MacFaddin, Jean F. 2000. Page 298 in *Biochemical Tests for Identification of Medical Bacteria,* 3rd ed. Lippincott Williams & Wilkins, Philadelphia.

Smibert, Robert M., and Noel R. Krieg. 1994. Page 630 in *Methods for General and Molecular Bacteriology*, edited by Philipp Gerhardt, R. G. E. Murray, Willis A. Wood, and Noel R. Krieg. American Society for Microbiology, Washington, DC.

Zimbro, Mary Jo, and David A. Power, eds. 2003. Page 606 in *Difco™ and BBL™ Manual—Manual of Microbiological Culture Media.* Becton Dickinson and Co., Sparks, MD.

NAME ______________________ DATE ______________________

LAB SECTION ______________________ I WAS PRESENT AND PERFORMED THIS EXERCISE (INITIALS) ________

Urea Hydrolysis

OBSERVATIONS AND INTERPRETATIONS

Using Table 5-13 as a guide, enter your results for the urea hydrolysis test after 24 hours.

Organism	Color Result	+ / −	Interpretation
Control			

Questions

1 Suppose you ran this test with *Providencia stuartii*, but it took 48 hours to turn pink. Do you think this is a false result? If so, is it a false positive or false negative? If not, why not? Give some possible explanations for this occurrence.

2 Explain why it is acceptable to record positive tests before the suggested incubation time is completed but it is not acceptable to record a negative result early

3 Typically, this medium is filter-sterilized because autoclaving breaks down the urea. Even unsterilized broth, however, rarely produces false-positive results. Why do you think this is true?

EXERCISE 5-13

Casein Hydrolysis Test

Theory

Many bacteria require proteins as a source of amino acids and other components for synthetic processes. Some bacteria have the ability to produce and secrete enzymes into the environment that catalyze the breakdown of large proteins to smaller peptides or individual amino acids, thus enabling their uptake across the membrane (Figure 5-43).

Casease is an enzyme that some bacteria produce to hydrolyze the milk protein **casein**, the molecule that gives milk its white color. When broken down into smaller fragments, the ordinarily white casein loses its opacity and becomes clear.

The presence of casease can be detected easily with the test medium Milk Agar (Figure 5-44). Milk Agar is an undefined medium containing pancreatic digest of casein, yeast extract, dextrose, and powdered milk. When plated Milk Agar is inoculated with a casease-positive organism, secreted casease will diffuse into the medium around the colonies and create a zone of clearing where the casein has been hydrolyzed. Casease-negative organisms do not secrete casease and do not produce clear zones in the medium.

Application

The casein hydrolysis test is used for the cultivation and differentiation of bacteria that produce the enzyme casease.

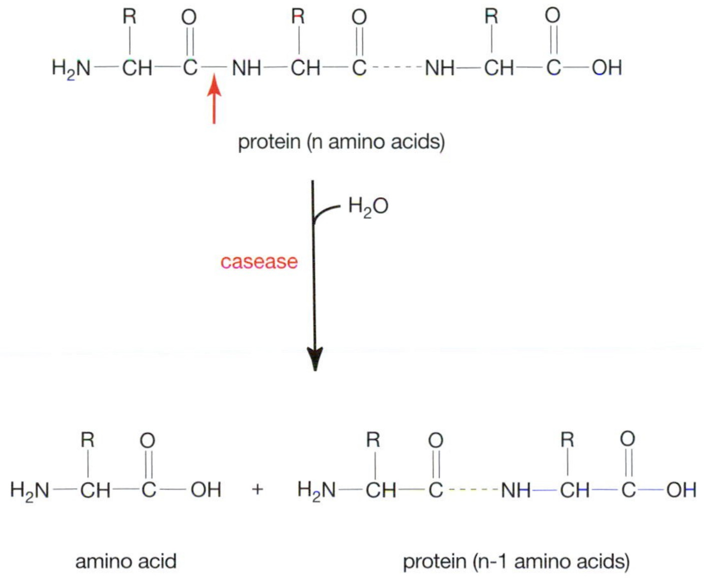

5-43 Casein Hydrolysis ◆ Hydrolysis of any protein occurs by breaking peptide bonds (red arrow) between adjacent amino acids to produce short peptides or individual amino acids.

In This Exercise

You will inoculate a single Milk Agar plate with a casease-positive and a casease-negative organism. Use Table 5-14 as a guide when making your interpretations.

Materials

Per Student Group

- ✓ one Milk Agar plate
- ✓ fresh cultures of:
  - *Bacillus cereus*
  - *Escherichia coli*

Medium Recipe

Skim Milk Agar

◆ Pancreatic digest of casein	5.0 g
◆ Yeast extract	2.5 g
◆ Powdered nonfat milk	100.0 g
◆ Glucose	1.0 g
◆ Agar	15.0 g
◆ Distilled or deionized water	1.1 L

pH 6.1 at 25°C

5-44 Casein Hydrolysis Test Results ◆ This Milk Agar plate was inoculated with a casease-positive organism above and a casease-negative organism below.

TABLE 5-14 Casease Test Results and Interpretations ◆

TABLE OF RESULTS

Result	Interpretation	Symbol
Clearing in agar	Casease is present	+
No clearing in agar	Casease is absent	–

Procedure

Lab One

1 Using a marking pen, divide the plate into three equal sectors. Be sure to mark on the bottom of the plate.

2 Label the plate with the names of the organisms, your name, and the date.

3 Spot-inoculate two sectors with the test organisms. Leave the third sector untouched.

4 Invert the plate and incubate it aerobically at 35 ± 2°C for 24 hours.

Lab Two

1 Examine the plates for clearing around the bacterial growth.

2 Record your results on the Data Sheet.

References

Atlas, Ronald M. 2004. Page 1563 in *Handbook of Microbiological Media,* 3rd ed. CRC Press LLC, Boca Raton, FL.

Holt, John G., Ed. 1994. *Bergey's Manual of Determinative Bacteriology,* 9th ed. Williams and Wilkins, Baltimore.

Smibert, Robert M., and Noel R. Krieg. 1994. Page 613 in *Methods for General and Molecular Bacteriology*, edited by Philipp Gerhardt, R. G. E. Murray, Willis A. Wood, and Noel R. Krieg. American Society for Microbiology, Washington, DC.

Zimbro, Mary Jo, and David A. Power, eds. 2003. Page 364 in *Difco™ and BBL™ Manual—Manual of Microbiological Culture Media.* Becton Dickinson and Co., Sparks, MD.

Name ______________________ Date ______________________

Lab Section ______________________ I was present and performed this exercise (initials) ________

Casein Hydrolysis Test

OBSERVATIONS AND INTERPRETATIONS

Using Table 5-14 as a guide, record your results in the chart below.

Organism	Result	+ / −	Interpretation
(Control) Sector			

Questions

1 How do we know that casease is an exoenzyme and not a cytoplasmic enzyme?

2 Is it acceptable to read a positive test before the incubation time is completed? How about an early negative result?

3 Why is the uninoculated sector relatively unnecessary in this test?

4 Why is it advisable to use a positive control along with organisms that you are testing?

EXERCISE 5-14

Gelatin Hydrolysis Test

Theory

Gelatin is a protein derived from collagen—a component of vertebrate connective tissue. **Gelatinases** comprise a family of extracellular enzymes produced and secreted by some microorganisms to hydrolyze gelatin. Subsequently, the cell can take up individual amino acids and use them for metabolic purposes. Bacterial hydrolysis of gelatin occurs in two sequential reactions, as shown in Figure 5-45.

The presence of gelatinases can be detected using Nutrient Gelatin, a simple test medium composed of gelatin, peptone, and beef extract. Nutrient Gelatin differs from most other solid media in that the solidifying agent (gelatin) is also the substrate for enzymatic activity. Consequently, when a tube of Nutrient Gelatin is stab-inoculated with a gelatinase-positive organism, secreted gelatinase (or gelatinases) will liquefy the medium. Gelatinase-negative organisms do not secrete the enzyme and do not liquefy the medium (Figures 5-46 and 5-47). A 7-day incubation period is usually sufficient to see liquefaction of the medium. However, gelatinase activity is very slow in some organisms. All tubes still negative after 7 days should be incubated an additional 7 days.

A slight disadvantage of Nutrient Gelatin is that it melts at 28°C (82°F). Therefore, inoculated stabs are typically incubated at 25°C along with an uninoculated control to verify that any liquefaction is not temperature-related.

Application

The Gelatin Hydrolysis Test is used to determine the ability of a microbe to produce gelatinases. *Staphylococcus aureus,* which is gelatinase-positive, can be differentiated from *S. epidermidis*. *Serratia* and *Proteus* species are positive members of *Enterobacteriaceae,* whereas most others in the family are negative. *Bacillus anthracis, B. cereus*, and several other members of the genus are gelatinase-positive, as are *Clostridium tetani* and *C. perfringens*.

In This Exercise

You will inoculate two Nutrient Gelatin tubes and incubate them for up to a week. Gelatin liquefaction is a positive result but is difficult to differentiate from gelatin that has melted as a result of temperature. Therefore, be sure to incubate a control tube along with the others. If the control melts as a result of temperature, refrigerate all tubes until it resolidifies. Use Table 5-15 as a guide for interpretation.

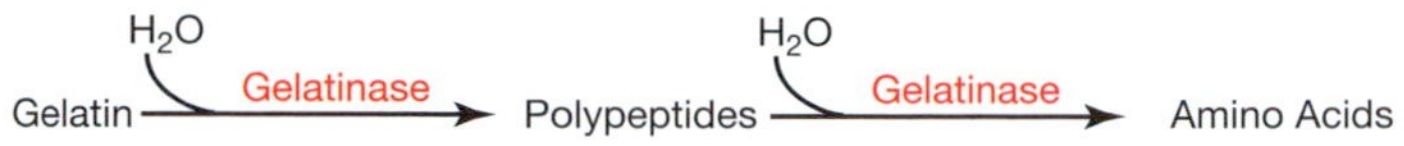

H_2O
H_2O
Gelatin
Gelatinase
Polypeptides
Gelatinase
Amino Acids

5-45 Gelatin Hydrolysis ◆ Gelatin is hydrolyzed by the gelatinase family of enzymes.

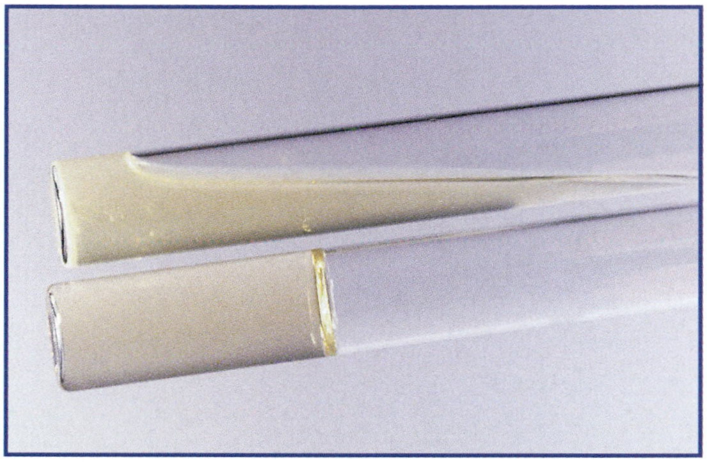

5-46 Nutrient Gelatin Stabs ◆ A gelatinase-positive (+) organism is above, and a gelatinase–negative (−) organism is below.

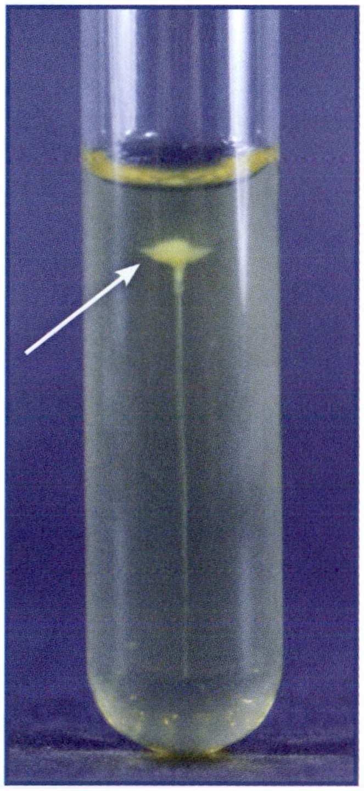

5-47 Crateriform Liquefaction ◆ This form of liquefaction also may be of diagnostic use because not all gelatinase-positive microbes liquefy the gelatin completely. Shown here is an organism liquefying the gelatin in the shape of a crater (arrow).

TABLE 5-15 Gelatinase Test Results and Interpretations ◆

TABLE OF RESULTS		
Result	**Interpretation**	**Symbol**
Gelatin is liquid (control is solid)	Gelatinase is present	+
Gelatin is solid	No gelatinase is present	−

Materials

Per Student Group

- ✓ three Nutrient Gelatin stabs
- ✓ ice bath
- ✓ fresh cultures of:
  - *Bacillus subtilis*
  - *Escherichia coli*

Medium Recipe

Nutrient Gelatin

♦ Beef Extract	3.0 g
♦ Peptone	5.0 g
♦ Gelatin	120.0 g
♦ Distilled or deionized water	1.0 L

pH 6.6–7.0 at 25°C

Procedure

Lab One

1. Obtain three Nutrient Gelatin stabs. Label each of two tubes with the name of the organism, your name, and the date. Label the third tube "control."
2. Stab-inoculate two tubes with heavy inocula of the test organisms. Do not inoculate the control.
3. Incubate all tubes at 25°C for up to 1 week.

Lab Two

1. Examine the control tube. If the gelatin is solid, the test can be read. If it is liquefied, place all tubes in the ice bath until the control has resolidified.
2. When the control has solidified, examine the inoculated media for gelatin liquefaction.
3. Record your results on the Data Sheet.

References

Collins, C. H., Patricia M. Lyne, and J. M. Grange. 1995. Page 112 in *Collins and Lyne's Microbiological Methods,* 7th ed. Butterworth-Heinemann, Oxford, United Kingdom.

Forbes, Betty A., Daniel F. Sahm, and Alice S. Weissfeld. 2002. Page 271 in *Bailey & Scott's Diagnostic Microbiology,* 11th ed. Mosby, St. Louis.

Lányi, B. 1987. Page 44 in *Methods in Microbiology,* Vol. 19, edited by R. R. Colwell and R. Grigorova. Academic Press, New York.

MacFaddin, Jean F. 2000. Page 128 in *Biochemical Tests for Identification of Medical Bacteria,* 3rd ed. Lippincott Williams & Wilkins, Philadelphia.

Smibert, Robert M., and Noel R. Krieg. 1994. Page 617 in *Methods for General and Molecular Bacteriology*, edited by Philipp Gerhardt, R. G. E. Murray, Willis A. Wood, and Noel R. Krieg. American Society for Microbiology, Washington, DC.

Zimbro, Mary Jo, and David A. Power, eds. 2003. Page 408 in *Difco™ and BBL™ Manual—Manual of Microbiological Culture Media.* Becton Dickinson and Co., Sparks, MD.

NAME ______________________ DATE ______________________

LAB SECTION ______________________ I WAS PRESENT AND PERFORMED THIS EXERCISE (INITIALS) ________

DATA SHEET 5-14

Gelatin Hydrolysis Test

OBSERVATIONS AND INTERPRETATIONS

Using Table 5-15 as a guide, record your liquefaction results in the chart below.

Organism	Result	+ / −	Interpretation
Control			

Questions

1 Some microbiologists recommend incubating this medium at 37°C, along with an uninoculated control, and then transferring all tubes to the refrigerator prior to reading them. Why might this be the preferred technique in some situations? What potential problems can you see with this method?

2 If the control is solid and an inoculated tube is liquid, is it acceptable to read the result before the complete incubation time has elapsed? Why or why not?

3 If the control is solid and a tube is also solid, is it acceptable to read the result before the complete incubation time has elapsed? Why?

4 Suggest some ways by which an organism could be a slow gelatin liquefier.

5 Suppose that after 7 days a tube inoculated with a slow liquefier shows no evidence of liquefaction. Is this a failure of the test system? If yes, why? If not, why not?

EXERCISE 5-15

DNA Hydrolysis Test

Theory

An enzyme that catalyzes the hydrolysis of DNA into small fragments (oligonucleotides) or single nucleotides is called a deoxyribonuclease or DNase (Figure 5-48). DNase is an exoenzyme; that is, an enzyme that is secreted by a cell and acts on the substrate extracellularly. Extracellular digestion typically allows utilization of a macro molecule too large to be transported into the cell. Ability to produce DNase can be determined by culturing and observing a suspected organism on a DNase Test Agar plate.

DNase Test Agar contains peptides derived from soybean and casein that serve as carbon and nitrogen sources, sodium chloride for osmotic balance, DNA as the substrate, and methyl green dye as an indicator of DNase activity. The dye forms a complex with polymerized DNA and gives the uninoculated agar a blue-green color. No complex is formed with nucleotides. Therefore, clearing that develops around bacterial growth on this medium is an indication of DNase activity (DNA hydrolysis) and is considered a positive result (Figure 5-49).

Application

DNase Test Agar is used to distinguish *Serratia* species (+) from *Enterobacter* species, *Moraxella catarrhalis* (+) from *Neisseria* species, and *Staphylococcus aureus* (+) from other *Staphylococcus* species.

In This Exercise

You will inoculate one DNase plate with two of the organisms the test was designed to differentiate. For best results, limit incubation time to 24 hours. Depending on the age of the plates and the length of incubation, the clearing around a DNase-positive organism may be subtle and difficult to see. When reading the plate, it might be helpful to look through it while holding it several inches above a white piece of paper. Use Table 5-16 to assist in interpretation.

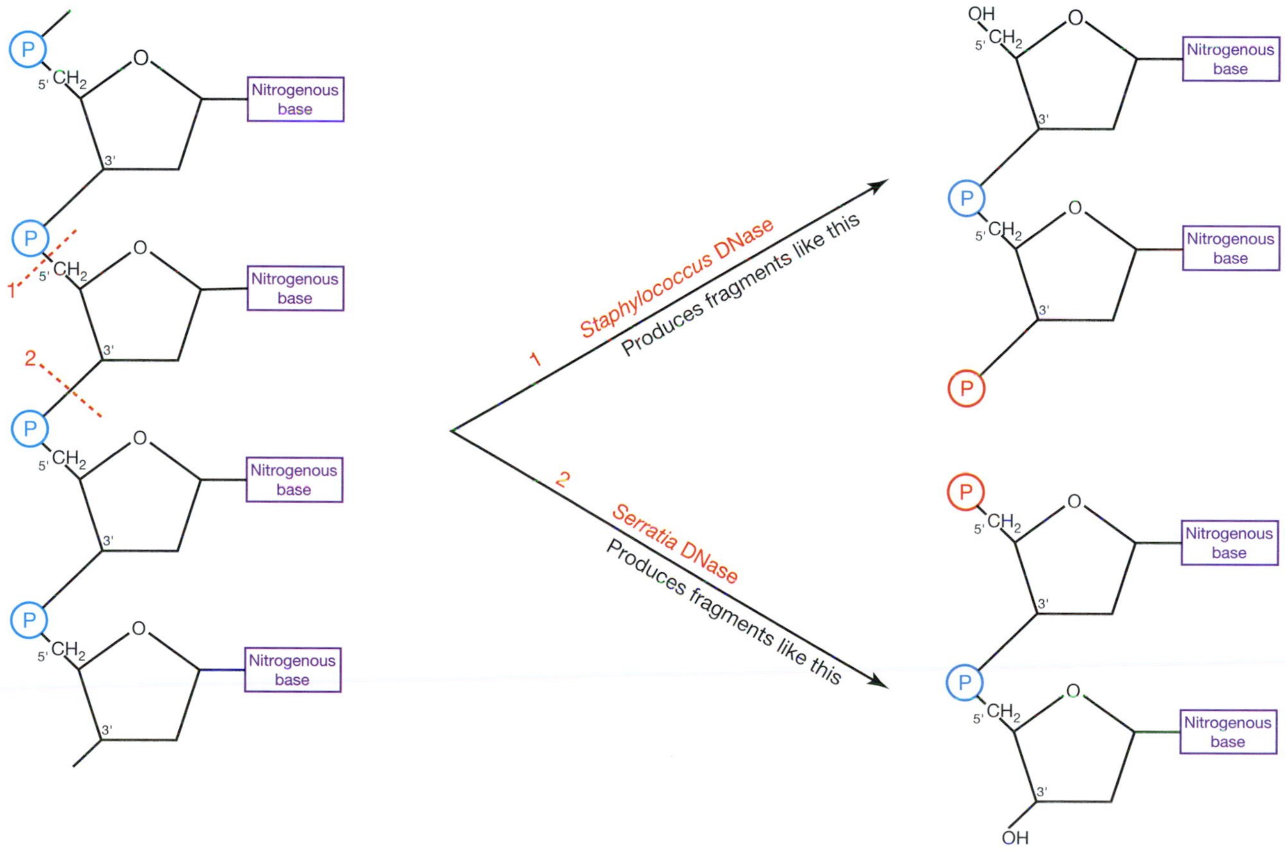

5-48 Two Patterns of DNA Hydrolysis ◆ DNase from *Staphylococcus* hydrolyzes DNA at the bond between the 5′-carbon and the phosphate (illustrated by line 1), thereby producing fragments with a free 3′-phosphate (shown in red on the upper fragment). Most fragments are one or two nucleotides long. A dinucleotide is shown here. *Serratia* DNase cleaves the bond between the phosphate and the 3′-carbon (illustrated by line 2) and produces fragments with free 5′-phosphates (shown in red on the lower fragment). Most fragments are two to four nucleotides in length. A dinucleotide is shown here.

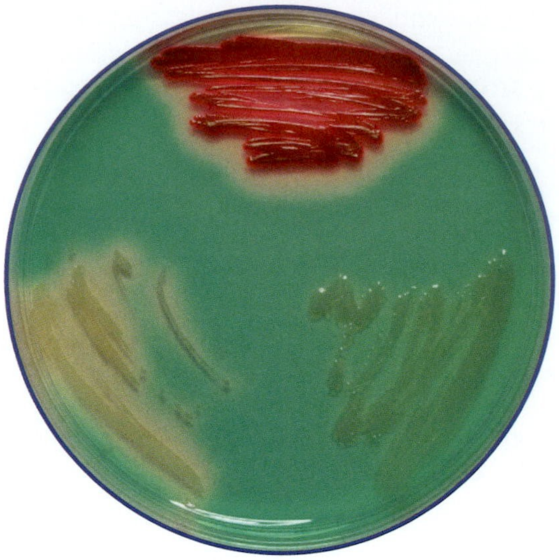

5-49 DNA Hydrolysis Test Results ◆ This DNase Test Agar plate was inoculated with three of the organisms chosen for this exercise. Note the clearing around the DNase-positive organisms on the top and left. The organism on the lower right is DNase-negative.

TABLE 5-16 DNA Hydrolysis Test Results and Interpretations ◆

TABLE OF RESULTS		
Result	**Interpretation**	**Symbol**
Clearing in agar around growth	DNAse is present	+
No clearing in agar around growth	DNAse is absent	–

Materials

Per Student Group

✓ one DNase Test Agar plate

✓ fresh cultures of:

- *Staphylococcus aureus* surrogate (*Aeromonas sobria* Carolina Biological #154828)
- *Serratia marcescens*
- *Enterobacter aerogenes*

Medium Recipe

DNase Test Agar with Methyl Green

- Tryptose 20.0 g
- Deoxyribonucleic acid 2.0 g
- Sodium chloride 5.0 g
- Agar 15.0 g
- Methyl green 0.05 g
- Distilled or deionized water 1.0 L

pH 7.1–7.4 at 25°C

Procedure

Lab One

1. Using a marking pen, divide the DNase Test Agar plate into four equal sectors. Be sure to mark the bottom of the plate.
2. Label the plate with the names of the organisms, your name, and the date.
3. Spot-inoculate three sectors with the test organisms and leave the fourth sector as a control.
4. Invert the plate and incubate it aerobically at 35 ± 2°C for 24 hours. (If your labs are scheduled more than 24 hours apart, arrange to have someone place your plates in a refrigerator until you can examine them.)

Lab Two

1. Examine the plates for clearing around the bacterial growth.
2. Record your results on the Data Sheet.

References

Collins, C. H., Patricia M. Lyne, and J. M. Grange. 1995. Page 114 in *Collins and Lyne's Microbiological Methods,* 7th ed. Butterworth-Heinemann, Oxford, United Kingdom.

Delost, Maria Dannessa. 1997. Page 111 in *Introduction to Diagnostic Microbiology*. Mosby, St. Louis.

Forbes, Betty A., Daniel F. Sahm, and Alice S. Weissfeld. 2002. Page 268 in *Bailey & Scott's Diagnostic Microbiology,* 11th ed. Mosby, St. Louis.

Lányi, B. 1987. Page 33 in *Methods in Microbiology,* Vol. 19, edited by R. R. Colwell and R. Grigorova. Academic Press, New York.

MacFaddin, Jean F. 2000. Page 136 in *Biochemical Tests for Identification of Medical Bacteria,* 3rd ed. Lippincott Williams & Wilkins, Philadelphia.

Zimbro, Mary Jo, and David A. Power, eds. 2003. Page 170 in *Difco™ and BBL™ Manual—Manual of Microbiological Culture Media.* Becton Dickinson and Co., Sparks, MD.

Name ______________________ Date ______________________

Lab Section ______________________ I was present and performed this exercise (initials) ________

DNA Hydrolysis Test

OBSERVATIONS AND INTERPRETATIONS

Using Table 5-16 as a guide, record your DNAse results in the chart below.

Organism	Result	+ / −	Interpretation
(Control) Sector			

Questions

1 In Theory on the first page of this exercise, the disassembly of DNA was described as a "depolymerization." What other term applies to the process?

2 A positive result for the DNA hydrolysis test does not distinguish between *Staphylococcus* DNase and the DNase produced by *Serratia*. If you had the expertise to correct this weakness in the system, would you improve the test's sensitivity or its specificity?

3 Suggest a reason why this test is read after only 24 hours while other tests (e.g., gelatinase test) may take a week.

4 Why is it advisable to use a positive control when testing unknown organisms?

EXERCISE 5-16

Lipid Hydrolysis Test

Theory

The word **lipid** is generally used to describe all types of fats. Bacterial enzymes that hydrolyze lipids are called **lipases**. Because lipases are secreted by the organism to digest lipids outside the cell, they are called extracellular enzymes or exoenzymes. Similar in function to other types of exoenzymes (Exercises 5-11, 5-13, and 5-14) lipases are necessary because intact lipids are macromolecules and too large to be transported into the cell. One purpose of this exercise is to demonstrate that bacteria can be differentiated based on their ability to produce and secrete lipases.

Simple fats, known as triglycerides or triacylglycerols, are composed of glycerol and three long-chain fatty acids. After lipolysis (lipid hydrolysis), the glycerol can be converted to dihydroxyacetone phosphate, which is an intermediate of glycolysis (Figures 5-50 and A-1). The fatty acids are catabolized by a process called **β-oxidation**. Two carbon fragments from the fatty acid are combined with Coenzyme A to produce Acetyl-CoA, which then may be used in the Krebs cycle to produce energy. Each Acetyl-CoA produced by this process also yields one NADH and one $FADH_2$ (important coenzymes in the electron transport chain). Alternatively, glycerol and fatty acids may be used in anabolic pathways.

A variety of simple fats can be used for the Lipid Hydrolysis Test, but tributyrin oil is a common choice because it is the simplest triglyceride found in natural fats. Tributyrin Agar is prepared as an emulsion that makes the agar appear opaque. When the plate is inoculated with a lipase-positive organism, clear zones will appear around the growth as evidence of lipolytic activity (Figure 5-51). If no clear zones appear, the organism is lipase-negative. Table 5-17 summarizes the results and interpretations.

Application

The lipase test is used to detect and enumerate lipolytic bacteria, especially in high-fat dairy products. A variety of other lipid substrates, including corn oil, olive oil, and

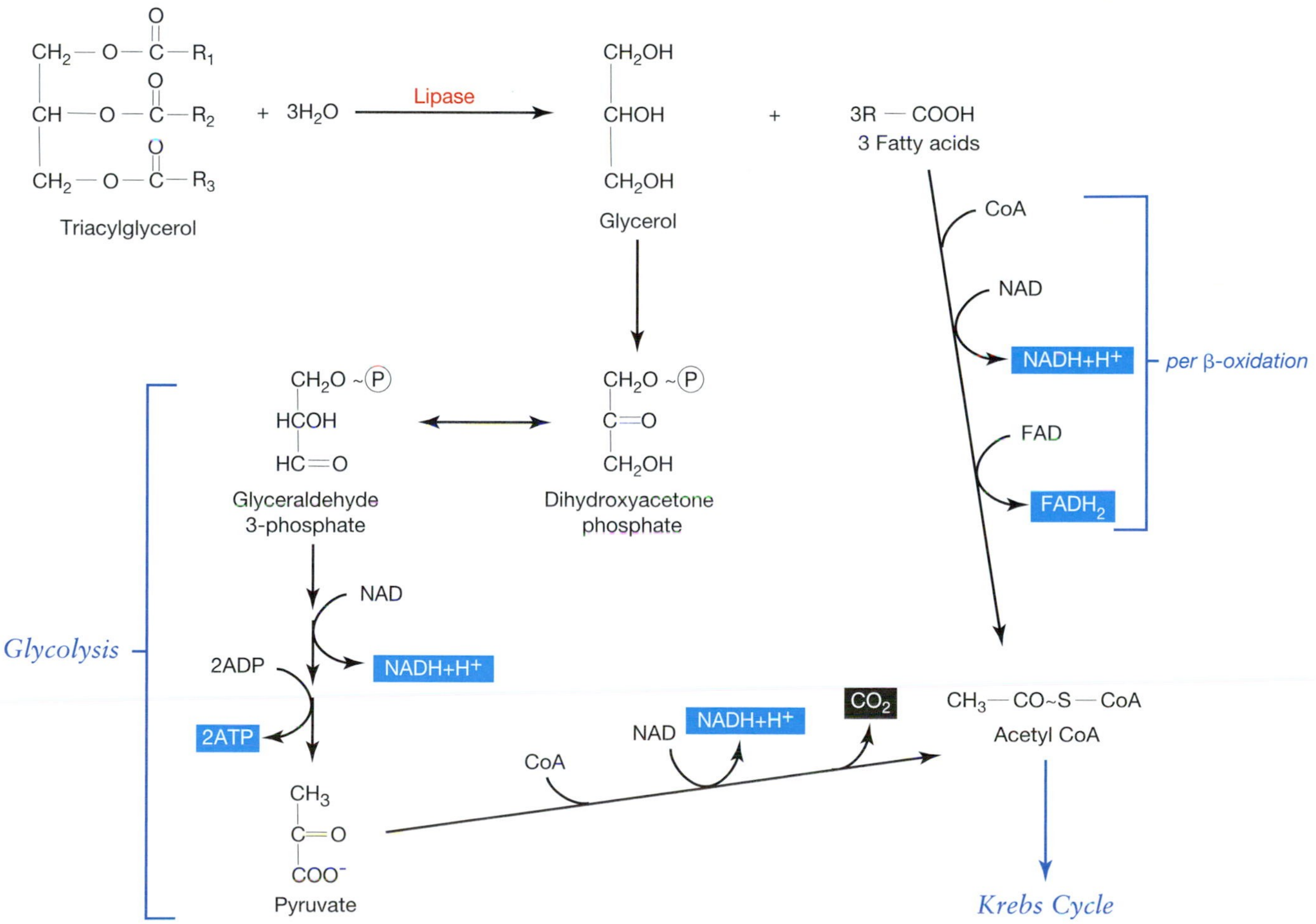

5-50 Lipid Metabolism ◆ A triacylglycerol is a simple fat molecule, composed of a three-carbon alcohol (glycerol) bonded to three fatty acid chains (represented by R_1, R_2, and R_3 in this diagram. The products of lipid catabolism can be used in glycolysis (glycerol) and the Krebs cycle (acetyl CoA).

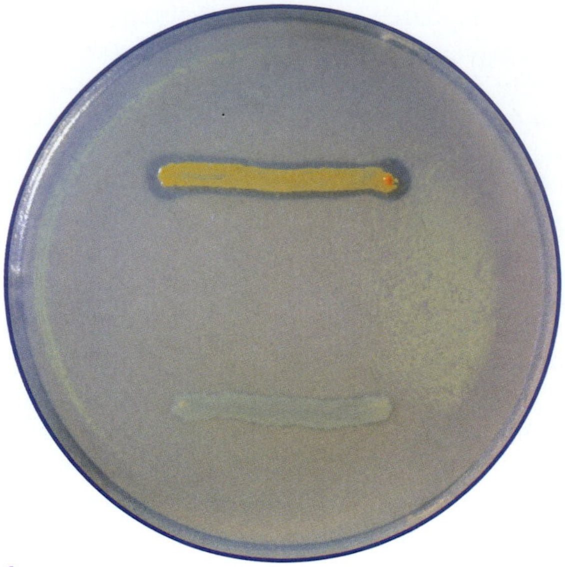

5-51 Lipid Hydrolysis Test Results ◆ This Tributyrin Agar plate was inoculated with a lipase-positive (+) organism above and a lipase-negative (−) organism below.

TABLE 5-17 Lipase Test Results and Interpretations ◆

TABLE OF RESULTS		
Result	**Interpretation**	**Symbol**
Clearing in agar around growth	Lipase is present	+
No clearing in agar around growth	Lipase is absent	−

soybean oil, are used to detect differential characteristics among members of *Enterobacteriaceae*, *Clostridium*, *Staphylococcus*, and *Neisseria*. Several fungal species also demonstrate lipolytic ability.

In This Exercise

You will inoculate a Tributyrin Agar plate with two organisms, one of which will produce a clearing in the medium surrounding the growth.

Materials

Per Student Group

✓ one Tributyrin Agar plate

✓ fresh cultures of:

- *Bacillus subtilis*
- *Escherichia coli*

Media Recipe

Tributyrin Agar

◆ Beef extract	3.0 g
◆ Peptone	5.0 g
◆ Agar	15.0 g
◆ Tributyrin oil	10.0 mL
◆ Distilled or deionized water	1.0 L

pH 5.8–6.2 at 25°C

Procedure

Lab One

1. Using a marking pen, divide the bottom of the plate into three equal sectors.
2. Label the plate with the names of the organisms, your name, and the date.
3. Spot-inoculate two sectors with the test organisms, leaving the third sector uninoculated as a control.
4. Invert the plate and incubate it aerobically at 35 ± 2°C for 24 to 48 hours.

Lab Two

Examine the plates for clearing around the bacterial growth, and record your results on the Data Sheet. If no clearing has developed, reincubate for 24 to 48 hours.

References

Collins, C. H., Patricia M. Lyne, and J. M. Grange. 1995. Page 114 in *Collins and Lyne's Microbiological Methods,* 7th ed. Butterworth-Heinemann, Oxford, United Kingdom.

Haas, Michael J. 2001. Chapter 15 in *Compendium of Methods for the Microbiological Examination of Foods,* 4th ed., edited by Frances Pouch Downes and Keith Ito. American Public Health Association, Washington, DC.

Knapp, Joan S., and Roselyn J. Rice. 1995. Page 335 in *Manual of Clinical Microbiology,* 6th ed., edited by Patrick R. Murray, Ellen Jo Baron, Michael A. Pfaller, Fred C. Tenover, and Robert H. Yolken. ASM Press, Washington, DC.

MacFaddin, Jean F. 2000. Page 286 in *Biochemical Tests for Identification of Medical Bacteria,* 3rd ed. Lippincott Williams & Wilkins, Philadelphia.

Zimbro, Mary Jo, and David A. Power, eds. 2003. Page 521 in *Difco™ and BBL™ Manual—Manual of Microbiological Culture Media.* Becton Dickinson and Co., Sparks, MD.

Name ______________________ Date ______________________

Lab Section ______________________ I was present and performed this exercise (initials) ________

Lipid Hydrolysis Test

OBSERVATIONS AND INTERPRETATIONS

Using Table 5-17 as a guide, record your results and interpretations in the table below.

Organism	Result	+ / −	Interpretation
(Control) Sector			

Questions

1 Many organisms possessing many different lipases produce positive results on Tributyrin Agar. Is the inability of this medium to distinguish between these different enzymes a weakness in its specificity or its sensitivity?

2 Tributyrin Agar has a shelf life of only a few days before it loses its opacity.

a. With this in mind, explain the importance of positive and negative controls in this test.

b. How would expired Tributyrin Agar affect the results of lipase (+) and lipase (−) organisms?

3 Why do you think Tributyrin Agar is prepared in the blender as an emulsion rather than simply stirred as a solution?

EXERCISE 5-17

PYR Test

Theory

Group A streptococci (*S. pyogenes*) and enterococci produce the enzyme L-pyrrolidonyl arylamidase. This enzyme hydrolyzes the amide pyroglutamyl-β-naphthylamide to produce L-pyrrolidone and β-naphthylamine, all of which are colorless. β-naphthylamine will react with *p*-dimethylaminocinnamaldehyde and form a red precipitate (Figure 5-52).

PYR may be performed as an 18-hour agar test, a four-hour broth test or, as used in this example, a rapid disc test. In each case the medium (or disc) contains pyroglutamyl-β-naphthylamide to which is added a heavy inoculum of the test organism. After the appropriate incubation or waiting period, a 0.01% *p*-dimethylaminocinnamaldehyde solution is added. Formation of a deep red color within a few minutes is interpreted as PYR-positive. Yellow or orange is PYR-negative (Figure 5-53).

Application

The PYR test is designed for presumptive identification of group A streptococci (*S. pyogenes*) and enterococci by determining the presence of the enzyme L-pyrrolidonyl arylamidase.

In This Exercise

You will inoculate two discs with test organisms to determine which one contains the enzyme L-pyrrolidonyl arylamidase.

Materials

Per Student Group

- ✓ one empty Petri dish
- ✓ PYR discs
- ✓ fresh slant cultures of:
  - *Enterococcus faecalis* (BSL-2)
  - *Staphylococcus epidermidis*

Procedure

1. Obtain a Petri dish, the PYR discs, and two sterile pipettes. (A clean microscope slide can be used in place of the Petri dish.)
2. Open the Petri dish and mark two spots where you will place the PYR discs. Label each spot with the name of its respective organism.
3. Place two PYR discs inside on the appropriate spots.
4. With a loop or other appropriate inoculating device, inoculate each disc with a large amount of its respective organism.
5. Add a drop of the *p*-dimethylaminocinnamaldehyde solution (PYR reagent) to each disc.
6. If the disc turns red within 5 minutes, it is positive. Orange or no color change are negative results.

1. L-Pyrrolidonyl–β–naphthylamide (colorless substrate) —L-pyrrolidonyl arylamidase→ L-Pyrrolidone (colorless) + β–naphthylamine (colorless)

2. β–naphthylamine (colorless substrate) + p–Dimethylaminocinnamaldehyde (PYR reagent) ⟶ Schiff base ↓ (red precipitate)

5-52 PYR Chemistry ◆ Reaction 1 shows the hydrolytic splitting of the substrate L–pyrrolidonyl–β–naphthylamide included in the medium. In this form of the test, the filter-paper discs were saturated with substrate. Both products of the reaction are colorless. Reaction 2 illustrates the reaction of the PYR reagent and β–naphthylamine. In the reaction, the amino group (H_2N) of β–naphthylamine reacts with the aldehyde group (–CH=O) of the reagent and produces a Schiff base (–CH=N–), which is red.

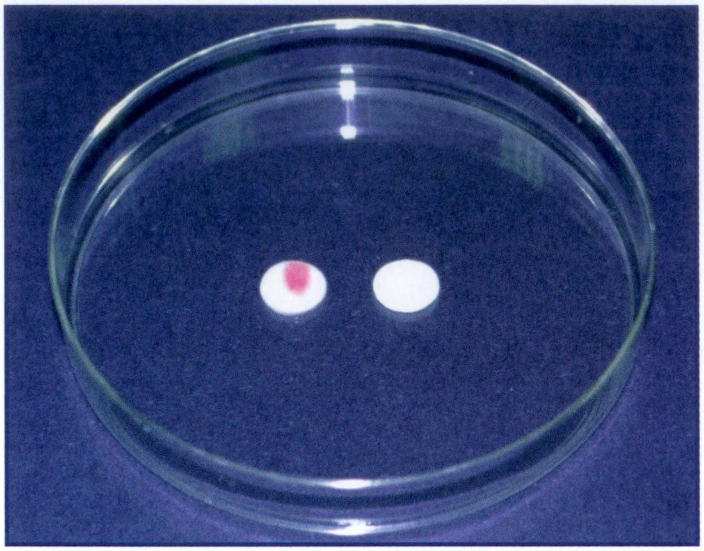

5-53 **PYR Disk Test** ◆ The disk on the left was inoculated with a PYR-positive organism; the disk on the right was inoculated with a PYR–negative organism.

TABLE **5-18** **PYR Test Results and Interpretations** ◆

TABLE OF RESULTS		
Result	**Interpretation**	**Symbol**
Red color formation	Organism produces L-pyrrolidonyl arylamidase	+
Orange color or no color change	Organism does not produce L-pyrrolidonyl arylamidase	−

References

Claus, G. William. 1989. Chapter 14 in *Understanding Microbes—A Laboratory Textbook for Microbiology*. W.H. Freeman and Company, New York, NY.

Collins, C. H., Patricia M. Lyne and J. M. Grange. 1995. Chapter 6 in *Collins and Lyne's Microbiological Methods*, 7th Ed. Butterworth-Heineman, Oxford.

Forbes, Betty A., Daniel F. Sahm, and Alice S. Weissfeld. 2002. Chapter 10 in *Bailey & Scott's Diagnostic Microbiology*, 11th Ed. Mosby-Yearbook, St. Louis, MO.

Koneman, Elmer W., Stephen D. Allen, William M. Janda, Paul C. Schreckenberger and Washington C. Winn, Jr. 1997. Chapter 2 in *Color Atlas and Textbook of Diagnostic Microbiology*, 5th Ed. J. B. Lippincott Company, Philadelphia, PA.

Name ____________________ Date ____________________

Lab Section ____________________ I was present and performed this exercise (initials) ________

PYR Test

OBSERVATIONS AND INTERPRETATIONS

Using Table 5-18 as a guide, enter your results and interpretations in the table below.

Organism	Result	+ / −	Interpretation

There are no questions for this exercise.

Combination Differential Media

Combination differential media combine components of several compatible tests into one medium, thus saving critical diagnostic time and money. Typically, they include core tests to differentiate members of specific bacterial groups. SIM medium, for example, tests for sulfur reduction, indole production, and motility—important characteristics of various members of *Enterobacteriaceae*.

As with most biochemical tests, combination media can be used as a followup to selective media. Selective media promote isolation of the organisms of interest and help determine the sequence of tests to follow. MacConkey Agar (Exercise 4-4), for example, may be streaked initially to encourage and isolate Gram-negative organisms, followed by Triple Sugar Iron Agar (TSIA) or Kligler Iron Agar (KIA). TSIA and KIA detect various fermentations and sulfur reduction.

EXERCISE 5-18

SIM Medium

Theory

SIM medium is used for determination of three bacterial activities: sulfur reduction, indole production from tryptophan, and motility. The semisolid medium includes casein and animal tissue as sources of amino acids, an iron-containing compound, and sulfur in the form of sodium thiosulfate.

Sulfur reduction to H_2S can be accomplished by bacteria in two different ways, depending on the enzymes present.

1. The enzyme **cysteine desulfurase** catalyzes the putrefaction of the amino acid cysteine to pyruvate (Figure 5-54).
2. The enzyme **thiosulfate reductase** catalyzes the reduction of sulfur (in the form of sulfate) at the end of the anaerobic respiratory electron transport chain (Figure 5-55).

Both systems produce hydrogen sulfide (H_2S) gas. When either reaction occurs in SIM medium, the H_2S that is produced combines with iron, in the form of ferrous ammonium sulfate, to form ferric sulfide (FeS), a black precipitate (Figure 5-56). Any blackening of the medium is an indication of sulfur reduction and a positive test. No blackening of the medium indicates no sulfur reduction and a negative reaction (Figure 5-57).

Indole production in the medium is made possible by the presence of tryptophan (contained in casein and animal protein). Bacteria possessing the enzyme **tryptophanase** can hydrolyze tryptophan to pyruvate, ammonia (by deamination), and indole (Figure 5-58).

The hydrolysis of tryptophan in SIM medium can be detected by the addition of Kovacs' reagent after a period of incubation. Kovacs' reagent contains **dimethylaminobenzaldehyde** (DMABA) and HCl dissolved in amyl alcohol. When a few drops of Kovacs' reagent are added to the tube, DMABA reacts with any indole present and produces a quinoidal compound that turns the reagent layer red (Figures 5-59 and 5-60). The formation of red color in the reagent layer indicates a positive reaction and the presence of tryptophanase. No red color is indole-negative.

Determination of motility in SIM medium is made possible by the reduced agar concentration and the method of inoculation. The medium is inoculated with a single stab from an inoculating needle. Motile organisms are able to move about in the semisolid medium and can be detected by the radiating growth pattern extending outward in all directions from the central stab line. Growth that radiates in *all* directions and appears slightly fuzzy is an indication

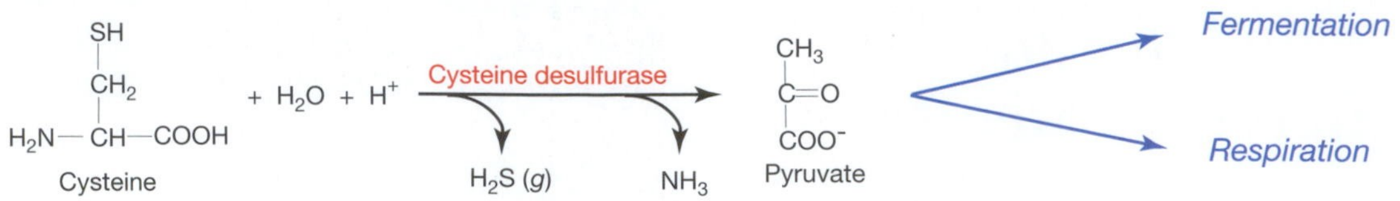

5-54 Putrefaction of Cysteine ◆ Putrefaction involving cysteine desulfurase produces H_2S. This reaction is a mechanism for getting energy out of the amino acid cysteine.

$$3S_2O_3^{=} + 4H^+ + 4e^- \xrightarrow{\text{Thiosulfate reductase}} 2SO_3^{=} + 2H_2S\ (g)$$

5-55 Reduction of Thiosulfate ◆ Anaerobic respiration with thiosulfate as the final electron acceptor also produces H_2S.

$$H_2S + FeSO_4 \longrightarrow H_2SO_4 + FeS\ (g)$$

5-56 Indicator Reaction ◆ Hydrogen sulfide, a colorless gas, can be detected when it reacts with ferrous sulfate in the medium to produce the black precipitate ferric sulfide.

5-57 Sulfur Reduction in SIM Medium ◆ The organism on the left is H_2S-negative. The organism on the right is H_2S-positive.

of motility (Figure 5-61). This should not be confused with the (seemingly) spreading growth produced by lateral movement of the inoculating needle when stabbing.

❖ Application

SIM medium is used to identify bacteria that are capable of producing indole, using the enzyme tryptophanase. The Indole Test is one component of the IMViC battery of tests (*I*ndole, *M*ethyl red, *V*oges-Proskauer, and *C*itrate) used to differentiate the *Enterobacteriaceae*. SIM medium also is used to differentiate sulfur-reducing members of *Enterobacteriaceae*, especially members of the genera *Salmonella, Francisella,* and *Proteus* from the negative *Morganella morganii* and *Providencia rettgeri*. In addition to the first two functions of SIM, motility is an important differential characteristic of *Enterobacteriaceae*.

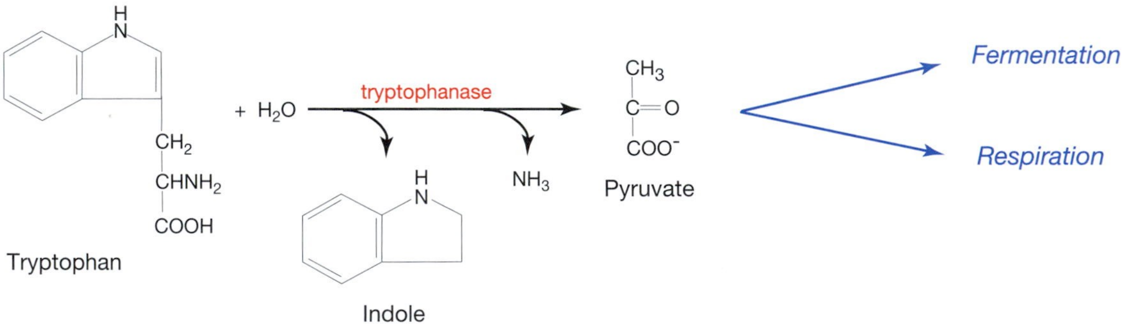

5-58 Tryptophan Catabolism in Indole-Positive Organisms ◆ This reaction is a mechanism for getting energy out of the amino acid tryptophan.

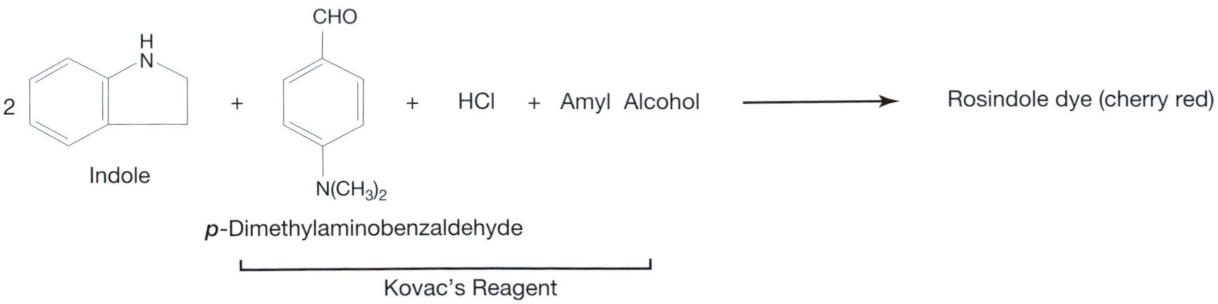

5-59 Indole Reaction with Kovacs' Reagent ◆ If present, indole reacts with the DMABA in Kovacs' reagent to produce a red color.

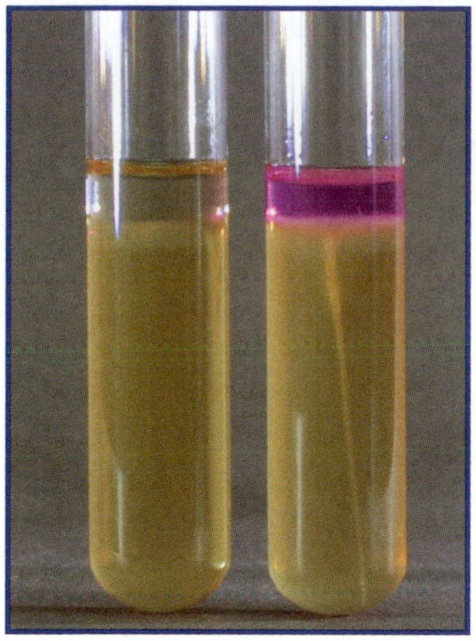

5-60 **Indole Test Results** ◆ These SIM tubes were inoculated with an indole-negative (−) organism on the left and an indole-positive (+) organism on the right.

5-61 **Motility in SIM** ◆ These SIM tubes were inoculated with a motile organism on the left and a nonmotile organism on the right.

In This Exercise

You will stab-inoculate SIM medium with three organisms demonstrating a variety of results. When reading your test results, make the motility and H_2S determinations before adding the indole reagent. Use Tables 5-19, 5-20 and 5-21, and Figures 5-57, 5-60, and 5-61 as guides when making your interpretations.

Materials

Per Student Group

- ✓ four SIM tubes
- ✓ Kovacs' reagent
- ✓ fresh cultures of:
  - ◆ *Escherichia coli*
  - ◆ *Citrobacter freundii* (Hardy Diagnostics #0315P)
  - ◆ *Klebsiella pneumoniae* surrogate (*Micrococcus luteus*)

Medium and Reagent Recipes

SIM (Sulfur-Indole-Motility) Medium

- ◆ Pancreatic digest of casein 20.0 g
- ◆ Peptic digest of animal tissue 6.1 g
- ◆ Ferrous ammonium sulfate 0.2 g
- ◆ Sodium thiosulfate 0.2 g
- ◆ Agar 3.5 g
- ◆ Distilled or deionized water 1.0 L

pH 7.1–7.4 at 25°C

Kovacs' Reagent

- ◆ Amyl alcohol 75.0 mL
- ◆ Hydrochloric acid, concentrated 25.0 mL
- ◆ *p*-dimethylaminobenzaldehyde 5.0 g

TABLE 5-19 **Sulfur Reduction Results and Interpretations** ◆

TABLE OF RESULTS

Result	Interpretation	Symbol
Black in the medium	Sulfur reduction (H_2S production)	+
No black in the medium	Sulfur is not reduced	−

TABLE 5-20 **Indole Production Results and Interpretations** ◆

TABLE OF RESULTS

Result	Interpretation	Symbol
Red in the alcohol layer of Kovacs' reagent	Tryptophan is broken down into indole and pyruvate	+
Reagent color is unchanged	Tryptophan is not broken down into indole and pyruvate	−

TABLE 5-21 **Motility Results and Interpretations** ◆

TABLE OF RESULTS

Result	Interpretation	Symbol
Growth radiating outward from the stab line	Motility	+
No radiating growth	Nonmotile	−

Procedure

Lab One

1 Obtain four SIM tubes. Label three with the names of the organisms, your name, and the date. Label one tube "control."

2 Stab-inoculate three tubes with the test organisms. Insert the needle into the agar to within 1 cm of the bottom of the tube. Be careful to remove the needle along the original stab line. Do not inoculate the control.

3 Incubate all tubes aerobically at 35 ± 2°C for 24 to 48 hours.

Lab Two

1 Examine the tubes for spreading from the stab line *and* formation of black precipitate in the medium. Record any H_2S production and/or motility on the Data Sheet.

2 Add Kovacs' reagent to each tube (to a depth of 2–3 mm). After several minutes, observe for the formation of red color in the reagent (upper) layer.

3 Record your results on the Data Sheet.

References

Delost, Maria Dannessa. 1997. Page 186 in *Introduction to Diagnostic Microbiology*. Mosby, St. Louis.

MacFaddin, Jean F. 1980. Page 162 in *Biochemical Tests for Identification of Medical Bacteria*, 2nd ed. Williams & Wilkins, Baltimore.

Zimbro, Mary Jo, and David A. Power, eds. 2003. Page 490 in *Difco™ and BBL™ Manual—Manual of Microbiological Culture Media*. Becton Dickinson and Co., Sparks, MD.

NAME ____________________ DATE ____________________

LAB SECTION ____________________ I WAS PRESENT AND PERFORMED THIS EXERCISE (INITIALS) ________

SIM Medium

OBSERVATIONS AND INTERPRETATIONS

1 Using Table 5-19 as a guide, record your results and interpretations in the chart below.

Sulfur Reduction			
Organism	Black PPT? Y / N	+ / −	Interpretation
Control			

2 Using Table 5-20 as a guide, record your results and interpretations in the chart below.

Indole Production			
Organism	Red Color? Y / N	+ / −	Interpretation
Control			

3 Using Table 5-21 as a guide, record your results and interpretations in the chart below.

	Motility		
Organism	Growth Pattern	+ / −	Interpretation
Control			

Questions

1 The sulfur reduction test is not able to differentiate H_2S produced by anaerobic respiration and H_2S produced by putrefaction. Is this inability the result of poor sensitivity or poor specificity of the test system?

2 What factors dictate the choice of tests included in a combination medium?

3 Virtually all sulfur-reducing members of *Enterobacteriaceae* are motile, as shown in Figure 5-57. Suppose for a moment that you inoculate a SIM medium with an indole-producing sulfur reducer that is nonmotile. After completing all aspects of the test, what would you expect the result to look like?

4 How would you expect this result to change if you eliminated the sodium thiosulfate? Ferrous ammonium sulfate?

EXERCISE 5-19

Triple Sugar Iron Agar / Kligler Iron Agar[1]

❖ Theory

Triple Sugar Iron Agar (TSIA) is a rich medium designed to differentiate bacteria on the basis of glucose fermentation, lactose fermentation, sucrose fermentation, and sulfur reduction. In addition to the three carbohydrates, it includes animal proteins as carbon and nitrogen sources, and both ferrous sulfate and sodium thiosulfate as sources of oxidized sulfur. Phenol red is the pH indicator, and ferrous sulfate is the hydrogen sulfide indicator.

The medium is prepared as a shallow agar slant with a deep butt, thereby providing both aerobic and anaerobic growth environments. It is inoculated by a stab in the agar butt followed by a fishtail streak of the slant. The incubation period is 18 to 24 hours for carbohydrate fermentation and up to 48 hours for hydrogen sulfide reactions. Many reactions in various combinations are possible (Figure 5-62 and Table 5-22).

When TSIA is inoculated with a glucose-only fermenter, acid products lower the pH and turn the entire medium yellow within a few hours. Because glucose is in short supply (0.1%), it will be exhausted within about 12 hours. As the glucose is used up, the organisms located in the aerobic region (slant) will begin to break down available amino acids, producing NH_3 and raising the pH. This process, which takes 18 to 24 hours to complete, is called a **reversion** and is limited to the slant because of the anaerobic conditions in the butt. Thus, a TSIA with a red slant and yellow butt after a 24-hour incubation period indicates that the organism ferments glucose but not lactose or sucrose.

Organisms that are able to ferment glucose *and* lactose *and/or* sucrose also turn the medium yellow throughout. However, because the lactose and sucrose concentrations are 10 times higher than that of glucose, resulting in greater acid production, both slant and butt will remain yellow after 24 hours. Therefore, a TSIA with a yellow slant and butt at 24 hours indicates that the organism ferments glucose and one or both of the other sugars. Gas produced by fermentation of any of the carbohydrates will appear as fissures in the medium or will lift the agar off the bottom of the tube.

Hydrogen sulfide (H_2S) may be produced by the reduction of thiosulfate in the medium or by the breakdown of cysteine in the peptone. Ferrous sulfate in the medium reacts with the H_2S to form a black precipitate, usually seen in the butt. Acid conditions must exist for thiosulfate reduction; therefore, black precipitate in the medium is an indication of sulfur reduction *and* fermentation. If the black precipitate obscures the color of the butt, the color of the slant determines which carbohydrates have been fermented (i.e., red slant = glucose fermentation, yellow slant = glucose and lactose and/or sucrose fermentation).

An organism that does not ferment any of the carbohydrates but utilizes animal proteins will alkalinize the medium and turn it red. If the organism can use the proteins aerobically and anaerobically, both the slant and butt will appear red. An obligate aerobe will turn only the slant red. (Note: The difference between a red butt and a butt unchanged by the organism may be subtle; therefore, comparison with an uninoculated control is always recommended.)

Timing is critical when reading results. An early reading could reveal yellow throughout the medium, leading one to believe that the organism ferments more than one sugar when it simply may not yet have exhausted the glucose. A reading after all of the sugars have been depleted could reveal a yellow butt and red slant leading one to falsely believe that the organism is a glucose-only fermenter. Tubes that are negative for sulfur reduction after 24 hours should be re-incubated an additional 24 hours before H_2S evaluation. Refer to Table 5-22 for information on the correct symbols and method of reporting the various reactions.

5-62 Kligler Iron Agar Slants ◆ These KIA slants from left to right illustrate: alkaline slant/acid butt (K/A); alkaline slant/no change in the butt (K/NC); uninoculated control; alkaline slant/acid butt, hydrogen sulfide present (K/A, H_2S); and acid slant/acid butt, gas (A/A, G). Refer to Table 5-22 for interpretations.

[1] This exercise is written for either Triple Sugar Iron Agar (TSIA) or Kligler Iron Agar (KIA). In form and function the two media are virtually identical except that KIA does not include sucrose. The results of both media look exactly the same. To avoid the awkwardness of repeatedly including both names, we refer only to TSIA. If you are using KIA, simply substitute the name where applicable. Table 5-22 contains result interpretations for both tests.

Application

TSIA and KIA are primarily used to differentiate members of *Enterobacteriaceae* and to distinguish them from other Gram-negative rods such as *Pseudomonas aeruginosa*.

In This Exercise

Today, you will inoculate four TSIA or KIA slants. Use a large inoculum and try not to introduce excessive air when stabbing the agar butt. Also be sure to remove all tubes from the incubator after no more than 24 hours to take fermentation readings.

Materials

Per Student Group

- ✓ five TSIA or KIA slants
- ✓ recommended organisms (grown on solid media):
  - *Alcaligenes faecalis*
  - *Escherichia coli*
  - *Citrobacter koseri* (Hardy Diagnostics #0106P)
  - *Citrobacter freundii* (Hardy Diagnostics #0315P)

Media Recipes

Triple Sugar Iron Agar

- Pancreatic Digest of Casein 10.0 g
- Peptic Digest of Animal Tissue 10.0 g
- Dextrose (glucose) 1.0 g
- Lactose 10.0 g
- Sucrose 1.0 g
- Ferrous Ammonium Sulfate 0.5 g
- Sodium chloride 5.0 g
- Sodium thiosulfate 0.2 g
- Agar 13.0 g
- Phenol red 0.025 g
- Distilled or deionized water 1.0 L

pH 7.1–7.5 at 25°C

Kligler Iron Agar

- Pancreatic Digest of Casein 10.0 g
- Peptic Digest of Animal Tissue 10.0 g
- Lactose 10.0 g
- Dextrose (glucose) 1.0 g
- Ferric Ammonium Citrate 0.5 g
- Sodium chloride 5.0 g

TABLE **5-22** TSIA/KIA Results and Interpretations

TABLE OF RESULTS

Result	Interpretation	Symbol
Yellow slant/yellow butt—KIA	Glucose and lactose fermentation with acid accumulation in slant and butt.	A/A
Yellow slant/yellow butt—TSIA	Glucose and lactose and/or sucrose fermentation with acid accumulation in slant and butt.	A/A
Red slant/yellow butt	Glucose fermentation with acid production. Proteins catabolized aerobically (in the slant) with alkaline products (reversion).	K/A
Red slant/red butt	No fermentation. Proteins catabolized aerobically and anaerobically with alkaline products. Not from *Enterobacteriaceae*.	K/K
Red slant/no change in the butt	No fermentation. Proteins catabolized aerobically with alkaline products. Not from *Enterobacteriaceae*.	K/NC
No change in slant/ no change in butt	Organism is growing slowly or not at all. Not from *Enterobacteriaceae*.	NC/NC
Black precipitate in the agar	Sulfur reduction. (An acid condition, from fermentation of glucose or lactose, exists in the butt even if the yellow color is obscured by the black precipitate.)	H_2S
Cracks in or lifting of agar	Gas production	

- Sodium thiosulfate 0.5 g
- Agar 15.0 g
- Phenol red 0.025 g
- Distilled or deionized water 1.0 L

pH 7.2 – 7.6 at 25°C

Procedure

Lab One

1 Obtain five slants. Label four of the slants with the names of the organisms, your name, and the date. Label the fifth slant "control."

2 Inoculate four slants with the test organisms. Using a heavy inoculum, stab the agar butt and then streak the slant. Do not inoculate the control.

3 Incubate all slants aerobically at 35 ± 2°C for 24 hours.

Lab Two

Examine the tubes for characteristic color changes and gas production. Use Table 5-22 as a guide while recording your results on the Data Sheet. The proper format for recording results is: slant reaction/butt reaction, gas production, hydrogen sulfide production. For example, a tube showing yellow slant, yellow butt, fissures in or lifting of the agar, and black precipitate would be recorded as: A/A, G, H_2S.

References

Delost, Maria Dannessa. 1997. Pages 184–185 in *Introduction to Diagnostic Microbiology.* Mosby, St. Louis.

Forbes, Betty A., Daniel F. Sahm, and Alice S. Weissfeld. 2002. Page 282 in *Bailey & Scott's Diagnostic Microbiology,* 11th ed. Mosby, St. Louis.

MacFaddin, Jean F. 2000. Page 239 in *Biochemical Tests for Identification of Medical Bacteria,* 3rd ed. Lippincott Williams & Wilkins, Philadelphia.

Zimbro, Mary Jo, and David A. Power, eds. 2003. Page 283 in *Difco™ and BBL™ Manual—Manual of Microbiological Culture Media.* Becton Dickinson and Co., Sparks, MD.

Name ______________________ Date ______________________

Lab Section ______________________ I was present and performed this exercise (initials) ________

Triple Sugar Iron Agar / Kligler Iron Agar

OBSERVATIONS AND INTERPRETATIONS

Refer to Table 5-22 when recording and interpreting your results.

Organism	Color Result	Symbol	Interpretation
Control			

Questions

1 As mentioned in Theory, the fermentation readings with TSIA and KIA must take place between 18 and 24 hours after inoculation. Why is this true? Is timing as critical with H_2S readings? Why or why not?

2 You learned in Theory that if the black precipitate obscures the color of the butt that it must be acidic and scored as "A." Why do you think this is true? *Hint:* See Figures 5-55 and 5-56.

3 TSIA and KIA are complex media with many ingredients. What would be the consequences of the following mistakes in preparing this medium? Consider each independently.

a. 1% glucose is added rather than the amount specified in the recipe.

b. Ferrous ammonium sulfate (or ferric ammonium citrate in KIA) is omitted.

c. Casein and animal tissue are omitted.

d. Sodium thiosulfate is omitted.

e. Phenol red is omitted.

f. The initial pH is 8.2.

g. The agar butt is shallow rather than deep.

Antimicrobial Susceptibility Testing

Antimicrobials are substances used to control or kill microorganisms. Some are produced naturally by microorganisms; others are synthetics produced in the laboratory. Most bacteria show susceptibility to one or more antimicrobial agents and, as illustrated in Exercise 7-2, may even react in predictable ways to the agent.

Susceptibility testing is frequently used in medicine to identify the antibiotic needed to treat an infection. But because of the predictable responses of some organisms to specific agents, susceptibility testing can also be helpful in microbial identification. All of the antibacterial susceptibility tests included in this unit apply the **disc-diffusion method** and are used to aid in bacterial identification.

In the disc-diffusion method, a commercially prepared filter paper disc impregnated with a specific concentration of antibiotic is placed on an agar plate immediately after it is inoculated to produce a "bacterial lawn." During incubation, two things happen simultaneously, 1) the antibacterial agent diffuses from the disc into the agar, becoming less concentrated the farther it travels, and 2) bacterial growth slowly develops on the agar surface.

In the area where antibiotic concentration in the agar is sufficient, growth of susceptible bacteria will be inhibited. This results in a clearing surrounding the disc called the **zone of inhibition**. The size of the zone relates to the **minimum inhibitory concentration (MIC)** of the antibacterial agent that is effective against the test organism.

The National Committee for Clinical Laboratory Standards (NCCLS) periodically publishes a Zone Diameter Interpretive Chart for use in testing and determining bacterial susceptibility or resistance to various antibacterial agents. The chart lists the concentrations of all currently used antibacterial agents and their applicable zone diameters.

It is important to understand that even resistant organisms may be inhibited by heavy concentrations of an antimicrobial agent. Therefore, the *size* of the zone, not the mere presence of a zone, is what determines susceptibility or resistance. Additionally, specificity of disc-diffusion tests is maintained largely by the amount of antibacterial agent added to the disc. For example, many streptococci are susceptible to heavy concentrations of optochin, but resistant to small concentrations. Conversely, *S. pneumoniae* is susceptible to very small concentrations of optochin. By preparing discs with a scant 5 µg of optochin, the test system is able to accurately differentiate *S. pneumoniae* from other organisms in the genus.

EXERCISE 5-20

Bacitracin, Novobiocin, and Optochin Susceptibility Tests

Theory[1]

Bacitracin, produced by *Bacillus licheniformis*, is a powerful peptide antibiotic that inhibits bacterial cell wall synthesis (Figure 5-63). Thus, it is effective only on bacteria that have cell walls and are in the process of growing. In this test, any zone of clearing 10 mm or greater around the disk is interpreted as bacitracin susceptibility (Figure 5-64). For more information on antimicrobial susceptibility, refer to Exercise 7-3, Antimicrobial Susceptibility Test.

Novobiocin is an antibiotic, produced by *Streptomyces niveus*. It interferes with ATPase activity and ultimately the production of ATP. In this test, a 5 µg disc should produce a zone of clearing 16 mm or more to be considered novobiocin susceptible (Figure 5-65).

Optochin is an antibiotic that interferes with ATPase activity and ATP production in susceptible bacteria. A 6 mm disc (containing 5 µg of optochin) should produce a zone of inhibition 14 mm or more to be considered optochin susceptible (Figure 5-66).

Application

The bacitracin test is used to differentiate and presumptively identify β-hemolytic group A streptococci (*Streptococcus pyogenes*) from other β-hemolytic streptococci. It also differentiates the genus *Staphylococcus* (resistant) from the susceptible *Micrococcus* and *Stomatococcus*.

The novobiocin test is used to differentiate coagulase-negative staphylococci. Most frequently it is used to presumptively identify the novobiocin-resistant *Staphylococcus saprophyticus*.

[1] Because of the repetitive nature of antibacterial susceptibility tests, information that applies to this and other tests appears only once in the introduction to this unit. Include that information as part of Theory in this exercise.

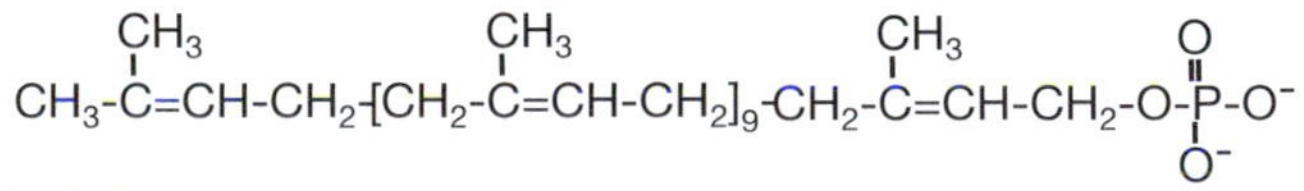

CH3-C=CH-CH2-[CH2-C=CH-CH2]9-CH2-C=CH-CH2-O-P-O⁻

5-63 Undecaprenyl Phosphate ◆ Undecaprenyl phosphate is involved in transporting peptidoglycan subunits across the cell membrane during cell wall synthesis. Bacitracin interferes with its release from the peptidoglycan subunit. It is a C_{55} molecule derived from 11 isoprene subunits plus a phosphate.

5-64 **Bacitracin Susceptibility on a Sheep Blood Agar Plate** ◆ The organism above has no clear zone and is resistant (R); the organism below has a clear zone larger than 10 mm and is susceptible (S).

5-65 **Novobiocin Disk Test** ◆ A novobiocin-resistant (R) organism is on the left; a susceptible (S) organism is on the right.

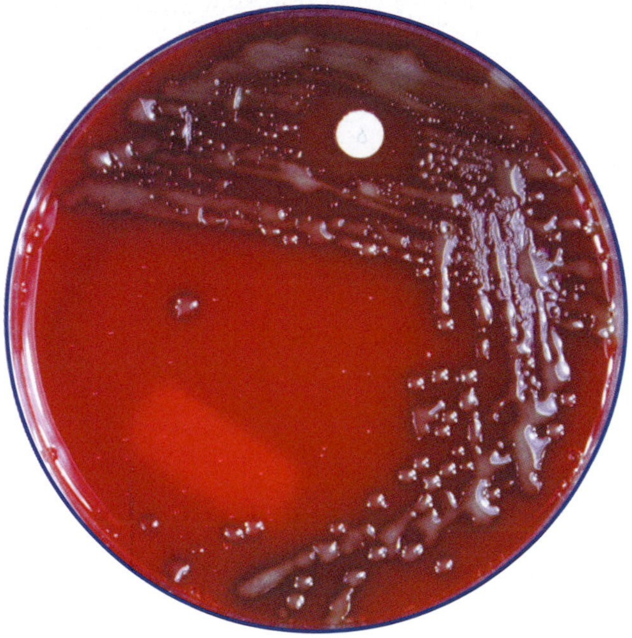

5-66 **Optochin Susceptibility Test** ◆ The zone of inhibition surrounding the disk indicates susceptibility to optochin and presumptive identification of *S. pneumoniae*.

The optochin test is used to presumptively differentiate *Streptococcus pneumoniae* from other α-hemolytic streptococci.

❖ In This Exercise

You will inoculate from one to three Blood Agar plates (depending on the tests chosen by your instructor) to produce confluent growth of the test organisms. You will then place specific antibiotic discs in the center of the inoculum. Following incubation, you will measure the clear zones surrounding the discs to determine susceptibility or resistance of the test organisms to the antibiotics.

❖ Materials

Per Student Group

- ✓ one Blood Agar plate per test (commercial preparation of TSA containing 5% sheep blood)
- ✓ sterile cotton applicators
- ✓ beaker of alcohol with forceps

Bacitracin Test

- ✓ 0.04 unit bacitracin discs
- ✓ fresh broth cultures of:
  - ◆ *Staphylococcus aureus* (BSL-2)
  - ◆ *Micrococcus luteus*

Novobiocin Test

- ✓ 5 µg novobiocin discs
- ✓ fresh broth cultures of:
  - ◆ *Staphylococcus epidermidis*
  - ◆ *Staphylococcus saprophyticus*

Optochin Test

- ✓ 5 µg (6 mm) optochin discs
- ✓ fresh broth cultures of:
  - ◆ *Streptococcus pneumoniae* (BSL-2)
  - ◆ *Streptococcus agalactiae* (BSL-2)

Procedure (Per Test)

Lab One

1 Obtain one Blood Agar plate for each test you are performing, label them accordingly, and draw a line on the bottom dividing the plate into two halves. Using a sterile cotton applicator, inoculate half of the plate with one of the two test organisms. (Make the inoculum as light as possible by wiping and twisting the wet cotton swab on the inside of the culture tube before removing it.) Inoculate the plate by making a single streak nearly halfway across its diameter. Turn the plate 90° and spread the organism evenly to produce a bacterial lawn covering nearly half the agar surface. Refer to the photo in Figure 5-64.

2 Being careful not to mix the cultures, repeat the process on the other half of the plate using the second test organism. Allow the broth to be absorbed by the agar for 5 minutes before proceeding to Step 3.

3 Sterilize the forceps by placing them in the Bunsen burner flame long enough to ignite the alcohol. (Note: Do not hold the forceps in the flame.) Once the alcohol has burned off, use the forceps to place an antibacterial disc in the center of the half containing organism 1. Gently tap the disk into place to ensure that it makes full contact with the agar surface. Return the forceps to the alcohol.

4 Repeat step 3, placing a disc on the other half of the plate—the area containing organism 2. Tap the disk into place and return the forceps to the alcohol.

5 Invert the plate and incubate it at 35 ± 2°C for 24 to 48 hours.

Lab Two

1 Remove all plates from the incubator and examine them for clearing around the disks.

2 Measure all zones of inhibition and record them on the Data Sheet.

TABLE 5-23 Bacitracin Test Results and Interpretations

TABLE OF RESULTS

Result	Interpretation	Symbol
Zone of clearing 10 mm or greater	Organism is sensitive to bacitracin	+
Zone of clearing less than 10 mm	Organism is resistant to bacitracin	−

TABLE 5-24 Novobiocin Test Results and Interpretations

TABLE OF RESULTS

Result	Interpretation	Symbol
Zone of clearing 16 mm or greater	Organism is sensitive to novobiocin	+
Zone of clearing less than 16 mm	Organism is resistant to novobiocin	−

TABLE 5-25 Optochin Test Results and Interpretations

TABLE OF RESULTS

Result	Interpretation	Symbol
Zone of clearing 14 mm or greater	Organism is sensitive to optochin	+
Zone of clearing less than 14 mm	Organism is resistant to optochin	−

References

Baron, Ellen Jo, Lance R. Peterson, and Sydney M. Finegold. 1994. Page 329 in *Bailey & Scott's Diagnostic Microbiology,* 9th ed. Mosby–Year Book, St. Louis.

Delost, Maria Dannessa. 1997. Page 107 in *Introduction to Diagnostic Microbiology.* Mosby, St. Louis.

DIFCO Laboratories. 1984. Page 292 in *DIFCO Manual,* 10th ed. DIFCO Laboratories, Detroit.

Forbes, Betty A., Daniel F. Sahm, and Alice S. Weissfeld. 2002. Page 290 in *Bailey & Scott's Diagnostic Microbiology,* 11th ed. Mosby, St. Louis.

Winn, Washington C., et al. 2006. Pages 645 and 1472 in *Koneman's Color Atlas and Textbook of Diagnostic Microbiology,* 6th ed. Lippincott Williams & Wilkins, Baltimore.

MacFaddin, Jean F. 2000. Page 3 in *Biochemical Tests for Identification of Medical Bacteria,* 3rd ed. Lippincott Williams & Wilkins, Philadelphia.

Name ______________________ Date ______________________

Lab Section ______________________ I was present and performed this exercise (initials) ________

Bacitracin, Novobiocin, and Optochin Susceptibility Tests

OBSERVATIONS AND INTERPRETATIONS

Refer to Tables 5-23, 24, and 25 when recording and interpreting your results in the chart below.

Antibacterial Agent	Organism	Susceptible (S)	Resistant (R)
Bacitracin			
Novobiocin			
Optochin			

Questions

1 The susceptibility tests included in this unit are typically used to differentiate between Gram-positive cocci. Would you predict them to be an effective differential test for Gram-negative organisms? Why or why not?

DATA SHEET 5-20

2 Why is it important to get a bacterial lawn rather than isolated colonies on the plate?

3 Does the zone of inhibition's edge indicate the limit of antibacterial agent diffusion into the agar? Give reasons or evidence to support your answer.

Other Differential Tests

This unit includes tests that do not fit elsewhere but are important to consider. Blood Agar is useful for detecting hemolytic ability of Gram-positive cocci, especially *Streptococcus* species. It also is used as a general-purpose growth medium appropriate for fastidious and nonfastidious microorganisms alike. The coagulase tests are commonly used to presumptively identify pathogenic *Staphylococcus* species. Motility Agar is used to detect bacterial motility, especially in differentiating *Enterobacteriaceae* and other Gram-negative rods.

EXERCISE 5-21

Blood Agar

❖ Theory

Several species of Gram-positive cocci produce exotoxins called **hemolysins**, which are able to destroy red blood cells (RBCs) and hemoglobin. Blood Agar, sometimes called Sheep Blood Agar because it includes 5% sheep blood in a Tryptic Soy Agar base, allows differentiation of bacteria based on their ability to hemolyze RBCs.

The three major types of hemolysis are β-hemolysis, α-hemolysis, and γ-hemolysis. β-hemolysis, the complete destruction of RBCs and hemoglobin, results in a clearing of the medium around the colonies (Figure 5-67). α-hemolysis is the partial destruction of RBCs and produces a greenish discoloration of the agar around the colonies (Figure 5-68). γ-hemolysis is actually nonhemolysis and appears as simple growth with no change to the medium (Figure 5-69).

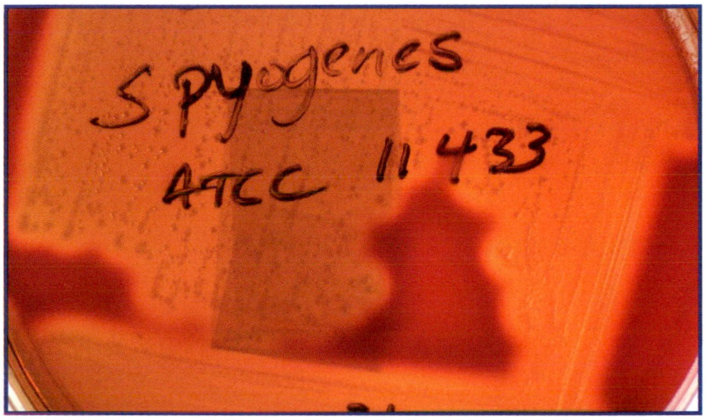

5-67 **β-Hemolysis** ◆ *Streptococcus pyogenes* demonstrates β-hemolysis. The clearing around the growth is a result of complete lysis of red blood cells. This photograph was taken with transmitted light.

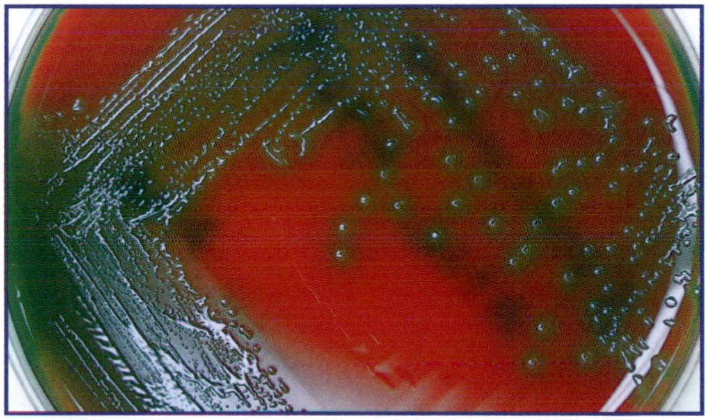

5-68 **α-Hemolysis** ◆ This is a streak plate of *Streptococcus pneumoniae* demonstrating α-hemolysis. The greenish zone around the colonies results from incomplete lysis of red blood cells.

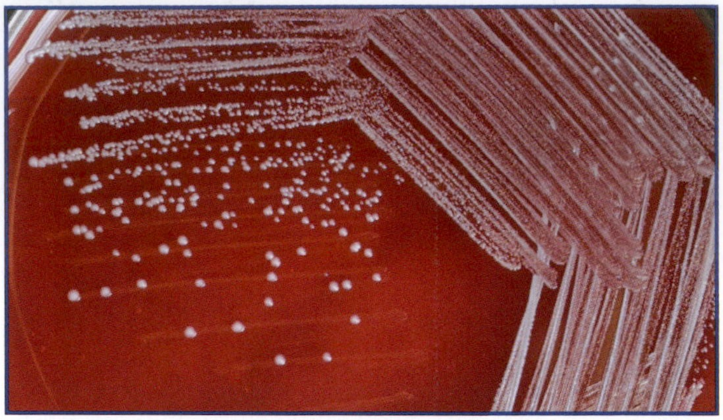

5-69 γ-**Hemolysis** ◆ This streak plate of *Staphylococcus epidermidis* on a Sheep Blood Agar illustrates no hemolysis.

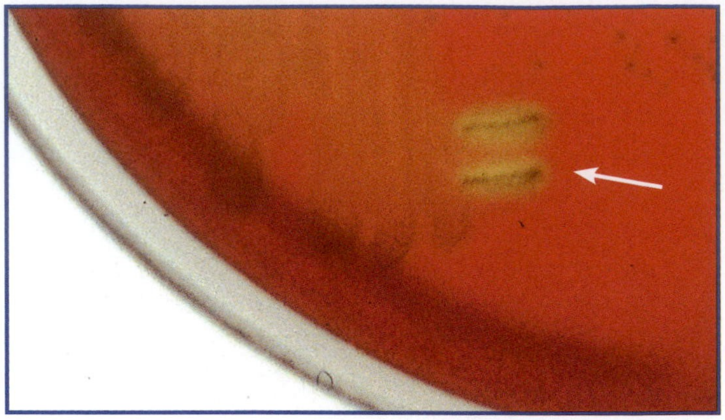

5-70 Aerobic Versus Anaerobic Hemolysis ◆ An unidentified throat culture isolate demonstrates α-hemolysis when growing on the surface, but β-hemolysis beneath the surface surrounding the stabs (arrow). This results from production of an oxygen-labile hemolysin.

Hemolysins produced by streptococci are called **streptolysins.** They come in two forms—type O and type S. **Streptolysin O** is oxygen-labile and expresses maximal activity under anaerobic conditions. **Streptolysin S** is oxygen-stable but expresses itself optimally under anaerobic conditions as well. The easiest method of providing an environment favorable for streptolysins on Blood Agar is what is called the **streak–stab technique.** In this procedure the Blood Agar plate is streaked for isolation and then stabbed with a loop. The stabs encourage streptolysin activity because of the reduced oxygen concentration of the subsurface environment (Figure 5-70).

Application

Blood Agar is used for isolation and cultivation of many types of fastidious bacteria. It also is used to differentiate bacteria based on their hemolytic characteristics, especially within the genera *Streptococcus*, *Enterococcus*, and *Aerococcus*.

In This Exercise

You will not be provided with organisms to inoculate your plate. Instead, your partner will take a swab of your throat. Then you will inoculate the Blood Agar plate in one quadrant to begin a streak for isolation. Isolation of the different colonies is the only way to observe the different forms of hemolysis properly. (Note: When you streak your plate, do only the first streak with the swab, and do streaks two, three, and four with your inoculating loop. If you are not sure how to do this, refer to Exercise 1-4. Because of the potential for cultivating a pathogen, cover the plate immediately after streaking and seal it with tape. Upon removing your plate from the incubator, do not open it until your instructor says it is safe to do so.)

Materials

Per Student Group

- ✓ one Sheep Blood Agar plate (commercially available—TSA containing 5% sheep blood)
- ✓ sterile cotton swabs
- ✓ sterile tongue depressors

Medium Recipe

5% Sheep Blood Agar (TSA ~ 5% Sheep Blood)

◆ Infusion from beef heart (solids)	2.0 g
◆ Pancreatic digest of casein	13.0 g
◆ Sodium chloride	5.0 g
◆ Yeast extract	5.0 g
◆ Agar	15.0 g
◆ Defibrinated sheep blood	50.0 mL
◆ Distilled or deionized water	1.0 L

pH 7.1–7.5 at 25°C

Procedure

Lab One

1. Obtain a Blood Agar Plate and label it with your name, the specimen source ("throat culture"), and the date.
2. Have your lab partner obtain a culture from your throat. Follow the procedure in Appendix B.
3. Immediately transfer the specimen to a Blood Agar plate. Use the swab to begin a streak for isolation. Refer to Exercise 1-4 if necessary.

4 Dispose of the swab in a container designated for autoclaving.

5 Finish the isolation procedure by streaking quadrants 2, 3, and 4 with your loop.

6 After completing the streak, use your loop to stab the agar in two or three places in the first streak pattern, and then in two or three places not previously inoculated.

7 Tape down the lid to prevent it from opening accidentally. Invert and incubate the plate aerobically at $35 \pm 2°C$ for 24 hours.

Lab Two

1 After incubation, do not open your plate until your instructor has seen it and given permission to do so.

2 Using transmitted light and referring to Table 5-26, observe for color changes and clearing around the colonies. This can be done using a colony counter or by holding the plate up to a light. Record your results on the Data Sheet. Refer to Figure 2-33 for two examples of throat cultures on Blood Agar.

References

Delost, Maria Dannessa. 1997. Page 103 in *Introduction to Diagnostic Microbiology.* Mosby, Inc. St. Louis.

Forbes, Betty A., Daniel F. Sahm, and Alice S. Weissfeld. 2002. Page 16 in *Bailey & Scott's Diagnostic Microbiology,* 11th ed. Mosby, St. Louis.

Krieg, Noel R. 1994. Page 619 in *Methods for General and Molecular Bacteriology,* edited by Philipp Gerhardt, R. G. E. Murray, Willis A. Wood, and Noel R. Krieg. American Society for Microbiology, Washington, DC.

Power, David A., and Peggy J. McCuen. 1988. Page 115 in *Manual of BBL™ Products and Laboratory Procedures,* 6th ed. Becton Dickinson Microbiology Systems, Cockeysville, MD.

Winn, Washington C., et al. 2006. Chapter 13 in *Koneman's Color Atlas and Textbook of Diagnostic Microbiology,* 6th ed. Lippincott Williams & Wilkins, Baltimore.

Zimbro, Mary Jo, and David A. Power, eds. 2003. *Difco™ and BBL™ Manual—Manual of Microbiological Culture Media.* Becton Dickinson and Co., Sparks, MD.

TABLE 5-26 Blood Agar Results and Interpretations

TABLE OF RESULTS		
Result	**Interpretation**	**Symbol**
Clearing around growth	Organism hemolyzes RBCs completely	β-hemolysis
Greening around growth	Organism hemolyzes RBCs partially	α-hemolysis
No change in the medium	Organism does not hemolyze RBCs	no (γ) hemolysis

Name ______________________ Date ______________________

Lab Section ______________________ I was present and performed this exercise (initials) ________

DATA SHEET 5-21

Blood Agar

OBSERVATIONS AND INTERPRETATIONS

Choose four different colonies (preferably with a diversity of hemolysis reactions) and fill in the chart. Refer to Table 5-26 when recording and interpreting your results.

Source of Culture	Colony Morphology	Hemolysis Result	Interpretation

Questions

1 The streak–stab technique, used to promote streptolysin activity, is preferred over incubating the plates anaerobically. Why do you think this is so? Compare and contrast what you see as the advantages and disadvantages of each procedure.

2 Assuming that all of the organisms cultivated in this exercise came from the throats of healthy students, why is it important to cover and tape the plates?

3 Why is the streak plate preferred over the spot inoculations in this procedure?

EXERCISE 5-22

CAMP Test

Theory

Group B *Streptococcus agalactiae* produces the CAMP factor—a hemolytic protein that acts synergistically with the β-hemolysin of *Staphylococcus aureus* subsp. *aureus*. When streaked perpendicularly to an *S. aureus* subsp. *aureus* streak on blood agar (Figure 5-71), an arrowhead-shaped zone of hemolysis forms and is a positive result.

Application

The CAMP test (an acronym of the developers of the test—Christie, Atkins, and Munch-Peterson) is used to differentiate Group B *Streptococcus agalactiae* (+) from other *Streptococcus* species (−).

In This Exercise

You will inoculate a Blood Agar plate as shown in Figure 5-72 with *Streptococcus agalactiae* and *Staphylococcus aureus*. Following incubation, you will examine your plate for the characteristic arrowhead pattern of clearing.

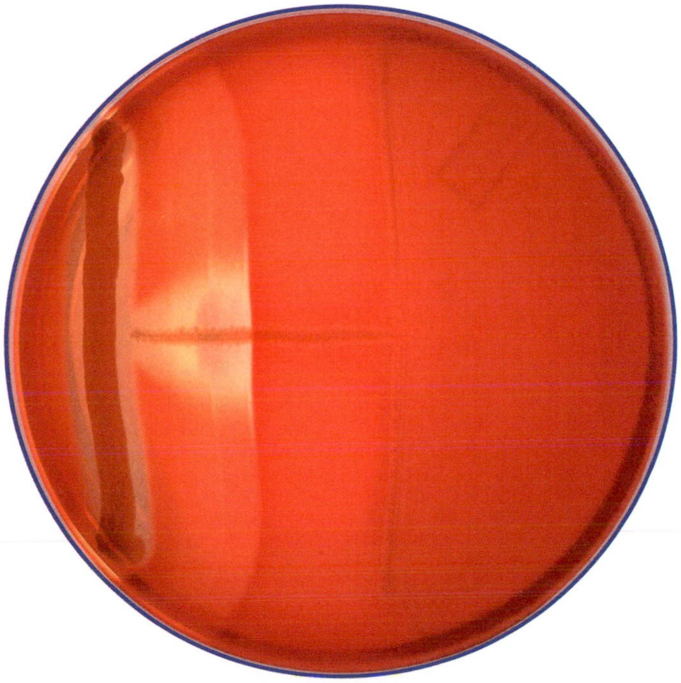

5-71 Positive CAMP Test Results ◆ Note the arrowhead zone of clearing in the region where the CAMP factor of *Streptococcus agalactiae* acts synergistically with the β-hemolysin of *Staphylococcus aureus* subsp. *aureus*.

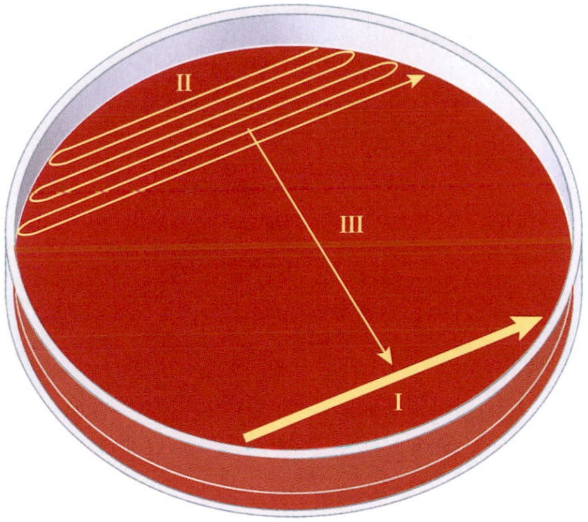

5-72 CAMP Test Inoculation ◆ Two inoculations are made. First *Staphylococcus aureus* subsp. *aureus* is streaked along one edge of a fresh Blood Agar plate (I). Then the isolate (when testing an unknown organism) is inoculated densely in the other half of the plate opposite *S. aureus* (II). Finally, a single streak is made from inside Streak II toward, but not touching, *S. aureus* (III).

Materials

- ✓ two Blood Agar plates (commercial preparation of TSA containing 5% sheep blood)
- ✓ sterile cotton applicators
- ✓ fresh broth cultures of:
  - ◆ *Staphylococcus aureus* subsp. *aureus* (BSL-2)
  - ◆ *Streptococcus agalactiae* (BSL-2)
  - ◆ *Streptococcus pneumoniae* (BSL-2)

Procedure

Lab One

1. Obtain two Blood Agar plates
2. Using a cotton swab, inoculate both plates as shown in Figure 5-72, each with a single streak of *S. aureus* along one edge (Streak I).
3. Inoculate each plate with a dense smear of one of the other two organisms across from the *S. aureus* streak, as shown in Streak II. Finish each plate with a single streak (as in Streak III) from the *Streptococcus* species toward, but not touching, the *S. aureus*. Note: It works best and saves material to do this third streak with the same swab used in Streak II before moving on to the other plate. Repeat procedure in the second plate with *S. aureus* and *S. pneumoniae*.

4 Label the plates with your name, the date, and the names of the organisms.

5 Invert the plates and incubate them at 35 ± 2°C for 24 hours.

Lab Two

1 Remove the plates from the incubator and observe for the characteristic arrowhead pattern.

2 Discard the plates in an appropriate biohazard or autoclave container.

TABLE 5-27 Test Results and Interpretations

TABLE OF RESULTS		
Result	**Interpretation**	**Symbol**
Arrowhead pattern	Organism produces hemolytic CAMP protein. Presumptively identified as *S. agalactiae*	+
No arrowhead pattern	Organism does not produce hemolytic CAMP protein. Presumptively identified as not *S. agalactiae*	–

NAME ____________________ DATE ____________________

LAB SECTION ____________________ I WAS PRESENT AND PERFORMED THIS EXERCISE (INITIALS) ________

DATA SHEET 5-22

CAMP Test

OBSERVATIONS AND INTERPRETATIONS

Refer to Table 5-27 when recording and interpreting your results in the chart below.

Organism (interacting with *S. aureus*)	Result + / −	Interpretation

Questions

1 You were instructed to not allow streak I and streak III to touch. Why do you think this is important?

2 Why do you think the clearing in a positive CAMP test is an arrowhead shape and not some other shape? Would it likely make an important difference if you began streak II near *S. aureus* and streaked across the plate in the other direction? Why?

3 What result would you expect to get if you accidentally reverse the organisms in the procedure?

EXERCISE 5-23

Coagulase Tests

Theory

Staphylococcus aureus is an opportunistic pathogen that can be highly resistant to both the normal immune response and antimicrobial agents. Its resistance is due, in part, to the production of a **coagulase** enzyme. Coagulase works in conjunction with normal plasma components to form protective fibrin barriers around individual bacterial cells or groups of cells, shielding them from phagocytosis and other types of attack.

Coagulase enzymes occur in two forms—**free coagulase** and **bound coagulase**. Free coagulase is an extracellular enzyme (released from the cell) that reacts with a plasma component called **coagulase-reacting factor** (CRF).The resulting reaction is similar to the conversion of prothrombin and fibrinogen in the normal clotting mechanism. Bound coagulase, also called clumping factor, is attached to the bacterial cell wall and reacts directly with fibrinogen in plasma. The fibrinogen then precipitates, causing the cells to clump together.

Two forms of the Coagulase Test have been devised to detect the enzymes: the Tube Test and the Slide Test. The Tube Test detects the presence of either bound or free coagulase, and the Slide Test detects only bound coagulase. Both tests utilize rabbit plasma treated with anticoagulant to interrupt the normal clotting mechanisms.

The Tube Test is performed by adding the test organism to rabbit plasma in a test tube. Coagulation of the plasma (including any thickening or formation of fibrin threads) within 24 hours indicates a positive reaction (Figure 5-73). The plasma typically is examined for clotting (without shaking) periodically for about 4 hours. After 4 hours coagulase-negative tubes can be incubated overnight, but no more than a total of 24 hours, because coagulation can take place early and revert to liquid within 24 hours.

In the Slide Test, bacteria are transferred to a slide containing a small amount of plasma. Agglutination of the cells on the slide within 1 to 2 minutes indicates the presence of bound coagulase (Figure 5-74). Equivocal or negative Slide Test results typically are given the Tube Test for confirmation.

Application

The Coagulase Test typically is used to differentiate *Staphylococcus aureus* from other Gram-positive cocci.

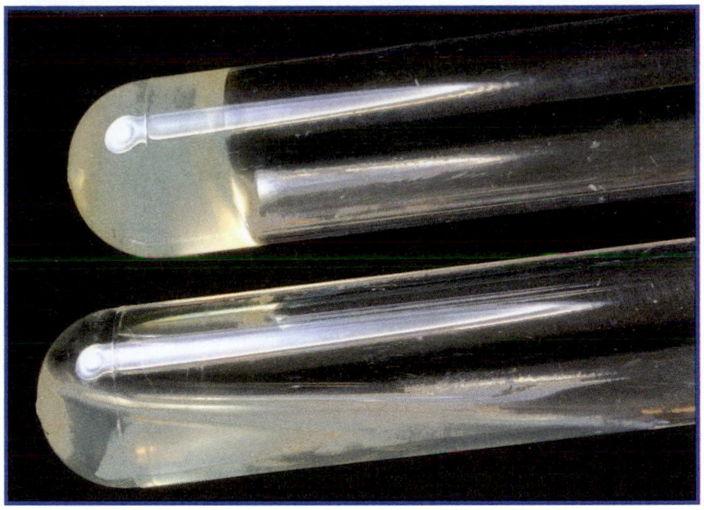

5-73 Coagulase Tube Test ◆ These coagulase tubes illustrate a coagulase-negative (−) organism, below, and a coagulase-positive (+) organism, above. The Tube Test identifies both bound and free coagulase enzymes. Coagulase increases bacterial resistance to phagocytosis and antibodies by surrounding infecting organisms with a clot.

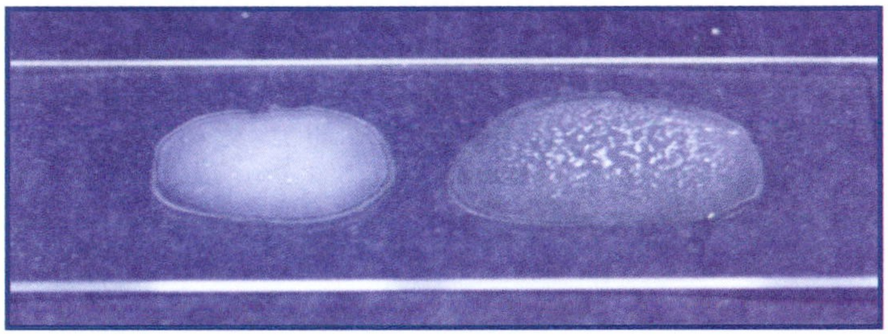

5-74 Coagulase Slide Test (Clumping Factor) ◆ This slide illustrates a coagulase-positive (+) organism on the right and a coagulase-negative (−) organism on the left. Agglutination of the coagulase plasma is indicative of a positive result for bound coagulase.

In This Exercise

You will perform both Tube and Slide Coagulase Tests. Immediately enter your Slide Test results on the Data Sheet. Read the Tube Test the following day.

Materials

Per Student Group

- ✓ three sterile rabbit plasma tubes (0.5 mL in 12 mm × 75 mm test tubes)
- ✓ sterile 1 mL pipettes
- ✓ sterile saline (0.9% NaCl)
- ✓ microscope slides
- ✓ fresh slant cultures of:
  - *Staphylococcus aureus* (BSL-2)
  - *Staphylococcus epidermidis*

Procedure

Lab One: Tube Test

1. Obtain three coagulase tubes. Label two tubes with the names of the organisms, your name, and the date. Label the third tube "control."
2. Inoculate two tubes with the test organisms. Mix the contents by gently rolling the tube between your hands. Do not inoculate the control.
3. Incubate all tubes at 35 ± 2°C for up to 24 hours, checking for coagulation periodically for the first 4 hours.

Lab One: Slide Test (Clumping Factor)

1. Obtain two microscope slides, and divide them in half with a marking pen. Label the sides A and B.
2. Place a drop of sterile saline on side A and a drop of coagulase plasma on side B of each slide.
3. Transfer a loopful of *S. aureus* to each half of one slide, making sure to completely emulsify the bacteria in the solutions. Observe for agglutination within 2 minutes. Clumping after 2 minutes is not a positive result.
4. Repeat step 3 using the other slide and *S. epidermidis*.
5. Record your results in the chart on the Data Sheet. Refer to Figure 5-74 and Table 5-28 when making your interpretations. Confirm any negative results by comparing with the completed Tube Test in 24 hours.

Lab Two

1. Remove all tubes from the incubator no later than 24 hours after inoculation. Examine for clotting of the plasma.
2. Record your results on the Data Sheet. Refer to Figure 5-73 and Table 5-29 when making your interpretations.

TABLE 5-28 Coagulase Slide Test Results and Interpretations ◆

TABLE OF RESULTS		
Result	**Interpretation**	**Symbol**
Clumping of cells	Plasma has been coagulated	+
No clumping of cells	Plasma has not been coagulated	−

TABLE 5-29 Coagulase Tube Test Results and Interpretations ◆

TABLE OF RESULTS		
Result	**Interpretation**	**Symbol**
Medium is solid	Plasma has been coagulated	+
Medium is liquid	Plasma has not been coagulated	−

References

Collins, C. H., Patricia M. Lyne, and J. M. Grange. 1995. Page 111 in *Collins and Lyne's Microbiological Methods,* 7th ed. Butterworth-Heinemann, Oxford, United Kingdom.

Delost, Maria Dannessa. 1997. Pages 98–99 in *Introduction to Diagnostic Microbiology*. Mosby, St. Louis.

DIFCO Laboratories. 1984. Page 232 in *DIFCO Manual,* 10th ed. DIFCO Laboratories, Detroit.

Forbes, Betty A., Daniel F. Sahm, and Alice S. Weissfeld. 2002. Pages 266–267 in *Bailey & Scott's Diagnostic Microbiology,* 11th ed. Mosby, St. Louis.

Holt, John G., Ed. 1994. *Bergey's Manual of Determinative Bacteriology,* 9th ed. Williams and Wilkins, Baltimore.

Lányi, B. 1987. Page 62 in *Methods in Microbiology,* Vol. 19, edited by R. R. Colwell and R. Grigorova. Academic Press, New York.

MacFaddin, Jean F. 2000. Page 105 in *Biochemical Tests for Identification of Medical Bacteria,* 3rd ed. Lippincott Williams & Wilkins, Philadelphia.

NAME ______________________ DATE ______________________

LAB SECTION ______________________ I WAS PRESENT AND PERFORMED THIS EXERCISE (INITIALS) ________

Coagulase Tests

OBSERVATIONS AND INTERPRETATIONS

Refer to Tables 5-28 and 5-29 when recording and interpreting your results in the charts below.

Slide Test Results			
Organism	**Slide**	**Result**	**Interpretation**
	A		
	B		
	A		
	B		

Tube Test Results		
Organism	**Result**	**Interpretation**
Control		

Questions

1 Why is it more important to use fresh cultures in the Coagulase Test than in a test medium such as Milk Agar?

2 How would you interpret a negative Slide Test and a positive Tube Test using the same organism?

3 Consider the Slide Test.

a. What is the role of sterile saline plus organism on the Slide Test?

b. Why is it advisable to run a known coagulase-positive organism along with your unknown organism?

c. How will the validity of the test be affected if clumping occurs on both smears of the known coagulase-positive organism?

4 List possible reasons why the Slide Test is not appropriate for detecting free coagulase.

EXERCISE 5-24

Motility Test

Theory

Motility Test Medium is a semisolid medium designed to detect bacterial motility. Its agar concentration is reduced from the typical 1.5% to 0.4%—just enough to maintain its form while allowing movement of motile bacteria. It is inoculated by stabbing with a straight transfer needle. Motility is detectable as diffuse growth radiating from the central stab line.

A tetrazolium salt (TTC) sometimes is added to the medium to make interpretation easier. TTC is used by the bacteria as an electron acceptor. In its oxidized form, TTC is colorless and soluble; when reduced it is red and insoluble (Figure 5-75). A positive result for motility is indicated when the red (reduced) TTC is seen radiating outward from the central stab. A negative result shows red only along the stab line (Figure 5-76).

Application

This test is used to detect bacterial motility. Motility is an important differential characteristic of *Enterobacteriaceae*.

In This Exercise

You will inoculate two Motility Stabs with an inoculating needle. Straighten the needle before you stab the medium. It is important, also, to stab straight into the medium and remove the needle along the same line. Lateral movement of the needle will make interpretation more difficult.

Materials

Per Student Group

- ✓ three Motility Test Media stabs
- ✓ fresh cultures of:
  - *Enterobacter aerogenes*
  - *Klebsiella pneumoniae* surrogate (*Micrococcus luteus*)

Medium Recipe

Motility Test Medium

♦ Beef extract	3.0 g
♦ Pancreatic digest of gelatin	10.0 g
♦ Sodium chloride	5.0 g
♦ Agar	4.0 g

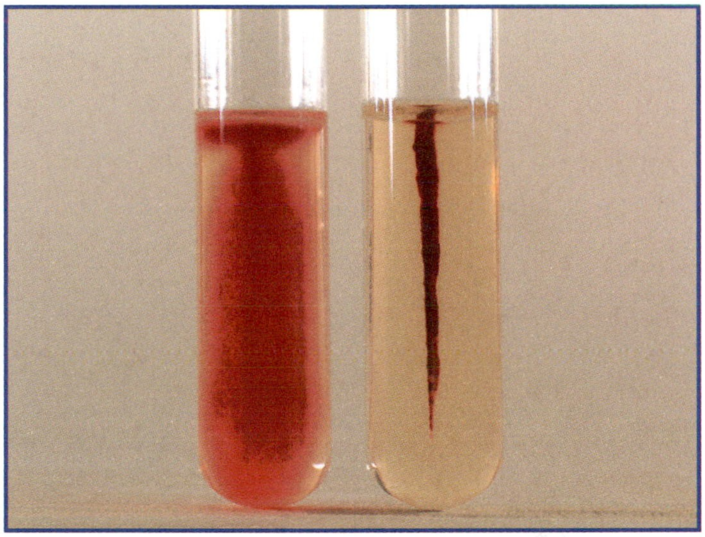

5-76 Motility Test Results ♦ These Motility Test Media were inoculated with a motile (+) organism on the left and a nonmotile (−) organism on the right.

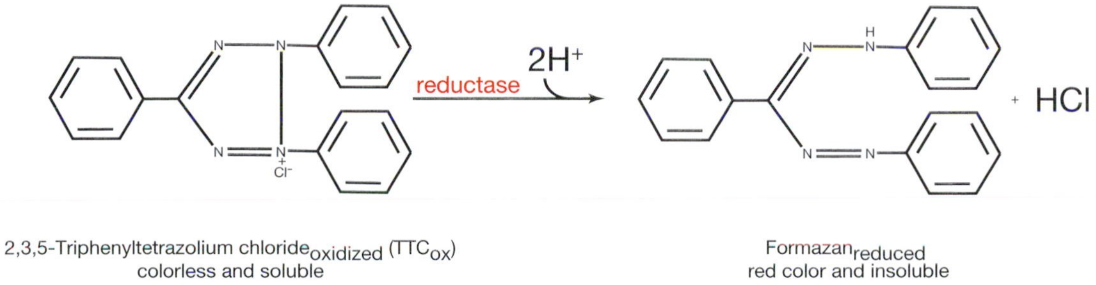
N
N
N
N
Cl⁻
reductase
2H⁺
H
N
N
N
N
+ HCl
2,3,5-Triphenyltetrazolium chloride oxidized (TTC ox)
colorless and soluble
Formazan reduced
red color and insoluble

5-75 Reduction of TTC ♦ Reduction of 2,3,5-Triphenyltetrazolium chloride by metabolizing bacteria results in its conversion from colorless and soluble to the red and insoluble compound formazan. The location of growing bacteria can be determined easily by the location of the formazan in the medium.

- Triphenyltetrazolium chloride (TTC) 0.05 g
- Distilled or deionized water 1.0 L

pH 7.1–7.4 at 25°C

Procedure

Lab One

1 Obtain three motility stabs. Label two tubes, each with the name of the organism, your name, and the date. Label the third tube "control."

2 Stab-inoculate two tubes with the test organisms. (Motility can be obscured by careless stabbing technique. Try to avoid lateral movement of the needle when performing this stab.) Do not inoculate the control.

3 Incubate the tubes aerobically at 35 ± 2°C for 24 to 48 hours.

Lab Two

1 Examine the growth pattern for characteristic spreading from the stab line.

2 Record your results on the Data Sheet.

TABLE 5-30 Motility Test Results and Interpretations

TABLE OF RESULTS		
Result	**Interpretation**	**Symbol**
Red diffuse growth radiating outward from the stab line	The organism is motile	+
Red growth only along the stab line	The organism is nonmotile	–

References

Forbes, Betty A., Daniel F. Sahm, and Alice S. Weissfeld. 2002. Page 276 in *Bailey & Scott's Diagnostic Microbiology,* 11th ed. Mosby, St. Louis.

MacFaddin, Jean F. 2000. Page 327 in *Biochemical Tests for Identification of Medical Bacteria,* 3rd ed. Lippincott Williams & Wilkins, Philadelphia.

Zimbro, Mary Jo, and David A. Power, eds. 2003. Page 374 in *Difco™ and BBL™ Manual—Manual of Microbiological Culture Media.* Becton Dickinson and Co., Sparks, MD.

Name ______________________ Date ______________________

Lab Section ______________________ I was present and performed this exercise (initials) ________

Motility Test

OBSERVATIONS AND INTERPRETATIONS

Refer to Table 5-30 when recording and interpreting your results in the chart below.

Organism	Result	+ / −	Interpretation

Questions

1 Why is it important to carefully insert and remove the needle along the same stab line?

2 Why is it essential that the reduced TTC be insoluble? Why is there less concern about the solubility of the oxidized form of TTC?

SECTION 6

Quantitative Techniques

Quantitative techniques in microbiology usually consist of counting cells or counting colonies formed by cells growing on agar plates. A typical result is a close estimate of the number of cells (or viruses) in one milliliter of sample—the **population density**, sometimes referred to as **cell density**, or **cell concentration.**

Knowledge of microbial population density is an important part of many areas of microbiology. Food and environmental microbiologists use population density for food production as well as the detection of food or water contamination. Medical microbiologists manipulate population density for use in standardized tests. Many researchers use population density to measure the effect of varying nutritional or environmental conditions. Industrial microbiologists maintain microbial populations at optimum levels in large-scale fermenters for the manufacture of many products including enzymes, antibiotics, beer, and wine.

In this section we include three bacterial counting techniques and one exercise to measure a viral population. We also included a Differential Blood Cell Count (Exercise 6-4) because, while not specifically designed to measure population density, it employs a direct microscopic counting technique that is useful as a clinical diagnostic tool. The Differential Blood Cell Count, as the only direct counting procedure, measures and records each cell type as a percentage of the total cells counted. The other four counts determine population density indirectly by means of dilutions and plate counts.

The standard dilution procedure used in microbial counts is the **serial dilution.** A serial dilution is a series of individual measured transfers, starting with an undiluted sample, one tube to the next, each reducing the cell density by a known amount, until it becomes dilute enough to be used in an appropriate counting procedure (Figure 6-1). The advantage of a serial dilution is that it not only reduces the cell density to a manageable size, but does so systematically so that each dilution tube contains a *known* volume of the undiluted original.

An important term you will see in the following exercises is **colony forming units**, or **CFU**. In plate counts, colonies are counted, but the count is usually recorded in CFU. CFU better describes the cell density, because colonies that grow on plates may start as single cells or as groups of cells, depending on the typical cellular arrangement of the organism. (Note: Viral samples are recorded in PFU [plaque forming units], which will be explained fully in Exercise 6-5.)

As you proceed through the exercises in this section, you will gain understanding and proficiency with serial dilutions. Because these exercises involve math, some of you will be a little (or a lot) intimidated. Take heart; these calculations are much simpler than they appear! If you are nervous about the math, we encourage you to see your instructor and to work as many practice problems as you can get. Before you know it, you will see these equations as a welcome solution to the very real problem of how to manage otherwise unmanageable numbers.

EXERCISE 6-1

Standard Plate Count (Viable Count)

❖ Theory

The standard plate count is a procedure that allows microbiologists to estimate the population density in a liquid sample by plating a very dilute portion of that sample and counting the number of colonies it produces. The inoculum that is transferred to the plate contains a *known* proportion of the original sample because it is the product of a **serial dilution**.

As shown in Figure 6-1, a serial dilution is simply a series of controlled transfers down a line of **dilution blanks** (tubes containing a known volume of sterile diluent—water, saline, or buffer). The series begins with a sample containing an unknown concentration (density) of cells and ends with a very dilute mixture containing only a few cells. Each dilution blank in the series receives a known volume from the mixture in the previous tube and delivers a known volume to the next, typically reducing the cell density to 1/10 or 1/100 at each step.

For example, if the original sample contains 1,000,000 cells/mL, following the first transfer, the 1/100 dilution (10 µL into 990 µL) in dilution tube 1 would contain 10,000 cells/mL. (Note that 10 µL added to 990 µL results in total volume of 1000 µL, 1/100th of which is original sample.) In the second dilution (tube 2) the 1/100 dilution would reduce it further to 100 cells/mL. Because the cell density of the original sample is not known at this time, only the dilutions (without mL units) are recorded on the dilution tubes. By convention, dilutions are expressed in scientific notation. Therefore, a 1/10 dilution is written as 10^{-1} and a 1/100 dilution is written as 10^{-2}.

A known volume of appropriate dilutions is then spread onto agar plates to produce at least one **countable plate**. A countable plate is one that contains between 30 and 300 colonies (Figure 6-2). A count lower than 30 colonies is considered statistically unreliable and greater than 300 is typically too many to be viewed as individual colonies.

In examining the procedural diagram, you can see that the first transfer in the series is a simple dilution, but that all successive transfers are compound dilutions. Compound dilutions are dilutions of previously diluted samples. Both types of dilutions can be calculated using the following formula,

$$V_1D_1 = V_2D_2$$

where V_1 and D_1 are the volume and dilution of the concentrated broth, respectively, while V_2 and D_2 are the volume and dilution of the completed dilution. Undiluted samples are always expressed as 1. Therefore, to calculate the dilution of a 100.0 μL sample added to 900.0 μL of diluent, the permuted formula would be used as follows.

$$D_2 = \frac{V_1 D_1}{V_2} = \frac{100.0\ \mu L \times 1}{1000.0\ \mu L} = \frac{1}{10} = 10^{-1}$$

As mentioned above, compound dilutions are calculated using the same formula. However, because D_1, in compound dilutions, no longer represents undiluted sample, but rather a fraction of the original density, it must be represented as something less than 1 (i.e., 10^{-1}, 10^{-2}, etc.) For example, if 100.0 μL of the 10^{-1} dilution from the last example were

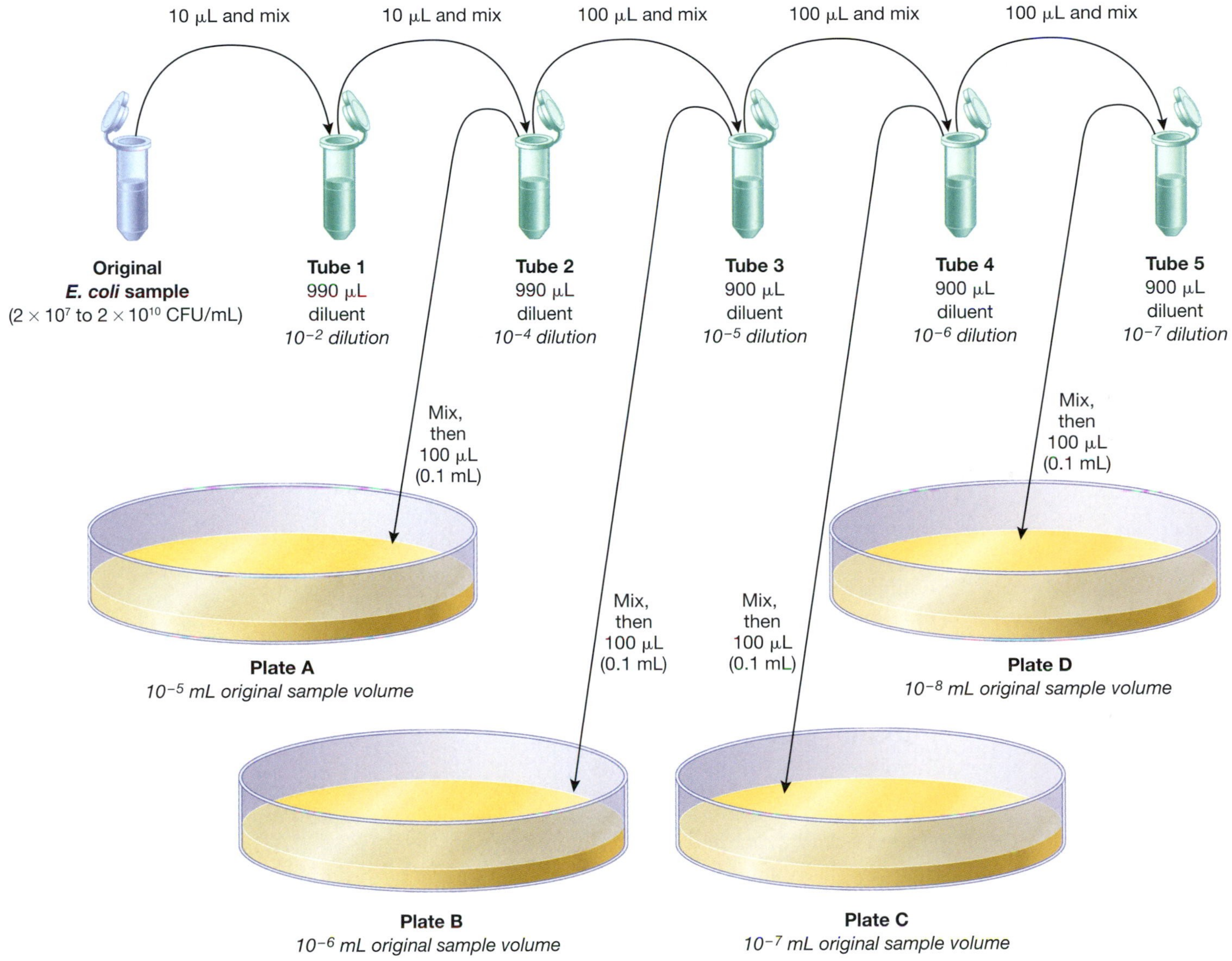

6-1 Serial Dilution Procedural Diagram ◆ This is an illustration of the dilution scheme outlined in the Procedure. The dilution assigned to each tube (written below the tube) represents the proportion of original sample inside that tube. For example, if the dilution is 10^{-4}, the proportion of original sample inside the tube would be 1/10000th of the total volume inside. When 0.1 mL (100.0 μL) of that solution is transferred to a plate, the volume of original sample in the plate is $0.1\ mL \times 10^{-4} = 10^{-5}\ mL$.

transferred to 900.0 µL of diluent, it would become a 10^{-2} dilution as follows.[1]

$$D_2 = \frac{V_1 D_1}{V_2} = \frac{100.0\ \mu L \times 10^{-1}}{1000.0\ \mu L} = \frac{10^{-1}}{10} = 10^{-2}$$

Spreading a known volume of this dilution onto an agar plate and counting the colonies that develop would give you all the information you need to calculate the original cell density (OCD). Below is the basic formula for this calculation.

$$OCD = \frac{CFU}{D \times V}$$

CFU (colony forming units) is actually the number of colonies that develop on the plate. CFU is the preferred term because colonies could develop from single cells or from groups of cells, depending on the typical cellular arrangement of the organism. D is the dilution as written on the dilution tube from which the inoculum comes. V is the volume transferred to the plate. (Note: The volume [in mL] is included in the formula because densities are expressed in CFU/mL, therefore a 100.0 µL [0.1 mL] inoculation [which would contain 1/10th as many cells as 1 mL] must be accounted for.)

As you can see in the formula, the volume of *original sample* being transferred to a plate is the product of the *volume transferred* and the *dilution* of the tube from which it came. Therefore 0.1 mL transferred from 10^{-2} dilution contains only 10^{-3} mL of the original sample (0.1 mL × 10^{-2}). The convention among microbiologists is to condense D and V in the formula into "Original sample volume" or simply "sample volume" (expressed in mL). The formula thus becomes,

$$OCD = \frac{CFU}{\text{Original sample volume}}$$

The sample volume is written on the plate at the time of inoculation. Following a period of incubation, the plates are examined, colonies are counted on the countable plates, and calculation is a simple division problem. Suppose, for example, you counted 37 colonies on a plate inoculated with 100.0 µL (0.1 mL) of a 10^{-5} dilution. Knowing that this plate now contains 10^{-6} mL of original sample, calculation would be as follows.

$$OCD = \frac{CFU}{\text{Sample volume}} = \frac{37\ CFU}{10^{-6}\ mL} = 3.7 \times 10^{7}\ CFU/mL$$

[1] Permutations of this formula work with all necessary dilution calculations. For calculations involving unconventional volumes or dilutions, the formula is essential, but for simple tenfold or hundredfold dilutions like the ones described in this exercise, the final compounded dilution in a series can be calculated by multiplying each of the simple dilutions by each other. For example, a series of three 10^{-1} dilutions would yield a final dilution of 10^{-3} ($10^{-1} \times 10^{-1} \times 10^{-1} = 10^{-3}$). Three 10^{-2} dilutions would yield a final dilution of 10^{-6} ($10^{-2} \times 10^{-2} \times 10^{-2} = 10^{-6}$). We encourage you to use whatever means is best for you. In time you will be doing the calculations in your head.

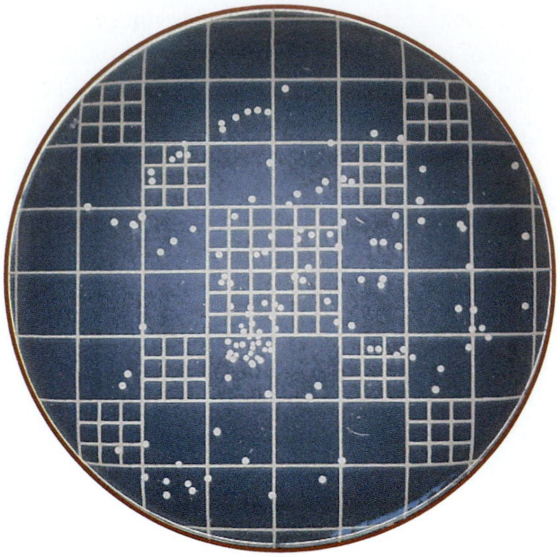

6-2 Countable Plate ◆ A countable plate has between 30 and 300 colonies. Therefore, this plate with approximately 130 colonies is countable and can be used to calculate cell density in the original sample. Plates with fewer than 30 colonies are TFTC ("too few to count"). Plates with more than 300 colonies are TNTC ("too numerous to count").

Application

The viable count is one method of determining the density of a microbial population. It provides an estimate of actual *living* cells in the sample.

In This Exercise

You will perform a dilution series and determine the population density of a broth culture of *Escherichia coli*. (Note: we have estimated your *E. coli* sample to contain 2×10^{7} to 2×10^{10} CFU/mL; therefore, you will plate dilutions from tube 2 through tube 5. If, when performing a plate count, you do not know the approximate cell density, you must plate all dilutions.) You will inoculate the plates using the **spread plate technique,** as illustrated in Exercise 1-5. As described in **Figure 1-31**, the inocula from the dilution tubes will be evenly dispersed over the agar surface with a bent glass rod. You will be sterilizing the glass rod between inoculations by immersing it in alcohol and igniting it. Be careful to organize your work area properly and *at all times keep the flame away from the alcohol beaker.*

Materials

Per Student Group

- ✓ micropipettes (10–100 µL and 100–1000 µL) with sterile tips
- ✓ five sterile microtubes
- ✓ sterile water
- ✓ eight Nutrient Agar plates

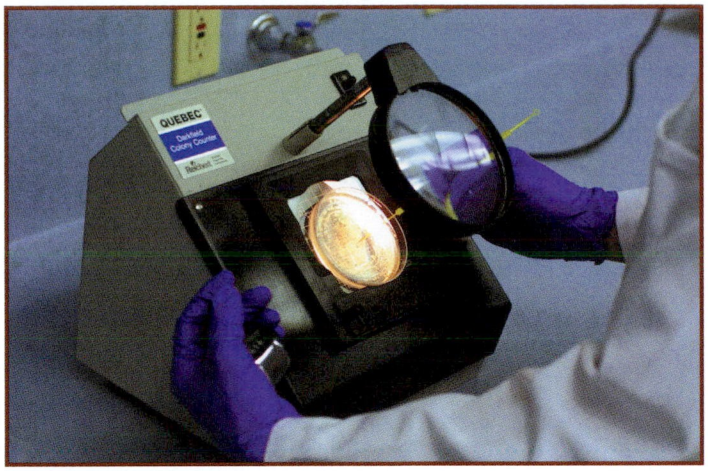

6-3 Counting Bacterial Colonies ◆ Place the plate upside down on the colony counter. Turn on the light and adjust the magnifying glass until all the colonies are visible. Using the grid in the background as a guide, count colonies one section at a time. Mark each colony with a felt-tip marker as you record with a hand tally counter.

- ✓ beaker containing ethanol and a bent glass rod
- ✓ hand tally counter
- ✓ colony counter
- ✓ 24-hour broth culture of *Escherichia coli* (This culture will have between 10^7 and 10^{10} CFU/mL.)

Procedure

Refer to the procedural diagram in Figure 6-1 as needed.

Lab One

1. Obtain eight plates, organize them into four pairs, and label them A_1, A_2, B_1, B_2, etc.
2. Obtain five microtubes, and label them 1–5. These are your dilution tubes; make sure they remain covered until needed.
3. Aseptically add 990 µL sterile water to dilution tubes 1 and 2. Cover when finished. Aseptically add 900 µL sterile water to dilution tubes 3, 4, and 5. Cover when finished.
4. Mix the broth culture, and aseptically transfer 10 µL to dilution tube 1; mix well. This is a 10^{-2} dilution.
5. Aseptically transfer 10 µL from dilution tube 1 to dilution tube 2; mix well. This is 10^{-4}.
6. Aseptically transfer 100 µL from dilution tube 2 to dilution tube 3; mix well. This is 10^{-5}.
7. Aseptically transfer 100 µL from dilution tube 3 to dilution tube 4; mix well. This is 10^{-6}.
8. Aseptically transfer 100 µL from dilution tube 4 to dilution tube 5; mix well. This is 10^{-7}.
9. Aseptically transfer 100 µL from dilution tube 2 to plate A_1. Using the spread plate technique (Exercise 1-5), dispense the diluent evenly over the entire surface of the agar. Repeat the procedure with plate A_2 and label both plates "10^{-5} mL original sample."
10. Following the same procedure, transfer 100 µL volumes from dilution tubes 3, 4, and 5 to plates B, C, and D, respectively. Label the plates with their appropriate original sample volumes.
11. Invert the plates, and incubate at 35 ± 2°C for 24 to 48 hours.

Lab Two

1. After incubation, examine the plates and determine the countable pair—plates with 30 to 300 colonies. Only one pair of plates should be countable (Figure 6-2).
2. Count the colonies on both plates and calculate the average (Figure 6-3). Record it on the Data Sheet. (Note: You may have more than one pair of plates that is countable. For the practice, count all plates that have between 30 and 300 colonies, and try to identify which pair you have the most confidence in. If no plates are in the 30–300 range, count the pair that is closest.)
3. Using the formula provided in Data and Calculations on the Data Sheet, calculate the density of the original sample, and record it in the space provided.

References

Collins, C. H., Patricia M. Lyne, and J. M. Grange. 1995. Page 149 in *Collins and Lyne's Microbiological Methods*, 7th ed. Butterworth-Heinemann, United Kingdom.

Koch, Arthur L. 1994. Page 254 in *Methods for General and Molecular Bacteriology*, edited by Philipp Gerhardt, R. G. E. Murray, Willis A. Wood, and Noel R. Krieg, American Society for Microbiology, Washington, DC.

Postgate, J. R. 1969. Page 611 in *Methods in Microbiology*, Vol. 1., edited by J. R. Norris and D. W. Ribbons. Academic Press, New York.

NAME ______________________ DATE ______________________

LAB SECTION ______________________ I WAS PRESENT AND PERFORMED THIS EXERCISE (INITIALS) ________

DATA SHEET 6-1

Standard Plate Count

OBSERVATIONS AND INTERPRETATIONS

1 Enter the number of colonies on each countable plate. Only one pair of plates should be countable, but for practice, record all countable plates anyway. For all plates containing more than 300 colonies, enter TNTC ("too numerous to count"). For plates containing fewer than 30, enter TFTC ("too few to count"). If no plates are countable, use the TFTC plate closest to 30 colonies for practice with the calculations. Make a note of this in Part 3 below.

2 Take the average number of colonies from the two (or more) countable plates and record it below.

Plate	A_1	A_2	B_1	B_2	C_1	C_2	D_1	D_2
Colonies Counted								
Average # Colonies								

3 Calculate the original density in CFU/mL using the following formula:

$$\text{OCD} = \frac{\text{CFU}}{\text{Original sample volume}}$$

Original density of *E. coli* in the broth	

Questions

(In the following questions, *cell density* and *cell concentration* are used interchangeably.)

1 Suppose your professor handed you a test tube with 2.0 mL of an *E. coli* broth culture in it and told you to make a 10^{-1} dilution of the entire culture. Explain how you would do this. Show your calculations.

2 Suppose your professor handed you a test tube with 2.0 mL of an *E. coli* broth culture in it and told you to make a 10^{-2} dilution of the entire culture. Explain how you would do this. Show your calculations.

3 How would you produce a 10^{-2} dilution of a 5 mL bacterial sample using the full 5 mL volume?

4 You have 0.05 mL of an undiluted culture at a concentration of 3.6×10^{6} CFU/mL. You then add 4.95 mL sterile diluent. What is the dilution, and what is the final concentration of cells?

5 What would be the dilution if 96 mL of diluent is added to 4 mL of a bacterial suspension?

6 You were instructed to add 1.0 mL out of 5.0 mL of an undiluted sample to 99 mL of sterile diluent. Instead, you add all 5.0 mL to the 99 mL. What was the intended dilution and what was the actual dilution?

7 Suppose you were instructed to add 0.2 mL of sample to 9.8 mL of diluent, but instead added 2.0 mL of sample. What was the intended dilution, and what was the actual dilution?

8 Plating 1.0 mL of a sample diluted by a factor of 10^{-3} produced 43 colonies. What was the original concentration in the sample?

Name ______________________ Date ______________________

Lab Section ______________________ I was present and performed this exercise (initials) ________

9 Plating 0.1 mL of a sample diluted by a factor of 10^{-3} produced 43 colonies. What was the original concentration in the sample?

10 The plate has 72 colonies, with a sample volume of 10^{-7} mL. What was the original concentration in the sample?

11 The plate has 259 colonies, with a sample volume of 10^{-6} mL. What was the original concentration in the sample?

12 How many colonies should be on the plate inoculated with a sample volume of 10^{-7} mL using the same sample as in Question #11?

13 How many colonies should be on the plate inoculated with a sample volume of 10^{-5} mL using the same sample as in Question #11?

14 A plate inoculated with a sample volume of 10^{-7} mL produced 170 colonies. What was the original concentration in the sample?

15 After incubation, how many colonies should be on the 10^{-8} mL plate from the dilution series in Question #14?

16 After incubation, how many colonies should be on the 10^{-6} mL plate from the dilution series in Question #14?

17 You have inoculated 100 μL of a sample diluted by a factor of 10^{-3} on a Nutrient Agar plate. After incubation, you count 58 colonies. What was the original cell density?

18 A plate that received 1000 μL of a bacterial sample diluted by a factor of 10^{-6} had 298 colonies on it after incubation. What was the original cell density?

19 A Nutrient Agar plate labeled 10^{-5} mL had 154 colonies after incubation. What was the cell density in the original sample? If the dilution tube used to inoculate the plate also was labeled 10^{-5}, what volume was used to inoculate this plate?

20 The original concentration in a sample is 2.79×10^{6} CFU/mL. Which sample volume should yield a countable plate? (Express your answer as 10^{x} mL.)

21 The original concentration in a sample is 5.1×10^{9} CFU/mL. Which sample volume should yield a countable plate?

22 A sample has a density of 1.37×10^{5} CFU/mL. What sample volume should yield a countable plate? Which two dilution tubes could be used to produce this sample volume? How?

NAME ______________________ DATE ______________________

LAB SECTION ______________________ I WAS PRESENT AND PERFORMED THIS EXERCISE (INITIALS) ________

23 A sample has a density of 7.9×10^9 CFU/mL. What sample volume should yield a countable plate? Which two dilution tubes could be used to produce this sample volume? How?

24 You are told that a sample has between 2.5×10^6 and 2.5×10^9 cells/mL. Devise a complete but efficient (that is, no extra plates!) dilution scheme that will ensure getting a countable plate.

25 A sample has between 3.3×10^4 and 3.3×10^8 CFU/mL. Devise a complete but efficient (that is, no extra plates!) dilution scheme that will ensure getting a countable plate.

26 Two plates received 100 μL from the same dilution tube. The first plate had 293 colonies, whereas the second had 158 colonies. Suggest reasonable sources of error.

27 Two parallel dilution series were made from the same original sample. The plates with sample volumes of 10^{-5} mL from each dilution series yielded 144 and 93 colonies. Suggest reasonable sources of error.

28 What are the *only* circumstances that would *correctly* produce countable plates from two different dilutions?

EXERCISE 6-2

Urine Culture

Theory

Urine culture is a semiquantitative CFU counting method that quickly produces countable plates without a serial dilution. The instrument used in this procedure is a volumetric loop, calibrated to hold 0.001 mL or 0.01 mL of sample. Urine culture procedures using volumetric loops are useful in situations where a rapid diagnosis is essential and approximations ($\pm 10^2$ CFU/mL) are sufficient to choose a course of action. Volumetric loops are useful in situations where population density is not likely to exceed 10^5 CFU/mL.

In this standard procedure, a loopful of urine is carefully transferred to a Blood Agar plate. The initial inoculation is a single streak across the diameter of the agar plate. Then the plate is turned 90° and (without flaming the loop) streaked again, this time across the original line in a zigzag pattern to evenly disperse the sample over the entire plate (Figure 6-4). Following a period of incubation, the resulting colonies are counted and population density, usually referred to as "original cell density," or OCD, is calculated (Figure 6-5).

OCD is recorded in "colony forming units," or CFU per milliliter (CFU/mL), as described in the introduction to this section. CFU/mL is determined by dividing the number of colonies on the plate by the volume of the loop. For example, if 75 colonies are counted on a plate inoculated with a 0.001 mL loop, the calculation would be as follows:

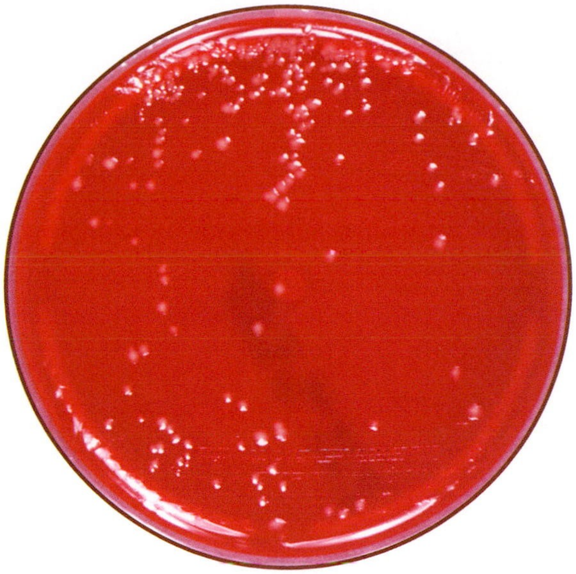

6-5 Urine Streak on Sheep Blood Agar ◆ This plate was inoculated with a 0.01 mL volumetric loop. The cell density can be determined by multiplying the number of colonies by 0.01.

$$OCD = \frac{CFU}{\text{loop volume}}$$

$$OCD = \frac{75\ CFU}{0.001\ mL}$$

$$OCD = 7.5 \times 10^4\ CFU/mL$$

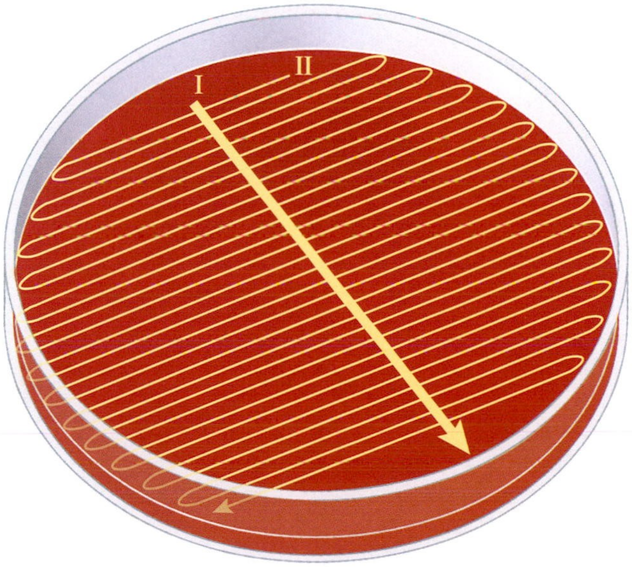

6-4 Semiquantitative Streak Method ◆ Streak 1 is a simple streak line across the diameter of the plate. Streak 2 is a tight streak across Streak 1 to cover the entire plate.

Application

Urine culture is a common method of detecting and quantifying urinary tract infections. It frequently is combined with selective media for specific identification of members of *Enterobacteriaceae* or *Streptococcus*.

In This Exercise

You will estimate cell density in a urine sample using a volumetric loop and the above formula. Be sure to hold the loop vertically and transfer slowly.

Materials

- ✓ one Blood Agar plate (TSA with 5% sheep blood)
- ✓ one sterile volumetric inoculating loop (either 0.01 mL or 0.001 mL)
- ✓ a fresh urine sample

Procedure

Lab One

1. Holding the loop vertically, immerse it in the urine sample. Then carefully withdraw it to obtain the correct volume of urine. This loop is designed to fill to capacity in the vertical position. Do not tilt it until you get it in position over the plate.
2. Inoculate the Blood Agar by making a single streak across the diameter of the plate.
3. Turn the plate 90° and, without flaming the loop, streak the urine across the entire surface of the agar, as shown in Figure 6-4.
4. Invert, label, and incubate the plate for 24 hours at 35 ± 2°C.

Lab Two

1. Remove the plate from the incubator and count the colonies. Also note any differing colony morphologies, which would indicate possible colonization by more than one species. Enter the data on the Data Sheet.
2. Calculate the original cell density of the sample using the formula on the Data Sheet. If two species are present, calculate each and enter both on the Data Sheet.

References

Forbes, Betty A., Daniel F. Sahm, and Alice S. Weissfeld. 2002. Pages 933–934 in *Bailey & Scott's Diagnostic Microbiology*, 11th ed. Mosby-Yearbook, St. Louis.

Winn, Washington C., et al. 2006. Page 85 in *Koneman's Color Atlas and Textbook of Diagnostic Microbiology*, 6th ed. Lippincott Williams & Wilkins, Baltimore.

Name ______________________ Date ______________________

Lab Section ______________________ I was present and performed this exercise (initials) ________

Urine Culture

OBSERVATIONS AND INTERPRETATIONS

Enter your colony count and loop volume data in the chart below. Then calculate the original cell density using the following formula.

$$OCD = \frac{CFU}{\text{loop volume}}$$

Use the extra rows to calculate cell densities of different colony types on your plate or for urine samples of other students. Label these appropriately.

Urine Sample	Colonies Counted	Loop volume (0.01 mL or 0.001 mL)	Original Cell Density (CFU/mL)

Questions

1 The plate pictured in Figure 6-5 was inoculated with a 0.01 mL volumetric loop and contains approximately 75 colonies. What was the original cell density?

2 The equation shown in the Theory explanation is used for calculating cell density in urine when using a 0.001 mL calibrated loop. The urine transferred in the loop is not literally diluted, yet its volume is equivalent to a dilution factor. What is the dilution factor (based on loop volume) expressed as a fraction? What is the dilution factor in scientific notation? (For help with dilutions, see Exercise 6-1.)

3 Calculation of original density in this exercise differs slightly from that offered in Exercise 6-1. Compare and contrast the formula used today with that used in Exercise 6-1. How do they differ? Could you have used the formula in Exercise 6-1 for today's calculations? Explain.

4 Using a volumetric loop is a semiquantitative technique. Why is it not quantitative? Design a procedure that would make it quantitative. (Hint: Refer to Exercise 6-1 if necessary.)

EXERCISE 6-3

Plaque Assay of Phage Titer

❖ Theory

Viruses that attack bacteria are called **bacteriophages,** or simply **phages** (Figure 6-6). Some viruses attach to the bacterial cell wall and inject viral DNA into the bacterial cytoplasm. The viral **genome** then commands the cell to produce more viral DNA and viral proteins, which are used for the assembly of more phages. Once assembly is complete, the cell lyses and releases the phages, which then attack other bacterial cells and begin the replicative cycle all over again. This process, called the **lytic cycle,** is shown in Figure 6-7.

Lysis of bacterial cells growing in a lawn on an agar plate produces a clearing that can be viewed with the naked eye. These clearings are called **plaques.** In the plaque assay, a sample of bacteriophage (generally diluted by means of a serial dilution) is added to a plate inoculated with enough bacterial **host** to produce a lawn of growth. After incubation, the number of plaques formed can be used to calculate the original phage **titer** (density). Refer to the procedural diagram in Figure 6-8.

Plaque assay technique is similar to the standard plate count, in that it employs a serial dilution to produce countable plates needed for later calculations. One key difference is that the plaque assay is done using the **pour-plate technique,** in which bacterial cells and viruses are first added

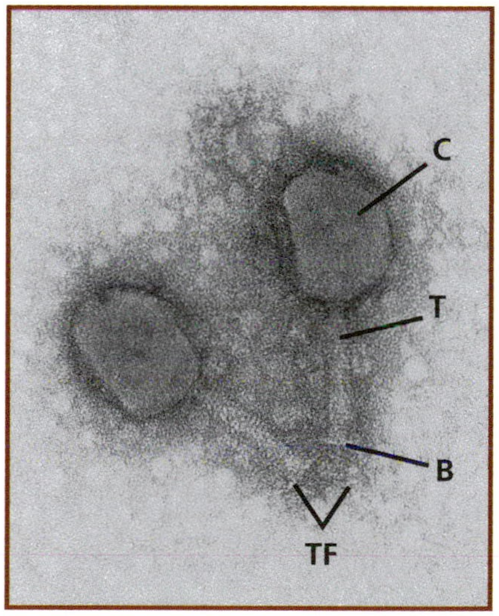

6-6 T4 Coliphage ◆ This is a negative stain of two T4 phage particles. Shown are the capsid (**C**), tail (**T**), and base plate (**B**). Tail fibers (**TF**) are visible only as broken pieces in the background. The length of this phage from base plate to tip of capsid is approximately 180 nm (0.18 µm).

(Photo by author taken at the San Diego State University Electron Microscope Facility.)

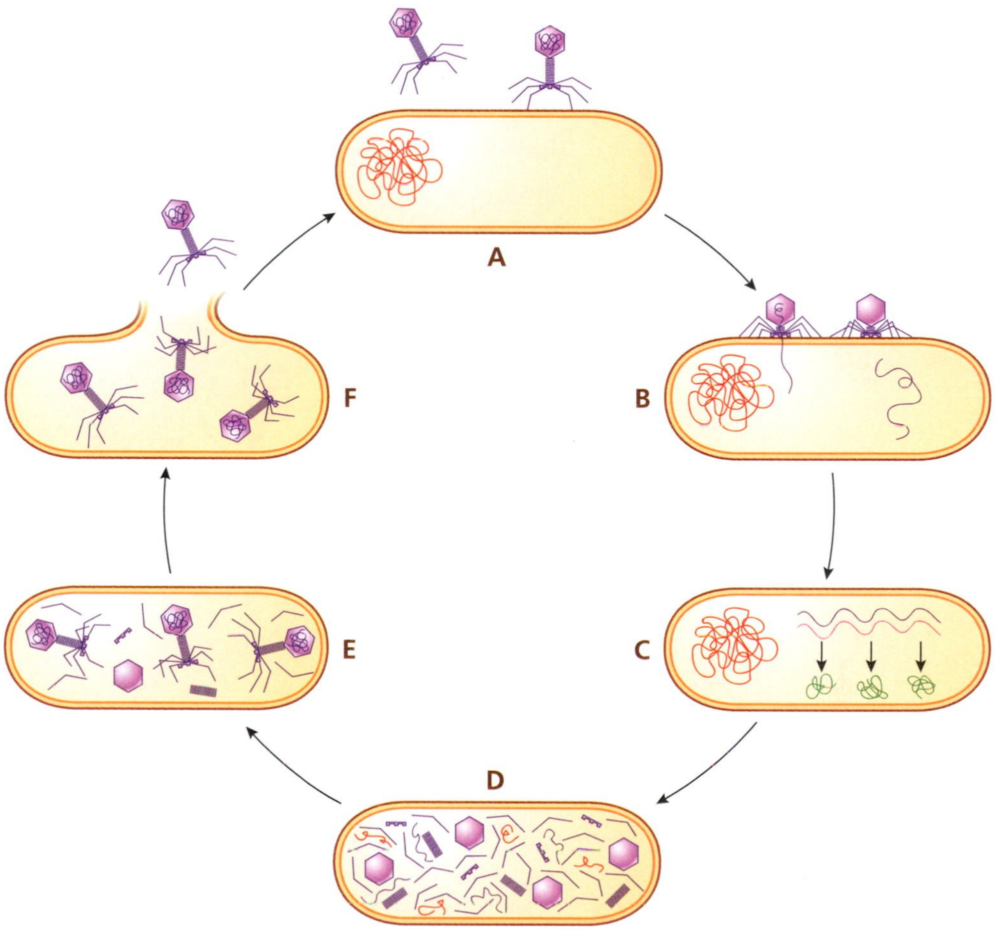

6-7 T4 Phage Lytic Cycle ◆ Shown is a simplified diagram of the T4 coliphage's replicative cycle. In **A**, an infective T4 phage (**left**) is approaching its host, *Escherichia coli*. On the right, the same phage is shown attached by its legs to specific receptors on *E. coli*. This is the **attachment** phase. The *E. coli* chromosome is the red tangled line at the left of the cell. On the left of **B**, T4 is in the process of injecting its DNA (purple line) into *E.coli*. This is the **penetration** phase. On the right, the entire phage genome is in the host. Removal of viral genome from its capsid is **uncoating**. In **C**, phage DNA is in the process of being transcribed into mRNA (red line), which is then translated into phage proteins (squiggly green lines). This is the **synthesis** phase and is performed by *E. coli* under the direction of the phage DNA. Simultaneously, phage DNA is being replicated, but this is not shown. Synthesis leads to **assembly** of the phage progeny, shown in **D** and **E**, where the capsid subunits come together to form the capsid into which the genome is inserted, and the tail with all its detailed parts comes together and attaches to the capsid. Note that host DNA has been degraded, but remains as short, red lines. (This is important because occasionally *E. coli* DNA is put into a viral capsid and is transferred to a new *E. coli* host, providing it with new genes in a process called **generalized transduction**). In **F**, the fully assembled phages are released as the cells bursts. At 37°C, this entire cycle can take as little as 25 minutes and release a few hundred phage progeny, each of which can repeat the process!

to molten agar and then poured into the plate. For help with serial dilutions and calculations, refer to Exercise 6-1, Standard Plate Count.

In this procedure (Figure 6-8), diluted phage is added directly to a small amount of *E. coli* culture (its host) and allowed a 15-minute **preadsorption period** to attach to the bacterial cells. Then this phage–host mixture is added to a tube of **soft agar**, mixed, and poured onto prepared Nutrient Agar plates as an **agar overlay**. The consistency of the solidified soft agar is sufficient to immobilize the bacteria while allowing the smaller bacteriophages, during incubation, to diffuse short distances and infect nearby cells. The end result is a plate covered with a lawn of *E. coli* bacterial growth in which plaques appear where contiguous cells have been lysed by the virus (Figure 6-9).

The procedure for counting plaques is the same as that for the standard plate count. To be statistically reliable, countable plates must have between 30 and 300 plaques. Calculating phage titer (original phage density) uses the same formula as other plate counts except that PFU (plaque

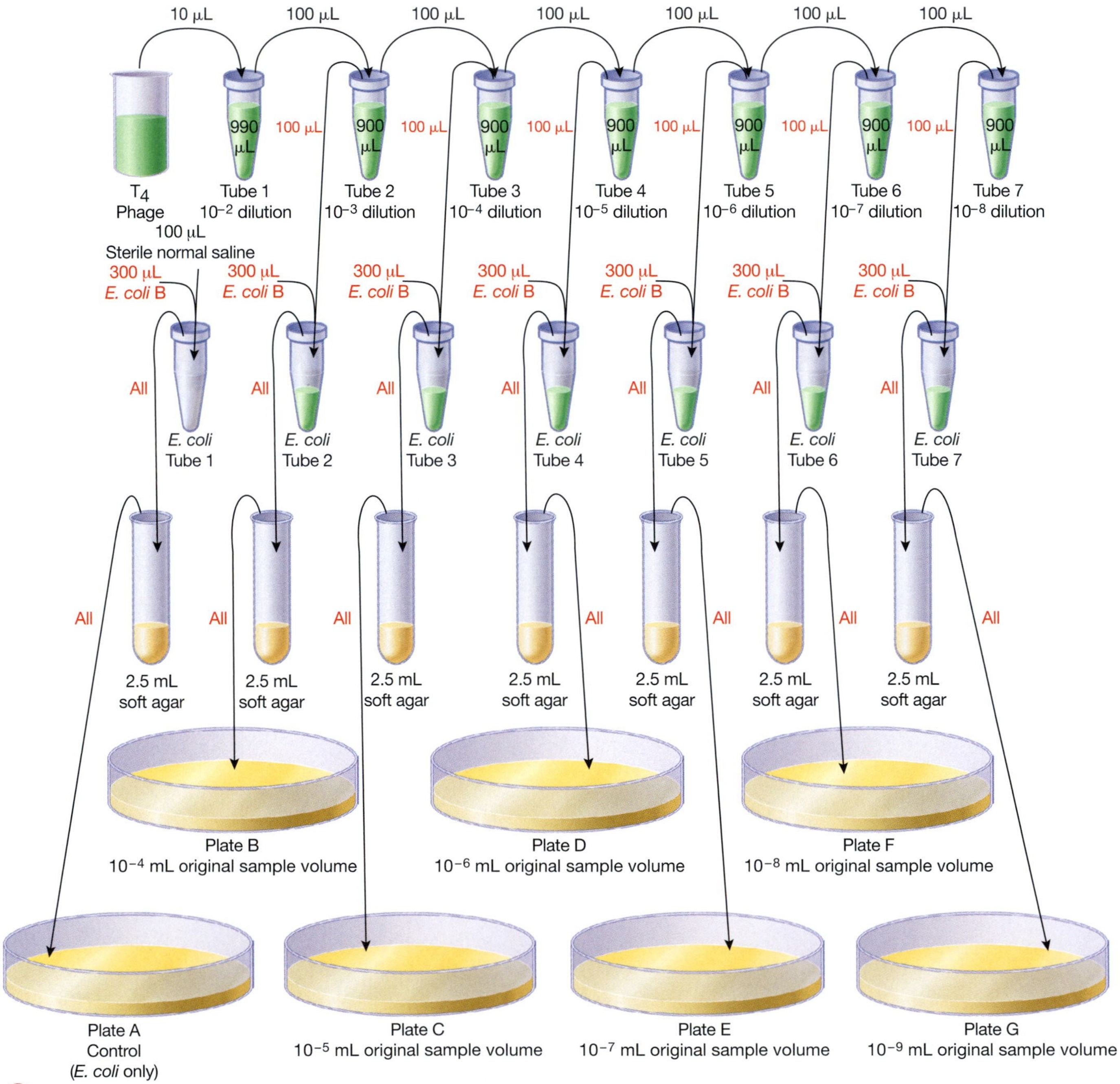

6-8 Procedural Diagram for Plaque Assay ◆ This is an illustration of the dilution scheme outlined in the Procedure. The dilution assigned to each tube (written below the tube) represents the proportion of original sample inside that tube. For example, if the dilution is 10^{-4}, the proportion of original sample inside the tube would be 1/10000th of the total volume inside. When 100µL (0.1 mL) of that solution is transferred to a plate, the volume of original sample in the plate is $0.1 \text{ mL} \times 10^{-4}$ or 10^{-5} mL. (Note: Do not be confused by the addition of *E. coli*; it does not affect the phage volume going to the plate.)

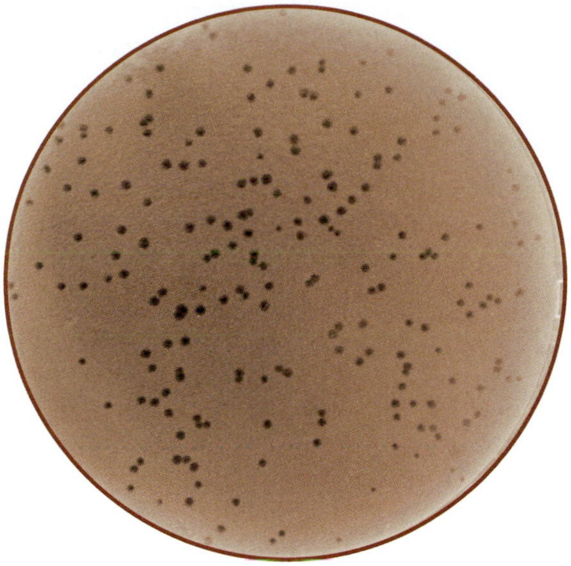

6-9 **Countable Plate** ◆ This plaque assay plate has between 30 and 300 plaques; therefore, it is countable.

forming unit) instead of CFU (colony forming unit) becomes the numerator in the equation. Phage titer, therefore, is expressed in PFU/mL and the formula is written as follows:

$$\text{Phage titer} = \frac{\text{PFU}}{\text{Volume plated} \times \text{Dilution}}$$

As with the standard plate count (Exercise 6-1), it is customary to condense *volume plated* and *dilution* into *original sample volume*. The formula then becomes:

$$\text{Phage titer} = \frac{\text{PFU}}{\text{Original sample volume}}$$

The original sample volume is written on the plate at the time of inoculation. Following a period of incubation, the plates are examined, plaques are counted on the countable plates, and calculation is a simple division problem.

Suppose, for example, you inoculated a plate with 100.0 µL (0.1 mL) of a 10^{-4} dilution. (Remember, you are calculating the *phage density;* the *E. coli* has nothing to do with the calculations.) This plate now contains 10^{-5} mL of original phage sample. If you subsequently counted 45 plaques on the plate, calculation would be as follows:

$$\text{Original phage density} = \frac{45 \text{ PFU}}{10^{-5} \text{ mL}} = 4.5 \times 10^{6} \text{ PFU/mL}$$

❖ Application

This technique is used to determine the concentration of viral particles in a sample. Samples taken over a period of time can be used to construct a viral growth curve.

❖ In This Exercise

You will be estimating the density (titer) of a T4 coliphage sample using a strain of *Escherichia coli* (*E. coli B*) as the host organism.

❖ Materials

Per Class

✓ 50°C hot water bath containing tubes of liquid soft agar (seven tubes per group)

Per Student Group

✓ micropipettes (10–100 µL and 100–1000 µL) with sterile tips

✓ fourteen sterile capped microtubes (1.5 mL or larger)

✓ seven Nutrient Agar plates

✓ seven tubes containing 2.5 mL soft agar

✓ small flask of sterile normal saline

✓ seven sterile transfer pipettes

✓ T4 coliphage

✓ 24-hour broth culture of *Escherichia coli B* (T-series phage host)

❖ Medium Recipe

Soft Agar

◆ Beef extract	3.0 g
◆ Peptone	5.0 g
◆ Sodium chloride	5.0 g
◆ Tryptone	2.5 g
◆ Yeast extract	2.5 g
◆ Agar	7.0 g
◆ Distilled or deionized water	1.0 L

Procedure

Refer to the procedural diagram as needed.

Lab One

1 Obtain all materials except for the soft agar tubes. To keep the agar tubes liquefied, leave them in the water bath and take them out one at a time as needed.

2 Label seven microtubes 1 through 7. Label the other seven microtubes *E. coli* 1–7. Place all tubes in a rack, pairing like-numbered tubes.

3 Label the Nutrient Agar plates A through G. Place them in the 35°C incubator to warm them. Take them out one at a time as needed. This will keep the soft agar (added at step 14) from solidifying too quickly and result in a smoother agar surface.

4 Aseptically transfer 990 µL sterile normal saline to dilution tube 1.

5 Aseptically transfer 900 µL sterile normal saline to dilution tubes 2–7.

6 Mix the *E. coli* culture and aseptically transfer 300 µL into each of the *E. coli* microtubes.

7 Mix the T4 suspension and aseptically transfer 10 µL to dilution tube 1. Mix well. This is a10^{-2} dilution.

8 Aseptically transfer 100 µL from dilution tube 1 to dilution tube 2. Mix well. This is 10^{-3}.

9 Aseptically transfer 100 µL from dilution tube 2 to dilution tube 3. Mix well. This is 10^{-4}.

10 Continue in this manner through dilution tube 7. The dilution in tube 7 should be 10^{-8}.

11 Aseptically transfer 100 µL sterile normal saline to *E. coli* tube 1. This will be mixed with 2.5 ml soft agar and used to inoculate a control plate.

12 Aseptically transfer 100 µL from dilution tube 2 to its companion *E. coli* tube. Repeat this procedure with the remaining five tubes.

13 This is the beginning of the preadsorption period. Let all seven tubes stand undisturbed for 15 minutes.

14 Remove one soft agar tube from the hot water bath and add the entire contents of *E. coli* tube 1. Using a sterile transfer pipette, mix well and immediately pour onto plate A. Gently tilt the plate back and forth until the soft agar mixture is spread evenly across the solid medium. Label the plate "Control."

15 Remove a second soft agar tube from the water bath and add the entire contents of *E. coli* tube 2. Using a sterile transfer pipette, mix well, and immediately pour onto plate B. Tilt back and forth to cover the agar and label it "10^{-4} mL original sample."

16 Repeat this procedure with dilutions 10^{-4} thru 10^{-8} and plates C through G. Label the plates with the appropriate original phage sample volumes.

17 Allow the agar to solidify completely.

18 Invert the plates and incubate aerobically at 35 ± 2°C for 24 to 48 hours.

Lab Two

1 After incubation, examine the control plate for growth and the absence of plaques.

2 Examine the remainder of your plates and determine which one is countable (30 to 300 plaques). Count the plaques and record the number on the Data Sheet. Record all others as either TNTC (to numerous to count) or TFTC (to few to count).

3 Using the sample volumes on the countable plate and the formula provided on the Data Sheet, calculate the original phage density. Record your results in the space provided on the Data Sheet.

References

Collins, C. H., Patricia M. Lyne, J. M. Grange. 1995. Page 149 in *Collins and Lyne's Microbiological Methods*, 7th ed. Butterworth-Heinemann, United Kingdom.

DIFCO Laboratories. 1984. Page 619 in *DIFCO Manual*, 10th ed. DIFCO Laboratories, Detroit.

Province, David L., and Roy Curtiss III. 1994. Page 328 in *Methods for General and Molecular Bacteriology*, edited by Philipp Gerhardt, R. G. E. Murray, Willis A. Wood, and Noel R. Krieg. American Society for Microbiology, Washington, DC.

NAME ______________________ DATE ______________________

LAB SECTION ______________________ I WAS PRESENT AND PERFORMED THIS EXERCISE (INITIALS) ________

Plaque Assay of Phage Titer

OBSERVATIONS AND INTERPRETATIONS

1 Enter the number of plaques counted on the countable plate. Only one plate should be countable. If there are more than 300 plaques, enter TNTC. If fewer than 30 plaques, enter TFTC.

Plate	A	B	C	D	E	F	G
Plaques Counted							
Sample volume							

2 Calculate the original density using the following formula, and enter the result below. Record your answer in PFU/mL.

$$\text{Phage titer} = \frac{\text{PFU}}{\text{Original sample volume}}$$

If you don't have any countable plates, use the TFTC plate closest to 30 in your calculation for practice. Make a note of this in your results.

Phage titer	

Questions

1 In this exercise, there must be enough bacteria inoculated to produce a lawn of growth. Why is that important?

DATA SHEET 6-3

2 How might the results be altered if you had skipped the preadsorption phase?

3 Suppose you followed all the necessary steps outlined in the Procedure and found no plaques on any of your plates after incubation. Suppose further that you knew with certainty that the bacteriophage was viable and had worked in other labs prior to yours. What possible explanations could there be for this occurrence?

4 Why was the water bath set at 50°C? What might be some consequences of changing the temperature?

5 Why was soft agar used for the agar overlay? What would you expect to see if standard Nutrient Agar had been used instead?

EXERCISE 6-4

Differential Blood Cell Count

Theory

Leukocytes (white blood cells or WBCs) are divided into two groups: **granulocytes** (which have prominent cytoplasmic granules) and **agranulocytes** (which lack these granules). There are three basic types of granulocytes: **neutrophils, basophils,** and **eosinophils**. The two types of agranulocytes are **monocytes** and **lymphocytes**.

Neutrophils (Figure 6-10) are the most abundant WBCs in blood. They leave the blood and enter tissues to phagocytize foreign material. An increase in neutrophils in the blood is indicative of a systemic bacterial infection. Mature neutrophils are sometimes referred to as **segs** because their nucleus is usually segmented into two to five lobes. Because of the variation in nuclear appearance, they are also called **polymorphonuclear neutrophils** (**PMNs**). Immature neutrophils lack this nuclear segmentation and are referred to as **bands**. This distinction is useful, because a patient with an active infection will be producing more neutrophils, meaning a higher percentage will be of the band (immature) type. Neutrophils are 12–15 µm in diameter, about twice the size of an erythrocyte (RBC).[1] Their cytoplasmic granules are neutral staining and thus do not have the intense color of other granulocytes when prepared with a Wright's or Giemsa stain.

Eosinophils are phagocytic, and their numbers increase during allergic reactions and parasitic infections (Figure 6-11). They are 12–15 µm in diameter (about twice the size of an RBC) and generally have two lobes in their nucleus. Their cytoplasmic granules stain red in typical preparations.

[1] It is convenient to discuss leukocyte size in terms of erythrocyte size because RBCs are so uniform in diameter. In an isotonic solution, erythrocytes are 7.5 µm in diameter.

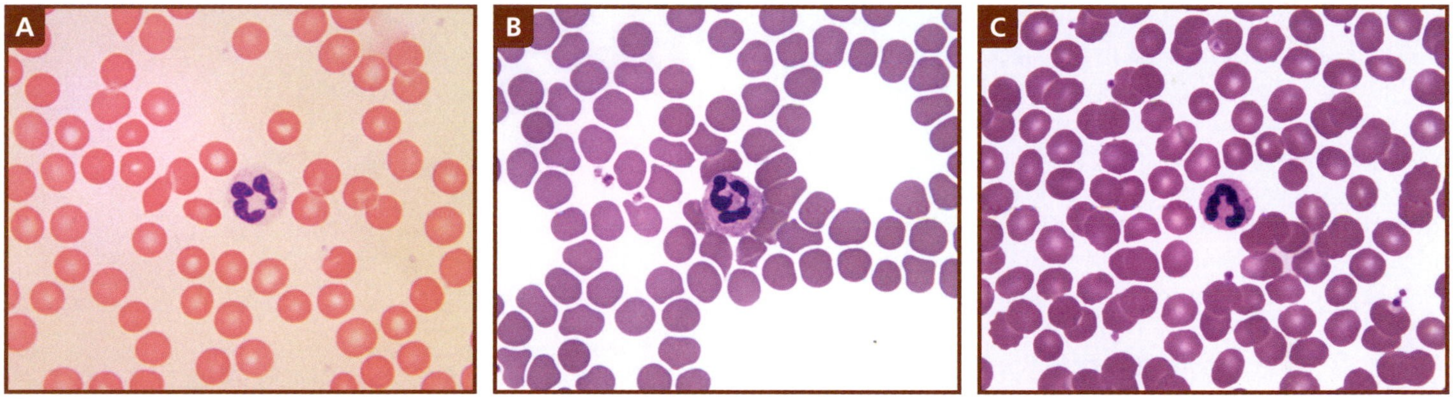

6-10 Neutrophils ◆ **A** and **B** The segmented nuclei of these cells identify them as mature neutrophils (segs). About 30% of neutrophils in blood samples from females demonstrate a "drumstick" protruding from the nucleus, as in **B**. This is the region of the inactive X chromosome. **C** This is an immature band neutrophil with an unsegmented nucleus. Micrograph **A** was stained with Giemsa stain; the others were prepared with Wright's stain. Neutrophils are approximately 12–15 µm in diameter.

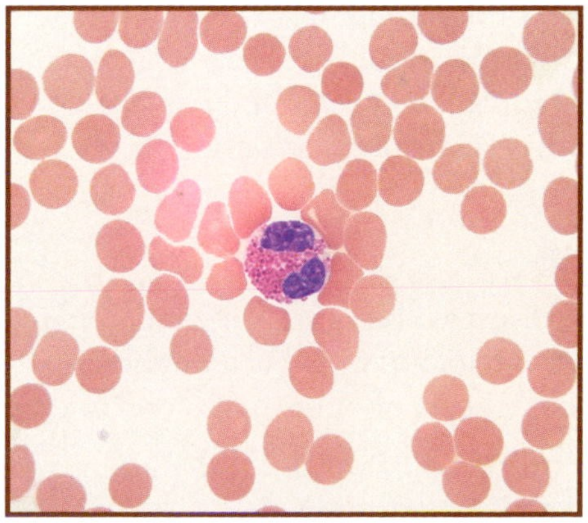

6-11 Eosinophil (Giemsa Stain) ◆ These granulocytes are relatively rare and are about twice the size of RBCs. Their cytoplasmic granules stain red, and their nucleus usually has two lobes.

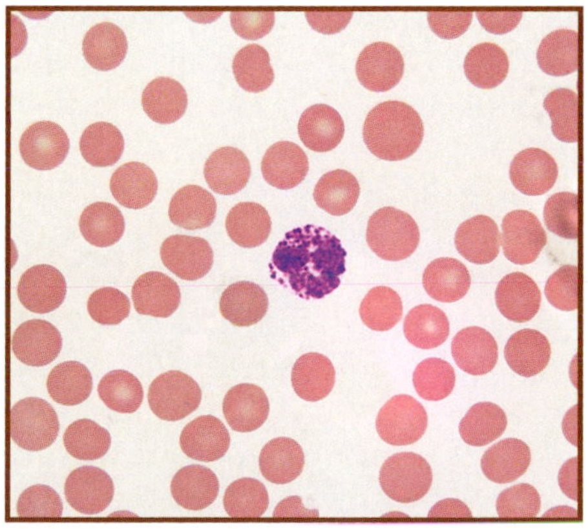

6-12 Basophil (Giemsa Stain) ◆ Basophils comprise only about 1% of all WBCs. They are up to twice the size of RBCs and have dark purple cytoplasmic granules that obscure the nucleus.

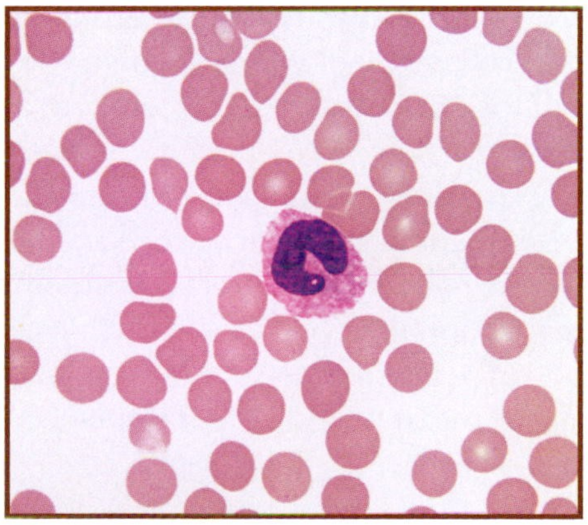

6-13 Monocyte (Giemsa Stain) ◆ Monocytes are the blood form of macrophages. They are about two to three times the size of RBCs and have a round or indented nucleus.

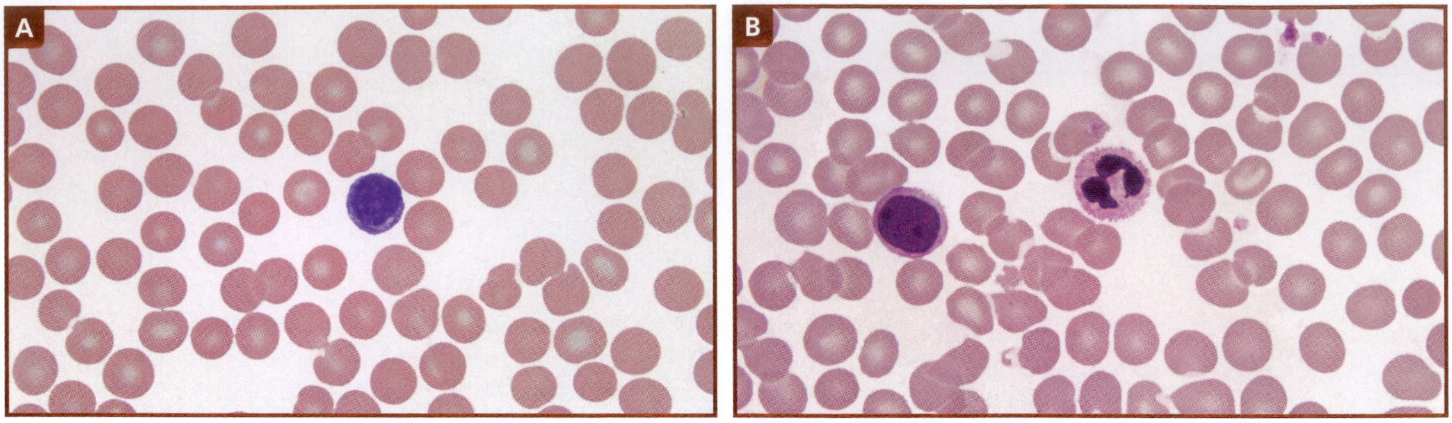

6-14 Lymphocytes ◆ Lymphocytes are common in the blood, comprising up to 33% of all WBCs. Most are about the size of RBCs and have only a thin halo of cytoplasm encircling their round nucleus. They belong to functional groups called B-cells and T-cells (which are morphologically indistinguishable). Micrograph **A** is a small lymphocyte and was prepared with Giemsa stain. Some lymphocytes are larger, as in micrograph **B**. These are natural killer (NK) cells or some other type of null cell. Also visible is a neutrophil.

Basophils (Figure 6-12) are the least abundant WBCs in normal blood. They are structurally similar to tissue mast cells and produce some of the same chemicals (histamine and heparin), but are derived from different stem cells in bone marrow. They are 12–15 µm in diameter. The nucleus is usually obscured by the dark staining cytoplasmic granules, but it either has two lobes or is unlobed.

Agranulocytes include monocytes and lymphocytes. Monocytes (Figure 6-13) are the blood form of **macrophages**. They are the largest of leukocytes, being two to three times the size of RBCs (12–20 µm). Their nucleus is horseshoe-shaped and the cytoplasm lacks prominent granules (but may appear finely granular).

Lymphocytes (Figure 6-14) are cells of the immune system. Two functional types of lymphocytes are the **T-cell**, involved in cell-mediated immunity, and the **B-cell**, which converts to a **plasma cell** (Figure 6-15) when activated and produces antibodies. The nucleus is usually spherical and takes up most of the cell. Lymphocytes are approximately the same size as RBCs or up to twice their size. The larger ones form a third functional group of lymphocytes, the **null cell**, many of which are **natural killer (NK) cells** that kill foreign or infected cells without antigen-antibody interaction (Figure 6-14b).

In a differential white cell count, a sample of blood is observed under the microscope and at least 100 WBCs are counted and tallied (this task is automated now). Approximate normal percentages for each leukocyte are as follows: 55–65% neutrophils (mostly segs), 25–33% lymphocytes, 3–7% monocytes, 1–3% eosinophils, and 0.5–1% basophils.

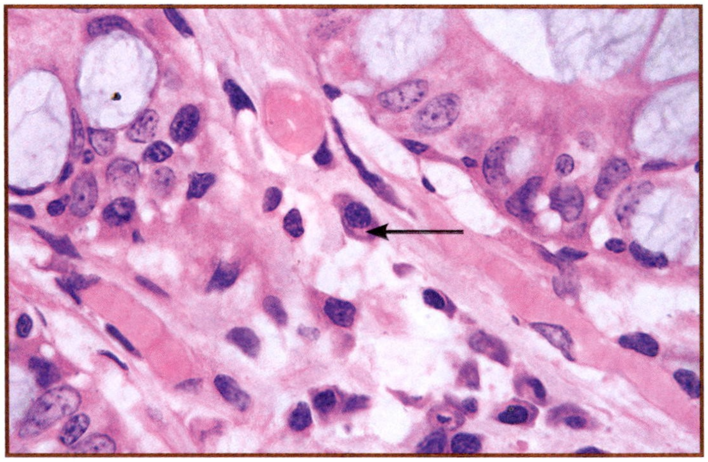

6-15 Plasma Cells ◆ B-lymphocytes develop into plasma cells when stimulated by the appropriate antigen. Plasma cells secrete protective antibodies against the antigen. This plasma cell (arrow) is recognizable because of its elongated shape, eccentric nucleus with "clock face" chromatin, and a pale region near the nucleus (which is the site of a Golgi apparatus). This specimen is from the colon.

Application

A differential blood cell count is done to determine approximate numbers of the various leukocytes in blood. Excess or deficiency of all or a specific group is indicative of certain disease states. Even though differential counts are automated now, it is good training to perform one "the old-fashioned way" using a blood smear and a microscope to get an idea of the principle behind the technique.

In This Exercise

You will be examining prepared blood smears and doing a differential count of white blood cells. As an optional activity, you may look at smears of abnormal blood and compare the differential count to normal blood.

Materials

Per Student Group

✓ commercially prepared human blood smear slides (Wright's or Giemsa stain)

TABLE **6-1** Typical Features of Human Leukocytes in Blood ◆

TABLE OF RESULTS					
Cell	**Abundance in Blood (%)**	**Diameter (μm)**	**Nucleus**	**Cytoplasmic Granules (Wright's or Giemsa Stain)**	**Functions**
Granulocyte					
Neutrophil	55–65	12–15	2–5 lobes	Present, but stain poorly; contain antimicrobial chemicals	Phagocytosis and digestion of (usually) bacteria
Eosinophil	1–3	12–15	2 lobes	Present and stain red; contain antimicrobial chemicals and histaminase	Present in inflammatory reactions and immune response against some multicellular parasites (worms)
Basophil	0.5–1	12–15	Unlobed or 2 lobes	Present and stain dark purple; contain histamine and other chemicals	Participate in inflammatory response
Agranulocyte					
Lymphocyte	25–33	7–18	Spherical (leaving little visible cytoplasm)	Absent	Active in specific acquired immunity (as T- and B-cells)
Monocyte	3–7	12–20	Horseshoe-shaped (cytoplasm is prominent)	Absent	Phagocytosis (as macrophages)

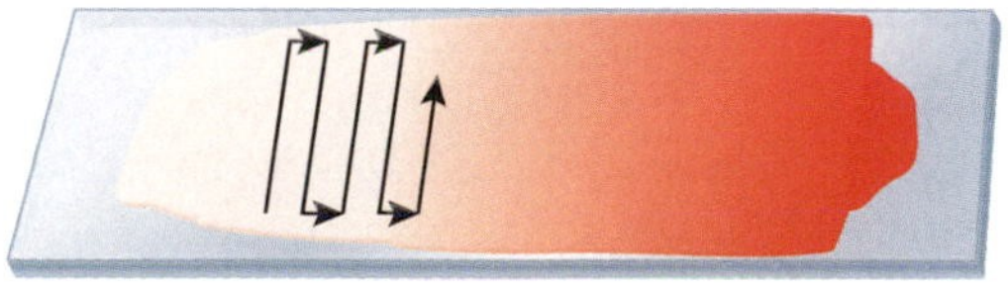

6-16 Following a Systematic Path ◆ A systematic scanning path is used to avoid wandering around the slide and perhaps counting some cells more than once. Remember that a microscope image is inverted. If you want the image to move left, you must move the slide to the right.

✓ (optional) commercially prepared abnormal human blood smear slides (e.g., infectious mononucleosis, eosinophilia, or neutrophilia)

Procedure

1 Obtain a blood smear slide and locate a field where the cells are spaced far enough apart to allow easy counting. (The cells should be fairly dense on the slide, but not overlapping.)

2 Using the oil immersion lens, scan the slide using the pattern shown in Figure 6-16. Be careful not to overlap fields when scanning the specimen. Choose a "landmark" blood cell at the right side of the field, and move the slide horizontally until that cell disappears off the left side. Avoid diagonal movement of the slide. As you scan, use the mechanical stage knobs separately to move the slide up and back or to the right and left in straight lines.

3 Make a tally mark in the appropriate box on the Data Sheet for the first 100 leukocytes you see.

4 Calculate percentages and compare your results with the accepted normal values.

5 Repeat with a pathological blood smear (if available).

References

Brown, Barbara A. 1993. *Hematology—Principles and Procedures*, 6th ed. Lea and Febiger, Philadelphia.

Diggs, L. W., Dorothy Sturm, and Ann Bell. 1978. *The Morphology of Human Blood Cells*, 4th ed. Abbott Laboratories, North Chicago, IL.

Junqueira, L. Carlos, and Jose Carneiro. 2003. *Basic Histology, Text and Atlas*, 10th ed. Lange Medical Books, McGraw Hill, New York.

Name ______________________ Date ______________________

Lab Section ______________________ I was present and performed this exercise (initials) ________

DATA SHEET 6-4

Differential Blood Cell Count

OBSERVATIONS AND INTERPRETATIONS

Record your data from the differential blood cell count in the chart below. As you count the 100 white blood cells, make tally marks in the appropriate boxes. Then calculate the percentages of each type and compare them to the expected values.

Normal Blood						
	Monocytes	Lymphocytes	Segmented Neutrophils	Band Neutrophils	Eosinophils	Basophils
Number						
Percentage						
Expected Percentage	3–7%	25–33%	55–65% (all neutrophils)	—	1–3%	0.5–1%

Abnormal Blood (Condition: ______)						
	Monocytes	Lymphocytes	Segmented Neutrophils	Band Neutrophils	Eosinophils	Basophils
Number						
Percentage						
Expected Percentage	3–7%	25–33%	55–65% (all neutrophils)	—	1–3%	0.5–1%

Questions

1 How do the percentages of each WBC compare to the published values? What might account for any differences you noted?

2 If you did a differential count on abnormal blood, how did the percentages compare to normal blood? How can any differences you noted be explained in the context of the disease and/or defense process?

3 What is the purpose of scanning the slide in a systematic pattern (as described in the text)?

EXERCISE 6-5

Environmental Sampling: The RODAC™ Plate

Theory

Monitoring of microbial surface contamination is an important practice in medical, veterinary, pharmaceutical, and food preparation settings. Often, the RODAC™ (Replicate Organism Detection and Counting) plate is used. It is a specially designed agar plate into which the medium is poured to produce a convex surface extending above the edge of the plate (Figure 6-17). The special design allows support of the lid above the agar without touching it. As a result, the sterile plate may be opened and pressed on a surface to be sampled. Most are 65 mm in diameter, which is smaller than standard-sized 100 mm Petri dishes. The smaller size makes it easier to apply uniform pressure across the entire plate when taking the sample. In addition, the base is marked in 16 1 cm squares, allowing an estimate of cell density on the surface (Figure 6-18).

The plate can be filled with a variety of media, the choice of which depends on the surface being sampled and the microbes to be recovered. For instance, in this lab you will be using plates with Trypticase™ Soy Agar, a good, general-purpose growth medium. Because the surface sampled may have been disinfected recently, Polysorbate 80 and Lecithin are added to counteract the effect of residual disinfectant. The medium can be supplemented with 5% sheep blood to improve recovery of fastidious bacteria. Monitoring yeast and mold contamination often employs Sabouraud Dextrose Agar.

The acceptable amount of growth on a RODAC™ plate is determined by the surface being sampled. It stands to reason that a surgical area would have a lower limit of acceptability than a food preparation area. Table 6-2 provides some guidelines.

6-17 RODAC™ Plate ◆ Notice that the agar extends above the edges of the plate to allow contact with the surface to be sampled.

6-18 Grid on the RODAC™ Plate ◆ Typically, RODAC™ plates are 65 mm in diameter. A grid of 16 squares is molded into the base, each with an area of 1 cm². Colonies growing in the grid can be counted and an average number of CFU per cm² of surface can be determined.

TABLE 6-2 Interpretation of RODAC™ Plate Colony Counts ◆

TABLE OF REPRESENTATIVE RESULTS[1] (Colonies in 25 cm)[2]

Interpretation	Critical Surfaces[2]	Floors
Good	0–5	0–25
Fair	6–15	26–50
Poor	>16	>50

[1] Adapted from BBL™ Trypticase™ Soy Agar with Lecithin and Polysorbate 80 package insert.

[2] Critical surfaces include those in operating rooms, nurseries, table tops, toilet seats, and other nonporous surfaces.

Application

The RODAC™ plate is used to monitor contamination of surfaces in food preparation, veterinary, pharmaceutical, and medical settings. The plates also can be used to assess the efficiency of decontamination of a surface by taking a sample with different plates before and after decontamination.

In This Exercise

You will sample your work surface before and after decontamination to evaluate your technique.

Materials

Per Student Group

✓ two Trypticase™ Soy Agar with Lecithin and Polysorbate 80 RODAC™ plates (available from BD Diagnostic Systems, http://www.bd.com/us/, catalog number 221288.) Alternatively, sterile RODAC™ plates may be purchased and filled with 16.5 to 17.5 mL TSA plus Lecithin and Polysorbate 80.

✓ disinfectant solution used in your laboratory

Medium Recipe

Trypticase™ Soy Agar with Lecithin and Polysorbate 80

♦ Pancreatic digest of casein	15.0 g
♦ Papaic digest of soybean meal	5.0 g
♦ Sodium chloride	5.0 g
♦ Lecithin	0.7 g
♦ Polysorbate 80	5.0 g
♦ Agar	15.0 g
♦ Distilled or deionized water	1.0 L

pH 7.1–7.5 at 25°C

Procedure

Lab One

1. **Important:** For today's lab only, **do not** disinfect your lab bench prior to beginning work.
2. Obtain two RODAC™ plates. Label the base of one plate "Before" and the base of the other "After." Also label the plate with your group and date.
3. Remove the lid of the "Before" plate and gently press the agar onto the surface of your lab bench. Apply uniform vertical pressure and do not slide the plate. Replace the lid.
4. Wipe down the bench surface with your lab's disinfectant. Allow it to dry on its own before proceeding.
5. Remove the lid of the "After" plate and gently press the agar onto the surface of your lab bench. (It is best not to sample the same spot as you did with the "Before" plate.) Apply uniform vertical pressure and do not slide the plate. Replace the lid.
6. Incubate both plates for 48–72 hours at 35 ± 2°C.

Lab Two

1. Examine the plates for growth. Count the number of colonies in the grid to get an estimate of cell density on the lab bench before and after decontamination.
2. Record your results on the Data Sheet.

References

American Public Health Association (Laboratory Section), Committee on Microbial Contamination of Surfaces. 1970. "A Cooperative Microbiological Evaluation of Floor Cleaning Procedures in Hospital Patient Rooms." *Health Lab. Sci.* 7 (4): 256–264.

BBL™ Trypticase™ Soy Agar with Lecithin and Polysorbate 80 package insert. BD Diagnostic Systems, http://www.bd.com/us/

Name ______ Date ______

Lab Section ______ I was present and performed this exercise (initials) ______

Environmental Sampling: The RODAC™ Plate

OBSERVATIONS AND INTERPRETATIONS

Record the results of each lab group and interpret them in the table below. To assess decontamination, compare the number of colonies on the "Before" and "After" plates. To interpret final colony count, apply the standards in Table 6-2 to the "After" plate.

TABLE OF RESULTS

Lab Group	Source of Sample	Colonies Per Plate Before Cleaning Lab Bench	Colonies Per Plate After Cleaning Lab Bench	Assessment of Decontamination	Interpretation of Final Colony Count

Questions

1 What information does sampling immediately after treatment of the surface with disinfectant provide?

2 What information would sampling several hours after treatment of the surface with disinfectant provide?

3 Why is it important to apply even pressure to the plate when sampling the surface?

4 Why is it important not to slide the plate when sampling the surface?

5 Why is it important that the plate be constructed so the lid is held above the agar (and not resting on it) during incubation?

6 Why is it important to counteract the effects of the disinfectant with Polysorbate 80 and Lecithin when incubating the sample?

7 Suppose a RODAC™ plate has no colonies on it after incubation. Can you assume that the surface is sterile? Explain your answer.

SECTION 7

Medical, Environmental, and Food Microbiology

Medical Microbiology

The study and application of many microbiological principles and those of medicine are inseparable. So, although this is not a medical microbiology manual, it is devoted largely to the study of microorganisms and their relationship to human health. The first four exercises in this section address specific medical issues and were chosen to augment exercises you have already completed from Sections 4, 5, and 6, many of which could have been, but for an editorial decision, included here.

Environmental Microbiology

Environmental microbiology is the study, utilization, and control of microorganisms in the environment. Many environmental organisms are used beneficially by industry, such as in the production of vinegar, leaching of low-grade ores, and sewage treatment. Other environmental organisms are of human or animal origin (usually fecal) and pose threats as food or water contaminants.

Food Microbiology

Food microbiology is devoted to the study and utilization of beneficial microbes, as well as the control of many common, and in some cases, potentially deadly, contaminants. Many molds, yeasts, and bacteria are responsible for the production, preservation, and flavoring of foods we enjoy, while others produce toxins so powerful that as little as one nanogram per kilogram of body weight can be lethal.

Medical Microbiology

So much of microbiology has to do with its medical applications that a strong case could be made that most of the exercises you have already completed fall under this heading. What we present in this first segment are two diagnostic tests and two epidemiological exercises. The Snyder Test is a simple and fun procedure that is used to detect susceptibility to dental decay. The Kirby-Bauer Test measures the effectiveness of various antibiotics on sample bacteria. In the first epidemiological exercise, you will use a standard resource—*Morbidity and Mortality Weekly Report* available from the Centers for Disease Control and Prevention (CDC) Web site—to follow the incidence of a disease of your choice over the course of two complete years. In the second exercise, you will simulate an epidemic outbreak in your class, determine the source of the outbreak, and perform epidemiological analyses of your class population.

EXERCISE 7-1

Snyder Test

Theory

Snyder Test medium is formulated to favor the growth of oral bacteria (Figure 7-1) and discourage the growth of other bacteria. This is accomplished by adjusting the starting pH of the medium to 4.8. Glucose is the fermentable carbohydrate and bromcresol green is the pH indicator. Lactobacilli and oral streptococci survive these harsh conditions, ferment the glucose, and lower the pH even further. The pH indicator, which is green at or above pH 4.8 and yellow below, turns yellow in the process. Development of yellow color in this medium, therefore, is evidence of fermentation and is highly suggestive of the presence of dental decay-causing bacteria (Figure 7-2).

The medium is autoclaved for sterilization, cooled to just over 45°C, and maintained in a warm water bath until needed. The molten agar then is inoculated with a small amount of saliva, mixed well, and incubated for up to 72 hours. The agar tubes are checked at 24-hour intervals for any change in color. High susceptibility to dental caries is indicated if the medium turns yellow within 24 hours. Moderate and slight susceptibility are indicated by a change within 48 and 72 hours, respectively. No change within 72 hours is considered a negative result. These results are summarized in Table 7-1.

Application

The Snyder Test is designed to measure susceptibility to dental caries (tooth decay), caused primarily by lactobacilli and oral streptococci.

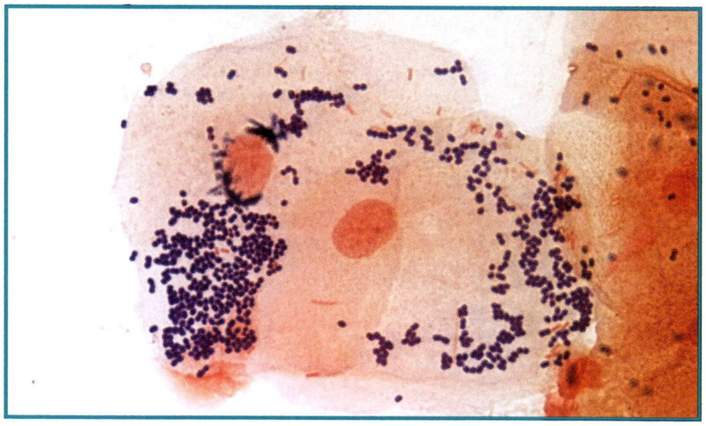

7-1 Gram Stain of Oral Bacteria (X1000) ◆ Note the predominance of Gram-positive cocci. Also present are a few Gram-negative rods and very few Gram-positive rods (*Lactobacillus*). Note also the size difference between the large epithelial cells in the center and the bacteria.

7-2 **Snyder Test Results** ◆
A positive result is on the left, and a negative result is on the right.

TABLE 7-1 **Snyder Test Results and Interpretations** ◆

TABLE OF RESULTS	
Result	**Interpretation**
Yellow at 24 hours	High susceptibility to dental caries
Yellow at 48 hours	Moderate susceptibility to dental caries
Yellow at 72 hours	Slight susceptibility to dental
Yellow at >72 hours	Negative

In This Exercise

You will check your own oral microflora for its potential to cause tooth decay.

Materials

Per Student Group

- ✓ hot water bath set at 45°C
- ✓ small sterile beakers
- ✓ sterile 1 mL pipettes with bulbs
- ✓ two Snyder Agar tubes

Medium Recipe

Snyder Test Medium

◆ Pancreatic digest of casein	13.5 g
◆ Yeast extract	6.5 g
◆ Dextrose	20.0 g
◆ Sodium chloride	5.0 g
◆ Agar	16.0 g
◆ Bromcresol green	0.02 g
◆ Distilled or deionized water	1.0 L

pH 4.6–5.0 at 25°C

Procedure

Lab One

1 Collect a small sample of saliva (about 0.5 mL) in the sterile beaker.

2 Remove two Snyder tubes from the water bath and label them with your name and the date. Aseptically add 0.2 mL of saliva to one tube and roll it between your hands until it is uniformly distributed throughout the agar. Do not inoculate the second tube.

3 Allow both tubes to cool to room temperature. Do not slant.

4 When the Snyder medium has solidified, incubate both tubes at 35 ± 2°C for up to 72 hours.

Lab Two

1 Examine the tubes at 24-hour intervals for color changes.

2 Record your results on the Data Sheet.

Reference

Zimbro, Mary Jo, and David A. Power, eds. 2003. Page 517 in *Difco™ and BBL™ Manual—Manual of Microbiological Culture Media*. Becton Dickinson and Co., Sparks, MD.

Name ______________________ Date ______________________

Lab Section ______________________ I was present and performed this exercise (initials) ________

Snyder Test

OBSERVATIONS AND INTERPRETATIONS

Refer to Table 7-1 when recording and interpreting your results below.

	TIME		
	24 Hrs.	48 Hrs.	72 Hrs.
Color			

Based on the results of this test, I have a ______________________ susceptibility to dental caries.

Questions

1 As demonstrated by this exercise, *Lactobacillus* species ferment glucose (dextrose). Considering the Snyder Test ingredients, suggest other metabolic functions that they might perform.

DATA SHEET 7-1

2 Considering, again, the Snyder Test, what kinds of dietary items would likely increase the number of lactobacilli in saliva?

3 Why isn't the molten Snyder Test Agar allowed to solidify as a slant?

EXERCISE 7-2

Antimicrobial Susceptibility Test (Kirby-Bauer Method)

Theory

Antibiotics are natural antimicrobial agents produced by microorganisms. One type of penicillin, for example, is produced by the mold *Penicillium notatum*. Today, because many agents that are used to treat bacterial infections are synthetic, the terms **antimicrobials** or **antimicrobics** are used to describe all substances used for this purpose.

The Kirby-Bauer Test, also called the disk diffusion test, is a valuable standard tool for measuring the effectiveness of antimicrobics against pathogenic microorganisms. In the test, antimicrobic-impregnated paper disks are placed on a plate that has been inoculated to form a bacterial lawn. The plates are incubated to allow growth of the bacteria and time for the agent to diffuse into the agar. As the drug moves through the agar, it establishes a concentration gradient. If the organism is susceptible to it, a clear zone will appear around the disk where growth has been inhibited (Figure 7-3).

The size of this **zone of inhibition** depends upon the sensitivity of the bacteria to the specific antimicrobial agent and the point at which the chemical's **minimum inhibitory concentration** (**MIC**) is reached. The zone may be produced as a result of killing the bacteria, in which case it is said to be **bactericidal**. In other cases, the drug may only stop bacterial growth without killing them, in which case it is said to be **bacteriostatic**. Table 7-2 outlines the mechanism of action and resistance for the antibiotics used in this exercise.

All aspects of the Kirby-Bauer procedure are standardized to ensure reliable results. Therefore, care must be taken to adhere to these standards. Mueller-Hinton Agar, which has a pH between 7.2 and 7.4, is poured to a depth of 4 mm in either 150 mm or 100 mm Petri dishes. The depth is important because of its effect upon diffusion. Thick agar slows lateral diffusion and thus produces smaller zones than plates held to the 4 mm standard. Inoculation is made with a broth culture diluted to match a 0.5 McFarland turbidity standard (Figure 7-4).

The disks, which contain a specified amount of the antimicrobial agent (printed on the disk) are dispensed onto the inoculated plate (Figure 7-5) and incubated at 35 ± 2°C. After 16 to 18 hours incubation, the plates are removed and the clear zone diameters are measured and recorded in millimeters (Figure 7-6).

Application

Antimicrobial susceptibility testing is a standardized method used to measure the effectiveness of antibiotics

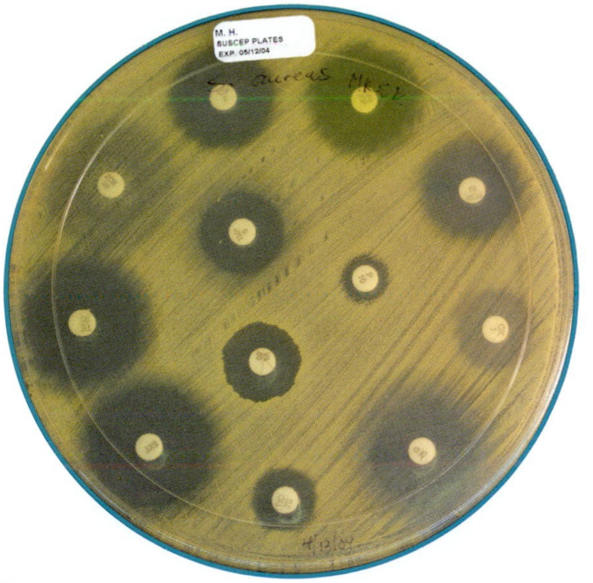

7-3 Disk Diffusion Test of Methicillin-Resistant *Staphylococcus aureus* (MRSA) ◆ This plate illustrates the effect of (clockwise from top outer right) Nitrofurantoin (F/M300), Norfloxacin (NOR 10), Oxacillin (OX 1), Sulfisoxazole (G 0.25), Ticarcillin (TIC 75), Trimethoprim-Sulfamethoxazole (SXT), Tetracycline (TE 30), Ceftizoxime (ZOX 30), Ciprofloxacin (CIP 5), and (inner circle from right) Penicillin (P 10), Vancomycin (VA 30), and Trimethoprim (TMP 5) on Methicillin-resistant *Staphylococcus aureus*.

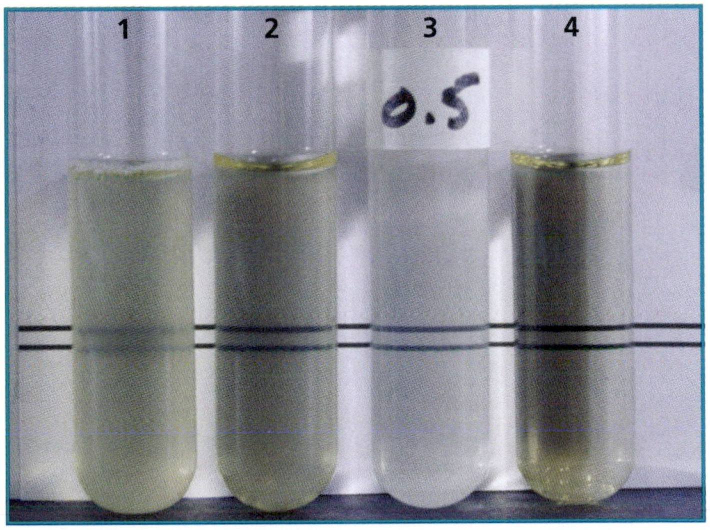

7-4 McFarland Standards ◆ This is a comparison of a McFarland turbidity standard to three broths with varying degrees of turbidity. There are 11 McFarland standards (0.5 to 10), each of which contains a specific percentage of precipitated barium sulfate to produce turbidity. In the Kirby-Bauer procedure, the test culture is diluted to match the 0.5 McFarland standard (roughly equivalent to 1.5×10^8 cells per mL) before inoculating the plate. Comparison is made visually by placing a card with sharp black lines behind the tubes. Tube 3 is the 0.5 McFarland standard. Notice that the turbidity of Tube 2 matches the McFarland standard exactly, whereas Tubes 1 and 4 are too turbid or too clear, respectively.

7-5 Disk Dispenser ◆ This antibiotic disk dispenser is used to uniformly deposit disks on a Mueller-Hinton Agar plate.

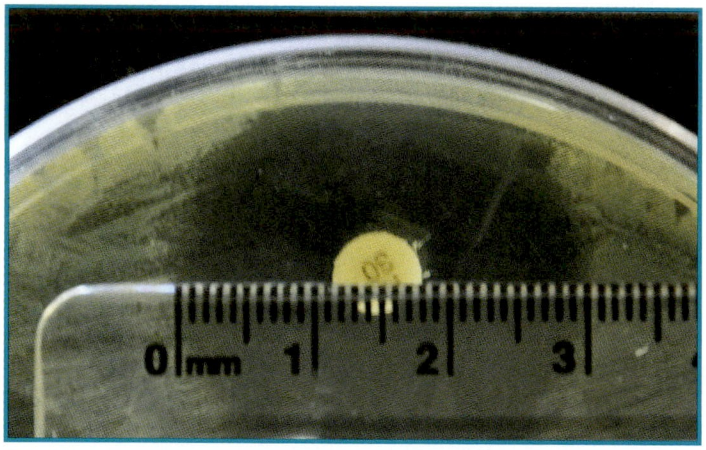

7-6 Measuring the Antimicrobial Susceptibility Zones ◆ Using a metric ruler and a dark, nonreflective background, measure the diameter of each clearing. Standards for comparison are given in millimeters (mm). This zone is 29 mm in diameter.

and other chemotherapeutic agents on pathogenic microorganisms. In many cases, it is an essential tool in prescribing appropriate treatment.

In This Exercise

You will test the susceptibility of *Escherichia coli* and *Staphylococcus aureus* strains to penicillin, ciprofloxacin, trimethoprim, and chloramphenicol. These antibiotics were chosen because they exhibit different modes of action on bacterial cells (Table 7-3).

Materials

Per Student Group

- ✓ two Mueller-Hinton Agar plates
- ✓ penicillin, chloramphenicol, trimethoprim, and ciprofloxacin antibiotic disks (and/or other disks as available)
- ✓ antibiotic disk dispenser or forceps for placement of disks
- ✓ small beaker of alcohol (for sterilizing forceps)
- ✓ two sterile cotton swabs
- ✓ one metric ruler
- ✓ sterile saline (0.85%)
- ✓ sterile transfer pipettes
- ✓ one McFarland 0.5 standard with card
- ✓ black, nonreflective posterboard (8.5" × 11")
- ✓ fresh broth cultures of:
  - *Escherichia coli*
  - *Staphylococcus epidermidis*

Medium Recipe

Mueller-Hinton II Agar

◆ Beef extract	2.0 g
◆ Acid hydrolysate of casein	17.5 g
◆ Starch	1.5 g
◆ Agar	17.0 g
◆ Distilled or deionized water	1.0 L

pH = 7.2–7.4 at 25°C

Procedure

Lab One

1. Gently mix the *E. coli* culture and the McFarland standard until they reach their maximum turbidity.
2. Holding the culture and McFarland standard upright in front of you, place the card behind them so you can see the black line through the liquid in the tubes. As you can see in Figure 7-4, the line becomes distorted by the turbidity in the tubes. Use the black line to compare the turbidity level of the two tubes. Dilute the broth with the sterile saline until it appears to have the same level of turbidity as the standard. (Add the saline a little at a time so you minimize the risk of diluting too much.) If it is not turbid enough, incubate it until it is.

3 Repeat Steps 1 and 2 with the *S. epidermidis* culture.

4 Label the plates with the organisms' names, your name, and the date.

5 Dip a sterile swab into the *E. coli* broth. As you remove it, press and rotate the cotton tip against the inside of the tube to remove excess broth.

6 Inoculate a Mueller-Hinton plate with *E. coli* by streaking the entire surface of the agar three times with the swab. Your goal is confluent growth, so make the streaks right next to each other. When you have covered the surface, rotate the plate 1/3 turn and repeat the streaking of the incoculum already on the plate, using the same technique to produce confluent growth. Then, rotate the plate another 1/3 turn and repeat.

7 Using a fresh sterile swab, inoculate the other plate with *S. epidermidis* in the same fashion to produce confluent growth.

8 Apply the penicillin, chloramphenicol, ciprofloxacin, and trimethoprim disks to the agar surface of each plate. You can apply the disks either singly using sterile forceps, or with a dispenser (Figure 7-5). Be sure to space the disks sufficiently (4 to 5 cm) to prevent overlapping zones of inhibition. Also, keep them away from the edge of the plate.

9 Press each disk gently with sterile forceps so it makes full contact with the agar surface. (Some disk dispensers are "self-tamping" and this step is unnecessary.)

10 Invert the plates and incubate them aerobically at 35 ± 2°C for 16 to 18 hours. Have a volunteer in the group remove and refrigerate the plates at the appropriate time.

Lab Two

1 Remove the plates from the incubator (or refrigerator). Hold the plate over the black, nonreflective posterboard and examine the plate with reflected light. The edge of a zone is where no growth is visible to the naked eye (Figure 7-6). Measure the diameter of each zone of inhibition in millimeters.

2 Using Table 7-3 and those provided with your antibiotic disks (if you used additional antibiotics), record your results in the table provided on the Data Sheet.

References

Clinical Laboratory Standards Institute (CLSI). 2009. *Performance Standards for Antimicrobial Disk Susceptibility Tests; Approved Standard*—10th Ed. CLSI document M02–A10. Wayne, PA.

Collins, C. H., Patricia M. Lyne, and J. M. Grange. 1995. Page 128 in *Collins and Lyne's Microbiological Methods*, 7th ed. Butterworth-Heinemann, UK.

Ferraro, Mary Jane, and James H. Jorgensen. 2003. Chapter 15 in *Manual of Clinical Microbiology*, 8th ed., edited by Patrick R. Murray, Ellen Jo Baron, James. H. Jorgensen, Michael A. Pfaller, and Robert H. Yolken, ASM Press, Washington, DC.

Forbes, Betty A., Daniel F. Sahm, and Alice S. Weissfeld. 2002. Pages 236–240 in *Bailey & Scott's Diagnostic Microbiology*, 11th ed. Mosby-Yearbook, St. Louis, MO.

Koneman, Elmer W., Stephen D. Allen, William M. Janda, Paul C. Schreckenberger, and Washington C. Winn, Jr. 1997. Pages 818–822 in *Color Atlas and Textbook of Diagnostic Microbiology*, 5th ed. J.B. Lippincott Company, Philadelphia, PA.

Mims, Cedric, Hazel M. Dockrell, Richard V. Goering, Ivan Roitt, Derek Wakelin, and Mark Zuckerman. 2004. Chapter 33 in *Medical Microbiology*, 3rd ed. Mosby, Philadelphia, PA.

Zimbro, Mary Jo, and David A. Power, eds. 2003. Page 376 in *Difco™ and BBL™ Manual—Manual of Microbiological Culture Media*. Becton Dickinson and Company, Sparks, MD.

TABLE 7-2 Antibiotic Targets and Resistance Medium ◆ Not all antibiotics affect cells in the same way. Some attack the bacterial cell wall, and others interfere with biosynthesis reactions. Resistance mechanisms can be broken down into three main categories: (a) altered target such that the antibiotic no longer can interact with the cellular process, (b) an alteration in how the drug is taken into the cell, and (c) enzymatic destruction of the drug.

Antibiotic	Cellular Target	Resistance Mechanism
Chloramphenicol	Prevents peptide bond formation during translation	1. Poor uptake of drug 2. Inactivation of drug
Ciprofloxacin	Interferes with DNA replication	1. Altered target 2. Poor uptake of drug
Trimethoprim	Inhibits purine and pyrimidine synthesis	1. Altered target
Penicillin	Inhibits cross-linking of the cell wall's peptidoglycan	One or more of: 1. Altered target 2. Poor uptake of drug 3. Production of β-lactamases

TABLE 7-3 Zone Diameter Interpretive Chart

				Zone Diameter Interpretive Standards		
Antibiotic	Organism(s)	Code	Disk Potency	Resistant	Intermediate	Susceptible
Chloramphenicol	Enterobacteriaceae (*E. coli*) and *Staphylococcus*	C 30	30 µg	≤12	13–17	≥18
Ciprofloxacin	Enterobacteriaceae (*E. coli*) and *Staphylococcus*	CIP 5	5 µg	≤15	16–20	≥21
Trimethoprim	Enterobacteriaceae (*E. coli*) and *Staphylococcus*	TMP 5	5 µg	≤10	11–15	≥16
Penicillin	*Staphylococcus*	P 10	10 U	≤28		≥29

This chart includes the antibiotics used in this exercise and contains data provided by the Clinical and Laboratory Standards Institute (CLSI). Permission to use portions (specifically Tables 2A and 2C) of M100-S19 (Performance Standards for Antimicrobial Susceptibility Testing; Nineteenth Informational Supplement) has been granted by CLSI. The interpretive data are valid only if the methodology in M02-A10 (Performance Standards for Antimicrobial Disk Susceptibility Tests—10th edition; Approved Standard) is followed. CLSI frequently updates the interpretive tables through new editions of the standard and supplements. Users should refer to the most recent editions. The current standard may be obtained from CLSI, 940 West Valley Road, Suite 1400, Wayne, PA 19087. Contact also may be made via the Web site (www.CLSI.org), email (customerservice@clsi.org), and by phone (1.877.447.1888).

Name ______________________ Date ______________________

Lab Section ______________________ I was present and performed this exercise (initials) ________

DATA SHEET 7-2

Antimicrobial Susceptibility Test

OBSERVATIONS AND INTERPRETATIONS

Record the zone diameters in mm. Using Table 7-3 as a guide, enter "S" if the organism is susceptible and "R" if it is resistant to the antibiotic. (Not all combinations of antibiotics and organisms are available.)

	Chloramphenicol		Ciprofloxacin		Penicillin		Trimethoprim	
Organism	Zone Diameter (mm)	S/R	Zone Diameter (mm)	S/R	Zone Diameter (mm)	S/R	Zone Diameter (mm)	S/R
Staphylococcus epidermidis								
Escherichia coli								

Questions

1 All aspects of the Kirby-Bauer Test are standardized to assure reliability of comparison with the published standards.

a. What might the consequence be of pouring the plates 2 mm deep instead of 4 mm deep?

b. The Mueller-Hinton II plates are supposed to be used within a specific time after their preparation and should be free of visible moisture. What negative effect(s) might moisture have on the test?

2 In clinical applications of the Kirby-Bauer Test, diluted cultures (for the McFarland standard comparison) must be used within 30 minutes. Why is this important?

3 *E. coli* and *S. epidermidis* were chosen to represent Gram-negative and Gram-positive bacteria, respectively. For a given antibiotic, is there a difference in susceptibility between the Gram-positive and Gram-negative bacteria? If so, what difference(s) do you see?

4 Suppose you do this test on a hypothetical *Staphylococcus* species with the antibiotics penicillin (P-10) and chloramphenicol (C30). You record zone identical diameters of 25 mm for the chloramphenicol and penicillin disks. Which antibiotic would be more effective against this organism? What does this tell you about comparing zone diameters to each other and the importance of the zone diameter interpretive chart?

5 How does the antibiotic get from the disk into the agar?

6 After incubation, does the antibiotic extend into the agar beyond the zone of inhibition? How does your answer relate to the concept of MIC?

EXERCISE 7-3

Morbidity and Mortality Weekly Report (MMWR) Assignment

Theory

Epidemiology is the study of the causes, occurrence, transmission, distribution, and prevention of diseases in a population. The Centers for Disease Control and Prevention (CDC) in Atlanta, GA (Figure 7-7), is the national clearinghouse for epidemiological data. The CDC receives reports related to the occurrence of over 60 notifiable diseases, some of which are listed in Table 7-4, from the United States and its territories, and compiles the data into tabular form in the publication *Morbidity and Mortality Weekly Report* (*MMWR*).

Two important disease measures that **epidemiologists** collect are **morbidity** (sickness) and **mortality** (death). Morbidity relative to a specific disease is the number of susceptible people who have the disease within a defined population during a specific time period. It usually is expressed as a rate. Because population size fluctuates constantly, it is conventional to use the population size at the midpoint of the study period. Also, the units for the rate fraction are "cases per person" and usually are small decimal fractions. To make the calculated rate more "user-friendly," it is multiplied by some power of 10 ("K") to achieve a value that is a whole number. Thus, a morbidity rate of 0.00002 is multiplied by 100,000 (10^5) so it can be reported as 2 cases per 100,000 people rather than 0.00002 cases per person. Morbidity rate is calculated using the following equation:

$$\text{Morbidity Rate} = \frac{\text{number of existing cases in a time period}}{\text{size of at-risk population at midpoint of time period}} \times K$$

Mortality, also expressed as a rate, is the number of people who die from a specific disease out of the total population *afflicted with that disease* in a specified time period. It, too, is multiplied by a factor "K" so the rate can be reported as a whole number of cases. The equation is:

$$\text{Mortality Rate} = \frac{\text{number of deaths due to a disease in a time period}}{\text{number of people with that disease in the time period}} \times K$$

7-7 Centers for Disease Control ◆ Shown is the entrance to the CDC campus in Atlanta, Georgia.

Minimally, an epidemiological study evaluates morbidity or mortality data in terms of **person** (age, sex, race, etc.), **place**, and **time**. Sophisticated analyses require training in biostatistics, but the simple epidemiological calculation you will be doing can be performed with little mathematical background. You will calculate **incidence rate,** which is the occurrence of new cases of a disease within a defined population during a specific period of time. As before, "K" is some power of 10 so the rate can be reported as a whole number of cases.

$$\text{Incidence Rate} = \frac{\text{number of new cases in a time period}}{\text{size of at-risk populations at midpoint of time period}} \times K$$

Because our focus is microbiology, we will deal only with **infectious diseases**—those caused by biological agents such as bacteria and viruses. **Noninfectious diseases**, such as stroke, heart disease, and emphysema, also are studied by epidemiologists but are not within the scope of microbiology.

Application

An understanding of the causes and distribution of diseases in a population is useful to health care providers in a couple of ways. First, awareness of what diseases are prevalent during a certain period of time aids in diagnosis. Second, an understanding of the disease and its cause(s) can be useful in implementing strategies for preventing it.

In This Exercise

First you will choose a disease from the list of notifiable diseases in Table 7-4. Then you will collect data for the

United States over the most recent complete year. Once you have collected data, you will calculate incidence values and construct graphs illustrating the cumulative totals and incidence values for the year.

Materials

✓ a computer with Internet access or access to copies of *Morbidity and Mortality Weekly Report*.

Procedure

1 Choose a disease from Table 7-4 and record its name on the Data Sheet.

2 Go to the CDC Web site http://www.cdc.gov. Then follow these links. (Note: Web sites often are revised, so if the site doesn't match this description exactly, it still is probably close. Improvise, and you'll find what you need. These links are current as of September 2011):

- Click on *Morbidity and Mortality Weekly Report* (*MMWR*) under "Publications" in the lower right of the page.
- Click on "State Health Statistics" in the menu bar at the left.
- On this new page, click on "Search Morbidity Tables."
- Read the "Note" below the menu window on the *MMWR* Tables (NNDSS) page. It describes how the data in the tables are collected and why the numbers are provisional.
- Choose the most recent complete year in the menu at the top of the *MMWR* Tables (NNDSS) page. Then, choose Week 1 under "*MMWR* Week" in the same menu. Click "Submit."
- Find your chosen disease in NNDSS Tables menu. Your chosen disease will be found in Table II. It is divided into parts with the diseases listed alphabetically, so you will have to try "hit and miss" until you find the part with your chosen disease. Once you find it, record the *cumulative* number of cases in the United States for Week 1 on the Data Sheet. This number is found in the top row (United States) under the column "Cum 20XX" (i.e., the most recent complete year).
- Return to the page with the *MMWR* Week and *MMWR* Year menu and continue the process for each week through *MMWR* Week 52. (Alternatively, you can just change the week number in the URL and press return.)[1] Record the cumulative totals on the Data Sheet for all 52 reporting weeks.

3 Answer the questions and complete the activities on the Data Sheet.

[1] Thanks to Joe Montes, Jr. for this shortcut tip!

TABLE 7-4 Notifiable Diseases in the United States and Its Territories Posted in Table II of *MMWR* (as of September 2011.)
Not all reportable diseases are listed in Table II. Table I lists "infrequently reported" diseases, whereas Table IV is used for notifiable diseases reported quarterly.

SELECTED NOTIFIABLE DISEASES	
Chlamydia trachomatis infection	Malaria
Coccidioidomycosis	Meningococcal diseases, invasive (all serogroups)
Cryptosporidiosis	Mumps
Dengue Fever	Pertussis
Ehrlichia chaffeensis	Salmonellosis
Anaplasma phagocytophilum	Shiga toxin-producing E. coli (STEC)
Giradiasis	Shigellosis
Gonorrhea	Spotted Fever Rickettsiosis (confirmed and probable) (including RMSF)
Haemophilus influenzae, invasive (all ages and serotypes)	*Streptococcus pneumoniae*, invasive disease (all ages)
Hepatitis A (viral, acute)	*Streptococcus pneumoniae*, invasive disease (<5 years)
Hepatitis B (viral acute)	Syphilis (primary and secondary)
Hepatitis C (viral, acute)	Varicella (chickenpox)
Legionellosis	West Nile Virus (neuroinvasive)
Lyme Disease	West Nile Virus (non-neuroinvasive)

Name ______________________ Date ______________________

Lab Section ______________________ I was present and performed this exercise (initials) ________

MMWR Report

DATA AND CALCULATIONS

1 Go to the *MMWR* Morbidity Tables at the CDC Web site (www.cdc.gov). Record the cumulative weekly totals for your selected disease for the last complete year in the Data Table on the next page. No calculation is necessary regarding these numbers at this point. Because they are cumulative totals, they should be greater than or equal to the preceding week. Nevertheless, because the reported numbers are provisional and subject to change, this may not always be the case. Simply record the numbers from the CDC Web site as they are reported.

2 Calculate the number of *new* cases during each 4-week period (weeks 1–4, weeks 5–8, etc.) by subtracting the total at the end of a 4-week period from the total at the end of the next 4-week period. That is, for the second four weeks subtract the total at the end of week 4 from the total at the end of week 8. Record these in the chart on page 406 in the column labeled "4-Week Totals." Show a sample calculation for the third 4-week total in the space below.

3 Calculate national incidence values of each 4-week period during the year using the calculated 4-week totals (weeks 1–4, weeks 5–8, etc.) in the numerator. Use the U.S. population size on July 1 in the denominator. This value can be obtained from the U.S. Census Bureau Web site at http://www.census.gov/popest/ and click on "Current Estimates Data" under "Quick Links." On the new page, click on the most recent national "Intercensal Estimates" link. At the bottom of this new page, under "Downloadable Datasets," click on "Data" for "Monthly Intercensal Resident Population Estimates for the United States." Find the year followed by a "7" (for July) to get the estimated population. Be sure to choose an appropriate value for K, and don't abuse the significant figures! Show a sample calculation for the third 4-week period in the space below.

Estimated population size for the year 20____:

$$\text{Incidence Rate} = \frac{\text{number of new cases in a time period}}{\text{size of at-risk population at midpoint of time period}} \times K$$

Week	Cumulative Totals by Week Year (20___)	4-Week Totals Year (20___)	Incidence Values Year (20___)	Week	Cumulative Totals by Week Year (20___)	4-Week Totals Year (20___)	Incidence Values Year (20___)
1				29			
2				30			
3				31			
4				32			
5				33			
6				34			
7				35			
8				36			
9				37			
10				38			
11				39			
12				40			
13				41			
14				42			
15				43			
16				44			
17				45			
18				46			
19				47			
20				48			
21				49			
22				50			
23				51			
24				52			
25							
26							
27							
28							

Name ______________________ Date ______________________

Lab Section ______________________ I was present and performed this exercise (initials) ________

DATA SHEET 7-3

Questions

1 Characterize the disease you have chosen. Be sure to include causative agent (by scientific name), symptoms, mode of transmission, and treatment. Cite your references.

2 Prepare a graph that illustrates the cumulative data for each week over the year. Be sure to include a title and axis labels with appropriate units in your graph.

3 Graph the calculated national incidence values over the year studied. In your graph, be sure to include a title and axis labels with appropriate units.

4 Briefly describe the national trend of your chosen disease over the year. Does the incidence appear to be seasonal?

EXERCISE 7-4

Epidemic Simulation

Theory

As mentioned in Exercise 7-3, epidemiology is the study of the causes, occurrence, transmission, distribution, and prevention of diseases in a population. Infectious diseases are transmitted by ingestion, inhalation, direct skin contact, open wounds or lesions in the skin, animal bites, direct blood-to-blood contact as in blood transfusions, and sexual contact. Infectious diseases can be transmitted by way of sick people or animals, healthy people or animals carrying the infectious organism or virus, water contaminated with human or animal feces, contaminated objects (**fomites**), aerosols, or biting insects (**vectors**).

When a disease is transmitted from an area such as the heating or cooling system of a building or from contaminated water that infects many people at once, it is called a **common source epidemic. Propagated transmission** is a disease transmitted from person to person. The first case of such a disease is called the **index case**. Determining the index case is the object of today's lab exercise.

A second objective of today's lab is to gather data to be used in performing simple epidemiological calculations. You will be calculating **incidence rate** and **prevalence** rate for the "disease" spreading through your class.

Incidence rate is the number of new cases of a disease reported in a defined population during a specific period of time. Incidence rate is calculated by the following equation:

$$\text{Incidence rate} = \frac{\text{number of new cases in a time period}}{\text{size of at-risk population at midpoint of time period}} \times K$$

Because population size fluctuates constantly, it is conventional to use the population size at the midpoint of the study period. Also, the usually small decimal fraction obtained for the rate is multiplied by some power of 10 (K in the equation) to get its value up to a number bigger than one. That is, an incidence rate of 0.00002 is multiplied by 100,000 (10^5) so it can be reported as two cases per 100,000 people, rather than 0.00002 cases per person (which is pretty meaningless).

You also will be calculating one form of prevalence rate called **point prevalence**, the number of cases of a disease at a specific point in time in a defined population. It is calculated by the following equation:

$$\text{Point prevalence} = \frac{\text{number of existing cases at a point in time}}{\text{size of at-risk population}} \times K$$

As with incidence rate, the calculated prevalence is multiplied by some power of 10 (K in the equation) to bring the number up to one or more.

Application

Epidemiologists have the task of identifying the source of a disease and establishing the mode of its transmission. Further, epidemiologists characterize diseases quantitatively, using measures such as incidence and prevalence. Hopefully, this information will allow them to make recommendations for prevention.

In This Exercise

One of you will become the "index case" in a simulation of an epidemic, and many of you who contact this person directly and indirectly will become "infected." Your job, besides having a little fun, will be to collect the data and determine which one of you is the index case. You will also use the data to calculate incidence and prevalence rates for your class over a hypothetical time period.

Materials

Per Student

- ✓ examination glove (latex or synthetic)
- ✓ numbered Petri dish containing a piece of hard candy (one will be contaminated with *Serratia marcescens*[1] or other microbe as chosen by your instructor; the others will be moistened with water)[2]
- ✓ one Nutrient Agar plate
- ✓ two sterile cotton applicators and sterile saline
- ✓ marking pen

Procedure

Lab One

1. Obtain a Petri dish containing candy, two sterile cotton applicators, a tube of sterile saline, and a Nutrient Agar plate. Record the number of your candy dish here and circle that number in the "case" column of the Data Sheet.
   Candy dish/case number: ______
2. Mark a line on the bottom of the agar plate to divide it into two halves. Label the sides "1" and "2." Also record the number of your candy dish on your plate.

---

[1] *S. marcescens* was chosen because of its obvious color. Other, less distinctive organisms can be substituted, if desired.

[2] Petri dishes should be numbered according to the number of students in the class. There should be no gaps in the sequence.

3 Open a package of sterile, cotton-tipped applicators so they can be easily removed.

4 Put a glove on your nondominant hand.

5 Using your ungloved hand, remove one cotton applicator, moisten it in the sterile saline, and sample the palm of the gloved hand. Be sure not to touch anything else with the palm of the glove. When finished, zigzag inoculate side 1 of the plate with the cotton applicator (Figure 1-32). Discard the applicator in an appropriate autoclave container.

6 Pick up the candy with your gloved hand and roll it around until a good amount of it is transferred to the palm of the glove. Drop the candy back into the Petri dish, close it with your ungloved hand, and touch nothing with your gloved hand. Just sit with your elbow on the table and your gloved hand in the air.

7 *Read this entire step before doing anything!* Using gloved hands only, case #1 rubs hands with case #2, making sure to transfer anything that may be on the palm of the glove to the palm of the other student's glove. After a few seconds of transfer, gently slide your hands apart, as shown in Figure 7-8. You musn't "snap" the gloves as you separate because this will produce aerosols. Also, hold on to the glove's cuff as you slide apart to keep it from coming off your hand.

8 Then, case #2 rubs gloved hands with #3, and #3 with #4, and so on until all students have contacted gloved hands. Continue to be careful not to snap the gloves when you separate. Hold on to the cuff when transferring and separating to prevent the glove from coming off your hand.

9 When all students have rubbed hands, the last (highest numbered) case should rub hands with case #1.

10 With a second sterile applicator, sample the palm of your gloved hand as before and inoculate side 2 of the plate. Be sure to sample the area of the glove that contacted other students' gloves. Dispose of the applicator in an appropriate autoclave container.

11 Remove your glove by inserting the thumb of your other hand under the cuff and rolling it off your hand, inverting it in the process. Dispose of it properly.

12 Wash your hands thoroughly. Use antiseptic soap, if available.

13 Discard all materials (except your inoculated plate) in appropriate containers. Be sure to tape the lid on your candy dish before you dispose of it.

14 Wipe down your work area with disinfectant.

15 Tape the lid on the inoculated plates. You may find it convenient to tape all plates from your group in one stack so they can be retrieved more easily.

16 Invert the inoculated plates and incubate at 25°C for 24 to 48 hours.

Lab Two

1 Remove the plates from the incubator, and examine them for characteristic reddish growth of *Serratia marcescens*.

2 Enter your results on the Data Sheet and follow the instructions for establishing the index case, incidence rate, and point prevalence.

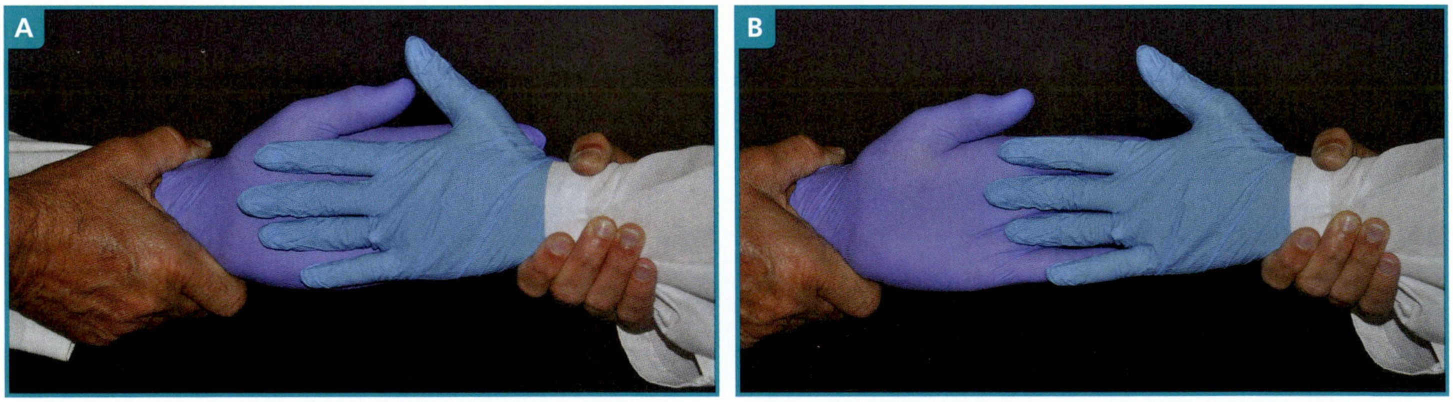

7-8 Sliding the Gloves Apart ◆ After rubbing palms of the hands, students should separate by gently sliding their gloved hands apart (as in **A** and **B**). Be sure to hold on to the glove's cuff to prevent it from coming off. These safety precautions must be followed and are designed to prevent the gloves from snapping apart (they will be sticky) and producing aerosols.

Name ______________________________ Date ______________________________

Lab Section ______________________________ I was present and performed this exercise (initials) ________

Epidemic Simulation

DATA AND CALCULATIONS

1 Circle the case number that matches your candy dish number.

2 After the plates have been incubated, compile data for your class population (each case is the student's candy number in the exercise). The organism spread during contact between student pairs was *Serratia marcescens,* which produces reddish-orange colonies. You will likely see growth on both sides of the plate, but red colonies are the only sign of disease. If there is reddish-orange growth on side 2 of the plate, consider the person to have the disease and record a " + " in the appropriate box. If there is no red growth on the plate or there is no difference between the two sides, consider the person to be healthy and record a " – " in the appropriate box.

Case	Week	Disease (+/−)
1	1	
2	1	
3	1	
4	1	
5	2	
6	2	
7	2	
8	2	
9	2	
10	3	
11	3	
12	3	
13	3	
14	3	
15	3	
16	4	
17	4	
18	4	
19	4	
20	4	
21	4	
22	4	
23	5	
24	5	
25	5	
26	5	
27	5	
28	6	
29	6	
30	6	
31	7	
32	7	
33	7	
34	7	
35	8	
36	8	
37	8	
38	9	
39	9	
40	10	

3 Identify the case number of the index patient.

4 Use the data obtained to calculate incidence and prevalence rates in your class population. For this purpose, assume that the cases have been identified over a 10-week period (or a shorter period, depending on how many students participated in the simulation). Also assume that the duration of the disease is 7 days, so new cases in one week are still diseased in the next week but are healthy by the following week.

Record the population size: ________________

Record incidence and point prevalence rates for each week in the chart at the right.

Week	Incidence	Prevalence
1		
2		
3		
4		
5		
6		
7		
8		
9		
10		

Questions

1 Explain how you determined the index case.

2 What was the purpose of side 1 on each plate?

3 Suggest possible reasons (based on the execution of the simulation) for cases showing no *S. marcescens* growth between cases that exhibited growth.

4 Suggest how the gaps could represent "subclinical infections" or "carriers" in this simulation.

5 Within the context of this simulation, how could you characterize any "non-*Serratia marcescens* growth on Side 1 or 2 of your plate? (Hint: "Contamination" is not an adequate answer!)

Environmental Microbiology

Environmental microbiology is the study, utilization, and control of microorganisms living in marine, freshwater, or terrestrial habitats. In this segment, you will focus on the detection and measurement of unwanted microorganisms in marine and freshwater habitats. You will be collecting water from an environmental source of your (or your instructor's) choosing and testing it for fecal contamination.

Fecal contamination frequently occurs in marine and freshwater habitats. To ensure public safety, governmental regulatory agencies and local water districts conduct daily tests of both ocean water and drinking water, using techniques similar to the two introduced in this section. Both procedures—the Membrane Filter Technique and Multiple Tube Fermentation Technique—detect and measure fecal contamination.

EXERCISE 7-5

Membrane Filter Technique

❖ Theory

Fecal contamination is a common pollutant in open water and a potential source of serious disease-causing organisms. Certain members of *Enterobacteriaceae*, (Gram-negative facultative anaerobes) such as *Escherichia coli*, *Klebsiella pneumoniae*, and *Enterobacter aerogenes,* are able to ferment lactose rapidly and produce large amounts of acid and gas. These organisms, called **coliforms,** are used as the indicator species when testing water for fecal contamination because they are relatively abundant in feces and easy to detect (Figure 7-9). Once fecal contamination is confirmed by the presence of coliforms, further tests will confirm or deny the presence of more dangerous fecal-borne pathogens.

In the Membrane Filter Technique, a water sample is drawn through a special porous membrane designed to trap microorganisms larger than 0.45 µm. After filtering the water sample, the membrane (filter) is applied to the surface of plated Endo Agar and incubated for 24 hours (Figure 7-10).

Endo Agar is a selective medium that encourages Gram-negative bacterial growth and inhibits Gram-positive growth. Endo Agar contains lactose for fermentation and a dye to indicate changes in pH. Because of their vigorous lactose fermentation resulting in acid and aldehyde formation, coliform colonies typically appear red and/or mucoid with a gold or green metallic sheen. Noncoliform bacteria (including several dangerous pathogens) tend to be pale pink, colorless, or the color of the medium.

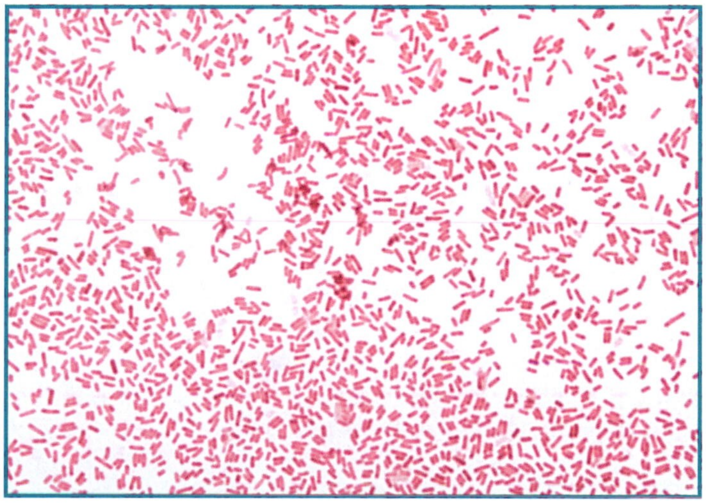

7-9 Gram stain of *Escherichia coli* ◆ *E. coli* is a principal coliform detected by the Membrane Filter Technique.

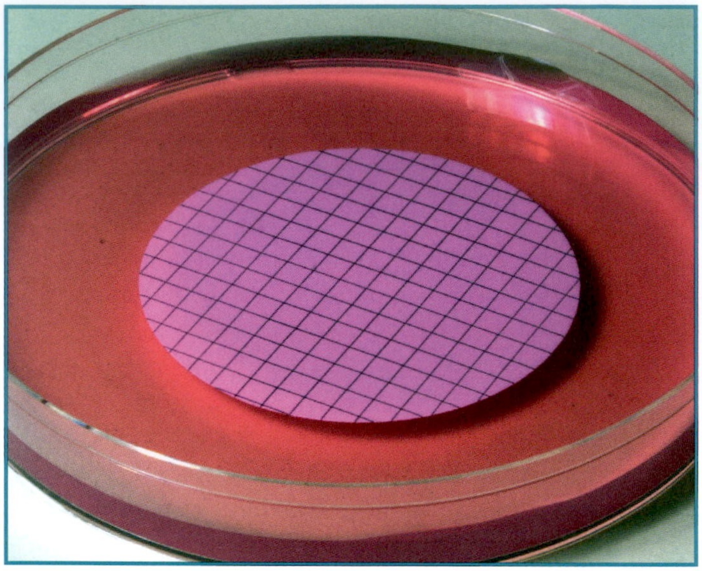

7-10 Membrane Filter ◆ This porous membrane will allow water to pass through but will trap bacteria and particles larger than 0.45 µm.

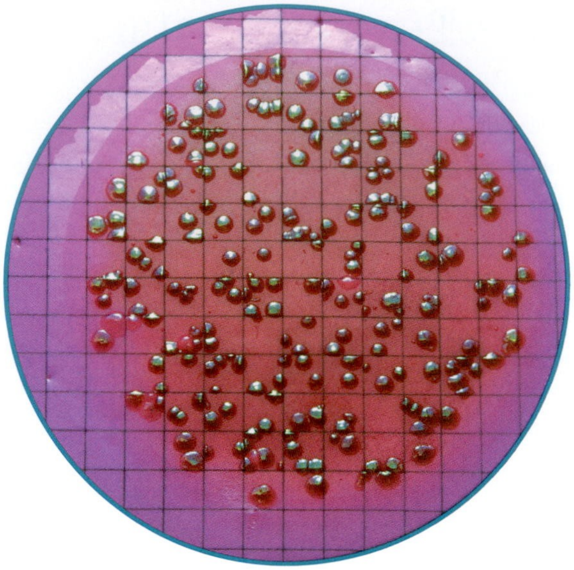

7-11 Coliform Colonies on a Membrane Filter ◆ Note the characteristic dark colonies with gold metallic sheen, indicating that this water sample is contaminated with fecal coliforms. Potable water has less than one coliform per 100 mL of sample tested.

7-12 Membrane Filter Apparatus ◆ Assemble the membrane filter apparatus as shown in this photograph. Use two suction flasks (as shown) to avoid getting water into the vacuum source. Secure the flasks on the table, as the tubing may make them top-heavy.

After incubation, all red or metallic colonies are counted and are used to calculate "coliform colonies/100 mL" using the following formula:

$$\frac{\text{Total coliforms}}{100 \text{ mL}} = \frac{\text{coliform colonies counted} \times 100}{\text{volume of original sample in mL}}$$

A "countable" plate contains between 20 and 80 coliform colonies with a total colony count no larger than 200 (Figure 7-11). To assure that the number of colonies will fall within this range, it is customary to dilute heavily polluted samples, thereby reducing the number of cells collecting on the membrane. When dilution is necessary, it is important to record the volume of *original sample* only, not any added water. **Potable water** contains less than one coliform per 100 mL of sample.

❖ Application

The Membrane Filter Technique is commonly used to identify the presence of fecal coliforms in water.

❖ In This Exercise

To determine total coliform population density, you will be using collected water samples and performing a membrane filter technique. Your results will be recorded in coliforms per 100 mL of sample.

❖ Materials

Per Student Group

- ✓ one m Endo Agar LES plate
- ✓ one sterile membrane filter (pore size 0.45 µm)
- ✓ sterile membrane filter suction apparatus (Figure 7-12)
- ✓ 100 mL water dilution bottle (to be distributed in the preceding lab)
- ✓ 100 mL water sample (obtained by student)
- ✓ 100 mL sterile water (for rinsing the apparatus)
- ✓ gloves
- ✓ household disinfectant and paper towels
- ✓ small beaker containing alcohol and forceps
- ✓ vacuum source (pump or aspirator)

Medium Recipe

m Endo Agar LES

Yeast extract	1.2 g
Casitone	3.7 g
Thiopeptone	3.7 g
Tryptose	7.5 g
Lactose	9.4 g
Dipotassium phosphate	3.3 g
Monopotassium phosphate	1.0 g
Sodium chloride	3.7 g
Sodium desoxycholate	0.1 g
Sodium lauryl sulfate	0.05 g
Sodium sulfite	1.6 g
Basic fuchsin	0.8 g
Agar	15.0 g
Distilled or deionized water	1.0 L

pH 7.3–7.7 at 25°C

Procedure

Prelab

1. Obtain a 100 mL water dilution bottle from your instructor.
2. Choose an environmental source to sample. (Your instructor may choose or provide samples.)
3. Visit the environmental site as close to your lab period as possible. Bring your water dilution bottle, a pair of gloves, some household disinfectant, and paper towels. While wearing gloves, fill the bottle to the white line (100 mL) and replace the cap.
4. Wipe the outside of the bottle with disinfectant. Dispose of the towels and gloves in the trash.
5. Store the sample in the refrigerator until your lab period. If the sample must sit for a while before your lab, leave the cap loose to allow some aeration.

Lab One

1. Alcohol-flame the forceps (Figure 7-13) and place the membrane filter (grid facing up) on the lower half of the filter housing (Figure 7-14).
2. Clamp the two halves of the filter housing together and insert the filter housing into the suction flask as shown in Figure 7-15. (This assembly can be a little top-heavy; so have someone hold it or otherwise secure it to prevent tipping.)

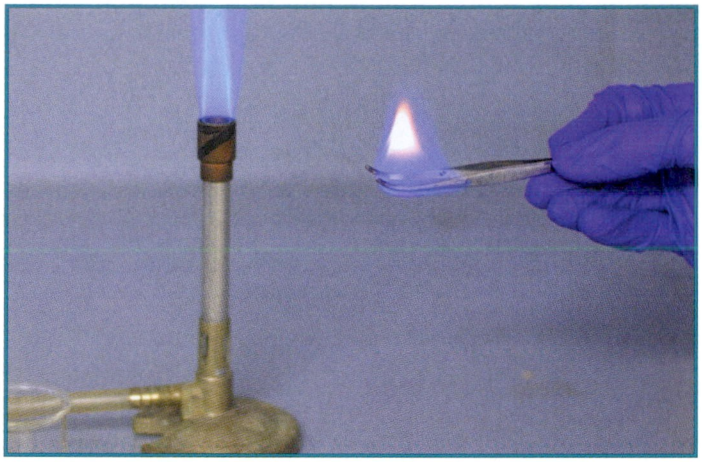

7-13 Alcohol-Flaming the Forceps ◆ Dip the forceps into the beaker of alcohol. Remove the forceps and quickly pass them through the Bunsen burner flame—just long enough to ignite the alcohol. **Be sure to keep the flame away from the beaker of alcohol.**

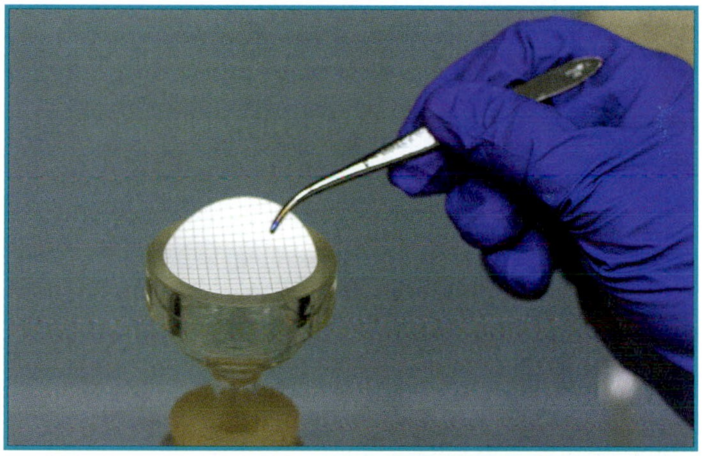

7-14 Placing the Filter on the Filter Housing ◆ Carefully place the filter on the bottom half of the filter housing. Clamp the filter funnel over the filter.

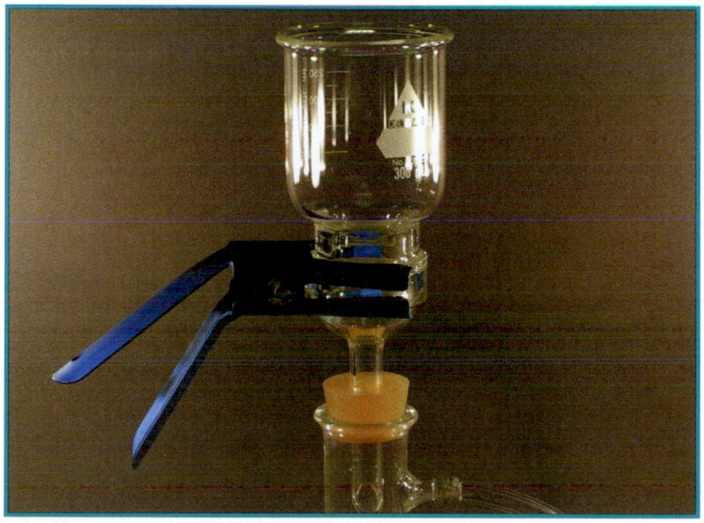

7-15 Membrane Filter Assembly ◆ The membrane filter assembly is made up of a two-piece funnel and clamp. The membrane filter is inserted between the two funnel halves, and the whole assembly is clamped together.

3 Pour the appropriate volume of water sample into the funnel. (Refer to Table 7-5 for suggested sample volumes. If the sample size is smaller than 10.0 mL, add 10 to 20 mL of sterile water before filtering, to help distribute the cells evenly on the surface of the membrane filter.)

4 Turn on the suction pump (or aspirator), and filter the sample into the flask.

5 Before removing the membrane, and with the suction pump still running, rinse the sides of the funnel two or three times with 20 or 30 mL of sterile water. This will wash off any cells adhering to the funnel and reduce the likelihood of contaminating the next sample.

6 Sterilize the forceps again, and carefully transfer the filter to the Endo Agar plate, being careful not to fold it or create air pockets (Figure 7-16).

7 Wait a few minutes to allow the filter to adhere to the agar, then invert the plate and incubate it aerobically at 35 ± 2°C for 22 to 24 hours.

Lab Two

1 Remove the plate and count the colonies on the membrane filter that are dark red, purple, have a black center, or produce a green or gold metallic sheen.

2 Record your data in the chart on the Data Sheet.

3 Calculate the coliform CFU per 100 mL using the following formula:

$$\frac{\text{Total coliforms}}{100\text{ mL}} = \frac{\text{coliform colonies counted} \times 100}{\text{volume of original sample in mL}}$$

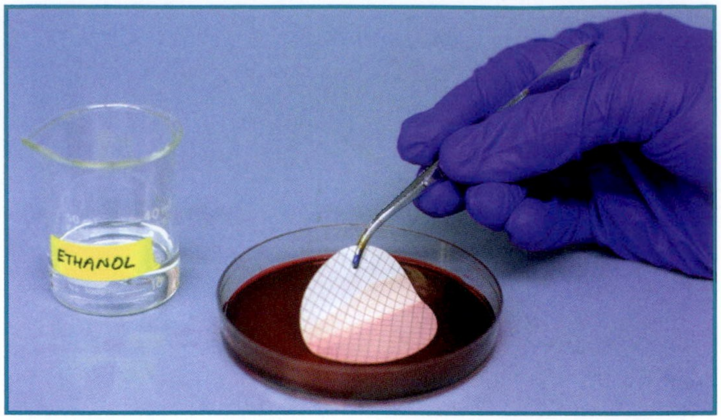

7-16 Placing the Filter on the Agar Plate ◆ Using sterile forceps, carefully place the filter onto the agar surface with the grid facing up. Try not to allow any air pockets under the filter, because contact with the agar surface is essential for growth of bacteria. Allow a few minutes for the filter to adhere to the agar before inverting the plate.

4 Record your results in the chart on the Data Sheet.

References

Collins, C. H., Patricia M. Lyne, and J. M. Grange. 1995. Page 270 in *Collins and Lyne's Microbiological Methods,* 7th ed. Butterworth-Heinemann, Oxford, UK.

Eaton, Andrew D. (AWWA), Lenore S. Clesceri (WEF), Eugene W. Rice (APHA), Arnold E. Greenberg (APHA), and Mary Ann H. Franson. 2005. Chapter 9 in *Standard Methods for the Examination of Water and Wastewater*, 21st ed. American Public Health Association, American Water Works Association, Water Environment Federation. APHA Publication Office, Washington, DC.

Zimbro, Mary Jo, and David A. Power, eds. 2003. Page 208 in *Difco™ and BBL™ Manual—Manual of Microbiological Culture Media.* Becton Dickinson and Co., Sparks, MD.

TABLE 7-5 Suggested Sample Volumes for Membrane Filter Total Coliform Test ◆

	Volume (●) To Be Filtered (mL)							
Water Source	100	50	10	1.0	0.1	0.01	0.001	0.0001
Drinking water	●							
Swimming pools	●							
Wells, springs	●	●	●					
Lakes, reservoirs	●	●	●					
Water supply intake			●	●	●			
Bathing beaches			●	●	●			
River water				●	●	●	●	
Chlorinated sewage				●	●	●		
Raw sewage					●	●	●	●

This standardized table is used to determine the volume of water to be tested using the Membrane Filter Technique. The volumes are based on the projected level of contamination in each of the sample sites. Note: The formula on page 414 includes the "volume of original sample" to compensate for any water volume filtered. (Reprinted with permission from the American Public Health Association)

Name ______________________ Date ______________________

Lab Section ______________________ I was present and performed this exercise (initials) ________

Membrane Filter Technique

OBSERVATIONS AND INTERPRETATIONS

1 Enter the number of colonies counted in each group's sample and calculate the total coliforms per 100 mL of water using the following formula:

$$\frac{\text{Total coliforms}}{100\text{ mL}} = \frac{\text{coliform colonies counted} \times 100\text{ mL}}{\text{volume of original sample in mL}}$$

2 Enter the results below.

Sample	Number of Colonies	Colonies/100 mL	Potable Y/N

Questions

1 For this test, why is Endo Agar used instead of Nutrient Agar?

2 How would adding glucose to this medium affect the results? Would it affect its sensitivity or specificity? Explain.

3 Suppose you were to count one coliform colony produced from a 10 mL water sample diluted with 90 mL of pure water. Is the water potable? Explain.

EXERCISE 7-6

Multiple Tube Fermentation Technique for Total Coliform Determination

Theory

The **Multiple Tube Fermentation Technique**, also called **most probable number**, or **MPN**, is a common means of calculating the number of coliforms present in 100 mL of a sample. The procedure determines both **total coliform** counts and *Escherichia coli* counts.

The three media used in the procedure are: Lauryl Tryptose Broth (LTB), Brilliant Green Lactose Bile (BGLB) Broth, and EC (*E. coli*) Broth. LTB, which includes lactose and lauryl sulfate, is selective for the coliform group. Because it does not screen out all noncoliforms, it is used to *presumptively* determine the presence or absence of coliforms. BGLB broth, which includes lactose and 2% bile, inhibits noncoliforms and is used to *confirm* the presence of coliforms. EC broth, which includes lactose and bile salts, is selective for *E. coli* when incubated at 45.5°C.

All broths are prepared in 10 mL volumes and contain an inverted Durham tube to trap any gas produced by fermentation. The LTB tubes are arranged in up to ten groups of five (see Figure 7-17, Procedural Diagram). Each tube in the first set of five receives 1.0 mL of the original sample. Each tube in the second group receives 1.0 mL of a 10^{-1} dilution. Each tube in group three receives 1.0 mL of 10^{-2}, etc.[1] (Note: The volume of broth in the tubes is not part of the calculation of the dilution factor. Dilutions of the water sample are made using sterile water prior to inoculating the broths. One milliliter of diluted sample is added to each tube in its designated group. (Refer to Exercise 6-1 for help with serial dilutions.)

After inoculation, the LTB tubes are incubated at 35 ± 2°C for up to 48 hours, then examined for gas production (Figure 7-18). Any positive LTB tubes then are used to inoculate BGLB tubes. Each BGLB receives one or two loopfuls from its respective positive LTB tube. Again, the cultures are incubated 48 hours at 35 ± 2°C and examined for gas production (Figure 7-19). Positive BGLB cultures then are transferred to EC broth and incubated at 45.5°C for 48 hours (Figure 7-20). After incubation the EC tubes with gas are counted. Calculation of BGLB MPN and EC MPN is based on the combinations of positive results in the BGLB and EC broths, respectively, using the following formula:

$$\text{MPN/100 mL} = \frac{100P}{\sqrt{V_n V_a}}$$

where

P = total number of positive results (BGLB or EC)

V_n = combined volume of sample in LTB tubes that produced negative results in BGLB or EC

V_a = combined volume of sample in all LTB tubes

It is customary to calculate and report *both* total coliform *and E. coli* densities. Total coliform MPN is calculated using BGLB broth results, and *E. coli* MPN is based on EC broth results.

Using the data from Table 7-6 and the formula above, the calculation would be as follows:

$$\text{MPN/100 mL} = \frac{100P}{\sqrt{V_n V_a}}$$

$$\text{MPN/100 mL} = \frac{100 \times 9}{\sqrt{0.24 \times 5.55}}$$

$$\text{MPN/100 mL} = 780$$

Application

This standardized test is used to measure coliform density (cells/100 mL) in water. It may be used to calculate the density of all coliforms present (total coliforms) or to calculate the density of *Escherichia coli* specifically.

In This Exercise

You will be using collected water samples to perform a multiple tube fermentation. The data collected then will be used to calculate a total coliform count and an *E. coli* count. Counts will be recorded in coliforms per 100 mL of sample and *E. coli* per 100 mL of sample, respectively.

Materials

Per Student Group

- ✓ 15 Lauryl Tryptose Broth (LTB) tubes (containing 10 mL broth and an inverted Durham tube)
- ✓ up to 15 Brilliant Green Lactose Bile (BGLB) Broth tubes (These tubes are needed for Lab Two. The number of tubes required will be determined by the results of the LTB test.)
- ✓ up to 15 EC Broth tubes (These tubes are needed for Lab Two. The number of tubes required will be determined by the results of the LTB test.)

[1] This exercise has been simplified for ease of instruction. The number of groups, number of tubes in each group, dilutions necessary, volume of broth in each tube, and volumes of sample transferred vary significantly, depending on the source and expected use of the water being tested.

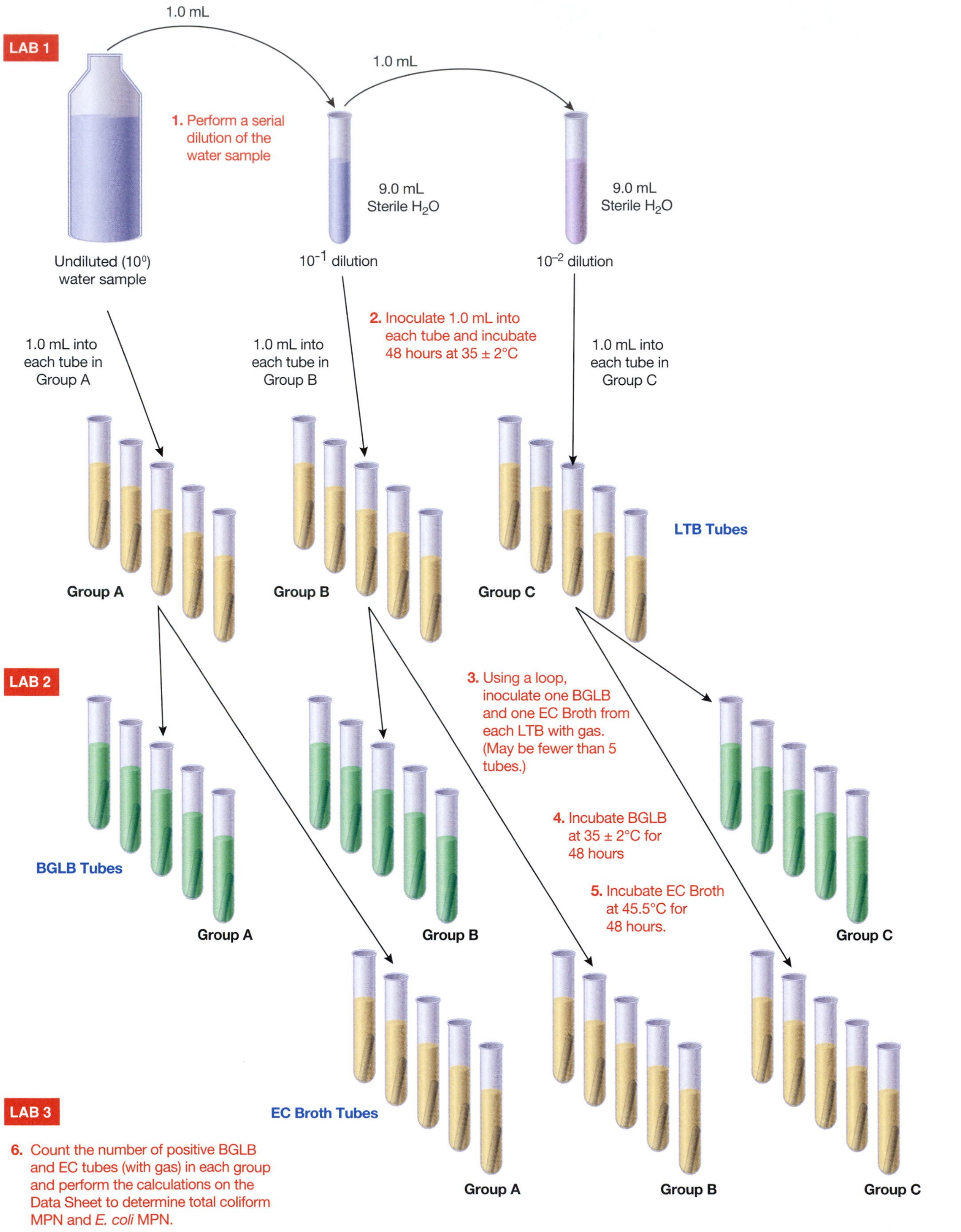
1.0 mL
LAB 1
1.0 mL
1. Perform a serial dilution of the water sample
9.0 mL Sterile H_2O
9.0 mL Sterile H_2O
Undiluted (10^0) water sample
10^{-1} dilution
10^{-2} dilution
2. Inoculate 1.0 mL into each tube and incubate 48 hours at 35 ± 2°C
1.0 mL into each tube in Group A
1.0 mL into each tube in Group B
1.0 mL into each tube in Group C
LTB Tubes
Group A
Group B
Group C
LAB 2
3. Using a loop, inoculate one BGLB and one EC Broth from each LTB with gas. (May be fewer than 5 tubes.)
4. Incubate BGLB at 35 ± 2°C for 48 hours
5. Incubate EC Broth at 45.5°C for 48 hours.
BGLB Tubes
Group A
Group B
Group C
LAB 3
EC Broth Tubes
6. Count the number of positive BGLB and EC tubes (with gas) in each group and perform the calculations on the Data Sheet to determine total coliform MPN and *E. coli* MPN.
Group A
Group B
Group C

7-17 Procedural Diagram of Multiple Tube Fermentation Technique for Determination of Total Coliform ◆

7-18 **LTB Tubes** ◆ The bubble in the Durham tube on the right is *presumptive* evidence of coliform contamination. The tube on the left is negative.

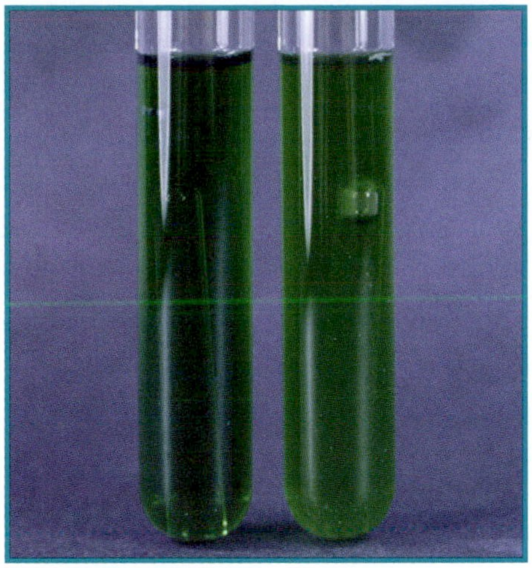

7-19 **BGLB Broth** ◆ The bubble in the Durham tube on the right is seen as *confirmation* of coliform contamination. The tube on the left is negative.

7-20 **EC Broth** ◆ The bubble in the Durham tube on the right is seen as *confirmation* of E. coli contamination. The tube on the left is negative.

- ✓ two 9.0 mL dilution tubes
- ✓ sterile 1.0 mL pipettes and pipettor
- ✓ water bath set at 45.5°C (Lab Two)
- ✓ test tube rack
- ✓ labeling tape
- ✓ 100 mL water dilution bottle (to be distributed in the preceding lab)
- ✓ 100 mL water sample (obtained by student)
- ✓ gloves
- ✓ household disinfectant and paper towels

Medium Recipes

Brilliant Green Lactose Bile Broth

◆ Peptone	10.0 g
◆ Lactose	10.0 g
◆ Oxgall	20.0 g
◆ Brilliant green dye	0.0133 g
◆ Distilled or deionized water	1.0 L

pH 7.0–7.4 at 25°C

Lauryl Tryptose Broth

◆ Tryptose	20.0 g
◆ Lactose	5.0 g
◆ Dipotassium phosphate	2.75 g
◆ Monopotassium phosphate	2.75 g
◆ Sodium chloride	5.0 g
◆ Sodium lauryl sulfate	0.1 g
◆ Distilled or deionized water	1.0 L

pH 6.6–7.0 at 25°C

EC Broth

◆ Tryptose	20.0 g
◆ Lactose	5.0 g
◆ Dipotassium phosphate	4.0 g
◆ Monopotassium phosphate	1.5 g
◆ Sodium chloride	5.0 g
◆ Distilled or deionized water	1.0 L

pH 6.7–7.1 at 25°C

Procedure

[Refer to the procedural diagram in Figure 7-17 as needed.]

Lab One

1 Mix the water sample. Then take a dilution by adding 1.0 mL of the water sample to one of the 9.0 mL dilution tubes. Mix well. This dilution is only 1/10 of the original sample; therefore, the dilution is 1/10 or 0.1 or 10^{-1} (preferred). For help with dilutions, refer to Exercise 6-1.

2 Make a second dilution by adding 1.0 mL of the 10^{-1} dilution to one of the 9.0 mL dilution tubes. Mix well. This dilution contains 1/100 original sample and is 10^{-2}.

3 Arrange the 15 LTB tubes into three groups of five in a test tube rack. Label the groups A, B, and C, respectively.

4 Aseptically transfer 1.0 mL of the (undiluted 10^{0}) water sample to each LTB tube in the group labeled A. Mix well. (Note: The dilutions are determined in the dilution tubes. The LTB is not part of the dilution.)

5 Add 1.0 mL of the 10^{-1} dilution to each of the LTB tubes in group B. Mix well.

6 Add 1.0 mL of the 10^{-2} dilution to each of the LTB tubes in the C group.

7 Incubate the LTB tubes at 35 to 37°C for 48 hours.

Lab Two

1 Remove the broths from the incubator and, one group at a time, examine the Durham tubes for accumulation of gas. Gas production is a positive result; absence of gas is negative if the broth is clear. (Note: If the LTB broth is turbid but contains no gas, the presence of coliforms is not yet ruled out. Therefore, record any turbidity as a positive result and include these tubes with those containing gas. Record positive and negative results in the chart provided on the Data Sheet. If necessary, refer to the example in Table 7-6.

2 Using an inoculating loop, inoculate one BGLB broth with each positive LTB tube. (Make sure that each BGLB tube is *clearly labeled* A, B, or C according to the LTB tube from which it is inoculated.)

3 Inoculate EC broths with the positive LTB tubes in the same manner as the BGLB above. Again, be sure to clearly label all EC tubes appropriately.

4 Incubate the BGLB at 35 to 37°C for 48 hours. Incubate the EC tubes in the 45.5°C water bath for 48 hours.

Lab Three

1 Remove all tubes from the incubator and water bath and examine the Durham tubes for gas accumulation. (Note: At this point, turbidity does not count as positive; the only positive result is the presence of gas in the Durham tube.) Count the positive BGLB tubes and enter your results in the chart on the Data Sheet.

2 Using Table 7-6 as a guide, complete the BGLB chart on the Data Sheet.

3 Using the data in the BGLB chart, calculate the total coliform MPN using the formula described in Theory. (This formula is used to calculate both Total Coliform MPN and *E. coli* MPN.)

4 Count the positive EC broth results in the same manner as the BGLB test. Record the results in the chart on the Data Sheet.

5 Determine the *E. coli* MPN in the same manner as described for total coliform count, and record your results in the chart on the Data Sheet.

Reference

Eaton, Andrew D. (AWWA), Lenore S. Clesceri (WEF), Eugene W. Rice (APHA), Arnold E. Greenberg (APHA), and Mary Ann H. Franson. 2005. Chapter 9 in *Standard Methods for the Examination of Water and Wastewater*, 21st ed. 2005. American Public Health Association, American Water Works Association, Water Environment Federation. APHA Publication Office, Washington, DC.

TABLE 7-6 BGLB Test Results ◆ The results shown here are of a hypothetical BGLB test using three groups (A, B, and C), each containing five tubes. (Note: On your data sheet, you are given two similar tables—one for BGLB results and one for EC broth results.) In this example, the original water sample was diluted to 10^0, 10^{-1}, and 10^{-2}. (Note: For purposes of calculation, undiluted sample is recorded as 10^0.) The three dilutions were used to inoculate the broths in groups A, B, and C, respectively. The first row contains the dilution of the inoculum used per group. The second row shows the actual amount of original sample that went into each broth. The third row contains the number of tubes in each group (five in this example). The fourth row shows the number of tubes from each group of five that showed evidence of gas production. This total (in red) inserts into the equation. The fifth row shows the number of tubes from each group that did not show evidence of gas production. The sixth row is used for calculating the "combined volume of sample in negative tubes" and refers to the inoculum that went into the LTB tubes that produced a negative result. This total (in red) inserts into the equation. The seventh row is used for calculating the "combined volume of sample in all tubes" and refers to the total volume of inoculum that went into all LTB tubes. This total (in red) inserts into the equation. As you can see, the undiluted inoculation (Group A) produced five positive results and zero negative results; the 10^{-1} dilution (Group B) produced three positive results and two negative; and the 10^{-2} dilution (Group C) produced one positive and four negative results. The total volume of original sample that went into LTB tubes was 5.55 mL, 0.24 mL of which produced no gas (shown in red in rows 7 and 6, respectively).

	Group	Group A	Group B	Group C	Totals (A+B+C)
1	Dilution (D)	10^0	10^{-1}	10^{-2}	NA
2	Portion of dilution added to each LTB tube that is original sample (1.0 mL + D)	1.0 mL	0.1 mL	0.01 mL	NA
3	# LTB tubes in group	5	5	5	NA
4	# positive (BGLB or EC) results	5	3	1	9
5	# negative (BGLB or EC) results	0	2	4	NA
6	Total volume of *original sample* in all LTB tubes that produced negative (BGLB or EC) results (D × 1.0 mL × # negative tubes)	0 mL	0.2 mL	0.04 mL	0.24 mL
7	Volume of *original sample* in all LTB tubes inoculated (D × 1.0 mL × # tubes)	5.0 mL	0.5 mL	0.05 mL	5.55 mL

NAME ______________________ DATE ______________________

LAB SECTION ______________________ I WAS PRESENT AND PERFORMED THIS EXERCISE (INITIALS) ________

DATA SHEET 7-6

Multiple Tube Fermentation Technique for Total Coliform Determination

DATA AND CALCULATIONS

1 Enter your data here.

BGLB Data				
Group	Group A	Group B	Group C	Totals (A + B + C)
Dilution (D)	10^0	10^{-1}	10^{-2}	NA
Portion of dilution added to each LTB tube that is *original sample* (1.0 mL × D)	1.0 mL	0.1 mL	0.01 mL	NA
# LTB tubes in group	5	5	5	NA
# Positive results (Gas)				
# Negative results (No gas)				NA
Total volume of *original sample* in all LTB tubes that produced negative results (D × 1.0 mL × # negative tubes)				
Volume of *original sample* in all LTB tubes inoculated (D × 1.0 mL × # tubes)	5.0 mL	0.5 mL	0.05 mL	5.55 mL

BGLB Data				
Group	Group A	Group B	Group C	Totals (A + B + C)
Dilution (D)	10^0	10^{-1}	10^{-2}	NA
Portion of dilution added to each LTB tube that is *original sample* (1.0 mL × D)	1.0 mL	0.1 mL	0.01 mL	NA
# LTB tubes in group	5	5	5	NA
# Positive results (Gas)				
# Negative results (No gas)				NA
Total volume of *original sample* in all LTB tubes that produced negative results (D × 1.0 mL × # negative tubes)				
Volume of *original sample* in all LTB tubes inoculated (D × 1.0 mL × # tubes)	5.0 mL	0.5 mL	0.05 mL	5.55 mL

2 Enter your final results here.

Total coliform MPN/100 mL (from BGLB data)	
E. coli MPN/100 mL (from EC data)	

Questions

1 In the following formula, V_n and V_a symbolize critical volumes in calculating the most probable number of coliforms in 100 mL of a sample. Do the symbols represent the amount of dilution added to tubes of LTB or the amount of original sample added to the tubes? Defend your answer in full and explain what would happen if you were to choose the wrong volumes in your calculation.

$$\text{MPN/100 mL} = \frac{100P}{\sqrt{V_n V_a}}$$

Name ______________________ Date ______________________

Lab Section ______________________ I was present and performed this exercise (initials) ________

2 Suppose you were to run this test on a water sample and, after 48 hours incubation of the LTB tubes, found turbidity but no gas bubbles. What should you do next?

3 What if you found gas in the LTB tubes but none in the BGLB? How does this affect the EC test?

4 All coliforms ferment glucose, but none of the media used for this test includes glucose. Why do you think glucose is not used?

5 Would adding glucose increase or decrease the sensitivity? Specificity?

Food Microbiology

Food microbiology is the study, utilization, and control of microorganisms in food. In Exercise 2-1 you learned that microorganisms exist virtually everywhere. Left undisturbed in their own habitat, most are beneficial in some way. Some microorganisms have been used successfully to produce our favorite foods, such as yogurt, wine, beer, sauerkraut, buttermilk, vinegar, bread, and cheeses. Generally speaking, however, the unintended introduction of microorganisms to our food (including otherwise beneficial microorganisms) is a serious health hazard. For example, *Escherichia coli*, one of the most common enterics in humans and other mammals, can cause mild to severe illness or even death if ingested.

It is not unusual to find unwanted microorganisms in unprocessed food. It is unavoidable and expected. This is why there are agencies such as the Food and Drug Administration (FDA) and the Centers for Disease Control and Prevention (CDC). These agencies were designed, in part, to protect consumers from the food-borne illnesses caused by improper management of food items. But these agencies exercise control only in the public arena; their influence over the practices of people in their homes typically is advisory only. This is why it is important to follow their recommendations and not only practice good personal hygiene, but to wash and cook food properly.

In the two exercises included in this unit, you will be performing a simple test of milk quality and then you will have a little fun by making yogurt. Don't forget to wash your hands before you start!

EXERCISE 7-7

Methylene Blue Reductase Test

Theory

Methylene blue dye is blue when oxidized and colorless when reduced. It can be reduced enzymatically either aerobically or anaerobically. In the aerobic electron transport system, methylene blue is reduced by cytochromes but immediately is returned to the oxidized state when it subsequently reduces oxygen. Anaerobically, the dye is in the reduced form, and in the absence of an oxidizing substance, remains colorless.

The reduction of methylene blue may be used as an indicator of milk quality. In the methylene blue reductase test, a small quantity of a dilute methylene blue solution is added to a sterilized test tube containing raw milk. The tube then is sealed tightly and incubated in a 35°C water bath. The time it takes the milk to turn from blue to white (because of methylene blue reduction) is a qualitative indicator of the number of microorganisms living in the milk (Figure 7-21). Good-quality milk takes longer than 6 hours to convert the methylene blue.

Application

This test is helpful in differentiating the *enterococci* from other *streptococci*. It also tests for the presence of coliforms in raw milk.

In This Exercise

You will test milk quality by measuring how long the indicator dye, methylene blue, takes to become oxidized by any bacterial contaminants present.

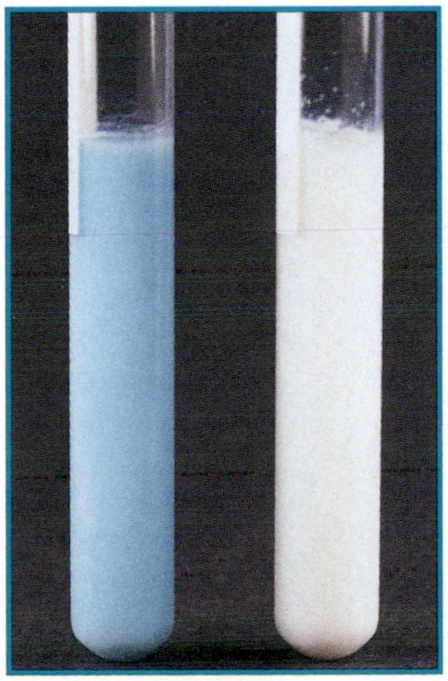

7-21 Methylene Blue Reductase Test ◆ The tube on the left is a control to illustrate the original color of the oxidized medium. The tube on the right indicates bacterial reduction of methylene blue after 20 hours. The speed of reduction is related to the concentration of microorganisms present in the milk.

Materials

Per Student Group

- ✓ milk samples (raw or processed; a variety is best)
- ✓ sterile screw-capped test tubes
- ✓ sterile 1 mL and 10 mL pipettes with mechanical pipettors
- ✓ hot water bath set at 35°C
- ✓ methylene blue reductase reagent
- ✓ overnight broth culture of *Escherichia coli*
- ✓ clock or wristwatch

Reagent

Methylene Blue Reductase Reagent

♦ Methylene blue dye	8.8 mg
♦ Distilled or deionized water	200.0 mL

Procedure

Lab One

1 Obtain sterile tubes for as many samples as you are testing, plus two more, to be used as positive and negative controls. Label them appropriately.

2 Using a sterile 10 mL pipette, aseptically add 10 mL milk to each tube.

3 Inoculate the tube marked "positive control" with 1.0 mL of *E. coli* culture.

4 Aseptically add 1.0 mL methylene blue solution to each test tube. Cap the tubes tightly and invert them several times to mix thoroughly.

5 Place the tube marked "negative control" in the refrigerator to prevent it from changing color.

6 Place all other tubes in the hot water bath, and note the time.

7 After 5 minutes, remove the tubes, invert them once to mix again, then return them to the water bath. Record the time in the chart in the Data Sheet under "Starting Time."

8 Using the control tubes for color comparison, check the tubes at 30-minute intervals, and record the time when each becomes white. Poor-quality milk takes less than 2 hours. Good-quality milk takes longer than 6 hours. If necessary, have someone check the tubes at 6 hours and record the results for you.

9 Using the chart provided, calculate the time it takes for each milk sample to become white.

References

Bailey, R. W., and E. G. Scott. 1966. Pages 114 and 306 in *Diagnostic Microbiology*, 2nd ed. Mosby, St. Louis.

Benathen, Isaiah. 1993. Page 132 in *Microbiology with Health Care Applications*. Star Publishing, Belmont, CA.

Power, David A., and Peggy J. McCuen. 1988. Page 62 in *Manual of BBL™ Products and Laboratory Procedures*, 6th ed. Becton Dickinson Microbiology Systems, Cockeysville, MD.

Richardson, Gary H., ed. 1985. *Standard Methods for the Examination of Dairy Products*, 15th ed. American Public Health Association, Washington DC.

NAME ______________________ DATE ______________________

LAB SECTION ______________________ I WAS PRESENT AND PERFORMED THIS EXERCISE (INITIALS) ________

Methylene Blue Reductase Test

OBSERVATIONS AND INTERPRETATIONS

Record your results and interpretations in the chart below.

Milk Sample (Brand and raw or processed)	Starting Time T_s (Milk is blue)	Ending Time T_e (Milk is white)	Elapsed Time $(T_e - T_s)$	Milk Quality

Questions

1 Why were you told to cap the tubes tightly? Explain.

2 What results would you expect if the tubes were inoculated with a strict aerobe? A strict anaerobe?

EXERCISE 7-8

Making Yogurt

Theory

Several species of bacteria are used in the commercial production of yogurt. Most formulations include combinations of two or more species to synergistically enhance growth and to produce the optimum balance of flavor and acidity. One common pairing of organisms in commercial yogurt is that of *Lactobacillus delbrueckii* subsp. *bulgaricus* and *Streptococcus thermophilus*.

Yogurt gets its unique flavor from acetaldehyde, diacetyl, and acetate produced from fermentation of the milk sugar lactose. The proportions of products, and ultimately the flavor, in the yogurt depend upon the types of enzyme systems possessed by the species used. Both species mentioned above contain **constitutive** β-galactosidase systems that break down lactose and convert the glucose to lactate, formate, and acetate via pyruvate in the Embden-Meyerhof-Parnas (glycolysis) pathway. (See Appendix A.)

As you may remember, lactose is a disaccharide composed of glucose and galactose. *S. thermophilus* does not possess the enzymes needed to metabolize galactose, and *L. delbrueckii* preferentially metabolizes glucose. This results in an accumulation of galactose, which adds sweetness to the yogurt. Acetaldehyde is produced directly from pyruvate by *S. thermophilus* and through the conversion of proteolysis products threonine and glycine by *L. delbrueckii*. Some strains of *S. thermophilus* also produce glucose polymers, which give the yogurt a viscous consistency.

Application

This exercise is designed for fun. Don't forget, it is unsafe to text while eating yogurt.

In This Exercise

You will produce yogurt using a simple home recipe with a commercial yogurt as a starter. Read the label to see which microorganisms are included. We hope you enjoy it.

Materials

Per Student Group

- ✓ whole, low-fat, or skim milk
- ✓ plain yogurt with active cultures (bring from home or supermarket)
- ✓ medium-size saucepan
- ✓ medium-size bowl
- ✓ wire whisk
- ✓ hotplate
- ✓ cooking thermometer
- ✓ measuring cup
- ✓ plastic wrap
- ✓ fresh fruit
- ✓ sugar (optional)
- ✓ pH meter or pH paper

Procedure[1]

Lab One

1. Obtain all materials, and set them up in a clean work area.
2. While stirring, slowly heat 5 cups milk in the saucepan to 185°F. Remove the milk from the heat, and let it cool to 110°F.
3. Place 1/4 cup starter yogurt in the bowl. Slowly, stir in cooled milk, about 1/3 to 1/2 cup at a time, mixing after each addition until smooth.
4. Cover the bowl with plastic wrap, and puncture several times to allow gases and excess moisture to escape.
5. Label the bowl with your name, the date, and the cultures present in your yogurt starter.
6. Incubate 5–6 hours at 30°–35°C. Remove the bowl from the incubator at the correct time, and place it in the refrigerator.

Lab Two

1. Remove your yogurt from the refrigerator.
2. Perform a simple stain on a smear from your yogurt and examine on the microscope.
3. Compare flavor, consistency, and starter cultures with other groups in the lab. With a pH meter or pH paper, measure the pH of your yogurt. Record your results in the chart provided on the Data Sheet.
4. Serve with fresh fruit and enjoy!

References

Downes, Frances Pouch, and Keith Ito. 2001. Chapter 47 in *Compendium of Methods for the Microbiological Examination of Foods*, 4th ed. American Public Health Association, Washington, DC.

Ray, Bibek. 2001. Chapter 13 in *Fundamental Food Microbiology*, 2nd ed. CRC Press LLC, Boca Raton, FL.

[1] Because food and beverages are ordinarily not allowed in microbiology labs, this exercise is best performed in an alternate location.

Name ______________________ Date ______________________

Lab Section ______________________ I was present and performed this exercise (initials) ________

DATA SHEET 7-8

Making Yogurt

OBSERVATIONS AND INTERPRETATIONS

Record yogurt made by students groups below.

Culture Organisms	Cell Morphology and Arrangement	Flavor	Consistency	pH

There are no questions for this exercise.

Microbial Genetics and Serology

SECTION

8

In the first part of this section you will perform three exercises dealing with DNA. First you will extract *E. coli* DNA. In the second exercise you will do an experiment to induce mutations—alterations of DNA. In it you will examine the effects of ultraviolet (UV) radiation on bacteria. UV illustrates how one (there are many) *mutagen* causes damage. You will also experiment with factors that affect its ability to cause mutation in the cell and how bacteria are able to repair that damage. The third exercise incorporates into one experiment, bacterial transformation, use of antibiotic selective media, genetic regulation, and genetic engineering. You will be transferring a jellyfish gene into *E. coli* cells, selecting for only those cells that have been transformed successfully, then manipulating the environment so the organisms produce the gene product only when you want them to.

In the second part of this section, you will have the opportunity to perform a variety of serological tests that are used to detect the presence of either antigen or antibody in a sample. Serological tests are useful as a diagnostic tool.

EXERCISE 8-1

Extraction of DNA from Bacterial Cells[1]

❖ Theory

DNA extraction from cells is surprisingly easy and occurs in three basic stages.

1. A detergent (e.g., sodium dodecyl sulfate—SDS) is used to lyse cells and release cellular contents, including DNA.
2. This is followed by a heating step (at approximately 65–70°C) that denatures protein (including DNases that would destroy the extracted DNA) and other cell components. Temperatures higher than 80°C will denature DNA, and this is undesirable. A protease also may be added to remove proteins. (Other techniques for purification may be used, but these will not be included in this exercise.)
3. Finally, the water-soluble DNA is precipitated in cold alcohol as a whitish, mucoid mass (Figure 8-1).

As an optional follow-up to extraction, an ultraviolet spectrophotometer (Figure 8-2) will be used to estimate DNA concentration in the sample by measuring absorbance at 260 nm, the optimum wavelength for absorption by DNA. An absorbance of A_{260nm} of 1 corresponds to 50 µg/mL of double-stranded DNA (dsDNA). Reading absorbance at 280 nm and calculating the following ratio can determine purity of the sample:

[1] *Note:* Thanks to the following individuals who offered helpful suggestions for this protocol: Dr. Melissa Scott, formerly of San Diego City College, Allison Shearer of Grossmont College, Donna Mapston and Dr. Ellen Potter of Scripps Institute for Biological Studies, and Dr. Sandra Slivka of Miramar College.

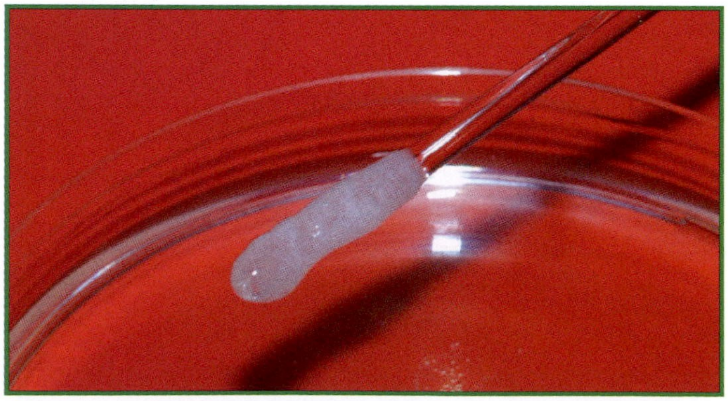

8-1 Precipitated DNA ◆ This onion DNA has been spooled onto a glass rod. Your DNA yield from *E. coli* will not be as high as shown here.

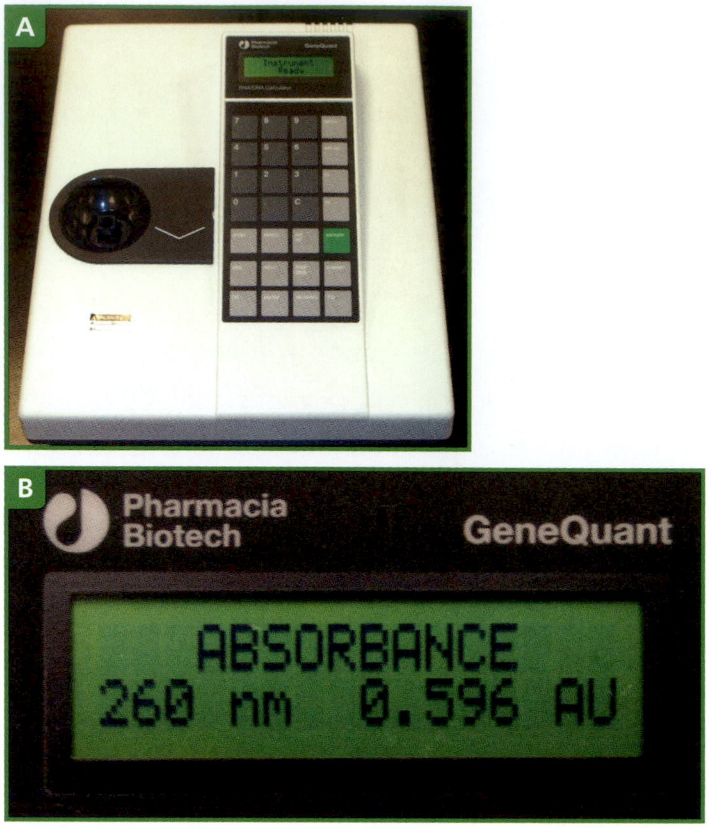
A
B
Pharmacia
Biotech
GeneQuant
ABSORBANCE
260 nm 0.596 AU

8-2 Ultraviolet Spectrophotometer ◆ **A** A UV spectrophotometer can be used to determine DNA concentration. A quartz cuvette is shown in the sample port. (Glass absorbs UV, so glass cuvettes can't be used.) **B** This specimen has an absorbance at 260 nm (A_{260nm}) of 0.596. Because an A_{260nm} of 1.0 is equal to 50 µg/mL of dsDNA, this specimen has a concentration of 29.8 µg dsDNA/mL (0.596 AU × 50 µg/mL). Absorbance also can be used to determine purity of the sample. A relatively pure DNA sample will have an A_{260nm}/A_{280nm} value of approximately 1.8.

$$\frac{\text{Absorbance}_{260nm}}{\text{Absorbance}_{280nm}}$$

If the sample is reasonably pure nucleic acid, the ratio will be about 1.8 (between 1.65 and 1.85). Because protein absorbs maximally at 280 nm, a ratio of less than 1.6 is likely because of protein contamination. If purity is crucial, the DNA extraction can be repeated. If the ratio is greater than 2.0, the sample is diluted and read again.

❖ Application

Extraction of DNA is a starting point for many lab procedures, including DNA sequencing and cloning.

❖ In This Exercise

You will extract DNA from *E. coli*. To improve yield, you first will concentrate an *E. coli* broth culture. The actual extraction involves cell lysis, denaturation of protein, and precipitation of the DNA in alcohol. Following extraction, an optional procedure may be used to determine DNA yield and the purity of your extract.

Materials

Per Pair of Students

- ✓ overnight culture of *Escherichia coli* in Luria-Bertani broth (young cultures work best)
- ✓ water bath set to 65°C
- ✓ microtube floats
- ✓ ice bath (small cups with crushed ice work well)
- ✓ 300 µL 10% sodium dodecyl sulfate (SDS)
- ✓ 50 µL 20 mg/mL Proteinase K solution (stored in freezer between uses) (Note: Meat tenderizer is an inexpensive substitute, though it may result in DNA hydrolysis if too much is used.)
- ✓ 300 µL 1.0 M sodium acetate solution (pH = 5.2)
- ✓ 3 mL 1X M Tris-Acetate-EDTA (TAE) buffer (dilute 10X TAE 9 + 1)
- ✓ 2 mL 90% isopropanol (stored in a freezer or an ice bath)
- ✓ two calibrated disposable transfer pipettes
- ✓ 100–1000 µL digital pipettor and tips
- ✓ 25 mL centrifuge tube
- ✓ tabletop centrifuge
- ✓ minicentrifuge
- ✓ disposable inoculating loop
- ✓ two microtubes (at least 1.5 mL in volume)
- ✓ vortex mixer (optional)
- ✓ ultraviolet spectrophotometer (optional)
- ✓ two quartz cuvettes (optional)

Medium Recipe

Luria-Bertani Broth

♦ Tryptone	10.0 g
♦ Yeast Extract	5.0 g
♦ NaCl	10.0 g
♦ Distilled or deionized water	1.0 L

pH 7.4 at 25°C

Procedure

Refer to the procedural diagram in Figure 8-3 as you read and follow this protocol.

1. Obtain the *E. coli* culture. Mix the suspension until uniform turbidity is seen, then transfer 5 mL to a clean, nonsterile centrifuge tube.
2. Spin the 5 mL sample in the tabletop centrifuge slowly for 10 minutes to produce a cell pellet.
3. After spinning and without disturbing the pellet, remove 4.5 mL of the supernatant with a transfer pipette. Dispose of the supernatant in the original culture tube.
4. Transfer the remaining 0.5 mL to a nonsterile microtube.
5. Add 200 µL of 10% SDS to the *E. coli*.
6. Add 30 µL of 20 mg/mL Proteinase K solution or a "pinch" of meat tenderizer.
7. Close the cap and gently mix the tube for 5 minutes by tipping it upside down every few seconds.
8. Place the tube in a float and incubate in a 65°C water bath for 5 minutes.
9. Add 200 µL mL 1 M sodium acetate solution and mix gently.
10. Place the tube in the ice bath until it is at or below room temperature.
11. When cooled, squirt 400 µL mL of cold 95% isopropanol into the preparation. (If there are bubbles on the surface, remove them with a nonsterile transfer pipette prior to adding the isopropanol.)
12. Use a disposable loop to mix the preparation, moving in and out and turning occasionally. A glob of mucoid DNA will begin to appear as you mix and will adhere to the loop.

Optional Procedure

(if your lab has an ultraviolet spectrophotometer)

Refer to the procedural diagram in Figure 8-3 as you read and complete the following protocol.

1. Remove the DNA from the original microtube and transfer it to a second microtube, using the loop.
2. Resuspend the DNA in 1000 µL isopropanol using a nonsterile transfer pipette, then spin it in a minicentrifuge for a few seconds. The DNA should be at the bottom of the tube.
3. With a nonsterile transfer pipette, remove the supernatant and allow the DNA pellet to air dry.
4. Resuspend the dried DNA in 1000 µL of TAE buffer. Mix vigorously to dissolve the DNA in the TAE. This mixing may be done by hand, or a vortex mixer may be used.
5. Transfer the suspended DNA solution into a quartz cuvette.

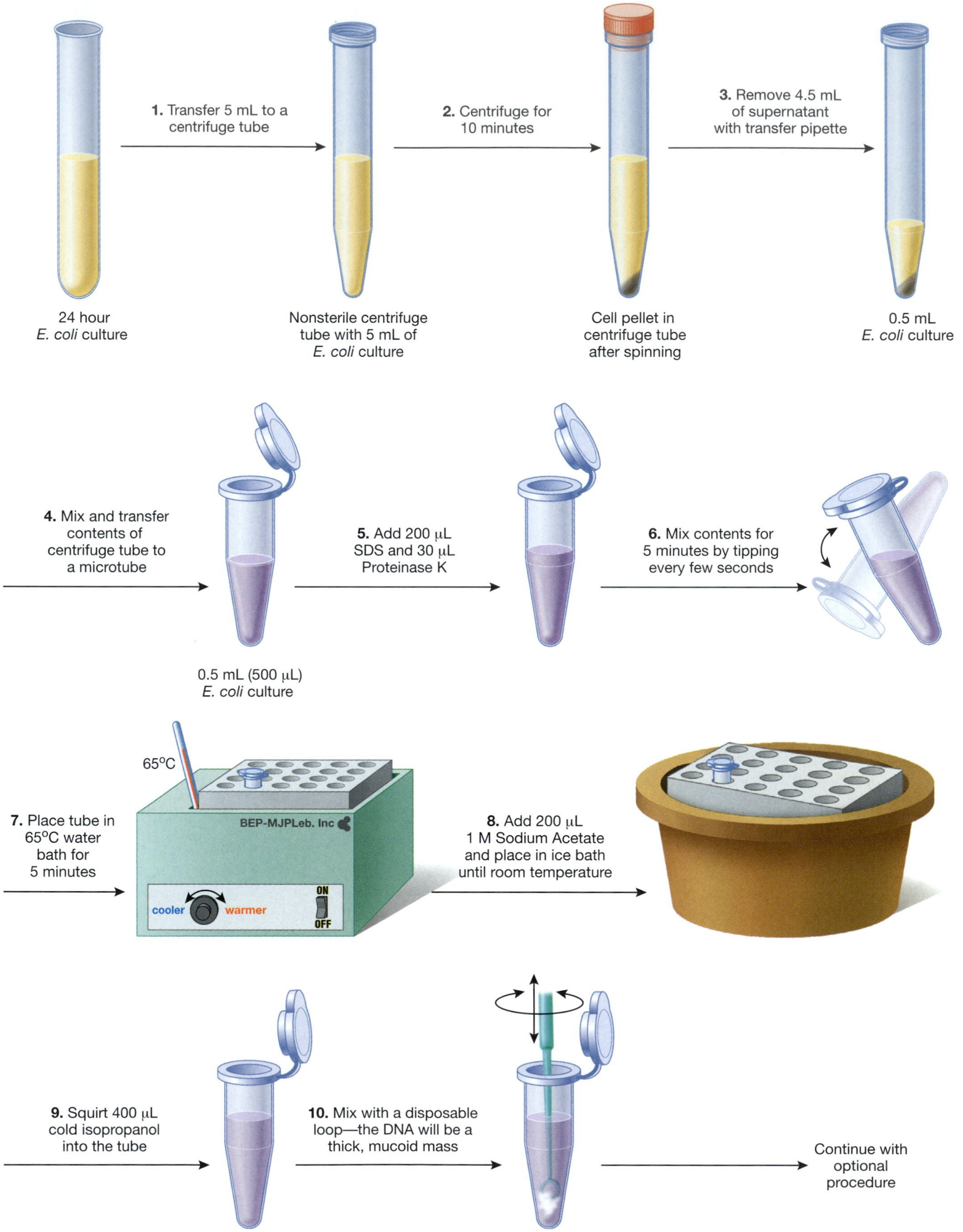
1. Transfer 5 mL to a centrifuge tube
2. Centrifuge for 10 minutes
3. Remove 4.5 mL of supernatant with transfer pipette
24 hour E. coli culture
Nonsterile centrifuge tube with 5 mL of E. coli culture
Cell pellet in centrifuge tube after spinning
0.5 mL E. coli culture
4. Mix and transfer contents of centrifuge tube to a microtube
5. Add 200 µL SDS and 30 µL Proteinase K
6. Mix contents for 5 minutes by tipping every few seconds
0.5 mL (500 µL) E. coli culture
65°C
7. Place tube in 65°C water bath for 5 minutes
BEP-MJPLeb. Inc
cooler
warmer
ON
OFF
8. Add 200 µL 1 M Sodium Acetate and place in ice bath until room temperature
9. Squirt 400 µL cold isopropanol into the tube
10. Mix with a disposable loop—the DNA will be a thick, mucoid mass
Continue with optional procedure

8-3 **Procedural Diagram for Bacterial DNA Extraction** *(continued on next page)* ◆

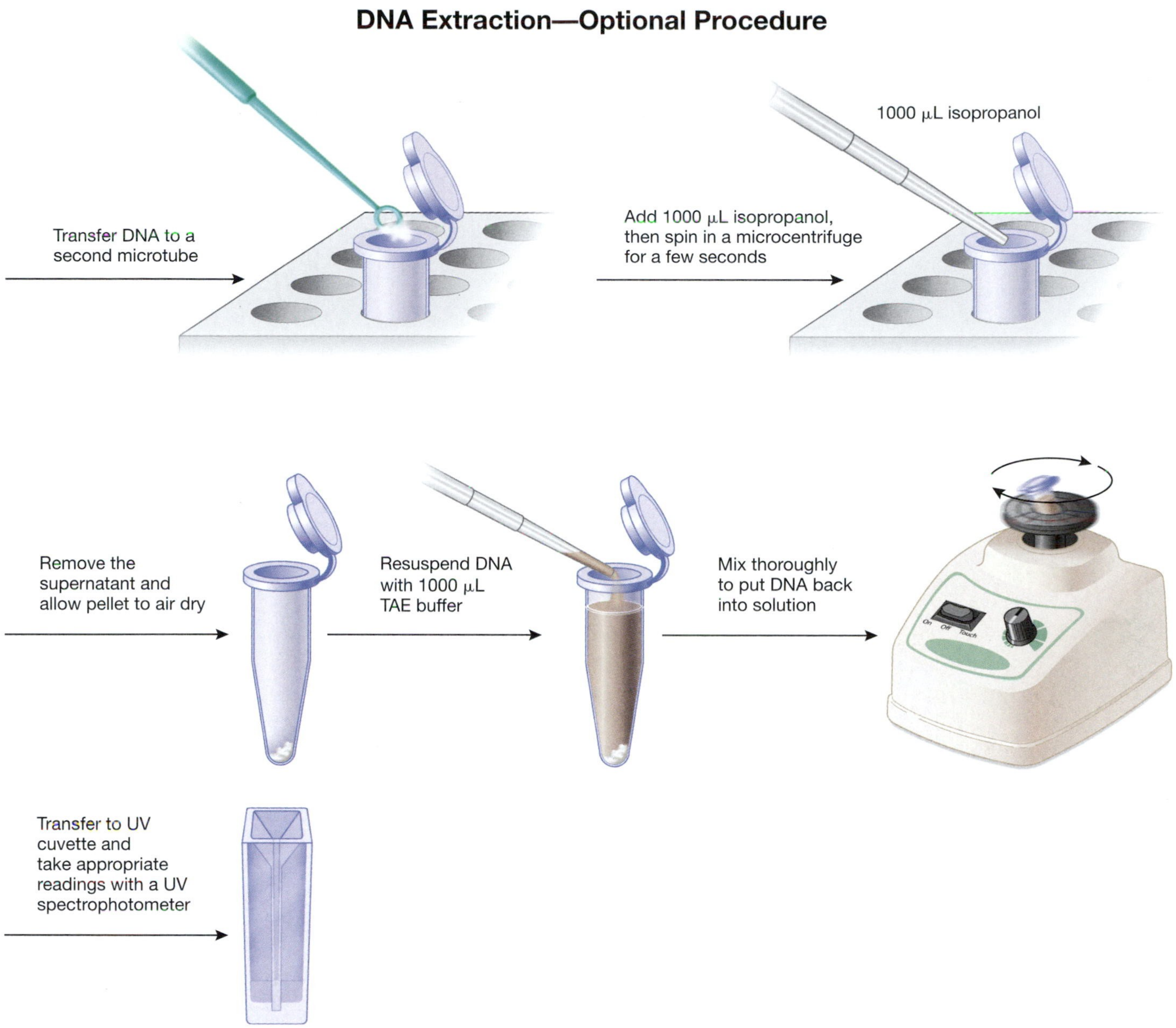
DNA Extraction—Optional Procedure
Transfer DNA to a second microtube
1000 µL isopropanol
Add 1000 µL isopropanol, then spin in a microcentrifuge for a few seconds
Remove the supernatant and allow pellet to air dry
Resuspend DNA with 1000 µL TAE buffer
Mix thoroughly to put DNA back into solution
Transfer to UV cuvette and take appropriate readings with a UV spectrophotometer

8-3 Procedural Diagram for Bacterial DNA Extraction *(continued)* ◆

6 Prepare a second cuvette containing 1000 µL of TAE as a blank.

7 Set the spectrophotometer to 260 nm wavelength. Follow the instructions for your UV spectrophotometer to check the absorbance of the extracted DNA, and record on the Data Sheet.

8 Set the spectrophotometer to 280 nm wavelength. Follow the instructions for your UV spectrophotometer to check the absorbance of the extracted DNA, and record on the Data Sheet.

9 Calculate the probable purity of your extracted DNA sample using the formula provided, and record on the Data Sheet.

References

Bost, Rod. 1989. *Down and Dirty DNA Extraction*. Carolina Genes. North Carolina Biotechnology, Research Triangle Park, NC.

Davis, Leonard G., Mark D. Dibner, and James F. Battey. 1986. *Basic Methods in Molecular Biology*. Elsevier Science Publishing, New York.

Freifelder, David. 1982. Pages 504–505 in *Physical Biochemistry*, 2nd ed. W. H. Freeman and Co, New York.

Kreuzer, Helen, and Adrianne Massey. 2001. *Recombinant DNA and Biotechnology—A Guide For Teachers*, 2nd ed. ASM Press, Washington, DC.

Zyskind, Judith W., and Sanford I. Bernstein. 1992. *Recombinant DNA Laboratory Manual*. Academic Press, San Diego.

Name ______________________ Date ______________________

Lab Section ______________________ I was present and performed this exercise (initials) ________

Extraction of DNA from Bacterial Cells

OBSERVATIONS AND INTERPRETATIONS

In the chart below, record the absorbance values for 260 and 280 nm.

Wavelength (nm)	Absorbance
260	
280	

1 Calculate the DNA concentration in your sample. Be sure to take any dilutions into account.

2 Determine the probable purity of your sample using the equation:

$$\frac{\text{Absorbance}_{260\text{nm}}}{\text{Absorbance}_{280\text{nm}}}$$

Questions

1 What is the importance of heating the cell lysate?

2 Why does the extraction work better with cold 95% isopropanol than with room-temperature 95% isopropanol? (Think in terms of how a change in temperature can affect physical properties of chemicals).

3 Based on your results, what wavelength gives maximum absorption of DNA? Does this match the accepted wavelength for maximum absorption? If not, suggest reasons for the discrepancy.

EXERCISE 8-2

Ultraviolet Radiation Damage and Repair

Theory

Ultraviolet radiation is part of the electromagnetic spectrum, but with shorter, higher energy wavelengths than visible light. Prolonged exposure can be lethal to cells because when DNA absorbs UV radiation at 254 nm, the energy is used to form new covalent bonds between adjacent pyrimidines: cytosine-cytosine, cytosine-thymine, or thymine-thymine. Collectively, these are known as **pyrimidine dimers**, with thymine dimers being the most common. These dimers distort the DNA molecule and interfere with DNA replication and transcription (Figure 8-4).

Many bacteria have mechanisms to repair such DNA damage. *Escherichia coli* performs **light repair** or **photoreactivation**, in which the repair enzyme, **DNA photolyase**, is activated by visible light (340–400 nm) and simply monomerizes the dimer by reversing the original reaction.

A second *E. coli* repair mechanism, **excision repair** or **dark repair**, involves a number of enzymes (Figure 8-5). The thymine dimer distorts the sugar-phosphate backbone of the strand. This is detected by an **endonuclease** (UvrABC) that breaks two bonds—one, eight nucleotides in the 5' direction from the dimer, and the other four nucleotides in the 3' direction. A **helicase** (UvrD) removes the 13-nucleotide fragment (including the dimer), leaving single-stranded DNA. **DNA polymerase I** inserts the appropriate complementary nucleotides in a 5' to 3' direction to make the molecule double-stranded again.

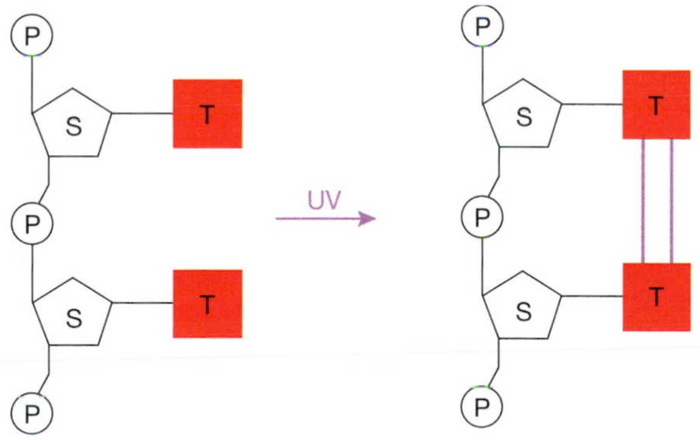

8-4 A Thymine Dimer in One Strand of DNA ◆ Thymine dimers form when DNA absorbs UV radiation with a wavelength of 254 nm. The absorbed energy results in the formation of two new covalent bonds between the thymines, resulting in distortion of the DNA strand. The enzyme DNA photolyase can break this bond to return the DNA to its normal shape and function. If it doesn't, and excision repair fails, the distortion interferes with DNA replication and transcription.

Finally, **DNA ligase** closes the gap between the last nucleotide of the new segment and the first nucleotide of the old DNA, and the repair is complete. Both mechanisms are capable of repairing a small amount of damage, but long and/or intense exposures to UV produce more damage than the cell can repair, making UV radiation lethal.

Application

Because ultraviolet radiation has a lethal effect on bacterial cells, it can be used in decontamination. Its use is limited,

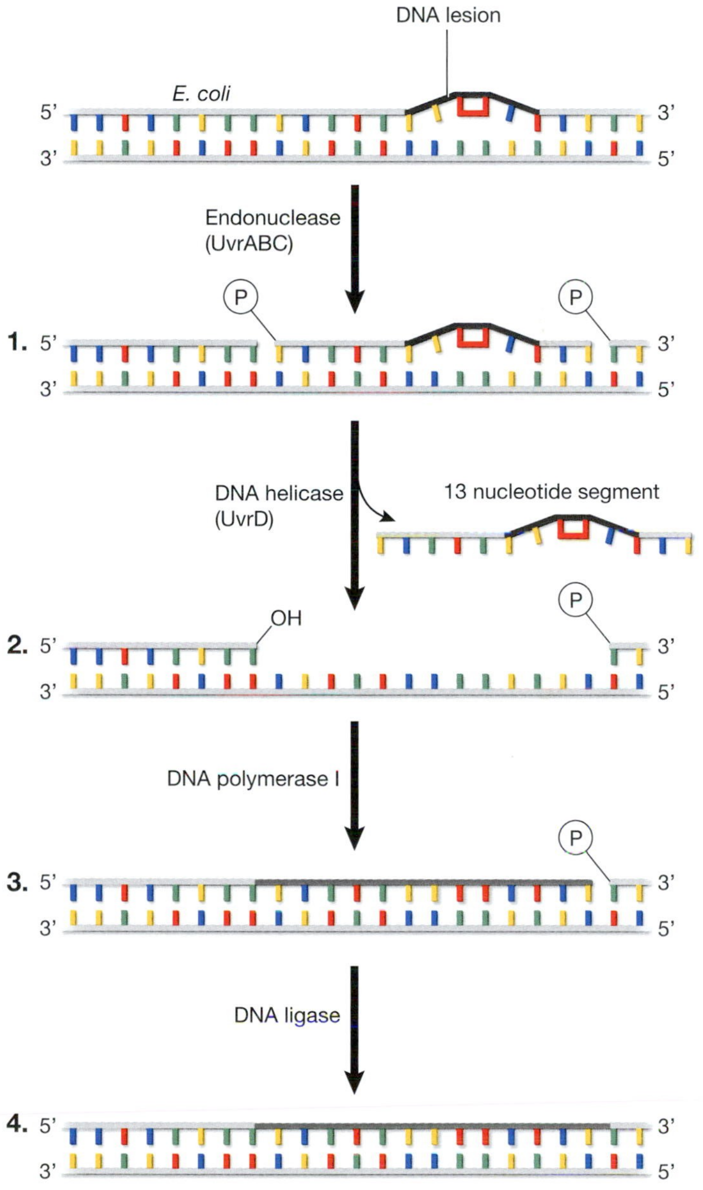

8-5 Excision or Dark Repair in *E. coli* ◆ In the repair process, four enzymes are used: (**1**) An endonuclease (UvrABC) to break two covalent bonds in the sugar-phosphate backbone of the damaged strand, (**2**) a helicase (UvrD) to remove the nucleotides in the damaged segment, (**3**) DNA polymerase I to synthesize a new strand, and (**4**) DNA ligase to form a covalent bond between the new and the original strands.

however, because it penetrates materials such as glass and plastic poorly. In addition, bacterial cells have mechanisms to repair UV damage.

In This Exercise

The lethal effects of ultraviolet radiation, its ability to penetrate various objects, and the cells' ability to repair UV damage will all be demonstrated. Do not be misled by the apparent simplicity of the experimental design—there's a lot going on!

Materials

Per Student Group

- ✓ seven Nutrient Agar plates
- ✓ disinfectant
- ✓ posterboard masks with 1" to 2" cutouts (Figure 8-6)
- ✓ UV light (shielded for eye protection)
- ✓ sterile cotton applicators
- ✓ 24-hour Trypticase Soy broth culture of *Serratia marcescens*

Procedure

Lab One

Refer to the Procedural Diagram in Figure 8-7 as you read and follow this procedure:[1]

1. Using a sterile cotton applicator, streak a Nutrient Agar plate to form a bacterial lawn over the entire surface by using a tight pattern of streaks. Rotate the plate one-third of a turn and repeat, then rotate it another one-third of a turn and streak one last time. Repeat this process for all remaining plates.
2. Number the plates 1, 2, 3, 4, 5, 6, and 7.
3. Remove the lid from plate 1 and set it on a disinfectant-soaked towel. Place the plate under the UV light and cover it with a mask.
4. Turn the UV lamp on, but do not look at it. After 30 seconds, turn off the UV lamp, remove the mask, and replace its lid. If space permits, you may combine this step with Step 5.

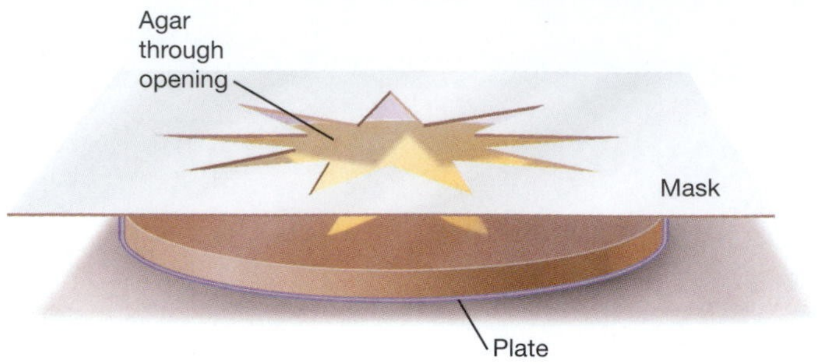

8-6 Posterboard Mask ◆ This is an example of a mask placed over a Petri dish. The cutout may be any shape but should leave the outer 25% of the plate masked.

5. Repeat the process for plate 2. If space permits, you may combine this step with Step 4.
6. Repeat the process for plate 3, but leave the UV lamp on for 3 minutes. If space permits, you may combine this step with Steps 7 and 8.
7. Irradiate plate 4 for 3 minutes, but leave the lid on and cover with the mask.
8. Repeat Step 7 for plate 5. If space permits, you may combine this step with Step 7.
9. Do not irradiate plates 6 and 7.
10. Incubate plates 1, 3, 4, and 6 for 24–48 hours at room temperature in an inverted position where they can receive natural light (e.g., a windowsill) for 24–48 hours. Do not stack them.

Lab Two

1. Examine the growth patterns on all plates. Observe for confluent-dense growth, confluent-sparse growth, individual colonies, or no growth in covered and exposed parts of each plate.
2. Record your results on the Data Sheet.

[1] Thanks to Roberta Pettriess of Wichita State University for her helpful suggestions to improve this exercise.

References

Moat, Albert G., John W. Foster, and Michael P. Spector. 2002. Chapter 3 in *Microbial Physiology,* 4th ed. Wiley-Liss, New York.

Nelson, David L., and Michael M. Cox. 2005. Chapter 25 in *Lehninger's Principles of Biochemistry.* W. H. Freeman and Co., New York.

White, David. 2000. Chapter 19 in *The Physiology and Biochemistry of Prokaryotes.* Oxford University Press, New York.

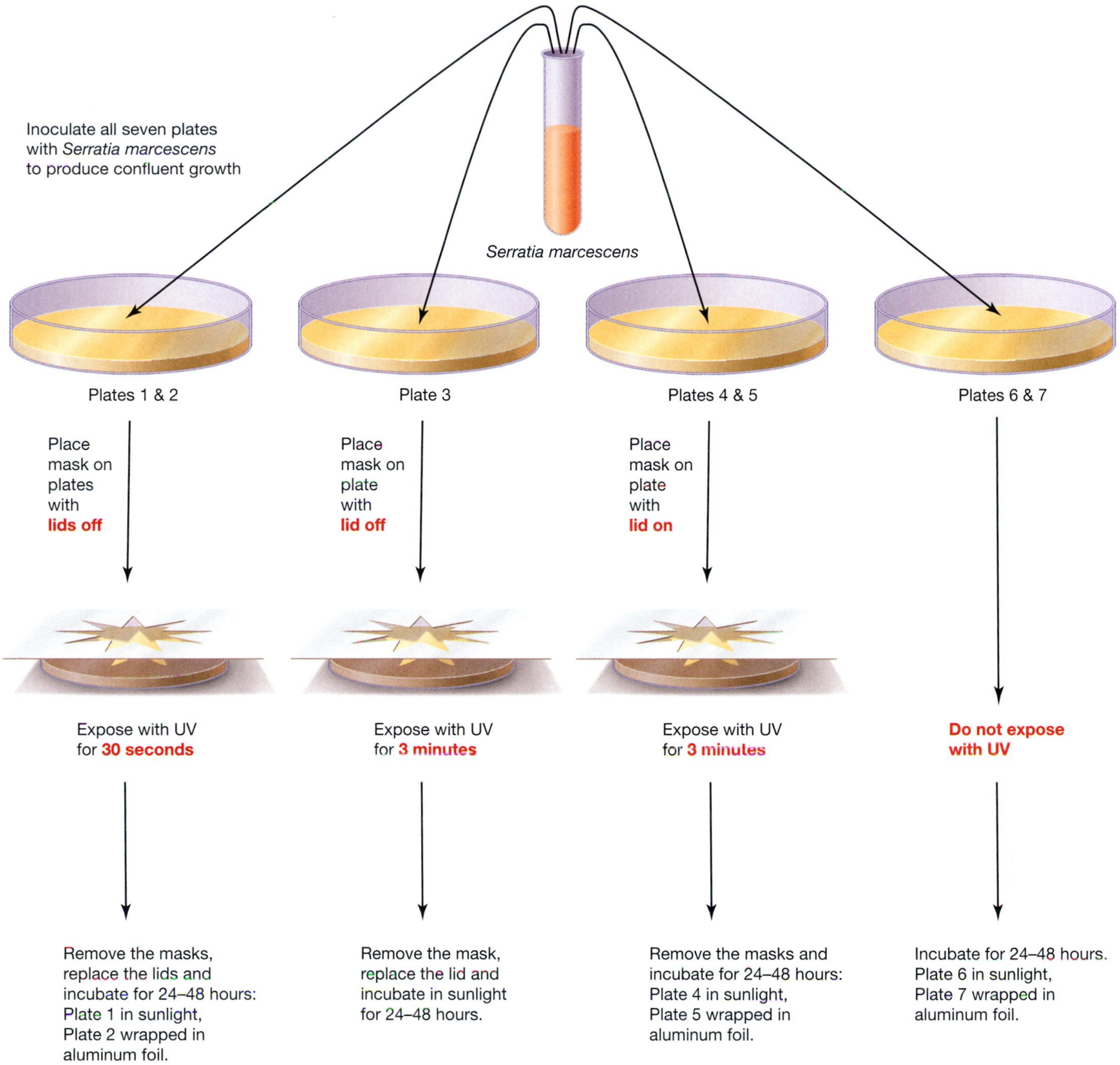

Inoculate all seven plates with *Serratia marcescens* to produce confluent growth
Serratia marcescens
Plates 1 & 2
Plate 3
Plates 4 & 5
Plates 6 & 7
Place mask on plates with **lids off**
Place mask on plate with **lid off**
Place mask on plate with **lid on**
Expose with UV for **30 seconds**
Expose with UV for **3 minutes**
Expose with UV for **3 minutes**
Do not expose with UV
Remove the masks, replace the lids and incubate for 24–48 hours: Plate 1 in sunlight, Plate 2 wrapped in aluminum foil.
Remove the mask, replace the lid and incubate in sunlight for 24–48 hours.
Remove the masks and incubate for 24–48 hours: Plate 4 in sunlight, Plate 5 wrapped in aluminum foil.
Incubate for 24–48 hours. Plate 6 in sunlight, Plate 7 wrapped in aluminum foil.

8-7 Procedural Diagram ◆ Inoculate and expose the plates as directed. Be sure to shield the UV light source adequately. Do not look at the light.

NAME ______________________ DATE ______________________

LAB SECTION ______________________ I WAS PRESENT AND PERFORMED THIS EXERCISE (INITIALS) ________

Ultraviolet Radiation Damage and Repair

OBSERVATIONS AND INTERPRETATIONS

Using the terms "confluent–dense growth," "confluent–sparse growth," "individual colonies," and "none," describe the growth of *Serratia marcescens* in both parts of each plate (i.e., where the mask shielded the growth from UV and where it did not). Record your observations in the chart below. Make a note as to the significance of each type of growth. (For example, unaffected by UV, probable photorepair, etc.)

Plate	Mask/Lid	UV Exposure Time (min.)	Incubation	Growth on Agar Surface Covered by Mask	Growth on Agar Surface Beneath Opening of Mask
1	Mask, no lid	0.5	Sunlight		
2	Mask, no lid	0.5	Dark		
3	Mask, no lid	3.0	Sunlight		
4	Mask and lid	3.0	Sunlight		
5	Mask and lid	3.0	Dark		
6	Lid only	0	Sunlight		
7	Lid only	0	Dark		

Questions

1 Consider the exposed part of the plate.

a. What does relative sparseness of growth tell you about the effect of UV radiation?

b. What do the colonies tell you about the cells from which they grew?

2 What does the relatively heavy growth on the plates covered by the mask and lid during exposure tell you about UV radiation?

3 What is the effect of longer UV exposure on *Serratia marcescens*? To answer this question, which plates should be compared?

4 What is the effect of the posterboard mask on UV radiation? To answer this question, which plates should be examined?

5 What is the effect of the plastic lid on UV radiation? To answer this question, which plates should be compared?

EXERCISE 8-3

Bacterial Transformation: the pGLO™ System

❖ Theory

This exercise utilizes a kit produced by Bio-Rad Laboratories[1] that efficiently illustrates the following principles of microbial genetics:

- bacterial transformation,
- use of an antibiotic selective medium to identify transformed cells, and
- the operon as a mechanism of microbial genetic regulation.

Bacterial **transformation** is the process by which **competent** bacterial cells pick up DNA from the environment and make use of the genes it carries. It was first demonstrated in a strain of pneumococcus (*Streptococcus pneumoniae*) in 1928 by Fred Griffith and since has been found to occur naturally in only certain genera. Modern techniques, however, have allowed biologists to make most cells (including eukaryotic cells) artificially competent, and this has made transformation a useful tool in genetic engineering. You will be using a $CaCl_2$ Transforming Solution and heat shock to make your *Escherichia coli* cells competent.

[1] Catalog Number 166-0003-EDU. Bio-Rad Laboratories Main Office 2000 Alfred Nobel Drive, Hercules, CA 94547; www.bio-rad.com, 1-800-424-6723.

Green fluorescent protein (GFP) is responsible for **bioluminescence** in the jellyfish *Aequorea victoria*. The GFP gene was isolated and altered so the GFP in this experiment fluoresces more than the natural version. You will be introducing the GFP gene into competent *E. coli* cells and will be using the ability of the cell to fluoresce as visual evidence of successful transformation and subsequent gene expression.

Operons are structural and functional genetic units of prokaryotes. Each operon minimally includes a **promoter site** (for binding **RNA polymerase**) and two or more **structural genes** coding for enzymes in the same metabolic pathway. In this exercise you will be using part of the arabinose operon.

The complete arabinose operon (Figure 8-8) is naturally found as part of the bacterial chromosome in some species and consists of the promoter (P_{BAD}) and three structural genes (*ara*B, *ara*A, and *ara*D) that code for enzymes used in arabinose digestion. *In vivo,* the arabinose enzymes are needed only when arabinose is present. After all, there is no point in the cell expending a lot of energy making the enzymes if the substrate is not there. A DNA binding protein called *ara*C attaches to the promoter of the arabinose operon and acts like a switch.

When arabinose is not present, *ara*C prevents RNA polymerase from binding to the promoter, so transcription

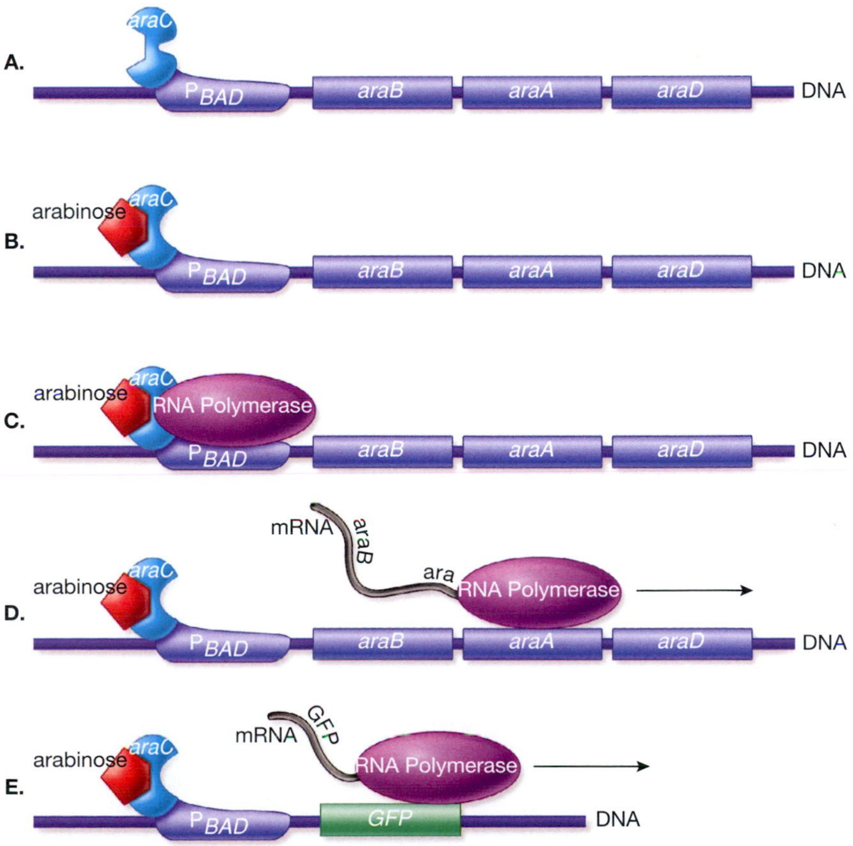

8-8 The Arabinose Operon—Normal and Genetically Engineered ◆ Operons are prokaryotic structural and functional genetic units: They carry genes for enzymes in the same metabolic pathway, and they are regulated together. Operons with genes for catabolic enzymes are transcribed only when the specific substrate is present. In this case, the substrate is the sugar arabinose. Figures A through D show how a normal arabinose operon works. Figure E shows how it has been modified for this experiment. **A** The arabinose operon consists of a promoter (P_{BAD}) and three structural genes (*ara*B, *ara*A, and *ara*D). The DNA binding protein (*ara*C) attaches to the promoter and acts like a switch. Without arabinose present, RNA polymerase is unable to bind to the promoter and begin transcription of the genes. **B** and **C** Arabinose binds to a receptor on *ara*C and causes it to change to a shape that allows binding of RNA Polymerase to the promoter. **D** Transcription of the structural genes occurs, the enzymes are produced, and arabinose is catabolized. **E** The pGLO™ plasmid has been engineered to contain the arabinose promoter and the gene for green fluorescent protein (GFP) instead of the genes for arabinose catabolism. If arabinose is present, the GFP gene will be transcribed. In addition, the pGLO™ plasmid has a replication origin and genes for antibiotic resistance and the DNA binding protein.

can't occur—the switch is "off." When arabinose is present, it binds to *ara*C and changes its shape so RNA polymerase *can* bind to the promoter and transcribe the genes—the switch is "on." Then this sequence of events occurs: The genes are transcribed, the enzymes are synthesized, they catalyze their reactions, and eventually the arabinose is consumed. Now, without arabinose, *ara*C returns to its "off" shape and the genes are no longer transcribed. In this exercise you will be using a modified arabinose operon that has been genetically engineered into the pGLO™ plasmid. It consists of the regulatory portion of the arabinose operon (the arabinose promoter and the *ara*C gene), but the structural genes have been replaced with the GFP gene.

All that remains is a means of carrying the GFP gene into the cell and replicating it—a **vector**. In this exercise you will be using a **plasmid** as a vector: the pGLO™ plasmid. Plasmids are small, naturally occurring, circular DNA molecules that possess only a few genes and replicate independent of the chromosome (because they have their own replication origin). Although they are nonessential, they often carry genes that are beneficial to the bacterium, such as antibiotic resistance. The antibiotic resistance gene (*bla*) used in this experiment produces an enzyme called β-lactamase, which hydrolyzes certain antibiotics, including penicillin and ampicillin.

The bottom line is this: The pGLO™ plasmid used in this experiment has been genetically engineered to contain the arabinose promoter (P_{BAD}), the gene for *ara*C, an antibiotic resistance gene (*bla)*, the gene for green fluorescent protein (GFP), and a replication origin (*ori*).

As you interpret the results of your experiment, you will see how these components come together and provide you with information about what is happening at the molecular level in your *E. coli* culture. Their functions are summarized in Table 8-1.

❖ Application

Introduction of foreign DNA into a cell, identification of transformed cells, and regulation of expression of the introduced genes are skills used in genetic engineering.

❖ In This Exercise

You will first make *E. coli* cells competent. Then, you will transform the competent *E. coli* cells with a plasmid containing the gene for green fluorescent protein. This technique of introducing a plasmid into cells is done routinely in genetic engineering protocols. You also will use arabinose as a genetic switch to regulate expression of the GFP gene by *E. coli*.

❖ Materials

Per Student Group

One Bio-Rad pGLO™ Transformation kit contains enough material for eight student workstations. Each workstation requires the following:

- ✓ one Luria-Bertani (LB) Agar plate of *E. coli*
- ✓ one sterile LB plate

TABLE 8-1 Cast of Characters and a Legend of Abbreviations ◆

Name	Symbol	Function in This Experiment
Green fluorescent protein	GFP	Used an indicator of successful transformation and gene transcription (expression) in this experiment.
Plasmid	pGLO™	Used as a vector to introduce the GFP gene into recipient *E. coli* cells.
Arabinose promoter	P_{BAD}	In this experiment it is the attachment site for RNA polymerase during transcription of the GFP gene on the pGLO™ plasmid. Normally, it is the attachment site for RNA polymerase during transcription of the *ara*B, *ara*A, and *ara*D genes.
DNA binding protein	*ara*C	It is a regulatory molecule for the arabinose promoter. In the presence of arabinose, *ara*C has a shape that allows RNA polymerase to bind to the promoter. Without arabinose, *ara*C's shape prevents RNA polymerase from binding to the promoter.
Antibiotic resistance gene	*bla*	It produces β-lactamase, an enzyme that hydrolyzes certain antibiotics, including ampicillin. The gene provides a means of differentiating cells that were transformed and those that were not.
Replication origin	*ori*	A replication origin is necessary for DNA replication. The pGLO™ plasmid is capable of replicating inside the cell because it has this. As a result of replication, copies of the plasmid can be distributed to the descendants of the original, transformed *E. coli* cell.

✓ two sterile LB + ampicillin plates
✓ one sterile LB + ampicillin + arabinose plate
✓ one microtube of $CaCl_2$ transformation solution
✓ two sterile microtubes
✓ one microtube of LB Broth
✓ seven disposable inoculating loops
✓ five disposable calibrated transfer pipettes
✓ one foam microtube holder/float
✓ one container of crushed ice
✓ one marking pen

Per Class

In addition, a class supply of the following is required:
✓ hydrated pGLO™ plasmid
✓ 42°C water bath and thermometer
✓ long-wave UV lamp

Procedure

Lab One

Refer to the procedural diagram in Figure 8-9 as you read and perform the following procedure.

1 Obtain two closed microtubes. Label them with your group's name, then label one "+DNA" and the other "−DNA." Put both tubes in the microtube holder/float.

2 Using a sterile calibrated transfer pipette, dispense 250 µL of $CaCl_2$ transformation solution into each tube. Close the caps. (The calibration marks are shown in Figure 8-10.)

3 Return the two tubes to the microtube holder/float and place them in the ice bath.

4 With a sterile loop, transfer one entire *E. coli* colony into the +DNA tube. Agitate the loop until all the growth is off of it and the cells are dispersed uniformly in the transformation solution. Close the lid and properly dispose of the loop.

5 Repeat Step 4 with a sterile loop and the −DNA tube.

6 Hold the UV light next to the vial of pGLO™ plasmid solution. Record your observation on the Data Sheet.

7 Using a sterile loop, remove a loopful of pGLO™ plasmid DNA solution. Be sure there is a film across the loop. Then transfer the loopful of plasmid solution to the +DNA tube and mix.

8 Leave the tubes on ice. Make sure they are far enough down in the microtube holder/float that they make good contact with the ice. Leave them on ice for 10 minutes.

9 As the tubes are cooling on ice for 10 minutes, label the four LB Agar plates.
- Label one LB/amp plate: +DNA
- Label the LB/amp/ara plate: +DNA
- Label the other LB/amp plate: −DNA
- Label the LB plate: −DNA

10 After 10 minutes on ice, transfer the microtube holder/float (still containing both tubes) to the 42°C water bath for exactly 50 seconds. This transfer from ice to warm water must be done rapidly. Also, make sure the tubes make good contact with the water.

11 After 50 seconds, quickly place both tubes back in the ice for 2 minutes. This process of heat shock makes the cell membranes more permeable to DNA; that is, you have made them "competent." (Timing is critical. According to Bio-Rad Laboratories, 50 seconds is optimal. No heat shock results in a 90% *reduction* in transformants, whereas a 90 second heat shock yields about half the transformants.)

12 After 2 minutes, remove the microtube holder/float from the ice and place it on the table.

13 Using a sterile calibrated transfer pipette, add 250 µL of LB broth to the +DNA tube. Repeat with another sterile pipette and the −DNA tube. Properly dispose of both pipettes.

14 Incubate both tubes for 10 minutes at room temperature. Then mix the tubes by tapping them with your fingers.

15 Using a different sterile calibrated transfer pipette for each, inoculate the "LB/amp/+DNA" plate and the "LB/amp/ara/+DNA" plate with 100 µL from the +DNA tube.

16 Using a different sterile calibrated transfer pipette for each, inoculate the "LB/amp/−DNA" plate and the "LB/−DNA" plate with 100 µL from the −DNA tube.

17 Using a different sterile loop for each, spread the inoculum over the entire surface of all four plates to get confluent growth. Use the "face" of the loop for this, not its edge. Properly dispose of the loops.

18 Tape the plates together in a stack so they face the same direction. Then label them with your group name and the date, and incubate them for 24 hours at 37°C in an inverted position.

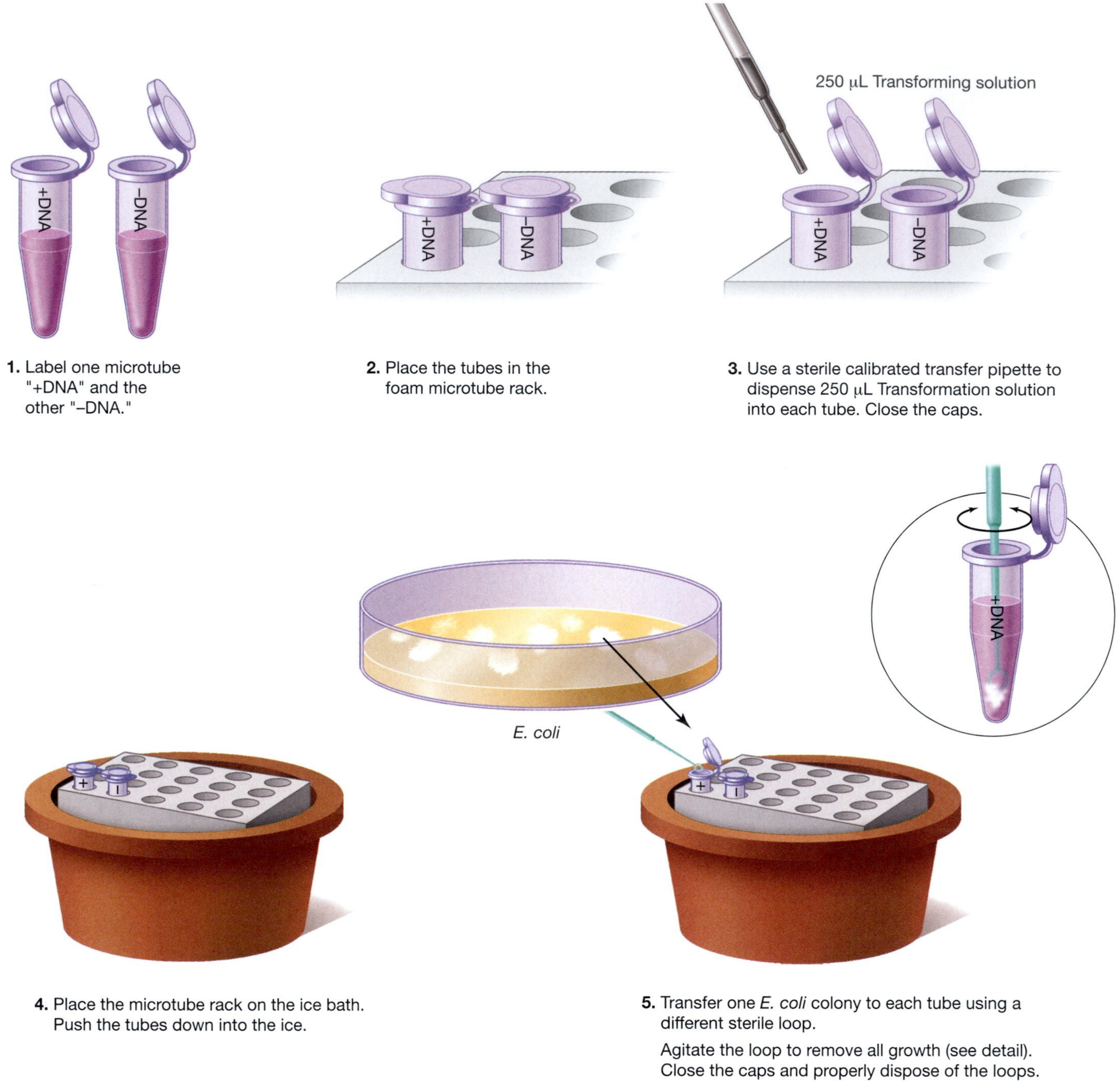
250 µL Transforming solution
+DNA
–DNA
E. coli
+
–
1. Label one microtube "+DNA" and the other "–DNA."
2. Place the tubes in the foam microtube rack.
3. Use a sterile calibrated transfer pipette to dispense 250 µL Transformation solution into each tube. Close the caps.
4. Place the microtube rack on the ice bath. Push the tubes down into the ice.
5. Transfer one E. coli colony to each tube using a different sterile loop.
Agitate the loop to remove all growth (see detail). Close the caps and properly dispose of the loops.

8-9 pGLO™ Transformation Procedural Diagram ◆ Be sure to dispose of all pipettes and loops properly. *(Continued on next page.)*

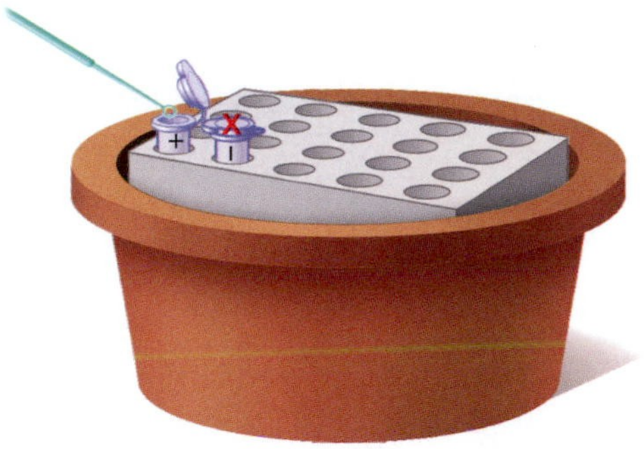
+
–

6. Transfer one loopful of pGLO™ plasmid DNA to the "+DNA" tube only.

Continue icing both tubes for 10 minutes.

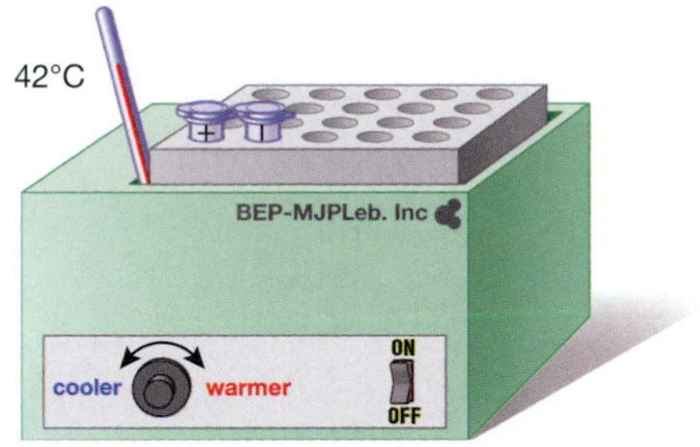
42°C
BEP-MJPLeb. Inc
cooler
warmer
ON
OFF

7. Quickly transfer the entire microtube rack to the 42°C water bath for exactly 50 seconds. Make sure the tubes contact the water.

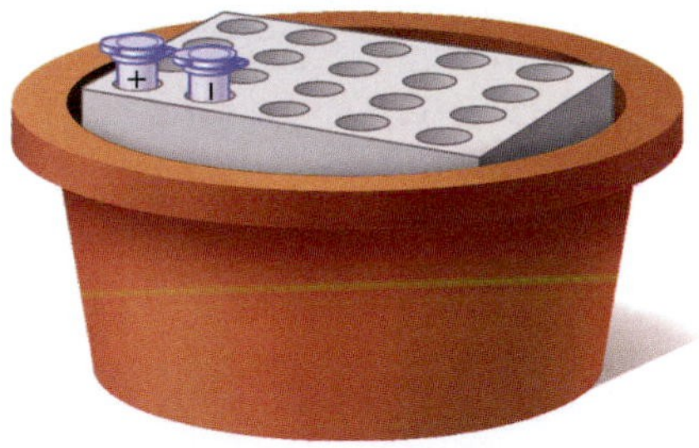

8. Quickly place the microtube rack back on ice for two minutes.

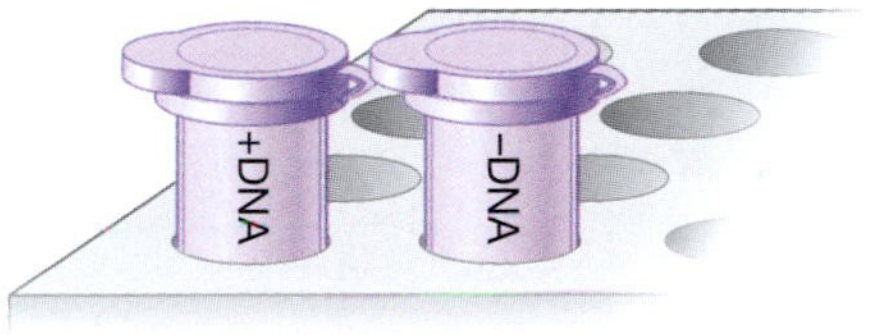
+DNA
–DNA

9. Place the microtube rack on the table.

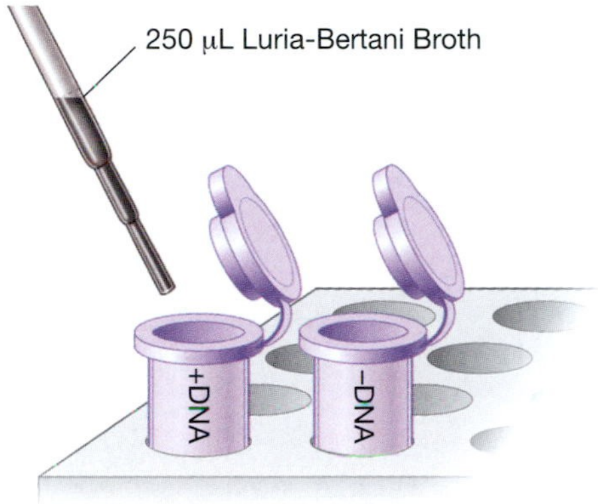
250 μL Luria-Bertani Broth
+DNA
–DNA

10. Add 250 μL LB broth to each tube with a different sterile transfer pipette.

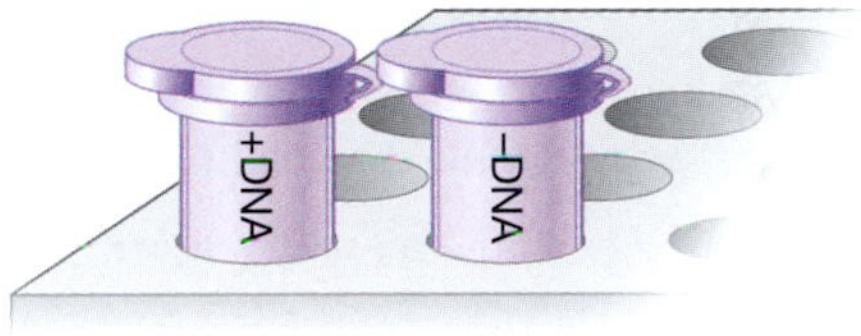
+DNA
–DNA

11. Incubate the tubes for 10 minutes at room temperature.

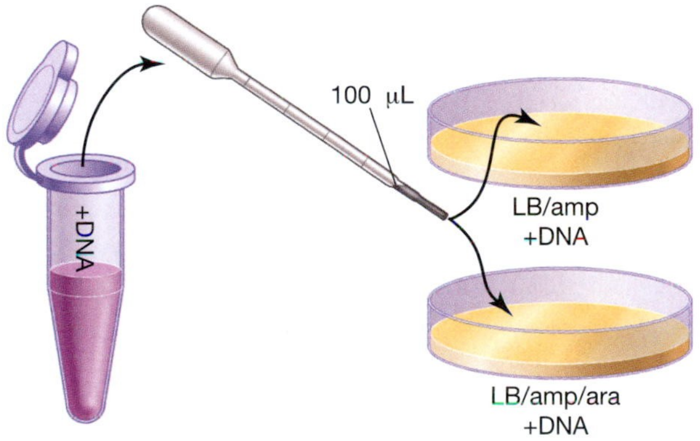
+DNA
100 μL
LB/amp
+DNA
LB/amp/ara
+DNA

12. Using a different pipette for each, transfer 100 μL of +DNA to an LB/amp and an LB/amp/ara plate. Using a different sterile loop for each, spread the inoculum over the entire agar surface. Use the "face" of the loop for this, not its edge. Incubate for 24 hours at 37°C.

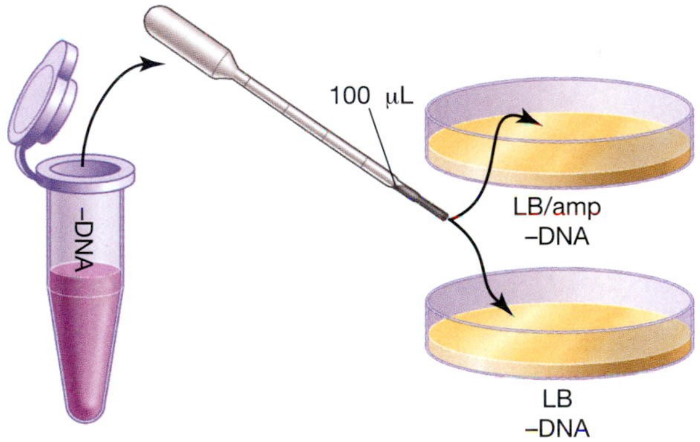
–DNA
100 μL
LB/amp
–DNA
LB
–DNA

13. Using a different pipette for each, transfer 100 μL of –DNA to an LB/amp and an LB plate. Using a different sterile loop for each, spread the inoculum over the entire agar surface. Use the "face" of the loop for this, not its edge. Incubate for 24 hours at 37°C.

8-9 pGLO™ Transformation Procedural Diagram *(continued)* ◆ Be sure to dispose of all pipettes and loops properly.

Lab Two

1 Retrieve your plates. Observe them in ambient room light, then in the dark with UV illumination. Record your observations on the Data Sheet, and answer the questions.

2 When finished, properly dispose of all plates.

Reference

Bio-Rad Laboratories. Instruction pamphlet for the *Bacterial Transformation—The pGLO™ System* kit (Catalog Number 166- 0003-EDU). Bio-Rad Laboratories, Hercules, CA.

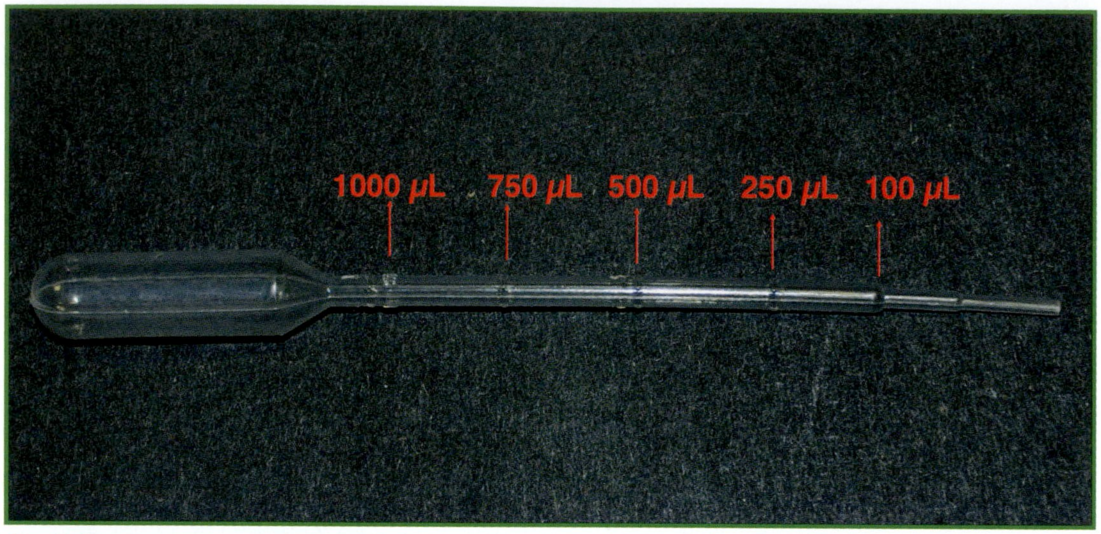

8-10 Disposable Transfer Pipette Calibrations ◆ The transfer pipettes supplied with the pGLO™ transformation kit are calibrated to deliver specific volumes. The volumes and their corresponding lines are shown here.

NAME ______________________ DATE ______________________

LAB SECTION ______________________ I WAS PRESENT AND PERFORMED THIS EXERCISE (INITIALS) ________

DATA SHEET 8-3

Bacterial Transformation: the pGLO™ System

OBSERVATIONS AND INTERPRETATIONS

1 Record the appearance of the pGLO™ plasmid solution with and without UV illumination.

2 In the chart below, record your observations of the plates after incubation.

Plate	Inoculum	Number of Colonies	Appearance in Ambient Light	Appearance with UV Light
LB	*E. coli* − DNA			
LB/amp	*E. coli* − DNA			
LB/amp	*E. coli* + DNA			
LB/amp/ara	*E. coli* + DNA			

Questions

1 In the chart below, fill in the genotype of *E. coli* prior to transformation and after transformation.

E. coli	GFP Gene (GFP* or GFP⁻)	*Bla* (*Bla** or *Bla*⁻)
Before transformation		
After transformation		

2 What was the purpose of examining the original pGLO™ solution with and without UV illumination?

3 What was the purpose of transferring the +DNA and −DNA tubes from ice, to hot water, to ice again?

4 Why were the vials incubated for 10 minutes in LB broth rather than transferring their contents directly to the plates?

5 What information is provided by the LB/−DNA plate?

6 Obviously, transformation could occur only if the pGLO™ plasmid was introduced into the solution. Which plate(s) exhibit transformation?

7 What information is provided by the LB/amp/−DNA plate?

8 Why does the LB/amp/ara/+DNA plate fluoresce when the LB/amp/+DNA plate does not?

9 Use the following information to calculate transformation efficiency.

a. You put 10 µL (one loopful—the loop is calibrated for this volume) of a 0.03 µg/µL pGLO™ solution into the +DNA tube. Calculate the µg of DNA you put into the tube.

b. The +DNA tube contained 510 µL of solution (10 µL of plasmid, 250 µL of Transforming Solution, and 250 µL of LB broth) prior to plating, but you did not use all of it. Calculate the fraction of the +DNA solution you used for each plate.

c. Use your answers to questions 9a and 9b to calculate the micrograms of DNA you plated.

d. Using the number of colonies on the LB/amp/ara/+DNA plate, calculate the transformation efficiency. (Hint: The units are transformants/µg of pGLO™ DNA.)

e. According to the manual supplied by Bio-Rad Laboratories, this protocol should yield a transformation efficiency between 8.0×10^2 and 7.0×10^3 transformed cells per microgram of pGLO™ DNA. How does your transformation efficiency compare? Account for any discrepancy.

EXERCISE 8-4

Slide Agglutination

Theory

Particulate antigens (such as whole cells) may combine with homologous antibodies to form visible clumps called **agglutinates. Agglutination** thus serves as evidence of antigen–antibody reaction and is considered a positive result. Agglutination reactions are highly sensitive and may be used to detect either the presence of antigen or antibody in a sample.

There are many variations of agglutination tests (Figure 8-11). **Direct agglutination** relies on the combination of antibodies and naturally particulate antigens. **Indirect agglutination** relies on artificially constructed systems in which agglutination will occur. These involve coating particles (such as red blood cells [RBCs] or latex microspheres) with either antibody or antigen, depending on what is being looked for in the sample. Addition of the homologous antigen or antibody then will result in clumping of the artificially constructed particles.

Application

Agglutination reactions may be used to detect the presence of either antigen or antibody in a sample. Direct agglutination reactions are used to diagnose some diseases, determine if a patient has been exposed to a certain pathogen, and are involved in blood typing. Indirect agglutination is used in some pregnancy tests as well as in diagnosing disease.

In This Exercise

You will perform a simple direct agglutination test to identify the presence of *Salmonella* H antigen in a sample. You will compare this reaction to a reaction between the nonhomologous *Salmonella* anti-H antiserum and the *Salmonella* O antigen.

Materials

Per Student Group

- ✓ one clean microscope slide
- ✓ two toothpicks
- ✓ marking pen
- ✓ *Salmonella* H antigen
- ✓ *Salmonella* O antigen
- ✓ *Salmonella* anti-H antiserum

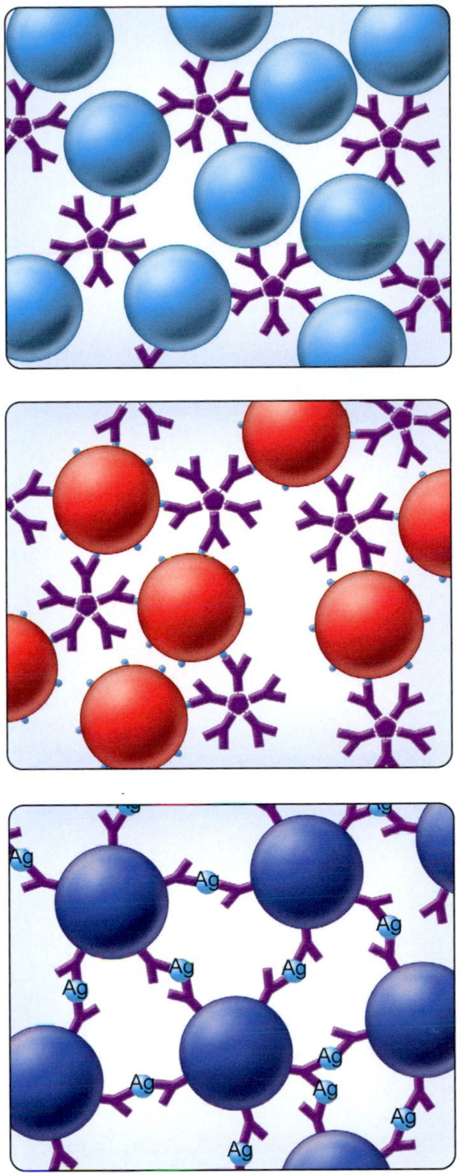

Direct agglutination occurs with naturally particulate antigens. Either antigen (blue spheres) or antibody (purple) may be detected in a sample using this style of agglutination test.

Detection of antibody in a sample can be done by indirect agglutination. A test solution is prepared by artificially attaching homologous antigen (blue) to a particle (red) such as red blood cells or latex beads and mixing with the sample suspected of containing the antibody (purple).

Another style of indirect agglutination detects antigen (light blue) in a sample. Antibodies (purple) are artificially attached to particles (dark blue), which are then mixed with the sample suspected of containing the antigen.

8-11 Direct and Indirect Agglutination Tests ❖ Direct agglutinations involve naturally particulate antigens. Indirect agglutination tests are artificially constructed systems that take advantage of the sensitivity of the agglutination reaction. Either the antigen or antibody is coated onto a particle, such as a latex bead or red blood cell, and allowed to react with the sample suspected of containing the antigen/antibody.

Procedure

1 Using a marking pen, draw two circles approximately the size of a dime on a microscope slide (Figure 8-12). Label one "O" and the other "H."

2 Place a drop of *Salmonella* anti-H antiserum in each circle.

3 Place a drop of *Salmonella* O antigen in the "O" circle and a drop of *Salmonella* H antigen in the "H" circle (Figure 8-13). **Be careful not to touch the dropper to the antiserum already on the slide.**

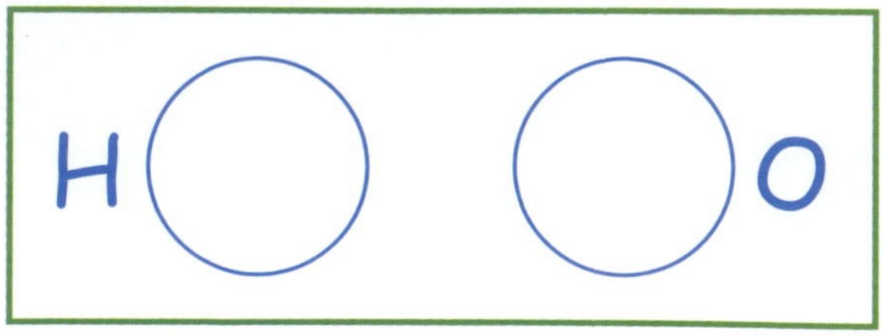

8-12 Prepare the Slide ◆ With your marking pen, draw two dime-sized circles on the slide. Label one circle "H" and the other "O." The circles will be where you check for the presence of *Salmonella* H and O antigens, respectively.

4 Using a *different* toothpick for each circle, mix until each of the antigen solutions is completely emulsified with the antiserum. **Do not over-mix.** Discard the toothpicks in a sharps container.

5 Gently rock this slide for a few minutes and observe for agglutination. (**When rocking, the fluid should not spread. If it does, you are rocking too hard.**) Record your observations on the Data Sheet.

❖ Alternative Test Procedure

Your instructor will cover the labels of the two *Salmonella* antigen bottles. You will use this procedure to identify which one contains the *Salmonella* H antigen.

1 Using a marking pen, draw two circles approximately the size of a dime on a microscope slide (Figure 8-12).

2 Place a drop of one *Salmonella* unknown antigen in one circle and a drop of the other *Salmonella* unknown antigen in the other circle (Figure 8-13).

3 Place a drop of *Salmonella* anti-H antiserum in each circle. **Be careful not to touch the dropper to the antigen solutions already on the slide.**

4 Using a *different* toothpick for each circle, mix until each of the antigen solutions is completely emulsified with the antiserum. **Do not over-mix.** Discard the toothpicks in a biohazard container.

5 Gently rock this slide for a few minutes, then observe for agglutination. (**When rocking, the fluid should not spread. If it does, you are rocking too hard.**) Record your observations on the Data Sheet.

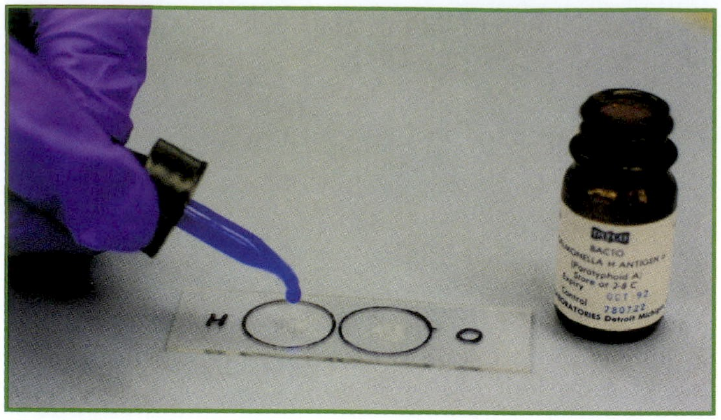

8-13 Adding the Antigen ◆ Carefully add the H and O antigens to the drops of Anti-H antiserum already on the slide. **Do not touch the dropper to the antiserum.**

References

Bopp, Cheryl A., Frances W. Brenner, Joy G. Wells, and Nancy A. Strockbine. 1999. Pages 467–471 in *Manual of Clinical Microbiology*, 7th ed., edited by Patrick R. Murray, Ellen Jo Baron, Michael A. Pfaller, Fred C. Tenover, and Robert H. Yolken. American Society for Microbiology, Washington, DC.

Collins, C. H., Patricia M. Lyne, J. M. Grange. 1995. Page 118 in *Collins and Lyne's Microbiological Methods*, 7th ed. Butterworth-Heinemann, Oxford, United Kingdom.

Constantine, Niel T., and Dolores P. Lana. 2003. Pages 222–223 in *Manual of Clinical Microbiology,* 8th ed., edited by Patrick R. Murray, Ellen Jo Baron, James H. Jorgensen, Michael A. Pfaller, and Robert H. Yolken. American Society for Microbiology, Washington, DC.

Forbes, Betty A., Daniel F. Sahm, and Alice S. Weissfeld. 2002. Pages 206–207 in *Bailey & Scott's Diagnostic Microbiology,* 11th ed. Mosby, St. Louis.

Lam, Joseph S., and Lucy M. Mutharia. 1994. Page 120 in *Methods for General and Molecular Bacteriology*, edited by Philipp Gerhardt, R. G. E. Murray, Willis A. Wood, and Noel R. Krieg. American Society for Microbiology, Washington, DC.

NAME ______________________ DATE ______________________

LAB SECTION ______________________ I WAS PRESENT AND PERFORMED THIS EXERCISE (INITIALS) ________

Slide Agglutination

OBSERVATIONS AND INTERPRETATIONS

Sketch and label your results in the diagram below.

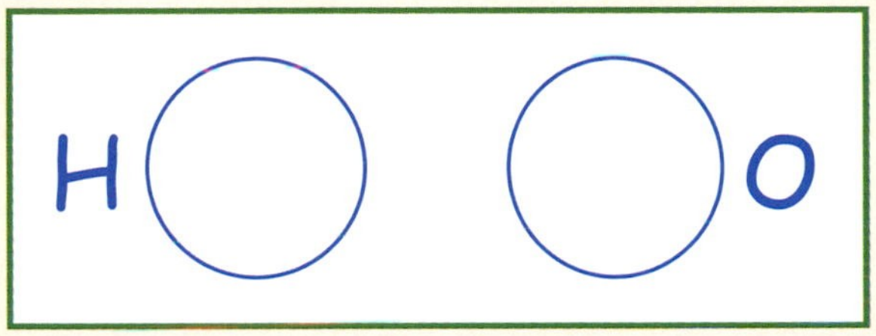

Questions

1 What is the purpose of performing this experiment with both H and O antigens?

2 Suppose you performed this test and got agglutination in both samples. Eliminating contamination of the antigen and antiserum samples as a possibility, provide an explanation of this hypothetical result. Is this attributable to poor sensitivity or specificity of the test system?

3 Suppose you performed this test and neither sample produced agglutination even though you know the antigens and antibodies should react. Would this be attributable to poor sensitivity or specificity of the test system?

4 Vertebrate animals (especially birds and mammals) produce antibodies; *Salmonella* is a bacterium. Given these facts, how can *Salmonella* antiserum (containing antibodies) be produced?

EXERCISE 8-5

Blood Typing

Theory

Hemagglutination is a general term applied to any agglutination test in which clumping of red blood cells (RBCs) indicates a positive reaction. Blood tests as well as a number of indirect diagnostic serological tests are hemagglutinations.

The most common form of blood typing detects the presence and absence of **A** and/or **B antigens** on the surface of red blood cells. A person's ABO blood type is genetically determined. An individual with type A blood has RBCs with the A antigen and produces anti-B antibodies. Conversely, an individual with type B blood has RBCs with the B antigen and produces anti-A antibodies. People with type AB blood have *both* A and B antigens on their RBCs and lack anti-A and anti-B antibodies. Type O individuals lack A and B antigens but produce *both* anti-A and anti-B antibodies.

ABO blood type is ascertained by adding a patient's blood separately to anti-A and anti-B antiserum and observing any signs of agglutination (Table 8-2 and Figure 8-14). Agglutination with anti-A antiserum indicates the presence of the A antigen and type A blood. Agglutination with anti-B antiserum indicates the presence of the B antigen and type B blood. If both agglutinate, the individual has type AB blood; lack of agglutination occurs in individuals with type O blood.

A similar test is used to determine the presence or absence of the **Rh factor** (antigen). If clumping of the patient's blood occurs when mixed with anti-Rh antiserum (also known as anti-D), the patient is Rh positive.

Application

Blood typing is a simple example of an agglutination test using blood. Clinically, it is used for cross-matching donor and recipient blood prior to transfusion.

In This Exercise

You will determine your blood type with respect to two different markers: the ABO group and the Rh factor. Then, you will compile class data and compare it to blood type frequencies in the United States (Table 8-3).

TABLE 8-2 Interpreting Blood Types

Reaction With				
Anti-A Antiserum	Anti-B Antiserum	Anti-Rh Antiserum	Interpretation	Symbol
Agglutination	No Agglutination	Agglutination	A antigen present B antigen absent Rh antigen present	A+
Agglutination	No Agglutination	No Agglutination	A antigen present B antigen absent Rh antigen absent	A−
No Agglutination	Agglutination	Agglutination	A antigen absent B antigen present Rh antigen present	B+
No Agglutination	Agglutination	No Agglutination	A antigen absent B antigen present Rh antigen absent	B−
Agglutination	Agglutination	Agglutination	A and B antigens present Rh antigen present	AB+
Agglutination	Agglutination	No Agglutination	A and B antigens present Rh antigen absent	AB−
No Agglutination	No Agglutination	Agglutination	A and B antigens absent Rh antigen present	O+
No Agglutination	No Agglutination	No Agglutination	A and B antigens absent Rh antigen absent	O−

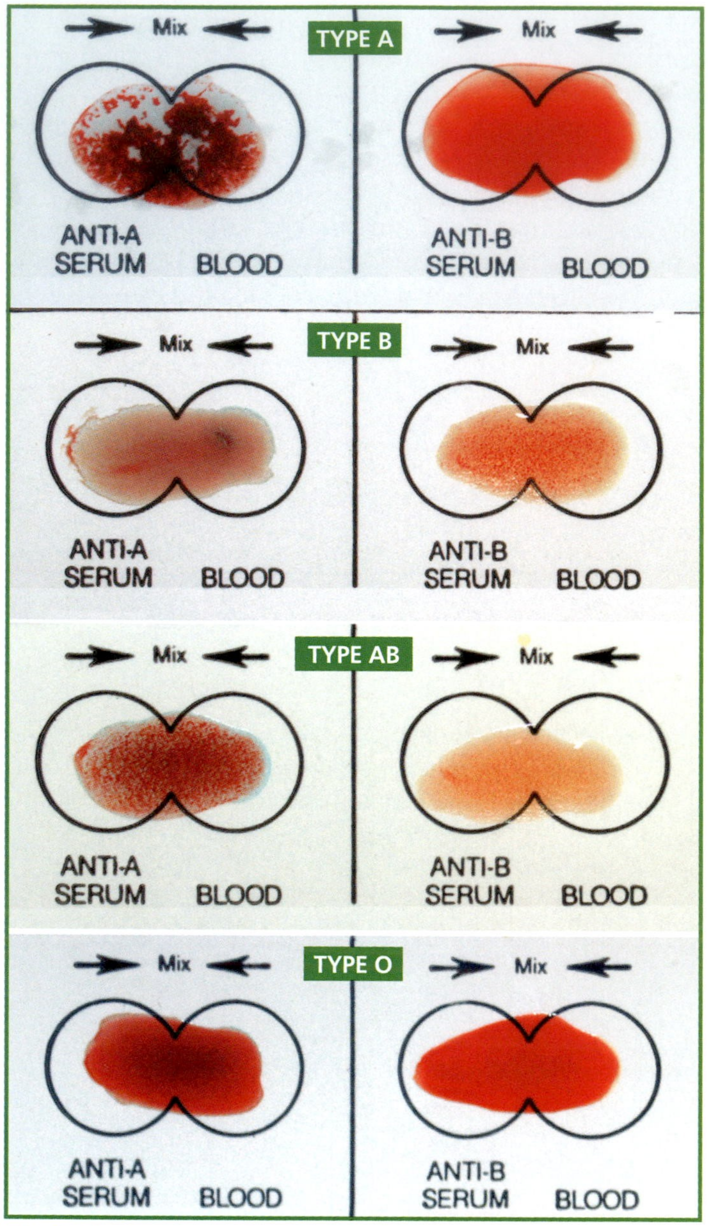

8-14 ABO Blood Groups ◆ Blood typing relies on agglutination of RBCs by Anti-A and/or Anti-B antisera. The blood types are as shown.

Materials

Per Student Group

- ✓ blood typing anti-A antiserum
- ✓ blood typing anti-B antiserum
- ✓ blood typing anti-Rh (anti-D) antiserum
- ✓ two microscope slides
- ✓ three toothpicks
- ✓ marking pen
- ✓ sterile lancets
- ✓ alcohol wipes
- ✓ small adhesive bandages
- ✓ sharps container
- ✓ disposable gloves

Procedure

1. With your marking pen, draw two circles about the size of a dime on one microscope slide. Label one circle "A" and the other "B" (Figure 8-15A).
2. Draw a single circle about the size of a dime with your marking pen in the center of a second microscope slide. Label it "Rh" (Figure 8-15B).
3. Place a drop of anti-A antiserum in the "A" circle.
4. Place a drop of anti-B antiserum in the "B" circle.
5. On the "Rh" microscope slide, place a drop of anti-Rh antiserum.
6. Clean the tip of your finger with an alcohol wipe. (The ring finger of your nondominant hand is a good choice.) Let the alcohol dry.
7. Shake your hand and "milk" blood down to the end of the finger you are going to prick.
8. Open a lancet package and remove the lancet, being careful not to touch the tip before you use it.
9. Prick the end of your finger and immediately place a drop of blood *beside* each drop of antiserum on both slides. Do not touch the antisera with your finger. It's okay to have someone else prick your finger, but make sure he or she wears protective gloves.
10. Discard the lancet in the sharps container.
11. Put an adhesive bandage on your wound.
12. Using a circular motion, mix each pair of drops with a toothpick. Be sure to *use a different toothpick* for each antiserum. Dispose of the toothpicks in the sharps container.
13. Gently rock the slides back and forth for a few minutes or until agglutination occurs.
14. After the agglutination reaction is complete, record the results in the table provided on the Data Sheet. Compare your results with the possible results in Table 8-2 to determine your blood type.
15. Collect class data and record these in the table provided on the Data Sheet. Compare class data with U.S. population data in Table 8-3.

TABLE 8-3 Approximate Percentages of Blood Types in the United States

Blood Type	White	African American	Hispanic	Asian
O+	37	47	53	39
O−	8	4	4	1
A+	33	24	29	27
A−	7	2	2	0.5
B+	9	18	9	25
B−	2	1	1	0.4
AB+	3	4	2	7
AB−	1	0.3	0.2	0.1

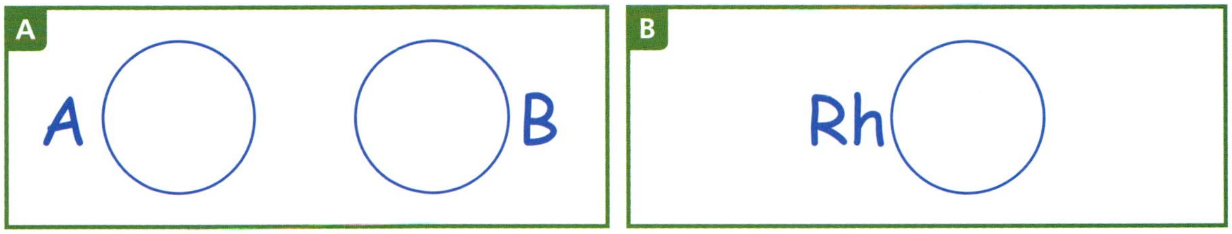

8-15 Prepare the Slides ◆ **A** Draw two circles on a slide and label them "A" and "B" for the type of antiserum each will receive. **B** Draw one circle on a second slide and label it "Rh."

Reference

American Red Cross. http://www.redcrossblood.org

NAME ______________________ DATE ______________________

LAB SECTION ______________________ I WAS PRESENT AND PERFORMED THIS EXERCISE (INITIALS) ________

Blood Typing

OBSERVATIONS AND INTERPRETATIONS

Record your results below.

Antiserum	Agglutination (+/−)
Anti-A	
Anti-B	
Anti-Rh (Anti-D)	

My blood type is: ______________________

Record class data in the chart below.

Blood Type	Percentage in U.S. Population	Number in Class	Percentage in Class	Deviation from National Values
O+	37.4			
O−	6.6			
A+	35.7			
A−	6.3			
B+	8.5			
B−	1.5			
AB+	3.4			
AB−	0.6			

Source: Stanford School of Medicine. Copyright © 2011.

Questions

1 Examine the blood type data obtained from your class. Attempt to explain any deviations from the national values.

2 People with Type "O" are said to be "Universal Donors" in blood transfusions. Explain the reasoning behind this. Which blood type is designated as the "Universal Recipient"?

3 What biological fact makes the concepts of "Universal Donor" and "Universal Recipient" potentially misleading?

4 Maternal-fetal Rh incompatibility (where the Rh− mother's Anti-Rh antibodies destroy the fetus's Rh+ red blood cells) is a well-known phenomenon. Less well known are situations of maternal-fetal ABO incompatibility. Suggest combinations of maternal and fetal blood types that could lead to this situation.

5 Why are red blood cells used in many indirect agglutination tests?

EXERCISE 8-6

ELISA For Detecting Antibodies in a Patient's Sample[1] (The Antibody Capture Method)

Theory

ELISA is an acronym for **Enzyme Linked Immunosorbant Assay**. As with other serological tests, ELISAs can be used to detect antigen in a sample or antibody in a sample. If antigen in a sample is to be identified, you must start with pure antibody to that antigen (homologous antibody). If the goal is to identify antibody in the sample, you must start with the homologous antigen. In either case, a secondary antibody with an attached, or **conjugated**, enzyme as an indicator of antigen-antibody reaction is used.

In this exercise, you will be looking for the **presence of antibody in a sample made to simulate patient's serum.** Follow along with Figure 8-18 (pages 468–471) as the general process is described. In this version of ELISA, the **purified antigen** (PA) homologous to the antibody being tested for is added to the wells of a microtiter strip. The antigen will **adsorb** (attach) to the plastic wall of the wells. After a wash with buffer to remove unbound antigen, a **blocking agent** (such as gelatin or other proteins) may be used to coat the well where antigens have not. This prevents antibodies used in subsequent steps from attaching to the well. Following another wash step, the "patient's serum" is added. If the antibodies are present, they will bind specifically to the purified antigen coating the well. Another wash step removes unbound antibody and equally importantly, other components in the patient's serum.

At this point, the test is either positive or negative, but we can't see it. What is required is an indicator, and that is where the enzyme-linked (**secondary** or **conjugate**) antibody comes in. This antibody is an **immunoglobin antibody**—its antigen is actually an antibody! The first step in the indicator reaction is adding secondary antibodies to the wells. They will bind to the patient's antibodies in the sample, if present. After allowing time for the secondary antibodies to react with their antigen (the patient's antibody), the well is washed again with buffer to remove any unbound secondary antibodies.

The second step in the indicator reaction is to use the enzyme on the secondary antibody to indicate its presence. The specific enzyme chosen for use in an ELISA is only important in that it catalyzes a reaction that can be detected easily. A commonly used enzyme is **horseradish peroxidase** (HRP). HRP catalyzes the conversion of hydrogen peroxide to water and oxygen ($2H_2O_2 \rightarrow O_2 + 2H_2O$). In this exercise, the hydrogen peroxide substrate is associated with a **chromogenic cosubstrate** (3,3',5,5'–tetramethylbenzidene –TMB). It turns a blue color when it becomes oxidized by the O_2 produced by peroxidase activity, indicating a positive result; a negative result shows no color change because the secondary antibody is not present and the peroxide is not broken down (Figure 8-16).

[1] Adapted from ELISA Immuno Explorer™ Kit (Catalog #166-2400EDU), Biotechnology Explorer™ instruction manual, Rev. D. Bio-Rad Laboratories, Life Science Education. 1-800-4-BIORAD (800-424-6723), www.explorer.bio-rad.com

Application

ELISA may be used to detect the presence and amount of either antigen or antibody in a sample. In an antibody capture ELISA, the test is used to screen patients for the presence of HIV antibodies (4–8 weeks after exposure), rubella virus antibodies, West Nile virus antibodies, and others. An antigen capture ELISA is used to detect hormones (such as HCG in some pregnancy tests and LH in ovulation tests), drugs, and viral (e.g., HIV p24 antigen within 4–8 weeks after exposure) and bacterial antigens.

In This Exercise

You will perform an ELISA to determine presence of an antibody in a "patient's serum" sample.

Materials Needed for this Exercise

Per Student Group

(The kit is designed for 12 groups of 4 students each)

- ✓ 4 "patient serum" test samples, each with 250 μL (one per student, with or without primary antibody) (yellow tubes, numbered from 1 to up to 48 depending on class size)
- ✓ 1 purified antigen containing 1.5 mL (green microtube—labeled "PA")

8-16 Positive and Negative Results ◆ A positive result in this ELISA produces a blue color within 5 minutes; no reaction is colorless. Waiting longer than 5 minutes may lead to a nonspecific (that is, not due to enzyme activity) color change.

✓ 1 positive control containing 500 µL (violet microtube—labeled "+")
✓ 1 negative control containing 500 µL (blue microtube—labeled "–")
✓ 1 secondary antibody containing 1.5 mL (orange microtube—labeled "SA")
✓ 1 substrate containing 1.5 mL (brown microtube—labeled "SUB")
✓ 1 microtube of 1% blocking agent containing 1.5 mL—labeled BA[2]
✓ 2 microtiter plate strips (one per pair of students)
✓ 2 (better with 4) micropipettors capable of dispensing 50 µL and tips
✓ 4 disposable transfer pipettes
✓ 2 beakers of 33 mL wash buffer labeled "WB" (one per pair of students)[3]
✓ large stack of paper towels
✓ 2 (up to 4) black marking pens
✓ microtube rack
✓ 2 receptacles (plastic cup or something similar–some use a sharps container for this) for pipette tip disposal labeled "tip disposal"

Procedure

Obtain all materials listed (Figure 8-17). From the yellow microtubes, choose a patient's "serum sample" to test. Record its number on the Data Sheet. Refer to the Procedural Diagram in Figure 8-18 as you read and perform this procedure. Dispose of pipettor tips in the container designated for them.

Step 1 Labeling the Microtiter Strip

With a permanent marker, label the sides of one 12–well microtiter plate strip per student pair as follows: The first three wells with " + ", the second three wells with " – ", the third three wells with one lab partner's sample number, and the fourth three wells with the other lab partner's sample number.

Step 2 Adding Purified Antigen

Using a digital pipettor with a fresh tip, dispense 50 µL of purified antibody (green tube labeled "PA") to each of the wells. Incubate at your desk for 5 minutes. Dispose of the pipette tip properly.

[2] Using blocking agent is an optional step in this protocol. Tubes for this are not supplied in the kit, so must come from instructor's stock. Blocking agent is 1% w/v gelatin, albumin, or nonfat dry milk.

[3] If the blocking step is used, 33 mL of Wash Buffer is just enough to carry out all the wash steps, but do not waste any! If it is not used, then another 5 mL may be added to the beaker. This presumes the prep is for the maximum 12 student groups of four.

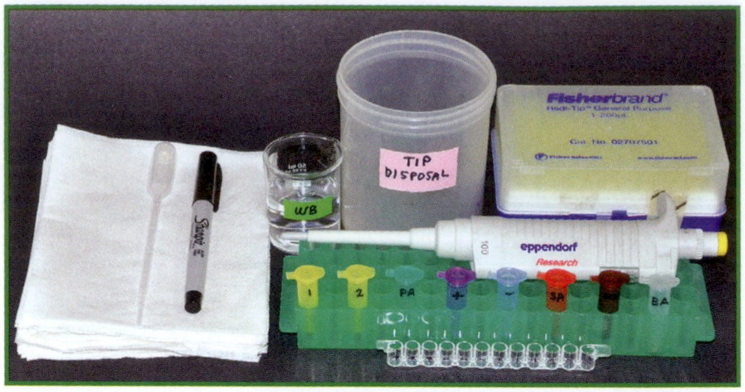

8-17 Materials Needed ◆ Shown are the materials needed for this procedure, but not necessarily in the proper quantity (due to space limitations). The tubes in the rack are as follows: yellow are numbered "patient serum" samples, green is purified antigen ("PA"), violet is the positive control ("+"), blue is the negative control ("–"), orange is the secondary antibody ("SA"), brown is the substrate ("SUB"), and clear is the optional 1% blocking agent ("BA"). Towels, a disposable transfer pipette, a permanent marker, wash buffer (WB), digital pipettor and tips, can for tip disposal, and microtiter strip are also shown.

Step 3 Removing Unbound Purified Antigen

Invert the strip and place it on the stack of paper towels (or a paper towel folded several times). Gently tap out the liquids onto the paper towel to remove unbound purified antigen. Avoid contaminating wells with "drainage" from other wells on the towel. Always begin with a fresh towel and replace it if it becomes so wet that there is the potential of contaminating wells with the "drainage."

Steps 4 and 5 Washing the Wells

Fill (but don't overflow) each well with wash buffer using a disposable pipette (which can be used for all subsequent wash steps). Remove the wash buffer onto fresh paper towels. Wash the wells a second time.

Steps 6, 7, 8, and 9 Optional: Using a Blocking Agent

Using a new tip, add 50 µL of blocking agent to each well. Incubate at your lab desk for 5 minutes. Remove unbound blocking agent and wash twice.

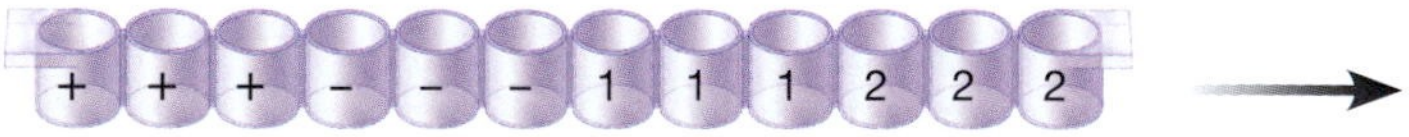

Step 1: Labeling the Microtiter Strip ◆ Using a permanent marker, label the outside of each well as shown. Replace numbers 1 and 2 with the "Patient Sample" numbers you have been assigned.

8-18 Procedural Diagram for the ELISA ◆ Pay special attention to volumes, solutions, and pipettes used in each step. Fresh pipette tips must be used for each solution and disposed of properly. The same disposable pipette should be used for transferring wash buffer throughout. Take care not to mix solutions from neighboring wells by overfilling or by emptying over a contaminated part of the paper towels. *(Continued)*

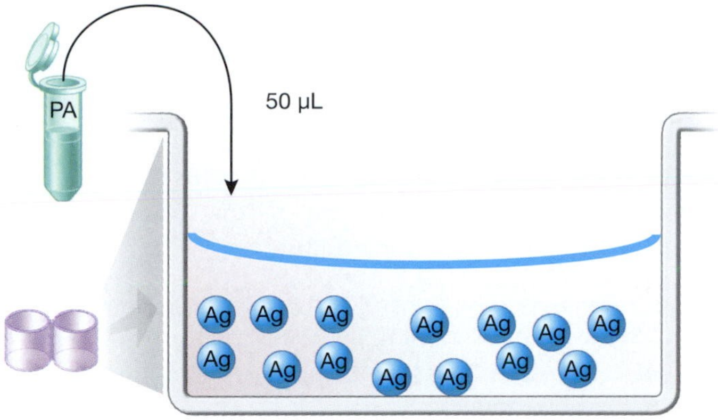

Step 2: Adding Purified Antigen (Ag) ◆ Add 50 µL of Purified Antigen (green microtube labled "PA") to each of the 12 wells. Incubate at your lab desk for 5 minutes. Dispose of the pipette tip properly.

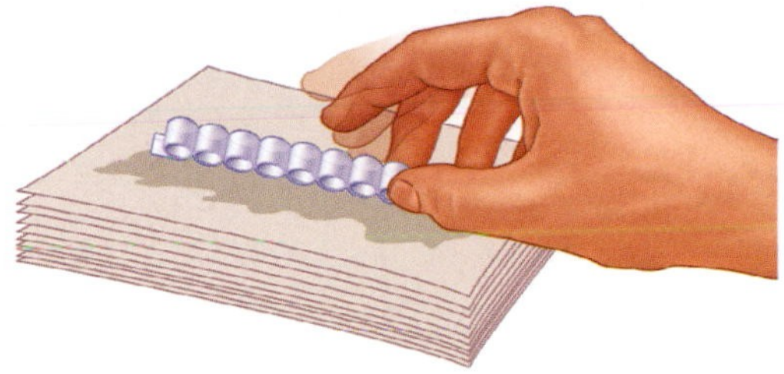

Step 3: Remove Unbound Purified Antigen ◆ Gently tap the strip on several thicknesses of fresh paper towels to remove unbound antigen. Be careful not to mix contents between wells. Dispose of wet towels.

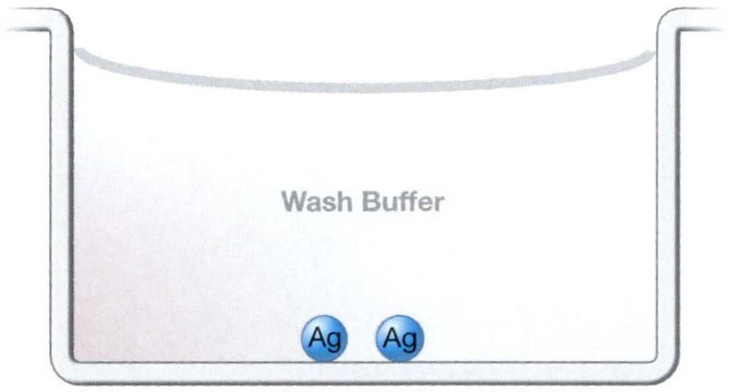

Step 4: Washing the Wells ◆ Using the disposable transfer pipette, fill (but don't overfill and mix neighboring well contents) all 12 wells with wash buffer. Save the disposable pipette for subsequent wash steps. Dispose of wet towels. (Note: In reality, at this point the entire lining of the well would be coated with Ag, but only two are shown to reduce cluttering in the art.)

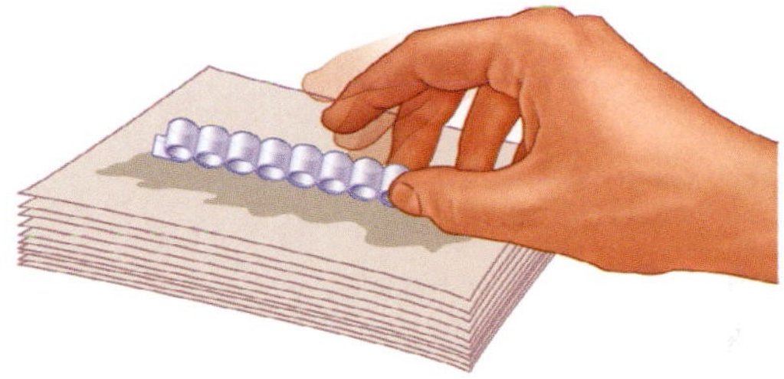

Step 5: Removing the Wash Buffer ◆ Drain the wells onto several thicknesses of fresh paper towels, and then gently tap out the remainder. Do not allow mixing between wells. Dispose of wet towels. Repeat Steps 4 and 5 to wash a second time. Continue with Step 6 (optional) or Step 10.

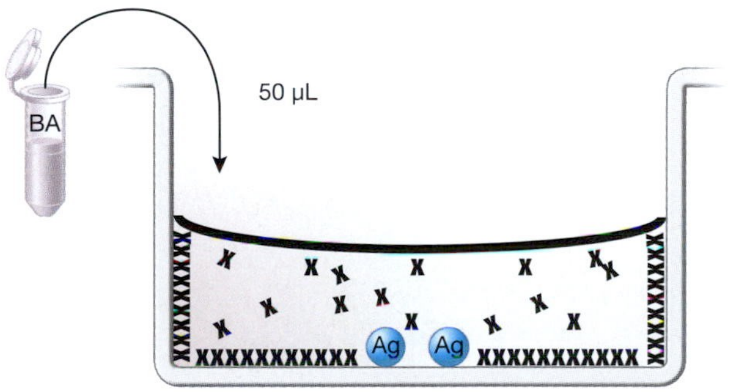

Step 6: Adding Blocking Agent (Optional) ◆ Add 50 µL of Blocking Agent (shown as x's in the drawing) to each well. Incubate at your lab desk for 5 minutes.

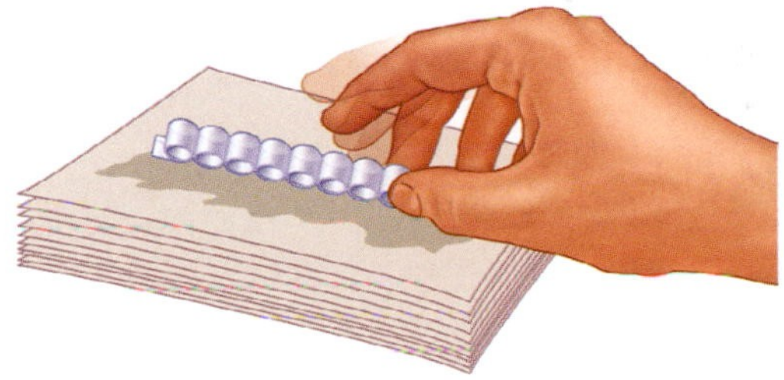

Step 7: Removing Unbound Blocking Agent ◆ Gently tap the strip on several thicknesses of fresh paper towels to remove unbound blocking agent. Be careful not to mix contents between wells.

Step 8: Washing the Wells ◆ Using the disposable transfer pipette, fill (but don't overfill and mix neighboring well contents) all 12 wells with wash buffer.

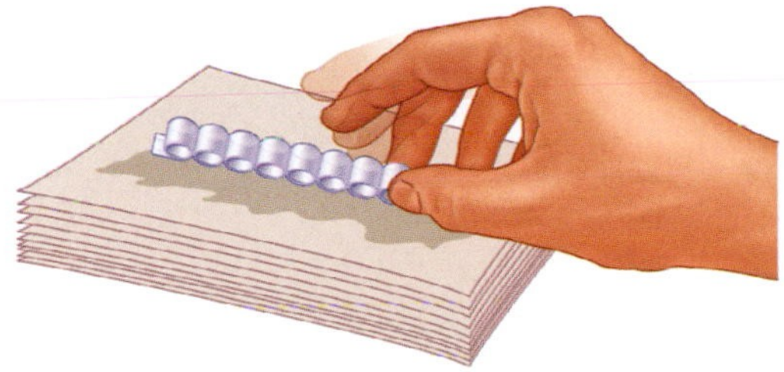

Step 9: Removing the Wash Buffer ◆ Drain the wells onto several thicknesses of fresh paper towels. Do not allow mixing between wells. Tap out the last remnants of wash buffer. Dispose of wet towels. **Repeat Steps 8 and 9 to wash a second time.**

8-18 **Procedural Diagram for the ELISA** *(Continued)* ◆

Step 10 Adding the Positive Control

Using the digital pipettor, add 50 μL of positive control (violet tube labeled " + ") to each of the wells labeled " + ." Dispose of the tip in the designated container. Continue immediately with Step 11.

Step 11 Adding the Negative Control

Using a new tip, add 50 μL of negative control (blue tube labeled " – ") to each of the wells labeled " – ." Dispose of the tip in the designated container. Continue immediately with Step 12.

Step 12 Adding Patient "Serum Samples"

Using a new tip, add 50 μl of one sample (yellow microtube) to the wells marked with its number. Dispose of the tip in the designated container. Using a new tip, add 50 μl of the second sample (yellow microtube) to the wells marked with its number. Dispose of the tip in the designated container. Incubate at your lab desk for 5 minutes.

Step 13 Removing the Control and Patient's "Serum Sample" Solutions and Washing the Wells

Drain the controls and parent samples out of the wells onto a fresh paper towel and gently tap out any residual as in Step 3. Be careful not to mix solutions between wells. Wash the wells as in Steps 8 and 9 twice.

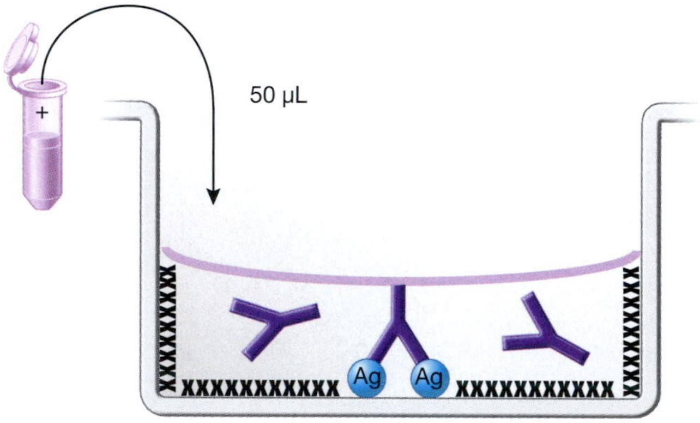

Step 10: Adding the Positive Control ◆ Add 50 μL of Positive Control (violet microtube labeled "+") **to the three "+" wells.** Continue immediately with Step 11.

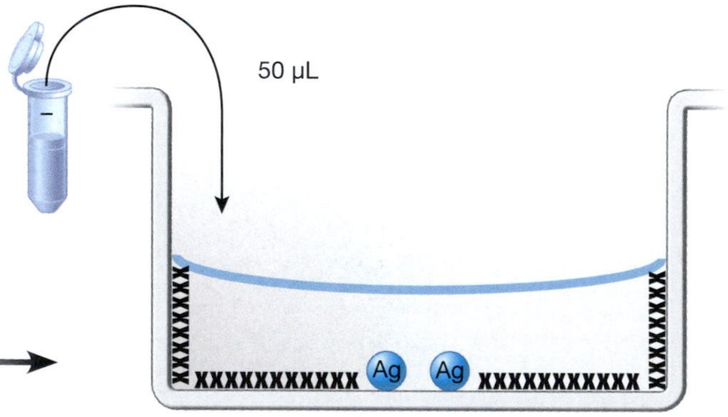

Step 11: Adding the Negative Control ◆ Add 50 μL of Negative Control (blue microtube labeled "–") **to the three "–" wells.** Continue immediately with Step 12.

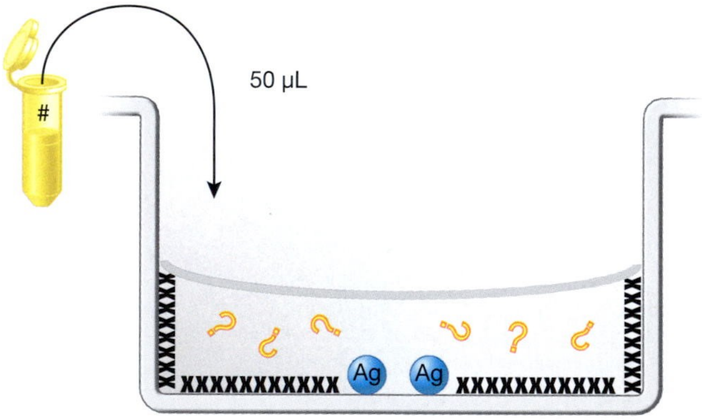

Step 12: Adding Patient's "Serum Samples" ◆ Add 50 μL of "Patient Serum Samples" (yellow microtubes labeled with numbers) to the appropriate remaining wells. Incubate for 5 minutes at your lab desk. The question marks indicate that the contents of the Patient's Sample are unknown.

Step 13: Removing the Control and "Patient's Serum Sample" Solutions and Washing the Wells ◆ Gently tap the strip on several thicknesses of fresh paper towels to remove the Control and "Patient Serum Sample" solutions. Be careful not to mix contents between wells. **Perform two washes with Wash Buffer as shown in Steps 8 and 9.** The illustration shows what a positive result looks like at this point. A negative result would not have the antibody attached to the antigens and would look like the art in Step 11.

8-18 **Procedural Diagram for the ELISA** *(Continued)* ◆

Step 14 Adding Secondary (Conjugate) Antibody

Using a new tip, dispense 50 µL of secondary antibody (orange microtube labeled "SA") to each of the wells. Incubate at your desk for 5 minutes.

Step 15 Removing Unbound Secondary Antibody

Remove the secondary antibody by gently tapping the strip on fresh towels as in Step 3, and then wash **three** times as in Steps 8 and 9. Replace towels as necessary.

Step 16 Adding Substrate

Using a new tip, dispense 50 µL of substrate (brown tube labeled "SUB") to all 12 wells. Incubate at your lab desk for 5 minutes. Observe for color changes and record your results on the Data Sheet. Take your readings within 5 minutes because there is the potential for a blue color developing (due to a reaction with light) if allowed to incubate longer.

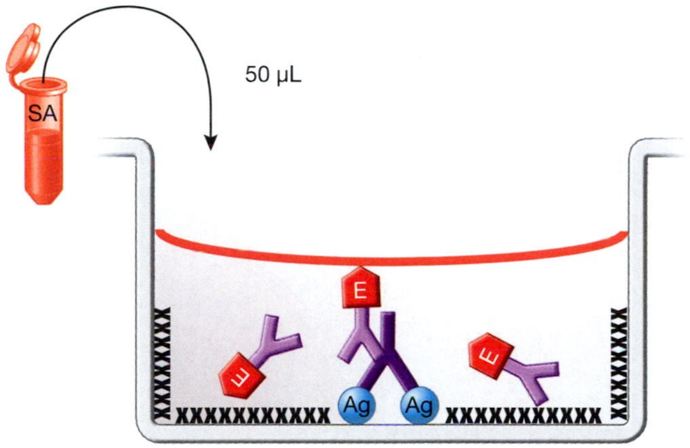

Step 14: Adding Secondary (Conjugate) Antibody ◆ Add 50 µL of Secondary Antibody (orange microtubes labeled with "SA") to the 12 wells. Incubate for 5 minutes at your lab desk.

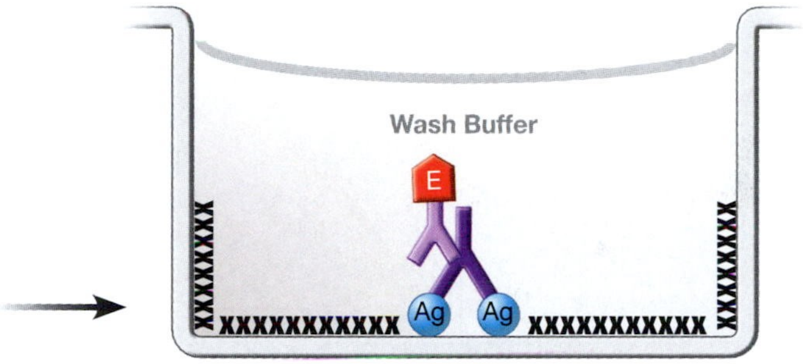

Step 15: Removing Unbound Secondary Antibody ◆ Gently tap the strip on several thicknesses of fresh paper towels to remove unbound Secondary Antibody solution. Be careful not to mix contents between wells. **Perform three washes with Wash Buffer as shown in Steps 8 and 9**. The illustration shows what a positive result looks like at this point. A negative result would not have either antibody present and would look like the art in Step 11.

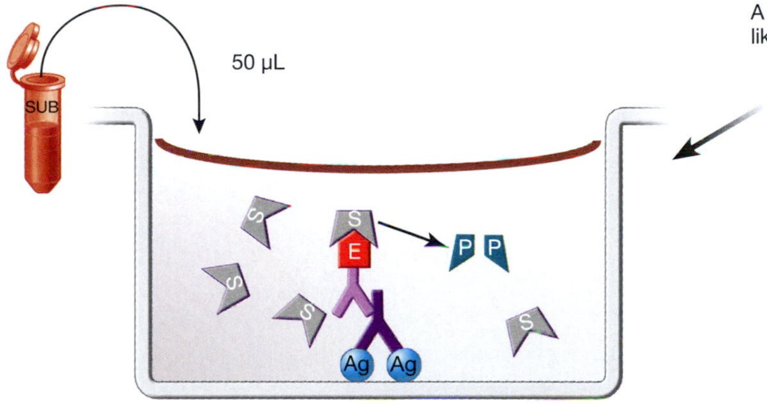

Step 16: Adding Substrate ◆ Add 50 µL of Substrate (brown microtube labeled "SUB") to the 12 wells. Incubate at your lab desk and read results within 5 minutes. A blue color is a positive result. **Waiting longer than 5 minutes to read results could lead to the substrate turning blue through its exposure to light, not enzyme.**

8-18 **Procedural Diagram for the ELISA** *(Continued)* ◆

References

Bio-Rad ELISA Immuno Explorer™ Kit (Catalog #166-2400EDU), Biotechnology Explorer™ Instruction Manual, Rev. D. Bio-Rad Laboratories, Life Science Education. 1-800-4-BIORAD (800-424-6723), www.explorer.bio-rad.com.

Lam, Joseph S., and Lucy M. Mutharia. 1994. Antigen-Antibody Reactions, Chapter 5 in *Methods for General and Molecular Bacteriology*, edited by Philipp Gerhardt, editor-in-chief, R.G.E. Murray, Willis A. Wood, and Noel. R. Krieg. ASM Press, Washington, DC.

Madigan, Michael T., John M. Martinko, Paul V. Dunlop, and David P. Clark. 2009. Pages 922-926 in *Brock's Biology of Microorganisms*, 12th ed. Pearson/Benjamin Cummings, San Francisco, CA.

Shores, Teri. 2009. Pages 102-103 in *Understanding Viruses*. Jones and Bartlett Publishers, LLC, Sudbury, MA.

Name ______________________ Date ______________________

Lab Section ______________________ I was present and performed this exercise (initials) ________

ELISA for Detecting Antibodies in a Patient's Sample (The Antibody Capture Method)

OBSERVATIONS AND INTERPRETATIONS

1 Record your Patient's Sample Number ____________.

2 Write the number of yours and your partner's Patient Sample Numbers in the Table below.

Tube Labels	+	+	+	−	−	−						
Results												

3 Record the results in each well of the microtiter strip in the Table above.

Questions

1 Why is it important not to mix antigen, antibody, or antigen/antibody solutions during the experiment or let solutions from one well overflow into another?

2 Consider the positive controls.

a. Given the design of this ELISA, what is the essential component in the positive control solution? Choose from antigen, primary antibody, secondary antibody, and substrate.

b. What purpose do the positive controls serve?

c. What are possible causes of negative results in all the positive controls?

d. How would negative results in *all* the positive controls affect your interpretation of results given by the patient samples?

3 Consider the negative controls.

a. Given the design of this ELISA, what component must be *absent* in the negative control solution? Choose from antigen, primary antibody, secondary antibody, and substrate.

b. What purpose do the negative controls serve?

c. What are possible causes of positive results in all the negative controls?

d. How would positive results in *all* the negative controls affect your interpretation of results given by the patient samples?

4 Consider the samples tested.

a. Why are the samples run in triplicate?

b. What might explain inconsistent results within each triplicate set and how might they affect your interpretations of the patient samples?

1) Positive controls

2) Negative controls

3) Patient sample

5 Refer to the Procedural Diagram for the following questions:

a. In Step 4, what could be the consequence of washing too vigorously with Wash Buffer? Of not washing enough?

b. What might be the consequence of not adding Blocking Agent in Step 6? (Omit this question if blocking agent wasn't used.)

d. In Step 13, what would be the consequence of washing too vigorously with Wash Buffer? Of not washing enough?

e. In Step 15, what would be the consequence of washing too vigorously with Wash Buffer? Of not washing enough?

Identification of Unknowns

SECTION

9

The exercises in this section are a culmination of all you have done up to this point. You will be using skills developed in earlier exercises and apply them as a practicing microbiologist would. That is, you will be identifying unknown microbes from samples supplied to you by your instructor.

The opportunity to apply what you have learned to a real problem with practical applications is both challenging and exciting for most students. A word of caution, though. Your instructor will know what your unknown is and can tell you if your identification is correct. But out of the classroom and in a professional laboratory, no one can tell the microbiologist if he or she is correct or not. In fact, "identification" is more about finding out what the unknown isn't, rather than finding out with certainty what it is. There is an unspoken understanding that the unknown's identity is based on the best match between the unknown's test results and the accepted results for the same tests of the named species. But the identification is provisional because other tests not run might lead to a different conclusion. Further, don't forget about those nasty false positives and false negatives. Now, with that in mind . . .

The first three exercises employ traditional methods of microbial identification. That is, you will use staining properties, cell morphology, cell arrangement, and biochemical test results to identify an unknown organism assigned to you. In each exercise, a flowchart will direct you in your choice of tests. Matching your unknown's results to the flowchart will lead you, by the process of elimination, to a provisional identification. This methodology will be applied to common species of Gram-negative rods (Exercise 9-1), Gram-positive cocci (Exercise 9-2), and Gram-positive rods (Exercise 9-3).

You have already done the tests in the flowcharts. Some, such as fermentation of sugars, use the same PR medium as Exercise 5-2, just with a different sugar. These are read as in Exercise 5-2.

In Exercise 9-4 and Exercise 9-5, you will identify unknown enterics using multitest systems. These take a shotgun approach but allow more rapid identification than traditional methods. Each has several biochemical tests that are inoculated simultaneously. These tests are based on the same principles as their tubed and plated forebears, but the combination of results produces a number that is compared to a database for identification.

Serological testing continues to be an important diagnostic tool for laboratorians. As an example of this approach, you will use the Streptex® Rapid Test system in Exercise 9-6. It identifies streptococci and related organisms using agglutination to detect group-specific antigens. In some cases, identification to species can be made; in others; only group identification is possible.

EXERCISE 9-1

Identification of *Enterobacteriaceae*

Theory

The *Enterobacteriaceae* (Figure 9-1) comprise a diverse family of Gram-negative rods that inhabit the human intestinal tract. The family has grown from 12 genera and 36 species in *Bergey's Manual of Determinative Bacteriology*, 8th ed. (1974) to 44 genera and 176 species in *Bergey's Manual of Systematic Bacteriology*, 2nd Ed., Volume 2, *Part B—The Gammaproteobacteria* (2005). This expansion has brought additional variability into the group and the recognition that the human intestines are not their main habitat. *Enterobacteriaceae* are also common in the environment (soil) and have hosts as diverse as plants and insects. Still, the following major characteristics are true for the majority of species.

- Growth on MacConkey agar
- Gram-negative rods
- Facultative anaerobes
- Oxidase negative
- Acid production from glucose (with or without gas)
- Most are catalase-positive
- Most reduce NO_3 to NO_2
- Peritrichous flagella, if motile

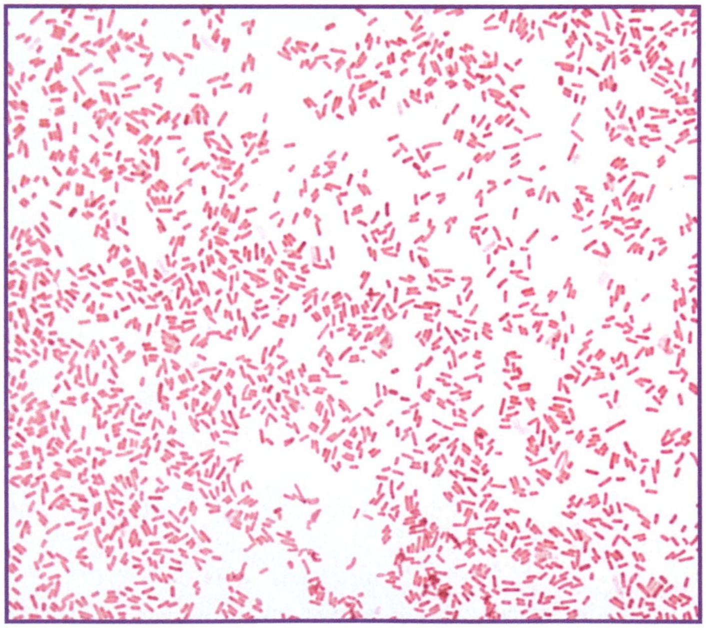

9-1 ***Escherichia coli*, a Common Enteric (X1000)** ◆ All *Enterobacteriaceae* are Gram-negative, oxidase-negative rods. They also ferment glucose to acid end products.

Many **enterics** are human pathogens (e.g., *Escherichia coli* 0157:H7—bloody diarrhea; *Klebsiella pneumoniae*—pneumonia of various types; *Proteus mirabilis*—urinary tract infections; *Salmonella typhi*—typhoid fever; *Shigella dysenteriae*—bacillary dysentery; and *Yersinia pestis*—plague) but just as many are harmless gut **commensals** or **opportunists**. The organisms for which the identification flowcharts in this exercise work are listed in Figure 9-2, and some are commonly associated with human infections. Your instructor will choose enteric unknowns appropriate to your microbiology course and facilities.

Application

Identification of enterics from human specimens using traditional methods requires a coordinated and integrated use of biochemical tests and stains. Although several multitest systems that utilize computer databases are available (see Exercises 9-4 and 9-5 for two examples), flowcharts are still a useful way to visualize the process of

List of *Enterobacteriaceae* in Identification Charts and IMViC Results

*Citrobacter freundii**
Citrobacter koseri
Enterobacter aerogenes
Enterobacter cloacae
Escherichia coli
*Hafnia alvei**
*Klebsiella pneumoniae** (BSL-2)
Morganella morganii (BSL-2)
*Proteus mirabilis** (BSL-2)
Proteus vulgaris (BSL-2)
Providencia alcalifaciens
Providencia stuartii
Salmonella enterica+ (BSL-2)
Serratia liquefaciens
*Serratia marcescens**
*Shigella flexneri (Group B)** (BSL-2)
Shigella sonnei (Group D) (BSL-2)

*Organism shows up in more than one figure due to variable results of one or more IMViC tests

+*Salmonella enterica* is *Salmonella enterica* subsp. *enterica* and includes *S. enteritidis* and *S. typhimurium*

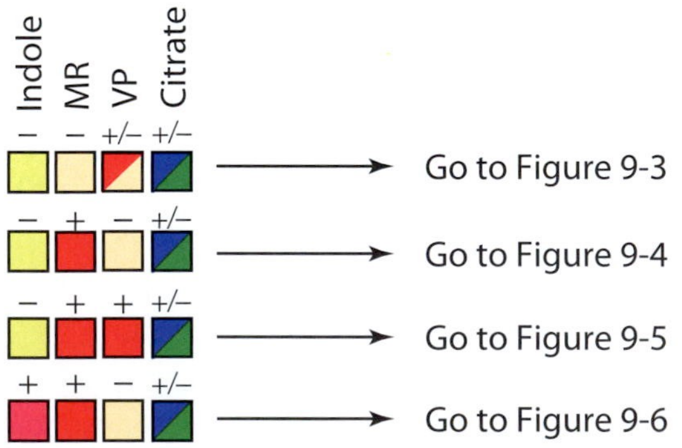

9-2 **IMViC Results of *Enterobacteriaceae*** ◆ The organisms listed are *Enterobacteriaceae* used in this exercise. Match the IMViC results of your isolate with one of the four shown, then proceed with the flowchart in the figure indicated. Some organisms are listed in more than one flowchart because of variable results for a particular test. A box divided diagonally means either result for that test is a match. (Most results after Farmer et al., 2007.) Note: Each flowchart can be used to identify only the organisms listed.

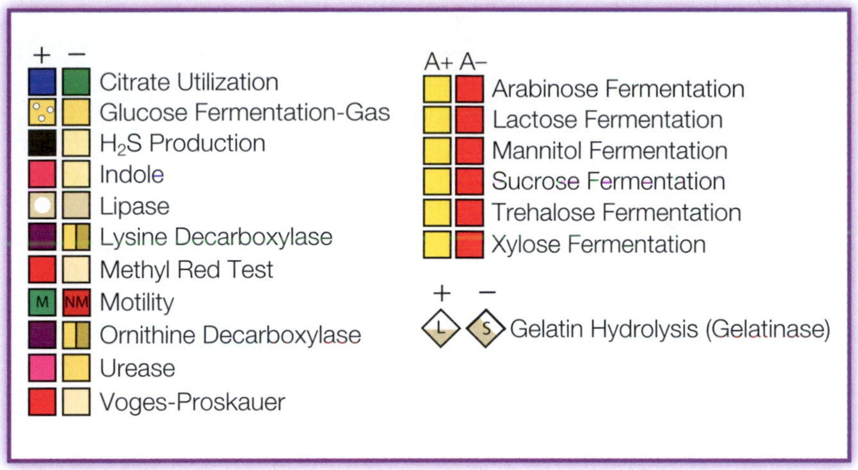
+ −
Citrate Utilization
Glucose Fermentation-Gas
H_2S Production
Indole
Lipase
Lysine Decarboxylase
Methyl Red Test
M NM Motility
Ornithine Decarboxylase
Urease
Voges-Proskauer
A+ A−
Arabinose Fermentation
Lactose Fermentation
Mannitol Fermentation
Sucrose Fermentation
Trehalose Fermentation
Xylose Fermentation
+ −
L S Gelatin Hydrolysis (Gelatinase)

TABLE 9-1 Key to Icons Used in This Exercise ◆ Colors are to remind you what results look like. They should not be used to compare with your results. ***Note:*** For the fermentation tests, "A+" means acid is produced; "A−" means no acid is produced.

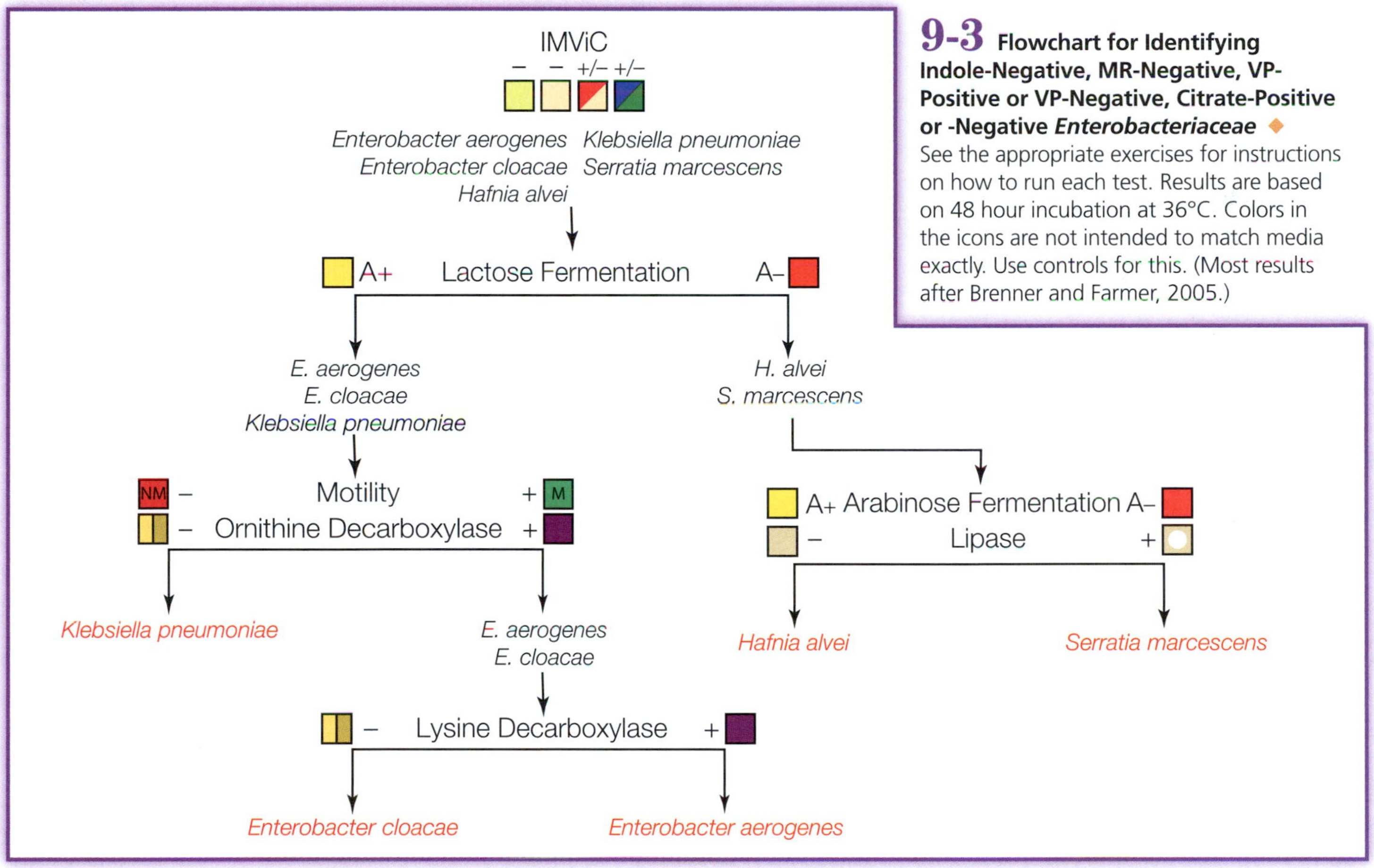
IMViC
− − +/− +/−
Enterobacter aerogenes
Enterobacter cloacae
Hafnia alvei
Klebsiella pneumoniae
Serratia marcescens
A+ Lactose Fermentation A−
E. aerogenes
E. cloacae
Klebsiella pneumoniae
H. alvei
S. marcescens
NM − Motility + M
− Ornithine Decarboxylase +
Klebsiella pneumoniae
E. aerogenes
E. cloacae
− Lysine Decarboxylase +
Enterobacter cloacae
Enterobacter aerogenes
A+ Arabinose Fermentation A−
− Lipase +
Hafnia alvei
Serratia marcescens

9-3 Flowchart for Identifying Indole-Negative, MR-Negative, VP-Positive or VP-Negative, Citrate-Positive or -Negative *Enterobacteriaceae* ◆ See the appropriate exercises for instructions on how to run each test. Results are based on 48 hour incubation at 36°C. Colors in the icons are not intended to match media exactly. Use controls for this. (Most results after Brenner and Farmer, 2005.)

identification by elimination. It is striking how often a few test results, when taken in combination, are necessary to identify an organism.

❖ In This Exercise

This exercise will span several lab periods. You will be assigned an unknown enteric and run biochemical tests as directed by the flowcharts in Figures 9-3 through 9-6 to identify it.

❖ Materials

Per Class

- ✓ media listed in Table 9-1 for biochemical testing
- ✓ pure cultures of organisms listed in Figure 9-2 to be used for positive controls and unknowns (All organisms are available from either Ward's or Carolina Biological Supply Company. BSL-2 organisms should not be chosen as unknowns by schools or classes not adequately prepared to handle them.)

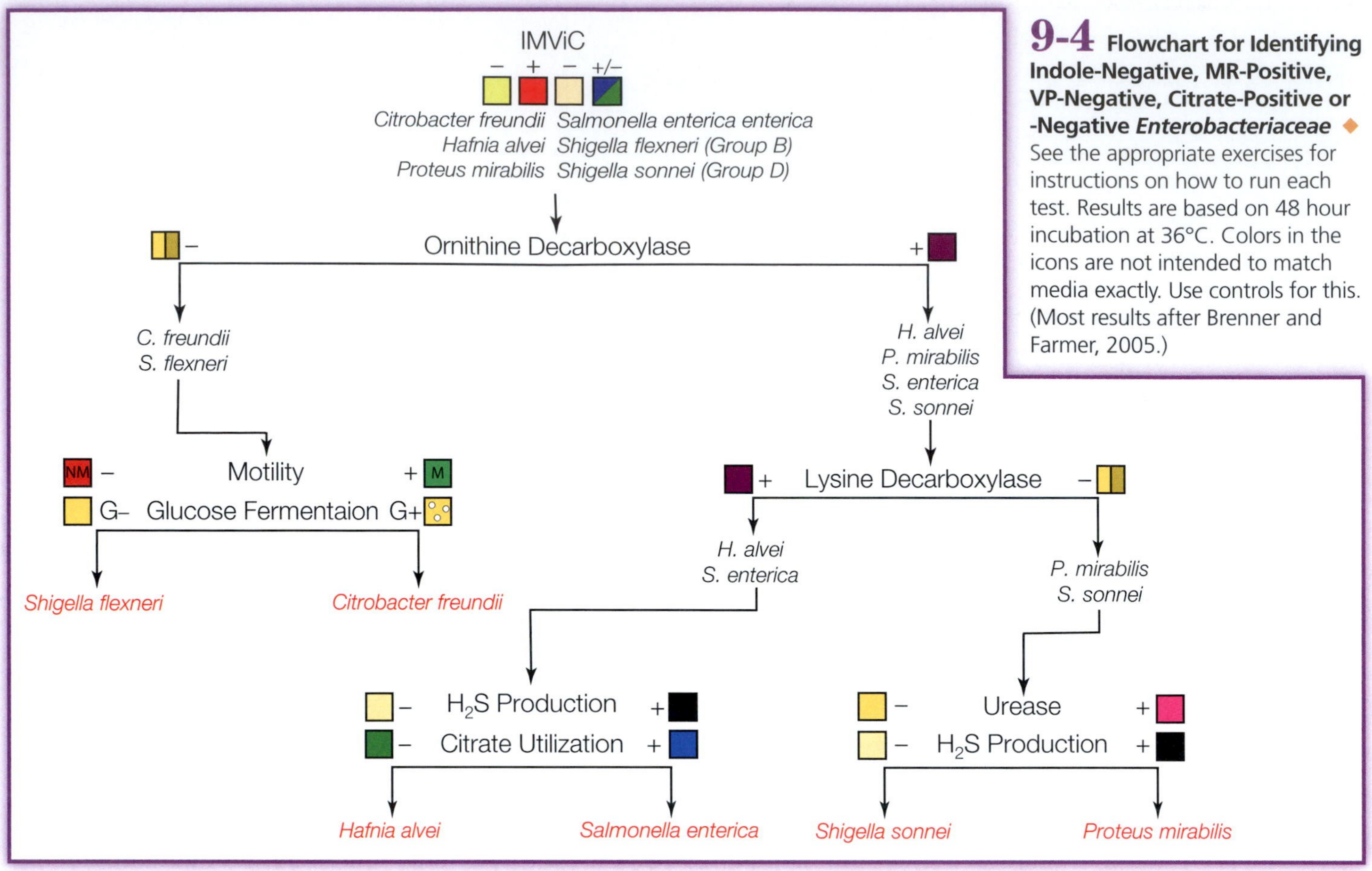
IMViC
– + – +/–
Citrobacter freundii
Hafnia alvei
Proteus mirabilis
Salmonella enterica enterica
Shigella flexneri (Group B)
Shigella sonnei (Group D)
– Ornithine Decarboxylase +
C. freundii
S. flexneri
H. alvei
P. mirabilis
S. enterica
S. sonnei
NM – Motility + M
G– Glucose Fermentaion G+
Shigella flexneri
Citrobacter freundii
+ Lysine Decarboxylase –
H. alvei
S. enterica
P. mirabilis
S. sonnei
– H2S Production +
– Citrate Utilization +
Hafnia alvei
Salmonella enterica
– Urease +
– H2S Production +
Shigella sonnei
Proteus mirabilis

9-4 Flowchart for Identifying Indole-Negative, MR-Positive, VP-Negative, Citrate-Positive or -Negative *Enterobacteriaceae* ◆ See the appropriate exercises for instructions on how to run each test. Results are based on 48 hour incubation at 36°C. Colors in the icons are not intended to match media exactly. Use controls for this. (Most results after Brenner and Farmer, 2005.)

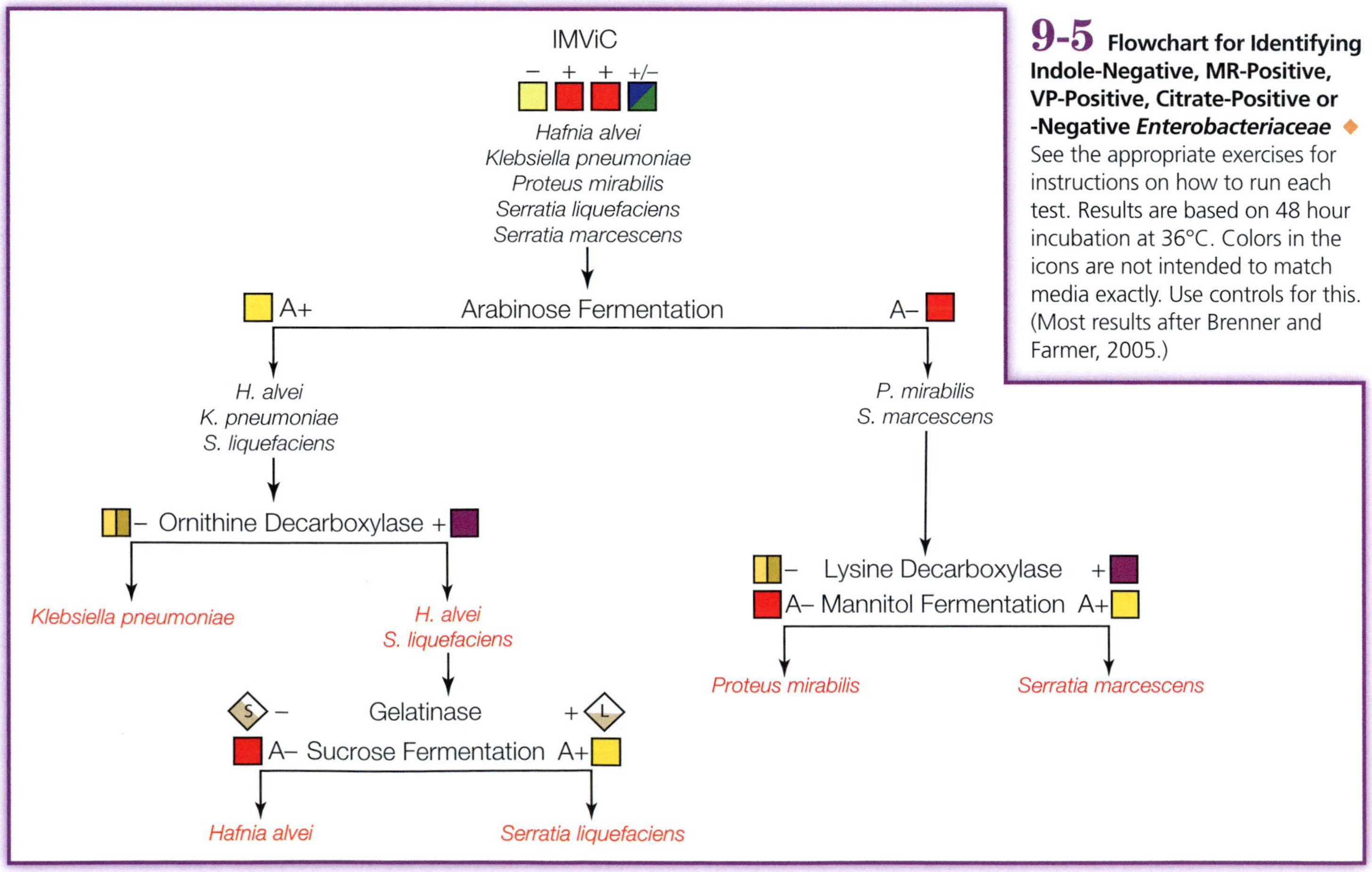
IMViC
– + + +/–
Hafnia alvei
Klebsiella pneumoniae
Proteus mirabilis
Serratia liquefaciens
Serratia marcescens
A+ Arabinose Fermentation A–
H. alvei
K. pneumoniae
S. liquefaciens
P. mirabilis
S. marcescens
– Ornithine Decarboxylase +
Klebsiella pneumoniae
H. alvei
S. liquefaciens
S – Gelatinase + L
A– Sucrose Fermentation A+
Hafnia alvei
Serratia liquefaciens
– Lysine Decarboxylase +
A– Mannitol Fermentation A+
Proteus mirabilis
Serratia marcescens

9-5 Flowchart for Identifying Indole-Negative, MR-Positive, VP-Positive, Citrate-Positive or -Negative *Enterobacteriaceae* ◆ See the appropriate exercises for instructions on how to run each test. Results are based on 48 hour incubation at 36°C. Colors in the icons are not intended to match media exactly. Use controls for this. (Most results after Brenner and Farmer, 2005.)

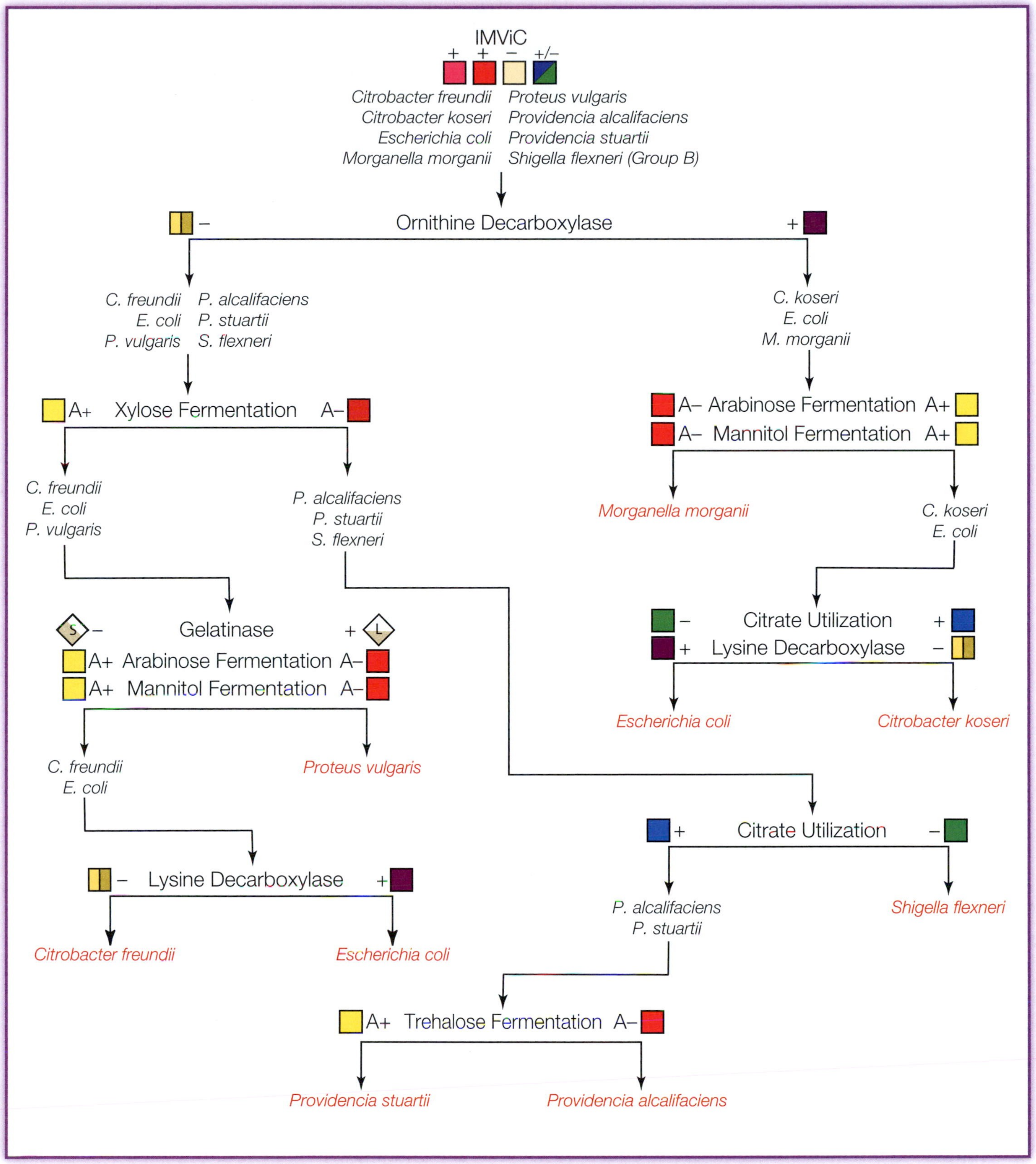

IMViC
+ + – +/–
Citrobacter freundii
Citrobacter koseri
Escherichia coli
Morganella morganii
Proteus vulgaris
Providencia alcalifaciens
Providencia stuartii
Shigella flexneri (Group B)
– Ornithine Decarboxylase +
C. freundii
E. coli
P. vulgaris
P. alcalifaciens
P. stuartii
S. flexneri
C. koseri
E. coli
M. morganii
A+ Xylose Fermentation A–
A– Arabinose Fermentation A+
A– Mannitol Fermentation A+
C. freundii
E. coli
P. vulgaris
P. alcalifaciens
P. stuartii
S. flexneri
Morganella morganii
C. koseri
E. coli
S – Gelatinase + L
A+ Arabinose Fermentation A–
A+ Mannitol Fermentation A–
– Citrate Utilization +
+ Lysine Decarboxylase –
Escherichia coli
Citrobacter koseri
C. freundii
E. coli
Proteus vulgaris
+ Citrate Utilization –
– Lysine Decarboxylase +
P. alcalifaciens
P. stuartii
Shigella flexneri
Citrobacter freundii
Escherichia coli
A+ Trehalose Fermentation A–
Providencia stuartii
Providencia alcalifaciens

9-6 Flowchart for Identifying Indole-Positive, MR-Positive, VP-Negative, Citrate-Positive or -Negative *Enterobacteriaceae* ◆
See the appropriate exercises for instructions on how to run each test. Results are based on 48 hour incubation at 36°C. Colors in the icons are not intended to match media exactly. Use controls for this. (Most results after Brenner and Farmer, 2005.)

Per Student

- ✓ two MacConkey Agar plates
- ✓ one unknown organism[1] from the list in Figure 9-2
- ✓ TSA slants (enough to keep the culture fresh over the time required for identification)
- ✓ Gram-stain kit
- ✓ microscope slides
- ✓ compound microscope with oil objective lens and ocular micrometer
- ✓ immersion oil
- ✓ lens paper
- ✓ gloves

Procedure

1 Obtain an unknown. Record its number on the Data Sheet.

2 Streak the sample for isolation on two MacConkey Agar plates. Incubate one at 25°C and the other 35 ± 2°C for 24 hours or more. It is best to check your plate for isolation at 24 hours. If you have isolation, continue with Step 3. If you don't have time to continue with Step 3, refrigerate your plates until you do. If you don't see isolation, ask your instructor if you should restreak for isolation, let the original plates continue to incubate, or do both. Record your isolation procedure as directed on the Data Sheet.

3 After incubation, record colony morphology (including medium) and optimum growth temperature under "preliminary observations" on the Data Sheet. Also include the result of the differential component of MacConkey Agar as Test #1 under "Differential Tests." Continue to record your isolation procedure on the Data Sheet.

4 Perform a Gram stain on a portion of one well-isolated colony. Record its Gram reaction and cell morphology, arrangement, and size on the Data Sheet under Preliminary Observations. If the isolate is a Gram-negative rod, transfer a portion of the *same* colony to a Trypticase Soy Agar slant (or another suitable growth medium as available in your lab) and incubate it at its optimum temperature. This is your pure culture to be used as a source of organisms for further testing. Complete the record of your isolation procedure on the Data Sheet.

5 Perform an oxidase test on what remains of the isolated colony, or wait until your pure culture has grown and do it then. Record the result as Test #2 under "Differential Tests" on the Data Sheet. If the isolate grew on MacConkey Agar and is an oxidase-negative, Gram-negative rod, it probably is a member of the *Enterobacteriaceae*. At the discretion of your instructor, further tests chosen from the characteristics listed may be run to assure that it is an organism from *Enterobacteriaceae*.

6 You now will begin biochemical testing to identify your organism. The tests in the flowcharts were chosen because of their uniformity of results as published in standard microbiological references, so they give the best chance of correct identification. Be aware, however, that the symbol " + " indicates that 90% or more of the strains tested give a positive result. This means that as many as 10% of the strains tested give a negative result. The same is true of the symbol " – "; that is, as many as 10% of the strains give a positive result for that test. Bottom line: There are no guarantees that your particular lab strains will behave in the majority, and if not, you will misidentify your unknown. This issue, if relevant to your unknown, will be addressed in Step 10.

7 Perform the IMViC series of tests (Indole Production, Methyl Red Test, Voges-Proskauer Test, and Citrate Utilization) at your isolate's optimum temperature.

- ♦ Record the relevant inoculation information and results on the Data Sheet under "Differential Tests."
- ♦ Use Test #3 for indole, Test #4 for MRVP, and Test #5 for citrate.
- ♦ Match your isolate's IMViC results with one shown in Figure 9-2, and proceed to the appropriate identification flowchart (Figures 9-3 through 9-6). Record on the Data Sheet the figure number of the flowchart you will use. (Note: If a result is shown as +/– [as it is for all citrate tests], either result for that test is a match.)

8 Follow the tests in the appropriate flowchart to identify your unknown. Use these guidelines.

- ♦ Do not run more than one test at a time unless your instructor tells you to do so.
- ♦ Where multiple tests are listed at a branch point, run only one, and then move to the next level in the flowchart as indicated by the result of that test.
- ♦ Continue to record the relevant inoculation information and results on the Data Sheet as you go. Keep accurate records of what you have done. Do not enter all your data at the end of the project.

---

[1] ***Note to instructor:*** This exercise can be combined with Exercises 9-2 or 9-3 by passing out a mixture of an enteric and a Gram-positive coccus or rod as unknowns. The student must isolate the two organisms from mixed culture and then proceed to identify the two independently. If this option is chosen, grow the two unknowns separately and then mix them immediately prior to handing them out in lab.

9 When you have identified your organism, use *Bergey's Manual* or another standard reference to find one more test to run for confirmation. The confirmatory test doesn't have to separate the final organisms on the flowchart. It only has to be a test for which you know the result and that you haven't run already as part of the identification process. Record this test on the Data Sheet and identify it as the confirmatory test. Record the *expected* result on the Comments line. After incubation, note whether the organism's result matches or differs from the expected result.

10 After the confirmatory test, write the name of the organism on the back of the Data Sheet under "My unknown is" and check with your instructor to see if you are correct. If you are, congratulations! If you aren't, your instructor will advise you as to which test(s) gave "incorrect" results by writing the test name(s) in the "rerun" space next to your identification. These should be rerun with appropriate controls from your school's inventory of organisms. Record the test(s) and control(s) on your Data Sheet. Checking the results with controls and your unknown will indicate if the "incorrect" test result was truly incorrect or if the strain of organism your school is using doesn't match the majority of strains for that test result.

References

Brenner, Don J. and J. J. Farmer, III. 2005. Order XIII. "Enterobacteriales" in *Bergey's Manual of Systematic Bacteriology*, Vol. 2, *The Proteobacteria, Part B—The Gammaproteobacteria*. Springer, New York.

Farmer III, J. J., K. D. Boatwright, and J. Michael Janda. 2007. Chapter 42 in *Manual of Clinical Microbiology*, 9th ed., edited by Patrick R. Murray, Ellen Jo Baron, James H. Jorgensen, Marie Louise Landry, and Michael A. Pfaller. ASM Press, Washington, DC.

Forbes, Betty A., Daniel F. Sahm, and Alice S. Weissfield. 2002. Chapter 25 in *Bailey & Scott's Diagnostic Microbiology*, 11th ed. Mosby, Inc., St. Louis, MO.

MacFaddin, Jean F. 2000. *Biochemical Tests for Identification of Medical Bacteria*, 3rd ed. Lippincott Williams & Wilkins, Philadelphia, PA.

Winn Jr., Washington, Stephen Allen, William Janda, Elmer Koneman, Gary Procop, Paul Schreckberger, and Gail Woods. 2006. Chapter 6 in *Koneman's Color Atlas and Textbook of Diagnostic Microbiology*, 6th ed., Lippincott Williams & Wilkins, Philadelphia, PA.

NAME ______________________ DATE ______________________

LAB SECTION ______________________ I WAS PRESENT AND PERFORMED THIS EXERCISE (INITIALS) ________

DATA SHEET 9-1

Identification of *Enterobacteriaceae*

ISOLATION PROCEDURE

Unknown Number ______________

Record all activities associated with isolation of your organisms—from mixed culture to pure culture. Always include the date, source of inoculum, destination, type of inoculation, incubation temperature, and any other relevant information. Also make note of transfers made to keep your pure culture fresh. (This log must be kept current.)

Preliminary Observations

Colony morphology (include medium and incubation temperature) ______________

Gram stain ________ Cell dimensions ________ Optimum temperature ________

Cellular morphology and arrangement ______________

Differential Tests

Begin recording with your MacConkey result, followed by the Oxidase test, and then your IMViC results. This log must be kept current.

Record the Identification Chart you are using. If you change charts, indicate the figure number of the new chart and the date of the change.

Test #1: ______ Date Begun: ______ Date Read: ______ Result: ______

Comments: ______

Test #2: ______ Date Begun: ______ Date Read: ______ Result: ______

Comments: ______

Test #3: ______ Date Begun: ______ Date Read: ______ Result: ______

Comments: ______

Test #4: ______ Date Begun: ______ Date Read: ______ Result: ______

Comments: ______

Test #5: ______ Date Begun: ______ Date Read: ______ Result: ______

Comments: ______

Test #6: ______ Date Begun: ______ Date Read: ______ Result: ______

Comments: ______

Test #7: ______ Date Begun: ______ Date Read: ______ Result: ______

Comments: ______

Test #8: ______ Date Begun: ______ Date Read: ______ Result: ______

Comments: ______

Test #9: ______ Date Begun: ______ Date Read: ______ Result: ______

Comments: ______

Test #10: ______ Date Begun: ______ Date Read: ______ Result: ______

Comments: ______

Test #11: ______ Date Begun: ______ Date Read: ______ Result: ______

Comments: ______

Test #12: ______ Date Begun: ______ Date Read: ______ Result: ______

Comments: ______

My Unknown is: ______ Rerun:[1] ______

______ Rerun: ______

______ Rerun: ______

[1] Your instructor will write what tests to rerun in this space if you misidentify your unknown.

EXERCISE 9-2

Identification of Gram-positive Cocci

Theory

Gram-positive cocci are frequent isolates in a clinical setting because they are common inhabitants of skin and mucous membranes. Five main genera, briefly described below, will be used in this lab exercise.

Staphylococcus (Figure 9-7)

- Gram-positive cocci in singles, pairs, tetrads, or clusters (especially when grown in broth)
- Catalase positive.
- With rare exception, facultatively anaerobic.
- Most are oxidase negative.
- G+C content within the genus ranges between 30 and 39%.
- Grow in 6.5% NaCl.
- Most produce acid from glucose.
- Most are resistant to bacitracin.
- Key pathogen is *Staphylococcus aureus* (toxic shock syndrome and a variety of other skin and deep organ infections, including bacterermia).

Kocuria

- A small genus; most species were formerly classified in the genus *Micrococcus*.
- Gram-positive cocci in pairs and tetrads
- Obligately aerobic.

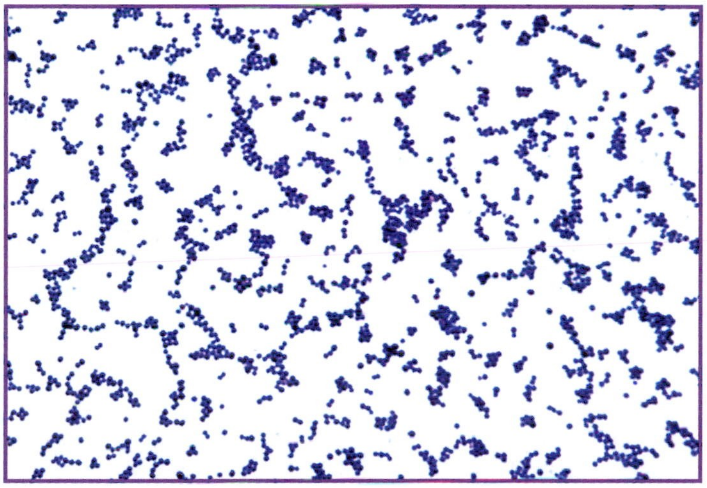

9-7 ***Staphylococcus aureus*** ◆ This specimen grown in broth illustrates the grape-like clusters of cells characteristic of the genus. Specimens grown on solid media may not show the clusters as clearly.

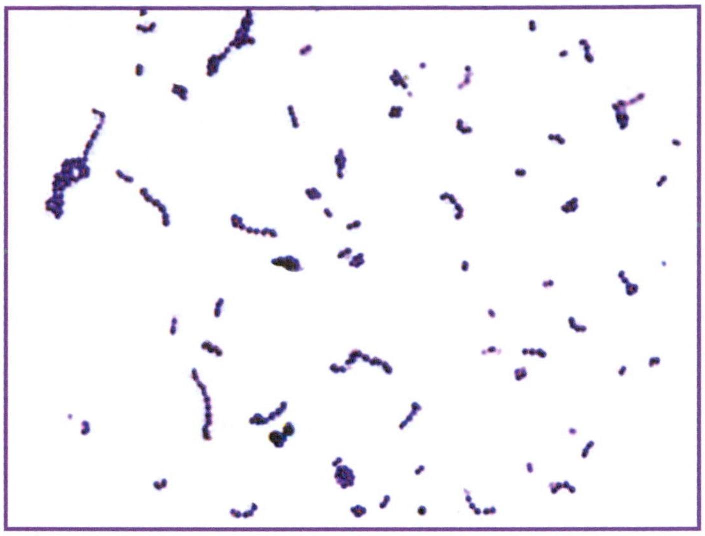

9-8 ***Streptococcus agalactiae*** ◆ This specimen grown in broth illustrates the streptococcal arrangement of cells characteristic of the genus. Specimens grown on solid media may not show the chains as clearly.

- Do not produce acid from glucose.
- Bacitracin susceptible.
- Catalase positive.
- Commensals or opportunistic pathogens, especially among immunocompromised patients.

Micrococcus

- Gram-positive cocci in pairs and tetrads
- Obligately aerobic.
- Oxidase positive.
- Do not produce acid from glucose.
- Bacitracin susceptible.
- Catalase positive.
- G+C content within the genus ranges between 66 and 75%.
- Grow in 6.5% NaCl.
- Commensals or opportunistic pathogens. *M. luteus* is a common skin commensal.

Streptococcus (Figure 9-8)

- Gram-positive cocci to ovoid cocci in singles, pairs, or short chains (especially when grown in broth)
- Gray to white, moist colonies are frequently observed.
- Catalase negative.
- Oxidase negative.
- Facultatively anaerobic.
- Nutritionally fastidious.
- Some require 5% CO_2 for growth.
- Ferment glucose and other carbohydrates, mostly to lactic acid.

- G+C content within the genus ranges between 34 and 46%.
- Many species produce hemolysins that either completely (β-hemolysins) or partially (α-hemolysins) destroy erythrocytes. Some species are nonhemolytic.
- Beta-hemolytic streptococci have traditionally been grouped by antigens first described by Rebecca Lancefield. Important Lancefield groups include Group A (*S. pyogenes*) and Group B (*S. agalactiae*).
- Key pathogens are *Streptococcus pyogenes* (strep throat, necrotizing fasciitis, scarlet fever) and *S. pneumoniae* (bacterial pneumonia, otitis media, and bacteremia).

Enterococcus

- Formerly members of the genus *Streptococcus*.
- Gram-positive cocci to ovoid cocci in singles, pairs, or short chains (especially when grown in broth); may be more rod-shaped if grown on solid media.
- Catalase negative
- Oxidase negative.
- Facultatively anaerobic.
- Lactic acid, but no gas, is the sole end product of fermentation.
- Grow in 6.5% NaCl broth.
- Grow in bile esculin.
- Most are PYR positive.
- G+C content within the genus ranges between 34 and 42%.
- Most express the Lancefield Group D antigen.
- Most species are commensals or opportunistic pathogens. The key opportunistic pathogen is *E. faecalis* (urinary tract infections, wound infections, and bacteremia in seriously ill elderly persons).

The organisms to be used as unknowns are listed in Table 9-2, as are the tests to be used in their identification. (Note: Your instructor will choose Gram-positive coccus unknowns appropriate to your microbiology course and facilities.)

Application

Identification of Gram-positive cocci from human specimens requires a coordinated and integrated use of biochemical tests and stains. Although several serological tests allow rapid identification, flowcharts are still useful way to visualize the process of identification by elimination.

In This Exercise

This exercise will span several lab periods. You will be given an unknown Gram-positive coccus from the organisms listed in Table 9-2, then run biochemical tests as directed by the flowcharts in Figure 9-9 or Figure 9-10 to identify it.

TABLE 9-2 List of Organisms and a Key to the Icons Used in This Exercise ◆ The organisms listed are Gram-positive cocci available from Ward's or Carolina Biological Supply Companies. Icon colors are to remind you what the results look like. They should not be used to compare with your results. ***Note:*** For the fermentation tests, "A+" means acid is produced; "A−" means no acid is produced.

Gram-Positive Cocci and Identification Tests

Catalase-Positive (Figure 9-9)

Kocuria rosea (= *Micrococcus roseus*)
Micrococcus luteus
Staphylococcus aureus (BSL-2)
Staphylococcus epidermidis (BSL-2)
Staphylococcus saprophyticus

Catalase-Negative (Figure 9-10)

Enterococcus faecalis (BSL-2)
Streptococcus agalactiae (BSL-2)
Streptococcus dysigalactae subsp. *equisimilis* (BSL-2) (= *Streptococcus equisimilis*)
Streptococcus mutans
Streptrococcus salivarius
Streptococcus sanguinis (BSL-2)
Streptococcus pneumoniae (BSL-2)
Streptococcus pyogenes (BSL-2)

+ −
Arginine Decarboxylase (Dihydrolase)
CAMP Test
Catalase
NO_3 reduced to NO_2
PYR Test
Starch Hydrolysis (Amylase)
Voges-Proskauer

β α γ
Hemolysis

A+ A−
Mannitol Fermentation

O F
Glucose O-F Test

S R
≥10 mm / <9 mm Bacitracin Susceptibility (0.04 U)
>16 mm / ≤16 mm Novobiocin Susceptibility (0.5 µg)
>14 mm / <14 mm Optochin Susceptibility (5 µg)

+ −
S / L Coagulase
L / S Gelatin Hydrolysis (Gelatinase)

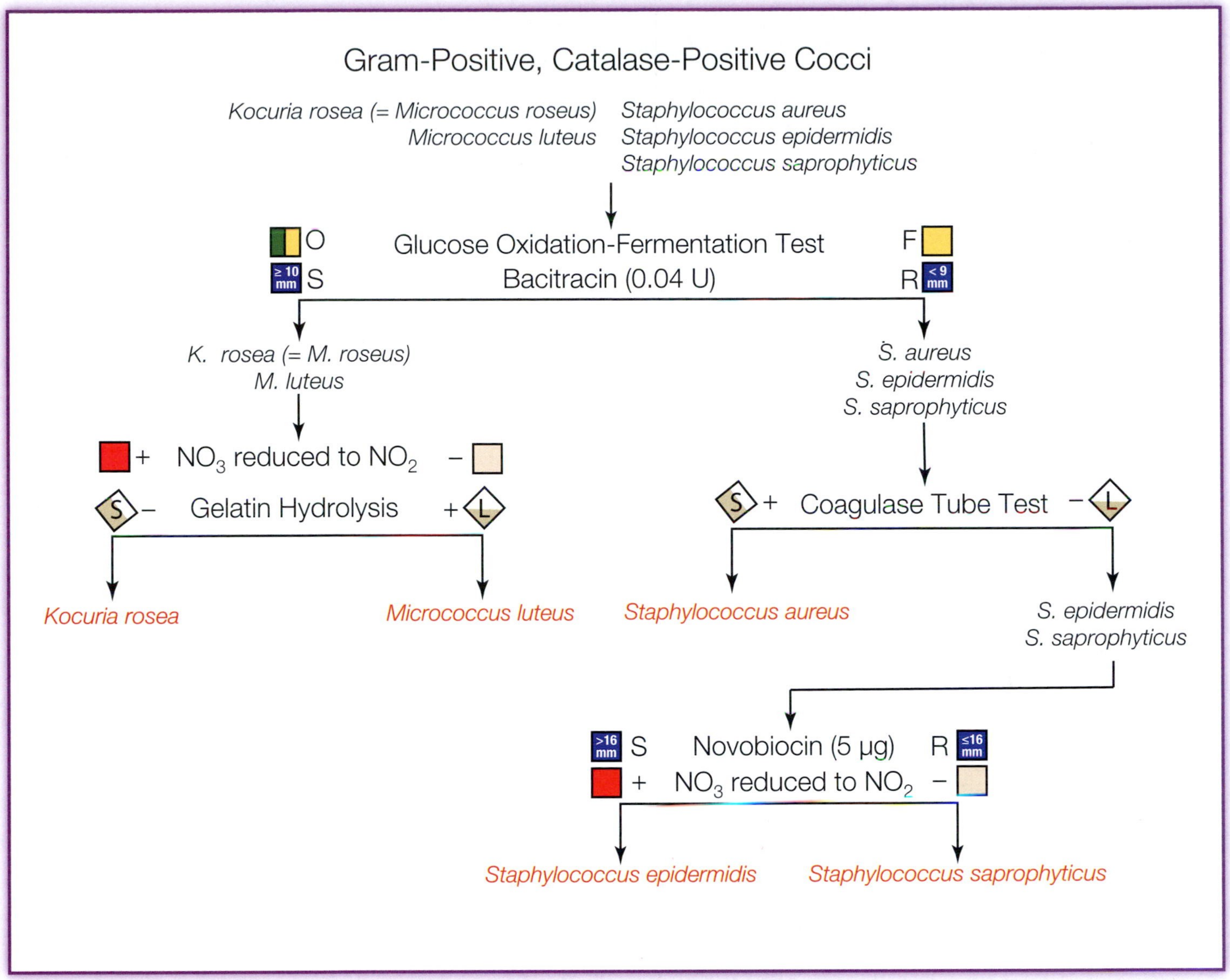
Gram-Positive, Catalase-Positive Cocci
Kocuria rosea (= Micrococcus roseus)
Micrococcus luteus
Staphylococcus aureus
Staphylococcus epidermidis
Staphylococcus saprophyticus
O
Glucose Oxidation-Fermentation Test
F
≥ 10 mm S
Bacitracin (0.04 U)
R < 9 mm
K. rosea (= M. roseus)
M. luteus
S. aureus
S. epidermidis
S. saprophyticus
+ NO3 reduced to NO2 −
S − Gelatin Hydrolysis + L
S + Coagulase Tube Test − L
Kocuria rosea
Micrococcus luteus
Staphylococcus aureus
S. epidermidis
S. saprophyticus
>16 mm S Novobiocin (5 µg) R ≤16 mm
+ NO3 reduced to NO2 −
Staphylococcus epidermidis
Staphylococcus saprophyticus

9-9 Identification Flowchart for Gram-positive, Catalase-Positive Cocci (Usually in Tetrads or Clusters) ◆ See the appropriate exercises for instructions on how to run each test. Novobiocin is covered in this exercise. In some cases, final identification to species is difficult. Try to get that far, but you may reach a point where your lab strains can't be differentiated any further. (Compiled from sources listed in the References.)

❖ Materials

Per Class

- ✓ media listed in Table 9-2 for biochemical testing
- ✓ pure cultures of organisms listed in Table 9-2 to be used for positive controls and unknowns (BSL–2 organisms should not be chosen by schools or classes not adequately prepared to handle them.)
- ✓ Todd-Hewitt Broth or Brain–Heart Infusion Broth
- ✓ candle jar set-up

Per Student

- ✓ two PEA, CNA, or 5% Sheep Blood Agar plates
- ✓ Gram-stain kit
- ✓ acid-fast stain kit
- ✓ gloves
- ✓ microscope slides
- ✓ 3% H_2O_2
- ✓ one unknown organism[1] from the list in Table 9-2
- ✓ compound microscope with oil objective lens and ocular micrometer
- ✓ lens paper
- ✓ immersion oil

[1] Note to instructor: This exercise can be combined with Exercise 9-1 by passing out a mixture of a Gram-negative enteric and a Gram-positive coccus as unknowns. The student then must isolate the two organisms from a mixed culture, and then proceed to identify the two independently. If this option is chosen, grow the two unknowns separately, and then mix them immediately prior to handing them out in lab.

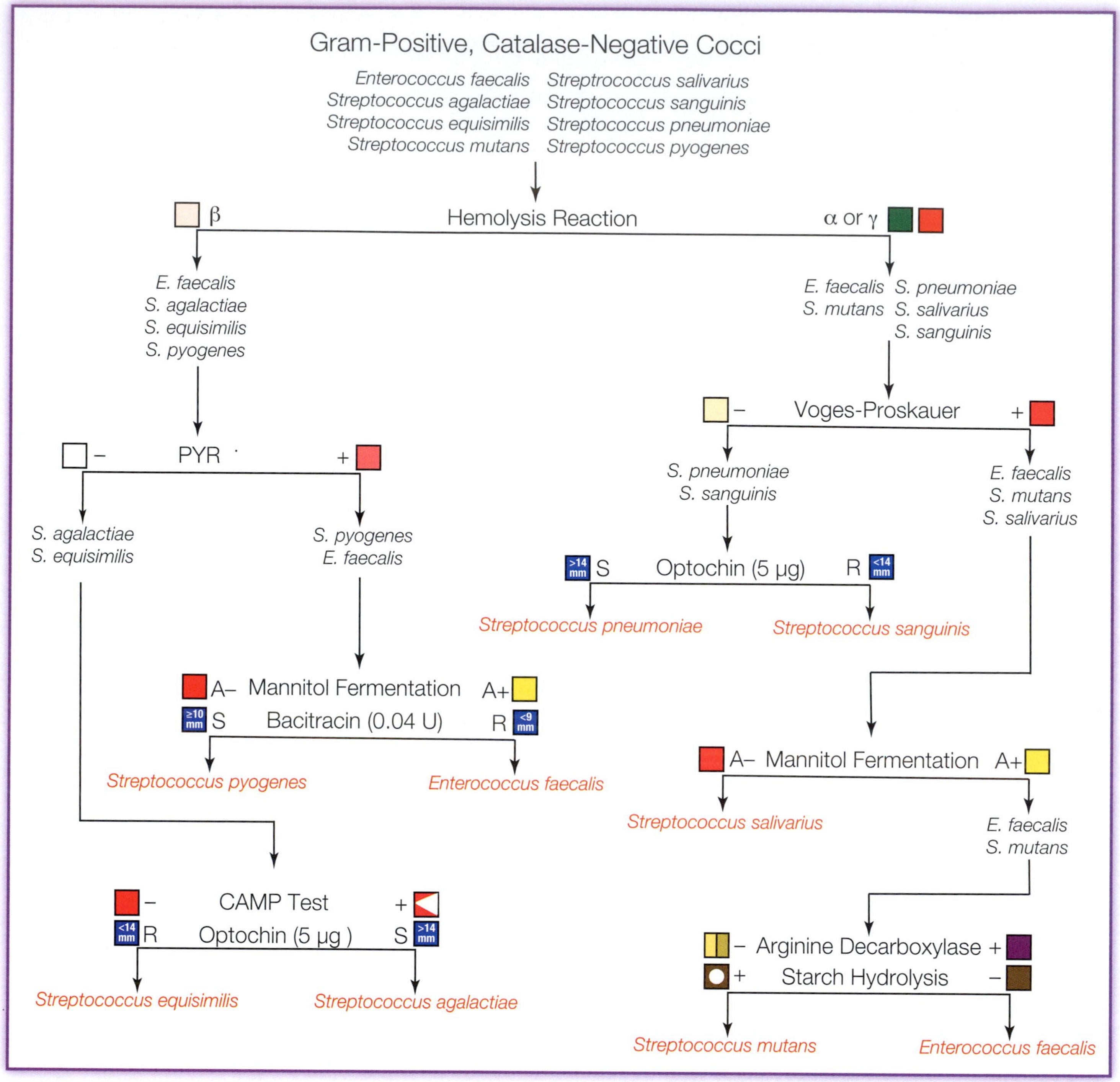

9-10 Identification Flowchart for Gram-positive, Catalase-Negative Cocci (Usually in Pairs or Chains) ◆ See the appropriate exercises for instructions on how to run each test. In some cases, final identification to species is difficult. Try to get that far, but you may reach a point where your lab strains can't be differentiated any further. (Compiled from sources listed in References.)

Procedure

1 Obtain an unknown. Record its number on the Data Sheet.

2 Streak the sample for isolation on two Sheep Blood Agar, PEA, or CNA plates. Incubate one in the candle jar and the other aerobically. Incubate both at 30–35°C. It is best to check your plate for isolation at 24 hours. If you have isolation, continue with Step 3. If you don't have time to continue with Step 3, refrigerate your plates until you do. If you don't see isolation, ask your instructor if you should restreak for isolation, let the original plates continue to incubate, or do both. Record your isolation procedure as directed on the Data Sheet.

3 After incubation, record colony morphology (including medium), optimum growth temperature, and CO_2 requirement under "Preliminary Results" on the Data Sheet. Also include the result of the differential component of blood agar as Test #1 if you used it. Continue to record your isolation procedure on the Data Sheet.

4 Perform a Gram stain on a portion of one well-isolated colony. Continue testing colonies until you find one that is a Gram-positive coccus.

- Record its Gram reaction and cell morphology, arrangement and size under "Preliminary Observations" on the Data Sheet.
- If the isolate is a Gram-positive coccus, transfer a portion of the *same* colony to a Todd-Hewitt Broth (or another suitable growth medium as available in your lab) and incubate it at 35±2°C. This is your pure culture to be used as a source of organisms for further testing. Complete the record of your isolation procedure on the Data Sheet.

5 Perform a catalase test on what remains of the isolated colony, or wait until your pure culture has grown and do it then. Record the result as Test #1 (or #2 if hemolysis reaction is already recorded) under Differential Tests on the Data Sheet. (Note: Be sure to use growth from the top of the colony if isolation was performed on a blood agar plate. This minimizes the possibility of a false positive from the catalase-positive erythrocytes in the medium. Alternatively, you can postpone the catalase test until your pure culture has grown.)

6 Use the chart in Figure 9-9 for identification of catalase-positive cocci (usually in tetrads or clusters). Use the chart in Figure 9-10 if the isolate is catalase-negative (usually with cocci in pairs or chains). Record the Figure number of the flowchart you will use on the Data Sheet. If your isolate is catalase-negative, ask your instructor if you would get better growth of your pure culture using Todd-Hewitt Broth or Brain–Heart Infusion Broth and incubating in a candle jar or CO_2 incubator.

7 Now you will begin biochemical testing to identify your organism. The tests in the flowcharts were chosen because of their uniformity of results as published in standard microbiological references, so they give the best chance for correct identification. Be aware, however, that the symbol "+" indicates that 90% or more of the strains tested give a positive result. This means that as many as 10% of the strains tested give a negative result. The same is true of the symbol "–". That is, as many as 10% of the strains give a positive result. Bottom line: There are no guarantees that your lab strains will behave in the majority, and if not, you will misidentify your unknown. This issue, if relevant to your unknown, will be addressed in Step 10.

8 Follow the tests in the appropriate flowchart to identify your unknown.[2] Use these guidelines.

- Do not run more than one test at a time unless your instructor tells you to do so.
- Where multiple tests are listed at a branch point, run only one, and then move to the next level in the flowchart as indicated by the result of that test.
- Continue to record the relevant inoculation information and results on the Data Sheet as you go. Keep accurate records of what you have done. *Do not* enter all your data at the end of the project.

9 When you have identified your organism, use *Bergey's Manual* or another standard reference to find one more test to run for confirmation. The confirmatory test doesn't have to separate the final organisms on the flowchart. It only has to be a test for which you know the result and that you haven't run already as part of the identification process. Record this test on the Data Sheet, and identify it as the confirmatory test. Record the *expected* result on the comments line. After incubation, note whether the organism's result matches or differs from the expected result.

10 After the confirmatory test, write the organism's name on the back of the Data Sheet under "My Unknown is" and check with your instructor to see if you are correct. If you are, congratulations! If you aren't, your instructor will advise you as to which test(s) gave "incorrect" results by writing the test name(s) in the "rerun" space next to your identification. These should be rerun with appropriate controls from your school's inventory of organisms. Record the test(s) and control(s) on your Data Sheet. Checking the results with controls and your unknown will indicate if the "incorrect" test result was truly incorrect, or if the strain of organism your school is using doesn't match the majority of strains for that test result.

[2] These flowcharts can be used only for the organisms listed in Table 9-2.

References

Becker, Karsten, and Christof von Eiff. 2011. "*Staphylococcus, Micrococcus, and Other Catalase-Positive Cocci,*" Chapter 19 in *Manual of Clinical Microbiology,* 10th ed. Edited by James Versalovic (Editor-in Chief), Karen C. Carroll, Guido Funke, James H. Jorgensen, Marie Louise Landry, and David W. Warnock. ASM Press, Washington, DC.

MacFaddin, Jean F. 2000. *Biochemical Tests for Identification of Medical Bacteria*, 3rd ed. Lippincott Williams & Wilkins, Philadelphia, PA.

Ruoff, Kathryn L. 2011. *Aerococcus*, *Abiotrophia*, and Other Aerobic Catalase-Negative, Gram-Positive Cocci, Chapter 22 in *Manual of Clinical Microbiology, 10th ed*. Edited by James Versalovic (Editor-in Chief), Karen C. Carroll, Guido Funke, James H. Jorgensen, Marie Louise Landry, and David W. Warnock. ASM Press, Washington, DC.

Spellerberg, Barbara, and Claudia Brandt. 2011. "*Streptococcus*," Chapter 20 in *Manual of Clinical Microbiology,* 10th ed. Edited by James Versalovic (Editor-in Chief), Karen C. Carroll, Guido Funke, James H. Jorgensen, Marie Louise Landry, and David W. Warnock. ASM Press, Washington, DC.

Švec, Pavel and Luc A. Devriese. 2009. *Enterococcus*, pages 594-607 in *Bergey's Manual of Systematic Bacteriology, 2nd ed., Vol. 3, The Firmicutes*. Springer, New York.

Teixeira, Lúcia Martins, Maria S Glória Siqueira Carvalho, Patricia Lynn Shewmaker, and Richard R. Facklam. 2011. "*Enterococcus*," Chapter 21 in *Manual of Clinical Microbiology*, 10th ed. Edited by James Versalovic (Editor-in Chief), Karen C. Carroll, Guido Funke, James H. Jorgensen, Marie Louise Landry, and David W. Warnock. ASM Press, Washington, DC.

Teuber, Michael. 2009. Lactococcus, pages 711-722 in Bergey's Manual of Systematic Bacteriology, 2nd Ed., Vol. 3, The Firmicutes. Springer, New York.

Whiley, Robert A. and Jeremy M. Hardie. 2009. *Streptococcus*, pages 655-711 in *Bergey's Manual of Systematic Bacteriology, 2nd Ed., Vol. 3, The Firmicutes*. Springer, New York.

NAME ______________________ DATE ______________________

LAB SECTION ______________________ I WAS PRESENT AND PERFORMED THIS EXERCISE (INITIALS) ________

DATA SHEET 9-2

Identification of Gram-positive Cocci

ISOLATION PROCEDURE

Unknown Number ______________

Record all activities associated with isolation of your organisms—from mixed culture to pure culture. Always include the date, source of inoculum, destination, type of inoculation, incubation temperature, and any other relevant information. Also make note of transfers made to keep your pure culture fresh. (This log must be kept current.)

Preliminary Observations

Colony morphology (include medium and incubation temperature) ______________

Gram stain ______________ Cell dimensions ______________ CO_2 Requirement ______________

Cell morphology and arrangement ______________

Differential Tests

Begin recording with your Catalase test (and hemolysis result, if determined). This log must be kept current.

Record the Identification Chart you are using. If you change charts, indicate the figure number of the new chart and the date of the change.

Test #1: ______ Date Begun: ______ Date Read: ______ Result: ______

Comments: ______

Test #2: ______ Date Begun: ______ Date Read: ______ Result: ______

Comments: ______

Test #3: ______ Date Begun: ______ Date Read: ______ Result: ______

Comments: ______

Test #4: ______ Date Begun: ______ Date Read: ______ Result: ______

Comments: ______

Test #5: ______ Date Begun: ______ Date Read: ______ Result: ______

Comments: ______

Test #6: ______ Date Begun: ______ Date Read: ______ Result: ______

Comments: ______

Test #7: ______ Date Begun: ______ Date Read: ______ Result: ______

Comments: ______

Test #8: ______ Date Begun: ______ Date Read: ______ Result: ______

Comments: ______

Test #9: ______ Date Begun: ______ Date Read: ______ Result: ______

Comments: ______

Test #10: ______ Date Begun: ______ Date Read: ______ Result: ______

Comments: ______

Test #11: ______ Date Begun: ______ Date Read: ______ Result: ______

Comments: ______

Test #12: ______ Date Begun: ______ Date Read: ______ Result: ______

Comments: ______

My Unknown is: ______ Rerun:[1] ______

______ Rerun: ______

______ Rerun: ______

[1] Your instructor will write what tests to rerun in this space if you misidentify your unknown.

EXERCISE 9-3

Identification of Gram-positive Rods

Theory

Several genera of Gram-positive rods have environmental, industrial, and clinical importance. Among these are *Bacillus*, *Clostridium*, *Corynebacterium*, *Lactobacillus* and *Mycobacterium*. Each is briefly characterized below.

Bacillus (Figure 9-11)

- A large, heterogeneous group comprising 162 species (as of 2011)
- Gram-positive rods (at least early in growth—Figure 9-11b), in singles or chains
- Aerobic or facultatively anaerobic
- Produce endospores aerobically; spore shape and position are variable, but never more than one per cell
- Most are catalase positive

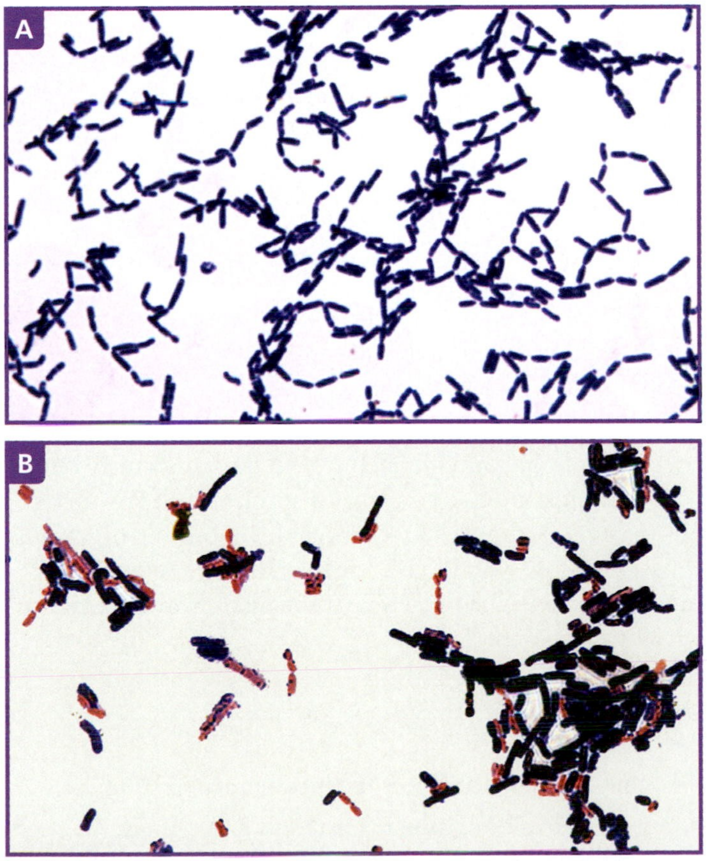

9-11 Gram Stains of *Bacillus cereus* ◆ **A** This Gram stain of a young (<24 hours old) *B. cereus* culture stains uniformly Gram-positive. **B** With age, many *Bacillus* species lose their ability to retain the crystal violet during a Gram stain and appear pink, some with purple spots, as in this specimen.

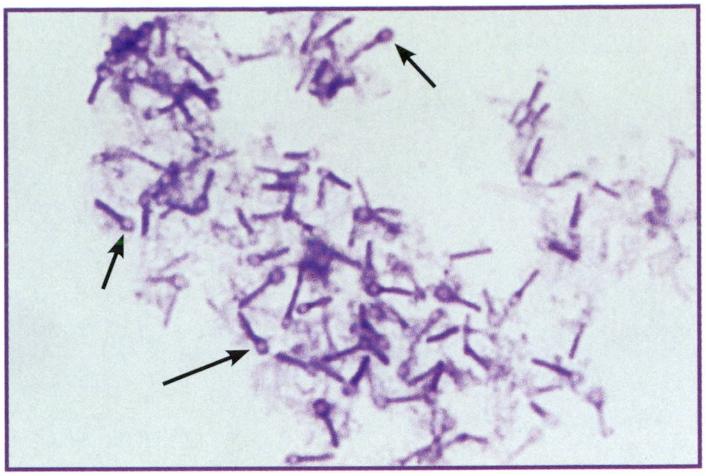

9-12 Gram Stain of *Clostridium tetani* ◆ Note the round, terminal endospores (**arrows**) that distend the cells.

- Most are motile
- Most are soil saprophytes and are chemoorganotrophic
- Key pathogens are *B. anthracis* (anthrax) and *B. cereus* (food poisoning and opportunistic infections)

Clostridium (Figure 9-12)

- A large, heterogeneous genus comprising over 160 species belonging to at least 12 different groups
- Gram-positive rods (at least early in growth), in singles, pairs, or chains
- Most are obligate anaerobes, but some are microaerophiles
- Produce endospores, but not aerobically; spore shape and position are variable, but usually cause distention (bulging) of the cell
- Most are catalase negative
- Most are chemoorganotrophic and are isolated from soil, sewage, or marine sediments
- Key pathogens are *Cl. tetani* (tetanus), *Cl. botulinum* (botulism), *Cl. perfringens* (food poisoning and gas gangrene), and *Cl. difficile* (pseudomembranous colitis)

Corynebacterium (Figure 9-13)

- Gram-positive rods, often club-shaped, in singles, pairs (V-forms), or arranged in stacks (palisades) or irregular clusters
- Metachromatic granules often present
- Facultatively anaerobic, but some are aerobes
- Catalase positive
- No endospores
- Nonmotile
- Some are found in the environment, and many are in the normal flora of humans (skin and mucous membranes)
- Key pathogen is *C. diphtheriae* (diphtheria), although many species are opportunistic pathogens

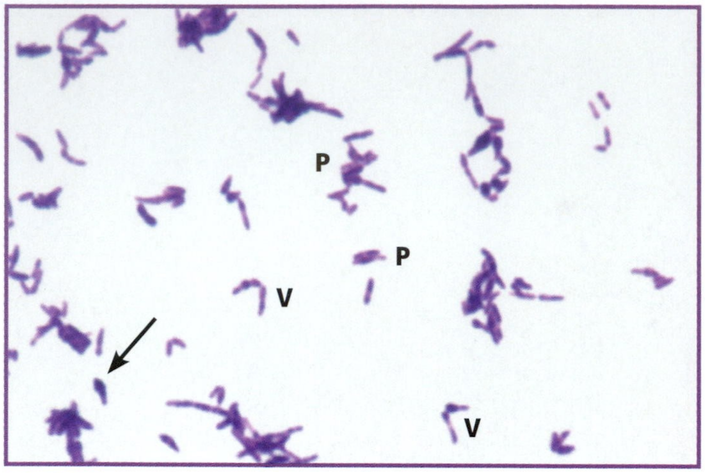

9-13 **Gram Stain of *Corynebacterium diphtheriae*** ◆ Note the club-shaped cells that might be mistaken for spore-producers (**arrow**). Cells of this genus often appear as "V-shaped" pairs (**V**) or palisades (**P**).

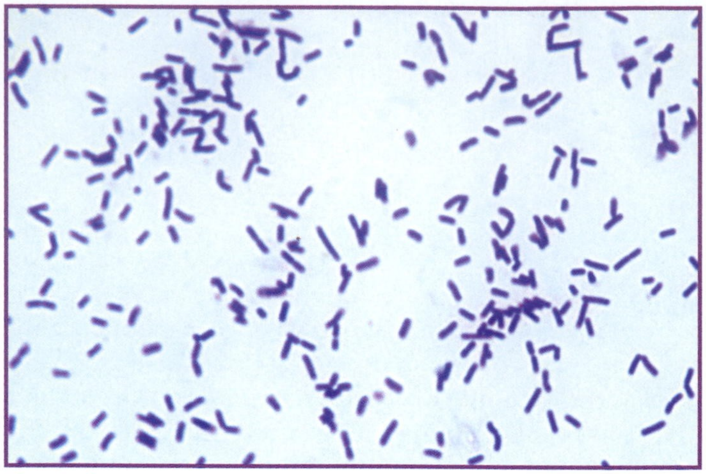

9-14 **Gram Stain of *Lactobacillus* spp.** ◆ Cells in the genus *Lactobacillus* range from long, slender rods to coocobacilli.

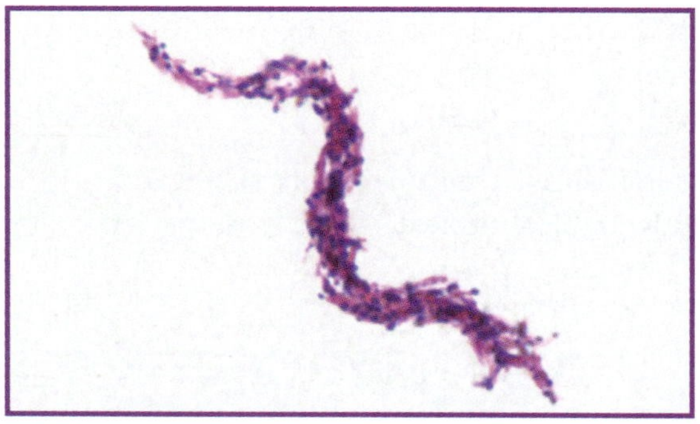

9-15 **Acid-fast Stain of *Mycobacterium tuberculosis*** ◆ Note the characteristic cord-like arrangement of cells.

Lactobacillus (Figure 9-14)

- ◆ Gram-positive rods, sometimes in chains
- ◆ Faculatively anaerobic with fermentative metabolism; lactose is an abundant product
- ◆ Catalase negative
- ◆ Oxidase negative
- ◆ No endospores
- ◆ Nonmotile
- ◆ Isolated from a variety of foods (dairy, fish, grain, meat), and many are in normal flora (mouth, intestines, and vagina)
- ◆ No key pathogens and therefore not usually identified to species in clinical setting

Mycobacterium (Figure 9-15)

- ◆ Over 130 recognized species divided into slow growers (>7 days to form colonies on Löwenstein-Jensen medium—about 70 species) and rapid growers (<7 days to form colonies on L-J medium—about 60 species)
- ◆ Weakly Gram-positive curved or straight rods, sometimes branched or filamentous
- ◆ Acid-fast (generally in early stages of growth)
- ◆ Aerobic
- ◆ Catalase positive
- ◆ Nonmotile
- ◆ No endospores
- ◆ Colony pigmentation with and without light is of taxonomic use
- ◆ Many are aquatic saprophytes
- ◆ Key pathogens are *M. tuberculosis* (tuberculosis) and *M. leprae* (leprosy or Hansen's disease). Many rapidly growing species are opportunists.

❖ Application

Staining reactions and biochemical testing are used widely in bacterial identification.

❖ In This Exercise

This exercise will span several lab periods. You will combine staining and biochemical testing to identify one of several unknown organisms to species level. For obvious reasons, pathogens are not included, but the organisms chosen will give you an overview of the characteristics of each genus as well as practice in the identification process.

❖ Materials

Per Class

- ✓ media listed in Table 9-3 for biochemical testing.
- ✓ pure cultures of organisms listed in Table 9-3 to be used for positive controls and unknowns (BSL–2 organisms should not be chosen as unknowns by schools or classes not adequately prepared to handle them).
- ✓ anaerobic jars and GasPaks to accommodate one plate per student

Gram-Positive Rods and Identification Tests

Bacillus cereus	*Clostridium sporogenes*
Bacillus coagulans	*Corynebacterium pseudodiphteriticum* (BSL-2)
Bacillus megaterium	*Corynebacterium xerosis*
Bacillus subtilis	*Lactobacillus plantarum*
Bacillus thuringiensis	*Lactobacillus delbrueckii*
Brevibacillus brevis	*Mycobacterium smegmatis*
Clostridium butyricum	

+	–	Test
		Acid-Fast Reaction (24-Hour Culture)
		Casein Hydrolysis
		Catalase
		Citrate Utilization
G/G	G/NG	Facultative Anaerobe (Thioglycollate)
		Lipid Hydrolysis (Lipase)
		Lysine Decarboxylase
M	NM	Motility
		Oxidation-Fermentation (Glucose O-F)
		Phenylalanine Deaminase
		Starch Hydrolysis (Amylase)
		VP
L	S	Gelatin Hydrolysis (Gelatinase)

A+	A–	Test
		Glucose Fermentation
		Mannitol Fermentation

TABLE 9-3 List of Organisms and a Key to the Icons Used in This Exercise ◆ The organisms listed are Gram-positive rods available from Ward's or Carolina Biological Supply Companies. Icon colors are to remind you what the results look like. They should not be used to compare your results. Note: For the fermentation tests, "A+" means acid is produced; "A−" means no acid is produced.

Per Student

- ✓ compound microscope with ocular micrometer and oil objective
- ✓ immersion oil
- ✓ clean glass slides
- ✓ Gram-stain kit
- ✓ gloves
- ✓ lens paper
- ✓ 3% H_2O_2
- ✓ one unknown organism in Thioglycollate Broth[1]
- ✓ two agar plates for streaking: instructor will choose from Trypticase Soy Agar, TSA plus 5% sheep blood, or Brain–Heart Infusion Agar
- ✓ TSA slant or Thioglycollate Broth

[1] *Note to instructor:* This exercise can be combined with Exercise 9-1 by passing out a mixture of a Gram-positive rod and an enteric. The student must isolate the two organisms from mixed culture and then proceed to identify the two independently. If this option is chosen, grow the two unknowns separately and then mix them immediately prior to handing them out in lab.

Procedure

1. Obtain an unknown and record its number on the Data Sheet.
2. Streak the sample for isolation on two agar plates (medium as provided by the instructor). Incubate both at 30°–35°C, one aerobically and the other in the anaerobic jar. It is best to check your plate for isolation at 24 hours. If you have isolation, continue with Step 3. If you don't have time to continue with Step 3, refrigerate your plates until you do. If you don't see isolation, ask your instructor if you should restreak for isolation, let the original plates continue to incubate, or do both. Record your isolation procedure as directed on the Data Sheet.
3. After incubation, record colony morphology (including medium) under "Preliminary Observations" and note any differences between aerobic and anaerobic growth. Continue to record your isolation procedure on the Data Sheet.

4 Perform a Gram stain on a portion of one well-isolated colony.

- Record its Gram reaction and cell morphology, arrangement, and size under "Preliminary Observations" on the Data Sheet.
- If the isolate is a Gram-positive rod (it should be!), transfer a portion of the same colony to a Trypticase Soy Agar slant (or another suitable growth medium as available in your lab) or Thioglycollate Broth (if anaerobic) and incubate it at 30°–35°C. This is your pure culture to be used as a source of organisms for further testing. Complete the record of your isolation procedure on the Data Sheet.

5 You now will begin biochemical testing to identify your organism. The tests in the flowchart (Figure 9-16) were chosen because of their uniformity of results, as published in standard microbiological references, so they give the best chance of correct identification. Be aware, however, that the symbol "+" indicates that 90% or more of the strains tested give a positive result. This means that as many as 10% of the strains tested give a negative result. The same is true of the symbol "−." That is, as many as 10% of the strains give a positive result for that test. Bottom line: There are no guarantees that your particular lab strains will behave in the majority and if not, you will misidentify your unknown. This issue, if relevant to your unknown, will be addressed in Step 10.

6 Follow the tests in Figure 9-16 to identify your unknown.[2] Use these guidelines.

- Do not run more than one test at a time unless your instructor tells you to do so.
- If your unknown is an obligate anaerobe, you will have to incubate media in an anaerobic jar. Coordinate your incubation with others who need the anaerobic jar to conserve GasPaks and incubator space.
- Where multiple tests are listed at a branch point, run only one, and then move to the next level in the flowchart as indicated by the result of that test.
- Continue to record the relevant information and results on the Data Sheet as you go. Keep accurate records of what you have done. Do *not* enter all of your data after you complete the project.

[2] This flowchart can be used only with the organisms listed in Table 9-3.

7 When you have identified your organism, use *Bergey's Manual* or another standard reference to find one more test to run for confirmation. The confirmatory test doesn't have to separate the final organisms on the flowchart. It only has to be a test for which you know the result and that you haven't run already as part of the identification process. Record this test on the Data Sheet and identify it as the confirmatory test. Record the *expected* result on the comments line. After incubation, note whether the organism's result matches or differs from the expected result.

8 After the confirmation test, write the organism's name on the back of the Data Sheet under "My Unknown is" and check with your instructor to see if you are correct. If you are, congratulations! If you aren't, your instructor will advise you as to which test(s) gave "incorrect" results, by writing the test name(s) in the "rerun" space next to your identification. These should be rerun with appropriate controls from your school's inventory of organisms. Record the test(s) and control(s) on your Data Sheet. Checking the results with controls and your unknown will indicate if the "incorrect" test result was truly incorrect or if the strain of organism your school is using doesn't match the majority of strains for that test result.

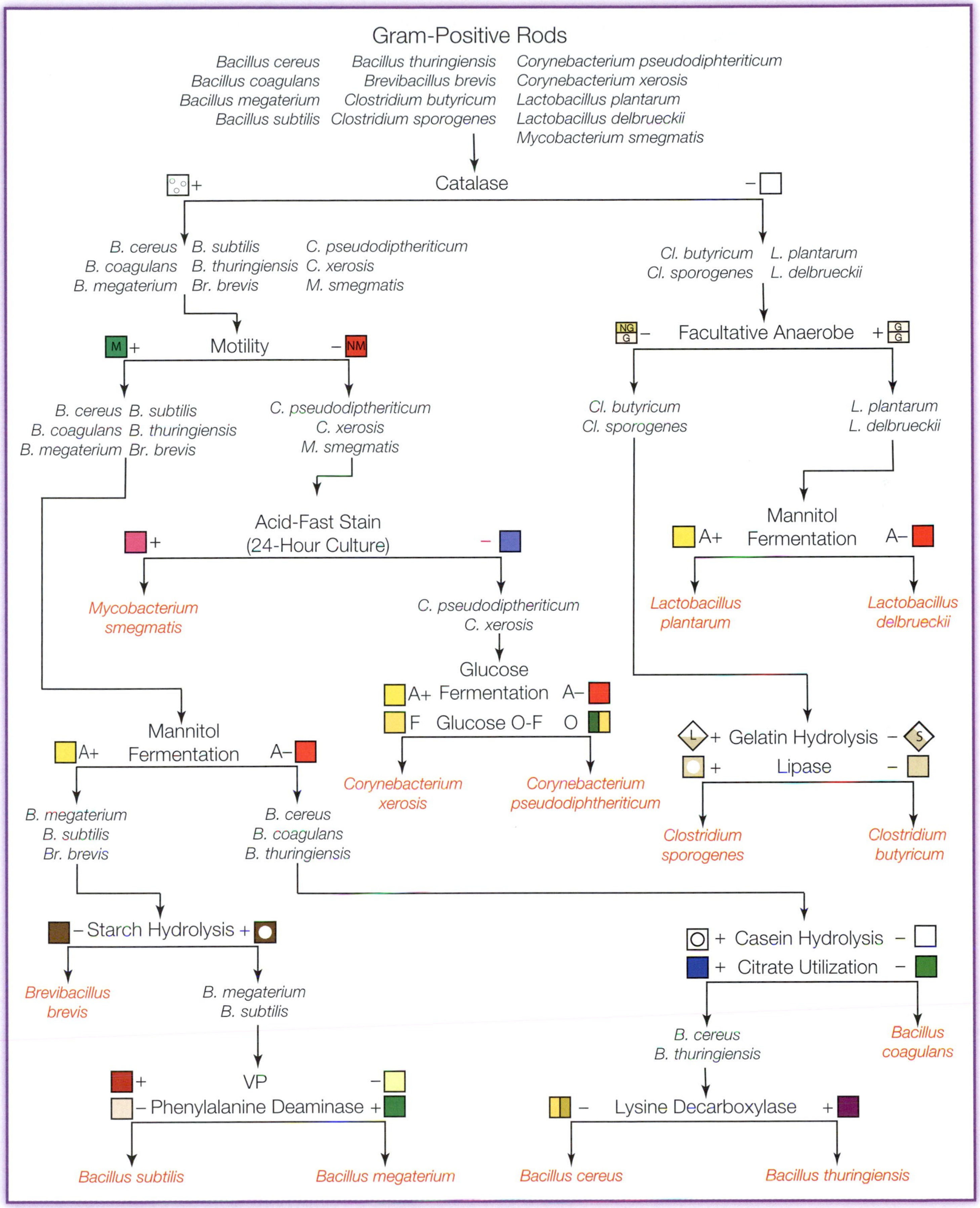

9-16 Identification Flowchart for Gram-positive Rods ◆ See the appropriate exercises for instructions on how to run each test.

(Compiled from sources listed in References)

References

Brown-Elliot, Barbara A, and Richard J. Wallace, Jr. 2011. "*Mycobacterium*: Clinical and Laboratory Characteristics of Rapidly Growing *Mycobacteria*," Chapter 30 in *Manual of Clinical Microbiology*, 10th ed. Edited by James Versalovic (Editor-in Chief), Karen C. Carroll, Guido Funke, James H. Jorgensen, Marie Louise Landry, and David W. Warnock. ASM Press, Washington, DC.

Collins, M. D., and C. S. Cummins. 1986. Section 15 (*Corynebacterium*) in *Bergey's Manual of Systematic Bacteriology*, Vol. 2, edited by John G. Holt. Williams and Wilkins, Baltimore, MD.

Funke, Guido, and Kathryn A. Bernard. 2007. "*Coryneform* Gram-Positive Rod," Chapter 34 in *Manual of Clinical Microbiology*, 9th ed., edited by Patrick R. Murray, Ellen Jo Baron, James H. Jorgensen, Marie Louise Landry, and Michael A. Pfaller. ASM Press, Washington, DC.

Hammes, Walter P. and Christian Hertel. 2009. *Lactobacillus*, pages 465-511 in *Bergey's Manual of Systematic Bacteriology*, 2nd ed., Vol. 3, *The Firmicutes*. Springer, New York.

Logan, Niall A. and Paul De Vos. 2009. *Bacillus*, pages 21-128 in *Bergey's Manual of Systematic Bacteriology*, 2nd ed., Vol. 3, *The Firmicutes*. Springer, New York.

Logan, Niall A., Alex R. Hoffmaster, Sean V. Shadomy, and Kendra E. Stauffer. 2011. "*Bacillus* and Other Aerobic Endospore-Forming Bacteria," Chapter 24 in in *Manual of Clinical Microbiology*, 10th ed. Edited by James Versalovic (Editor-in Chief), Karen C. Carroll, Guido Funke, James H. Jorgensen, Marie Louise Landry, and David W. Warnock. ASM Press, Washington, DC.

MacFaddin, Jean F. 2000. *Biochemical Tests for Identification of Medical Bacteria*, 3rd ed. Lippincott Williams & Wilkins, Philadelphia, PA.

Pfyffer, Gaby E. and Frantiska Palicova. 2011. "*Mycobacterium*: General Characteristics, Laboratory Detection, and Staining Procedures," Chapter 28 in *Manual of Clinical Microbiology*, 10th ed. Edited by James Versalovic (Editor-in Chief), Karen C. Carroll, Guido Funke, James H. Jorgensen, Marie Louise Landry, and David W. Warnock. ASM Press, Washington, DC.

Rainey, Fred A., Becky Jo Hollen, and Alanna Small. 2009. *Clostridium*, pages 738-828 in *Bergey's Manual of Systematic Bacteriology*, 2nd ed., Vol. *3*, *The Firmicutes*. Springer, New York.

Stevens, Dennis L., Amy E. Bryant, Anja Berger, and Christoph von Eichel-Streiber. 2011. "*Clostridium*," Chapter 50 in *Manual of Clinical Microbiology*, 10th ed. Edited by James Versalovic (Editor-in Chief), Karen C. Carroll, Guido Funke, James H. Jorgensen, Marie Louise Landry, and David W. Warnock. ASM Press, Washington, DC.

Wayne, Lawrence G., and George P. Kubrica. 1986. Section 15 (*Mycobacterium*) in *Bergey's Manual of Systematic Bacteriology*, Vol. 2, edited by John G. Holt. Williams and Wilkins, Baltimore, MD.

NAME ______________________ DATE ______________________

LAB SECTION ______________________ I WAS PRESENT AND PERFORMED THIS EXERCISE (INITIALS) ________

Identification of Gram-positive Rods

ISOLATION PROCEDURE

Unknown Number ______________

Record all activities associated with isolation of your organisms—from mixed culture to pure culture. Always include the date, source of inoculum, destination, type of inoculation, incubation temperature, and any other relevant information. Also make note of transfers made to keep your pure culture fresh. (This log must be kept current.)

Preliminary Observations

Colony morphology (include medium, incubation temperature, and differences between aerobic and anaerobic growth, if applicable) ______________________

Gram stain ______________ Cellular morphology, arrangement and dimensions ______________

Differential Tests

Begin recording with your catalase result. This log must be kept current.

Record the Identification Chart you are using. If you change charts, indicate the figure number of the new chart and the date of the change.

Test #1: ______ Date Begun: ______ Date Read: ______ Result: ______

Comments: ______

Test #2: ______ Date Begun: ______ Date Read: ______ Result: ______

Comments: ______

Test #3: ______ Date Begun: ______ Date Read: ______ Result: ______

Comments: ______

Test #4: ______ Date Begun: ______ Date Read: ______ Result: ______

Comments: ______

Test #5: ______ Date Begun: ______ Date Read: ______ Result: ______

Comments: ______

Test #6: ______ Date Begun: ______ Date Read: ______ Result: ______

Comments: ______

Test #7: ______ Date Begun: ______ Date Read: ______ Result: ______

Comments: ______

Test #8: ______ Date Begun: ______ Date Read: ______ Result: ______

Comments: ______

Test #9: ______ Date Begun: ______ Date Read: ______ Result: ______

Comments: ______

Test #10: ______ Date Begun: ______ Date Read: ______ Result: ______

Comments: ______

Test #11: ______ Date Begun: ______ Date Read: ______ Result: ______

Comments: ______

Test #12: ______ Date Begun: ______ Date Read: ______ Result: ______

Comments: ______

My Unknown is: ______ Rerun:[1] ______

______ Rerun: ______

______ Rerun: ______

[1] Your instructor will write what tests to rerun in this space if you misidentify your unknown.

Multiple Test Systems

Multiple test systems are systems designed to run an entire battery of tests simultaneously. They employ the same biochemical principles that you encountered in Section 5 and are read essentially in the same manner. All conditions possible with standard tubed media can be achieved in a fraction of the time with multiple test media, including aerobic or anaerobic growth conditions and the addition of reagents for confirmatory testing.

These exercises will give you an opportunity to have fun while using some of the skills you have learned. The organisms chosen for these tests will be provided as unknowns for you to identify. Alternatively, your instructor may allow you to isolate an environmental sample to identify. Finally, the system you use may be used in conjunction with your bacterial unknown project or with one of the identification exercises in this section.

EXERCISE 9-4

API 20 E Identification System for *Enterobacteriaceae* and Other Gram-negative Rods

Theory

The API 20 E system is a plastic strip containing 20 microtubes and cupules, partially filled with different dehydrated substrates. Bacterial suspension is added to the microtubes, which simultaneously rehydrates and inoculates the media. As with the other biochemical tests in this section, color changes take place in the tubes either during incubation or after addition of reagents. These color changes indicate positive or negative results (Figure 9-18).

Following incubation, reactions that do not require addition of reagents are evaluated and recorded on the Result Sheet (Figure 9-19). After these "spontaneous results" have been recorded, reagents are added to those microtubes requiring them and the results are recorded on the Results Sheet. An oxidase test is performed separately and constitutes the 21st test.

As shown in Figure 9-19, the Result Sheet divides the tests into seven groups of three, with the members of a group having numerical values of 1, 2, and 4, respectively. These numbers are assigned (entered in the circle at the top of the microtube) for positive (+) results only. Negative (−) results are not counted. After the three values for a group have been entered, they are added together to produce a number from 0 to 7. This value is entered in the oval below the group. The totals from each group are combined sequentially to produce a seven-digit code, which can then be interpreted on the Analytical Profile Index* (Figure 9-20).

In rare instances, information from the 21 tests (and the seven-digit code) is not discriminatory enough to identify an organism. When this occurs, the organism is grown and examined on MacConkey Agar and supplemental tests are performed for nitrate reduction, oxidation/reduction of glucose, and motility. The results are entered separately in the supplemental spaces on the Result Sheet and used for final identification.

*The index is now available only as a subscription through BioMérieux-USA.com.

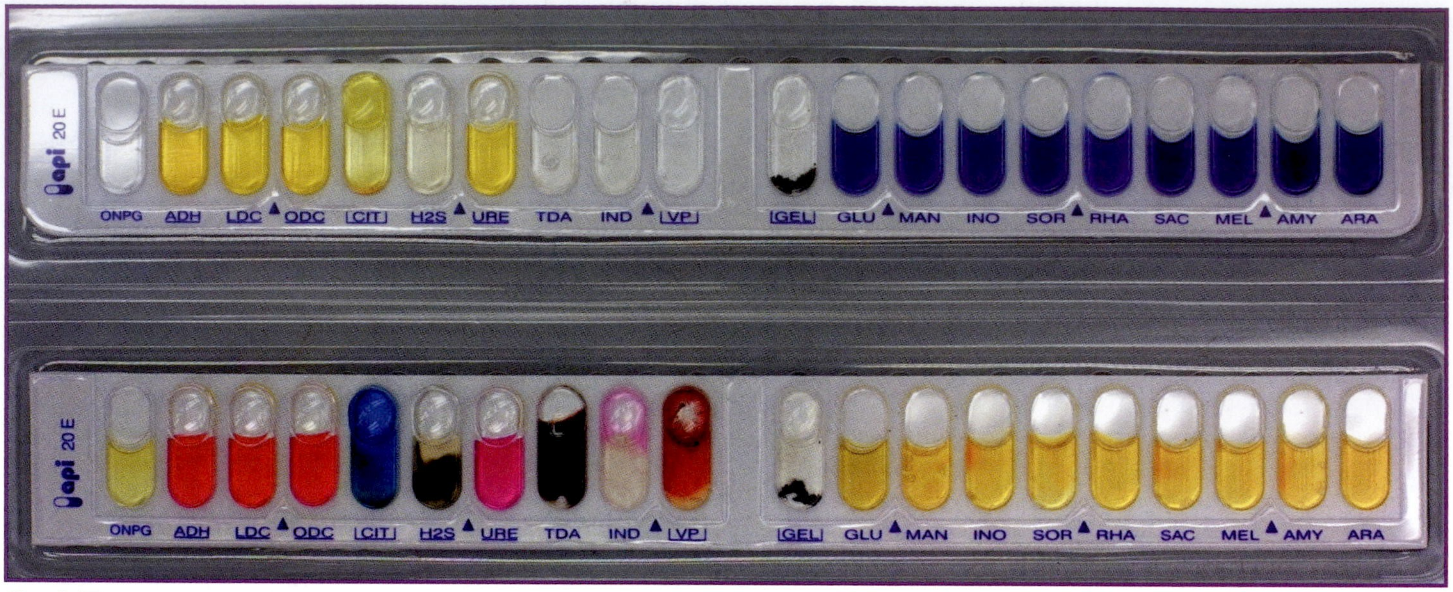

9-17 API 20 E Test Strips ◆ The top strip is uninoculated. The bottom strip (with the exception of GEL) illustrates all positive results.

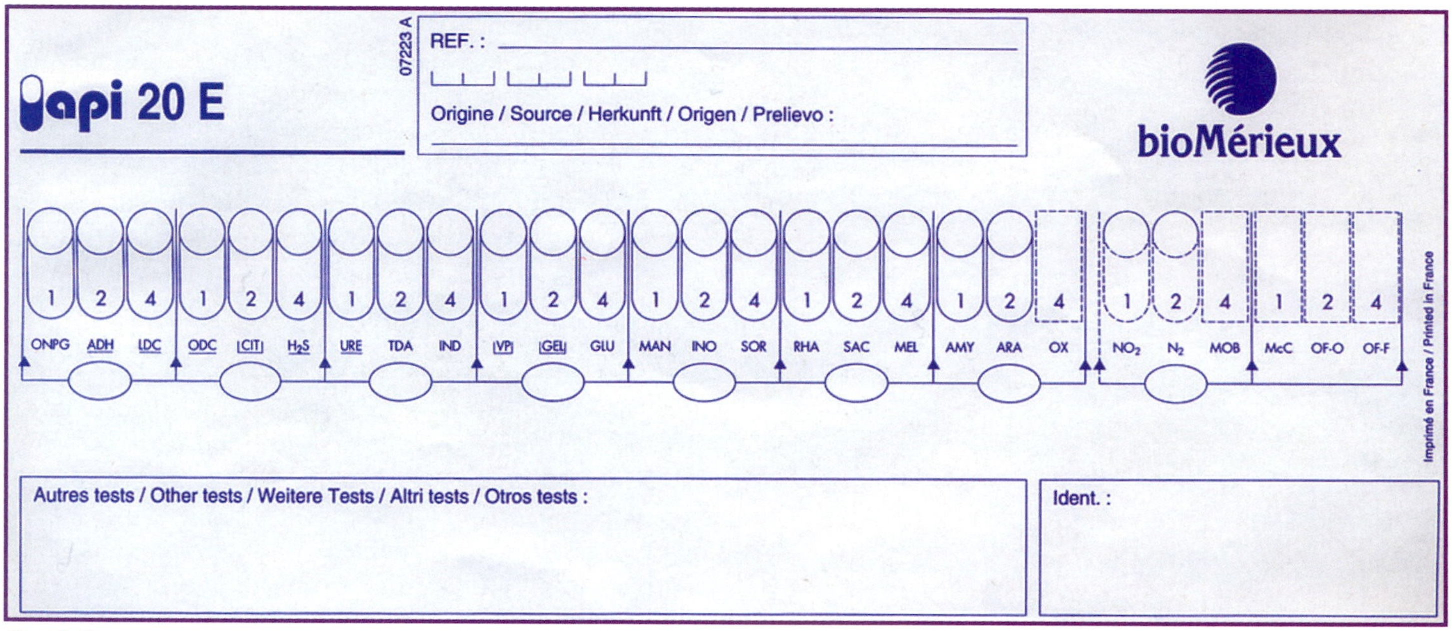

9-18 API 20 E Result Sheet ◆ The API 20 E Result Sheet divides the tests into seven groups of three. Each *positive* result is given the numerical value printed on the cupule (1, 2, or 4, respectively). Negative results are zero. The sum of each triplet becomes the number entered in the oval below. The numbers in the first seven ovals combine to form a seven-digit code that can be interpreted using the Analytical Profile Index supplied by the company (visit http://www.BioMerieux-USA.com).

Application

The API 20 E multitest system (available from BioMérieux, Inc.) is used clinically for the rapid identification of *Enterobacteriaceae* (more than 5,500 strains) and other Gram-negative rods (more than 2,300 strains).

In This Exercise

Today you will use a multitest system to run an entire battery of tests designed to identify a member of the *Enterobacteriaceae*.

Materials

- ✓ API 20 E identification system for *Enterobacteriaceae* and other Gram-negative rods
- ✓ API Analytical Profile Index or online subscription
- ✓ sterile suspension medium
- ✓ ferric chloride reagent (see Exercise 5-9)
- ✓ Kovacs' reagent (see Exercise 5-18)
- ✓ potassium hydroxide reagent (see Exercise 5-3)
- ✓ alpha-naphthol reagent (see Exercise 5-3)

```
7 046 124      TRES BONNE IDENTIFICATION       VERY GOOD IDENTIFICATION        SEHR GUTE IDENTIFIZIERUNG       7 046 124
               A L'ESPECE                      TO THE SPECIES LEVEL            AUF SPEZIES EBENE
                                                           |                       GLUCOSEg  ESC (HYD.)  O/129 R     RM/MR
  Aer.hydrophila gr. 1   %id=56.1 T=0.77 (LDC   25%)(AMY   75%) |Aeromonas caviae        -        +          +          +
                                         (ARA   75%)            |Aeromonas hydrophila    +        +          +          86%
  Aer.hydrophila gr. 2   %id=43.1 T=0.73 (CIT   80%)(VP    80%) |Vibrio fluvialis        0%       NT         -          NT
                                         (AMY   75%)            |Aeromonas sobria        +        -          +          -
  -POSSIBILITE DE Vibrio fluvialis        -POSSIBILITY OF Vibrio fluvialis       -MOEGLICHKEIT VON Vibrio fluvialis
--------------------------------------------------------------------------------------------------------------------------
7 046 125      EXCELLENTE IDENTIFICATION       EXCELLENT IDENTIFICATION        AUSGEZEICHNETE IDENTIFIZIERUNG  7 046 125
               A L'ESPECE                      TO THE SPECIES LEVEL            AUF SPEZIES EBENE
                                                           |                       GLUCOSEg  ESC (HYD.)  O/129 R     RM/MR
  Aer.hydrophila gr. 1   %id=56.5 T=0.84 (LDC   25%)(ARA   75%) |Aeromonas caviae        -        +          +          +
  Aer.hydrophila gr. 2   %id=43.4 T=0.80 (CIT   80%)(VP    80%) |Aeromonas hydrophila    +        +          +          86%
                                                                |Vibrio fluvialis        0%       NT         -          NT
                                                                |Aeromonas sobria        +        -          +          -
  -POSSIBILITE DE Vibrio fluvialis        -POSSIBILITY OF Vibrio fluvialis       -MOEGLICHKEIT VON Vibrio fluvialis
--------------------------------------------------------------------------------------------------------------------------
7 046 126      BONNE IDENTIFICATION            GOOD IDENTIFICATION             GUTE IDENTIFIZIERUNG            7 046 126
                                                           |                       GLUCOSEg  ESC (HYD.)  O/129 R     RM/MR
  Aer.hydrophila gr. 1   %id=97.9 T=0.84 (LDC   25%)(AMY   75%) |Aeromonas caviae        -        +          +          +
                                                                |Aeromonas hydrophila    +        +          +          86%
                                                                |Vibrio fluvialis        0%       NT         -          NT
                                                                |Aeromonas sobria        +        -          +          -
  -POSSIBILITE DE Vibrio fluvialis        -POSSIBILITY OF Vibrio fluvialis       -MOEGLICHKEIT VON Vibrio fluvialis
```

9-19 A Portion of the Analytical Profile Index (API) ◆ Identification is made by locating the seven-digit number in the Analytical Profile Index. This information is also available through online subscription.

- ✓ sulfanilic acid reagent (see Exercise 5-6)
- ✓ N,N-Dimethyl-1-naphthylamine reagent (see Exercise 5-6)
- ✓ oxidase test slides (see Exercise 5-5)
- ✓ hydrogen peroxide
- ✓ zinc dust
- ✓ sterile mineral oil
- ✓ sterile Pasteur pipettes
- ✓ distilled water
- ✓ agar plates containing unidentified bacterial colonies

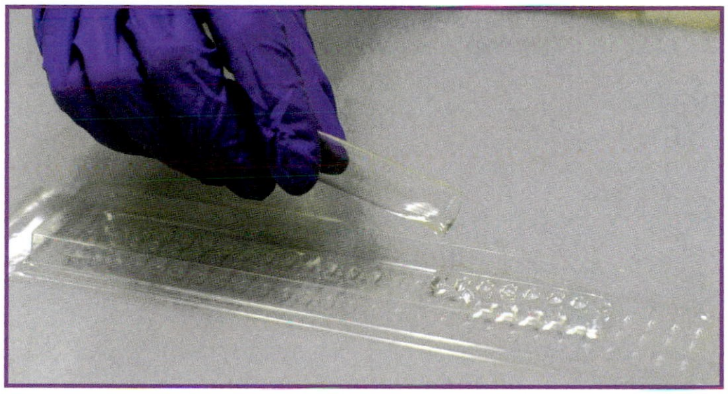

9-20 Add Water to Honeycombed Wells ◆ The purpose of the water is to humidify the strip during incubation. Spread the water as uniformly as possible in the honeycomb, but it is not necessary to add identical amounts to each well.

Procedure

Lab One

1 Perform a Gram stain on your unknown organism to confirm that it is a Gram-negative rod.

2 Open an incubation box and distribute about 5 mL of distilled water into the honeycombed wells of the tray to create a humid atmosphere (Figure 9-20).

3 Record the unknown number on the elongated flap of the tray.

4 Remove the strip from its packaging and place it in the tray.

5 Perform the oxidase test on a colony identical to the colony that will be tested. If you need help with this, refer to Exercise 5-5. Record the result on the score sheet as the 21st test.

6 Using a sterile Pasteur pipette, remove a single, well-isolated colony from the plate (Figure 9-21) and fully emulsify it in a tube of suspension medium or sterile 0.85% saline (Figure 9-22). Try not to pick up any agar.

7 Using the same pipette, fill both tube and cupule of test CIT, VP, and GEL with bacterial suspension (Figure 9-23). Fill only the tubes (not the cupules) of all other tests.

8 Overlay the ADH, LDC, ODC, H_2S, and URE microtubes by filling the cupules with sterile mineral oil.

9 Close the incubation box and incubate at 35 ± 2°C for 24 hours. (Note: If the tests cannot be read at 24 hours, have someone place the incubation box in a refrigerator until the tests can be read.)

9-21 **Pick a Colony** ◆ Being careful not to disturb the agar surface, pick up one isolated colony with a Pasteur pipette.

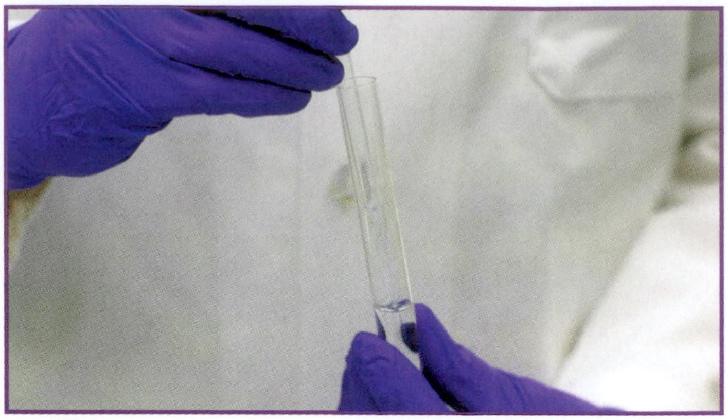

9-22 **Emulsify the Colony** ◆ Emulsify the colony in 5 mL suspension medium or sterile 0.85% saline. Do this by filling and emptying the pipette several times until uniform turbidity is achieved in the medium.

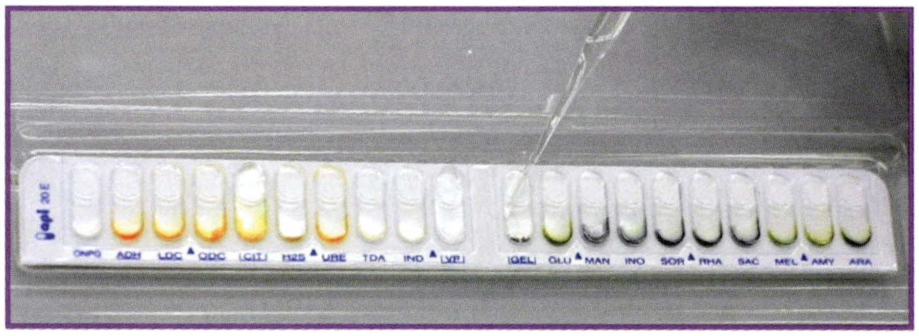

9-23 **Inoculate the Tubes** ◆ Inoculate the tubes by placing the pipette at the side of the tube and gently filling them with the suspension. If necessary to avoid creating bubbles, tilt the strip slightly. Tap gently with your index finger if necessary to remove bubbles.

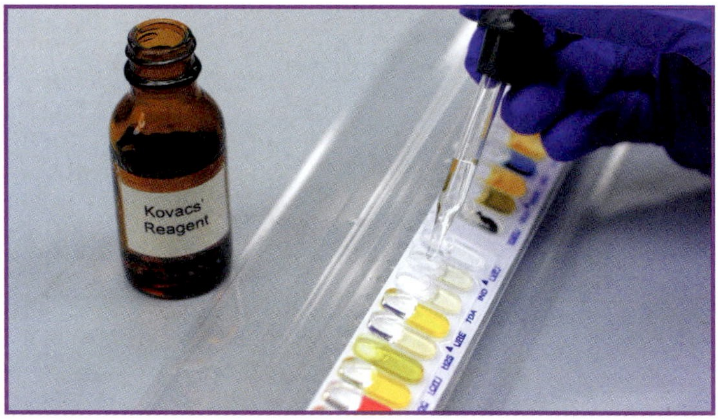

9-24 **Add Reagents** ◆ Follow the instructions in the text.

Lab Two

1 Refer to Table 9-4 (Table 2 on the package insert) as you examine the test results on the strip.

2 Record all spontaneous reactions on the Result Sheet. (Spontaneous reactions are reactions completed without the addition of reagents.)

3 Add 1 drop of ferric chloride to the TDA microtube (Figure 9-24). A dark brown color indicates a positive reaction to be recorded on the Result Sheet.

4 Add 1 drop of potassium hydroxide reagent and 1 drop of alpha-naphthol reagent to the VP microtube. A pink or red color indicates a positive reaction to be recorded on the Result Sheet. If a slightly pink color appears in 10 to 12 minutes, the action should be considered negative.

5 Add 2 drops of sulfanilic acid reagent and 2 drops of N,N-dimethyl-1-naphthylamine reagent to the GLU microtube. Wait 2 to 3 minutes. A red color indicates a positive reaction ($NO_3 \rightarrow NO_2$). Enter this on the Result Sheet as positive. A yellow color at this point is inconclusive because the nitrate may have been reduced to nitrogen gas ($NO_3 \rightarrow N_2$, sometimes evidenced by gas bubbles). If the substrate in the tube remains yellow after adding reagents, add 2 to 3 mg of zinc dust to the tube. A yellow tube after 5 minutes is positive for N_2 and is recorded on the Result Sheet as positive. If the test turns pink-red after the addition of zinc, this is a negative reaction; the nitrates *still present* in the tube have been reduced by the zinc.

6 Add 1 drop of Kovacs' reagent to the IND microtube. Wait 2 minutes. A red ring indicates a positive reaction to be recorded on the Result Sheet.

7 Add the positive results within each group on the Result Sheet, using the values given, and enter the number in the circle below each group's results (Figure 9-19).

8 Locate the seven-digit code in the Analytical Profile Index, and identify your organism. Enter the name on the Result Sheet and tape it on the Data Sheet. Answer the questions.

References

API 20 E Identification System for *Enterobacteriaceae* and Other Gram-Negative Rods package insert.

API 20 E Analytical Profile Index.

TABLE **9-4** **API 20 E Results and Interpretations** ◆

TABLE OF RESULTS					
#1	Test	Substrate/Activity	Result	Interpretation	Symbol
1	ONPG	O-nitophenyl-β-D-galactopyranoside	Yellow[1]	Organism produces β-galactosidase	+
			Colorless	Organism does not produce β-galactosidase	–
2	ADH	Arginine	Red-Orange	Organism produces arginine dehydrolase	+
			Yellow	Organism does not produce arginine dehydrolase	–
3	LDC	Lysine	Red-Orange[2]	Organism produces lysine decarboxylase	+
			Yellow	Organism does not produce lysine decarboxylase	–
4	ODC	Ornithine	Red-Orange[2]	Organism produces ornithine decarboxylase	+
			Yellow	Organism does not produce ornithine decarboxylase	–
5	CIT	Sodium citrate	Blue-green/Blue[3]	Organism utilizes citrate as sole carbon source	+
			Pale green/Yellow	Organism does not utilize citrate	–
6	H_2S	Sodium thiosulfate	Black	Organism reduces sulfur	+
			Colorless/grayish	Organism does not reduce sulfur	–
7	URE	Urea	Red-Orange[2]	Organism produces urease	+
			Yellow	Organism does not produce urease	–
8	TDA	Tryptophan	Brown-red	Organism produces tryptophan deaminase	+
			Yellow	Organism does not produce tryptophan deaminase	–
9	IND	Tryptophane	Red ring	Organism produces indole	+
			Yellow	Organism does not produce indole	–
10	VP	Creatine Sodium pyruvate	Pink-red[4]	Organism produces acetoin	+
			Colorless	Organism does not produce acetoin	–
11	GEL	Kohn's charcoal gelatin	Diffusion of black pigment	Organism produces gelatinase	+
			No black pigment diffusion	Organism does not produce gelatinase	–
12	GLU	Glucose[5]	Yellow	Organism ferments glucose	+
			Blue/blue-green	Organism does not ferment glucose	–
13	MAN	Mannitol[5]	Yellow	Organism ferments mannitol	+
			Blue/blue-green	Organism does not ferment mannitol	–
14	INO	Inositol[5]	Yellow	Organism ferments inositol	+
			Blue/blue-green	Organism does not ferment inositol	–
15	SOR	Sorbitol[5]	Yellow	Organism ferments sorbitol	+
			Blue/blue-green	Organism does not ferment sorbitol	–

[1] A very pale yellow is also positive.
[2] Orange after 36 hours is negative.
[3] Reading made in the cupule.
[4] A slightly pink color after 5 minutes is negative.
[5] Fermentation begins in the lower portion of the tube; oxidation begins in the cupule.

TABLE 9-4 API 20 E Results and Interpretations ◆ *(continued)*

TABLE OF RESULTS					
#1	Test	Substrate/Activity	Result	Interpretation	Symbol
16	RHA	Rhamnose[5]	Yellow	Organism ferments rhamnose	+
			Blue/blue-green	Organism does not ferment rhamnose	–
17	SAC	Sucrose (saccharose)[5]	Yellow	Organism ferments sucrose	+
			Blue/blue-green	Organism does not ferment sucrose	–
18	MEL	Melibiose[5]	Yellow	Organism ferments melibiose	+
			Blue/blue-green	Organism does not ferment melibiose	–
19	AMY	Amygdalin[5]	Yellow	Organism ferments amygdalin	+
			Blue/blue-green	Organism does not ferment amygdalin	–
20	ARA	Arabinose[5]	Yellow	Organism ferments arabinose	+
			Blue/blue-green	Organism does not ferment arabinose	–
21	OX	Separate test done on paper test strip	Violet	Organism possesses cytochrome-oxidase	+
			Colorless	Organism does not possess cytochrome-oxidase	–
22	GLU (nitrate reduction)	Potassium nitrate	Red after addition of reagents	Organism reduces nitrate to nitrate	+
			Yellow after addition of zinc	Organism does not reduce nitrate to nitrate	–
			Yellow after addition of reagents	Organism reduces nitrate to N_2 gas	+
			Orange-red after addition of zinc	Organism does not reduce nitrate	–
23	MOB	Motility Medium or wet mount slide	Motility	Organism is motile	+
			Nonmotility	Organism is not motile	–
24	McC	MacConkey Medium	Growth	Organism is probably *Enterobacteriaceae*	+
			No growth	Organism is not *Enterobacteriaceae*	–
25	OF	Glucose	Yellow under mineral oil	Organism ferments glucose	+
			Green under mineral oil	Organism does not ferment glucose (not *Enterobacteriaceae*)	–
			Yellow without mineral oil	Organism either ferments glucose or utilizes it oxidatively	+
			Green without mineral oil	Organism does not utilize glucose (not *Enterobacteriaceae*)	–

Name ______________________ Date ______________________

Lab Section ______________________ I was present and performed this exercise (initials) ________

API 20 E Identification System for *Enterobacteriaceae* and Other Gram-negative Rods

OBSERVATIONS AND INTERPRETATIONS

Tape your API 20 E Result Sheet here.

Questions

1 Why is it important to perform the reagent tests last?

2 In clinical applications of this test system, reagents are added only if the glucose (oxidation/fermentation) test result is yellow or at least three other tests are positive. If these conditions are not met, a MacConkey Agar plate is streaked and additional tests are performed confirming glucose metabolism, nitrate reduction, and motility. Why do you think this is so? Be specific.

3 Suppose, after 24 hours incubation, you notice no growth in the tubes containing mineral oil. Assuming that it is behaving properly under these conditions, what do you know about the organism and what predictions can you safely make about its performance in the decarboxylase tests, fermentation tests, and nitrate reduction test? Is it a member of *Enterobacteriaceae*?

EXERCISE 9-5

Enterotube® II

Theory

The Enterotube® II is a multiple test system designed to identify enteric bacteria based on: glucose, adonitol, lactose, arabinose, sorbitol, and dulcitol fermentation, lysine and ornithine decarboxylation, sulfur reduction, indole production, acetoin production from glucose fermentation, phenylalanine deamination, urea hydrolysis, and citrate utilization.

The Enterotube® II, as diagramed in Figure 9-25, is a tube containing 12 individual chambers. Inside the tube, running lengthwise through its center, is a removable wire. After the end-caps are removed (aseptically), one end of the wire is touched to an isolated colony (on a streak plate) and drawn back through the tube to inoculate the media in each chamber (Figures 9-26 through 9-31).

After 18 to 24 hours' incubation, the results are interpreted, the indole test is performed (Figure 9-31), and the tube is scored on an Enterotube® II Results Sheet (Figure 9-32). As shown in the figure, the combination of positives entered on the score sheet results in a five-digit numeric code. This code is used for identification in theEnterotube® II Computer Coding and Identification System (CCIS) (Figure 9-33).

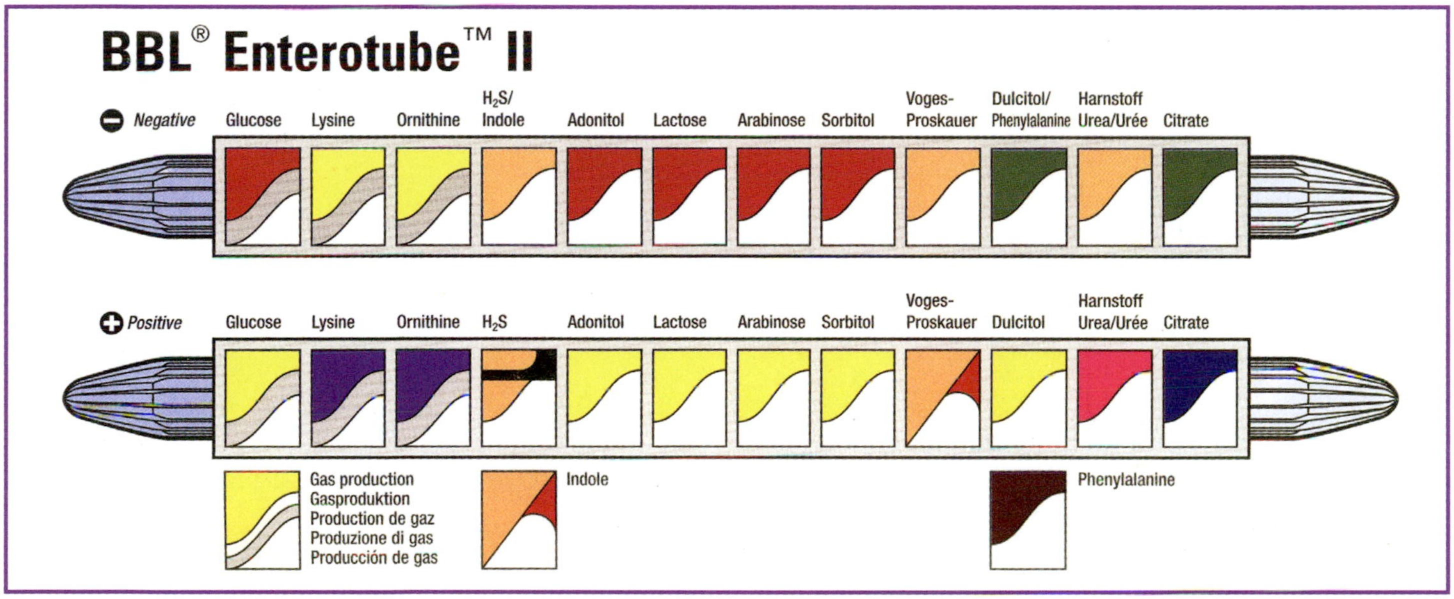

Illustration courtesy Becton Dickinson and Co.

9-25 **Enterotube® II Test Result Diagram**

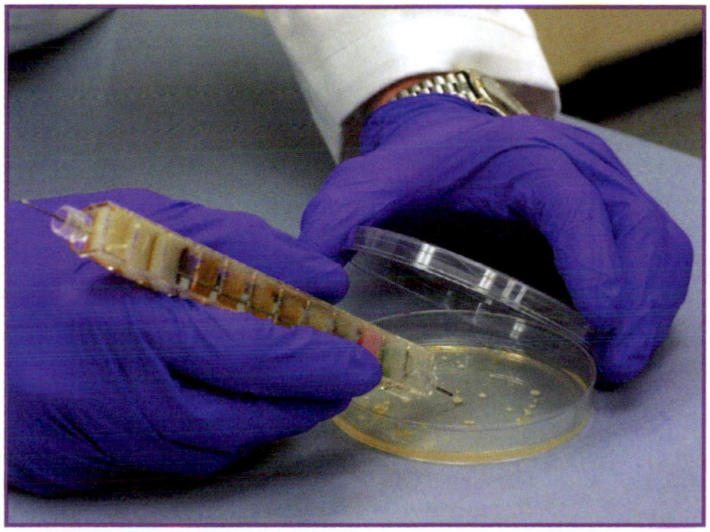

9-26 **Harvest Growth** ◆ Using the sterile tip of the wire, aseptically remove growth from a colony on the agar surface. Do not dig into the agar. The inoculum should be large enough to be visible.

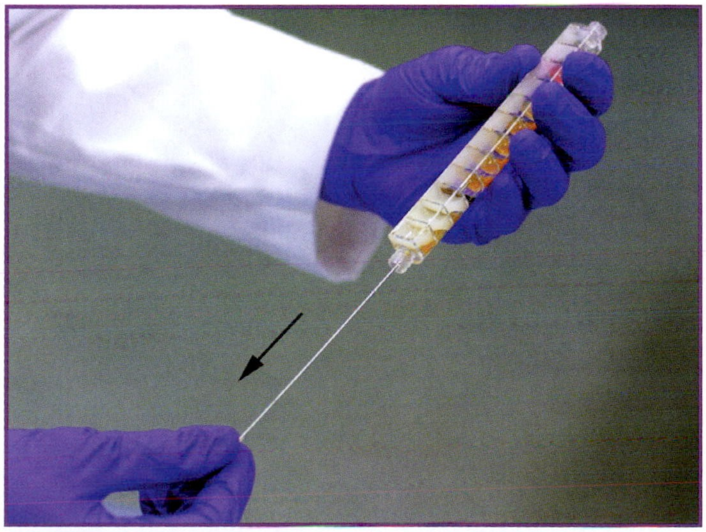

9-27 **Pull Wire Through** ◆ Loosen the wire by turning it slightly. While continuing to rotate it withdraw the wire until its tip is inside the last compartment (glucose).

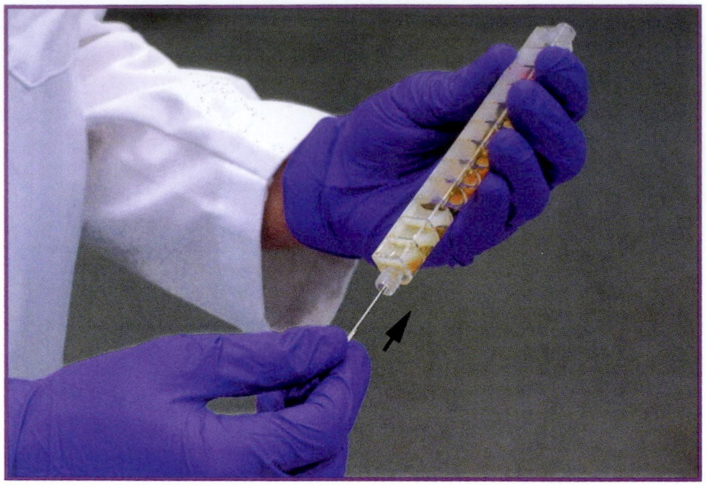

9-28 Reinsert Wire ◆ Slide the wire back into the Enterotube® II until you can see the tip inside the citrate compartment. The notch in the wire should be lined up with the end of the tube nearest the glucose compartment.

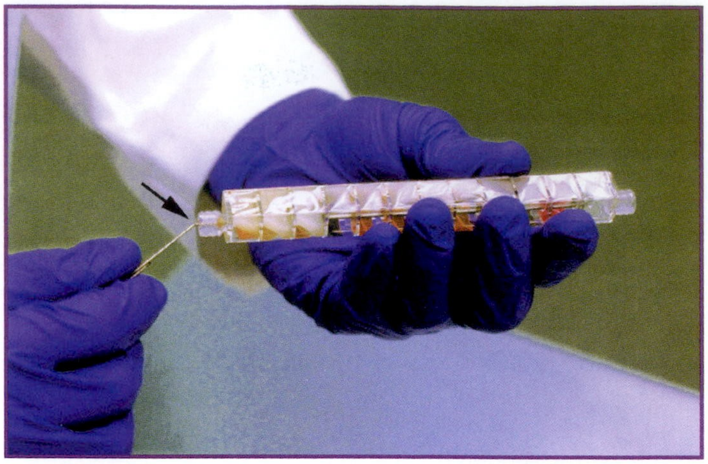

9-29 Break Wire ◆ Bend the wire until it breaks off at the notch.

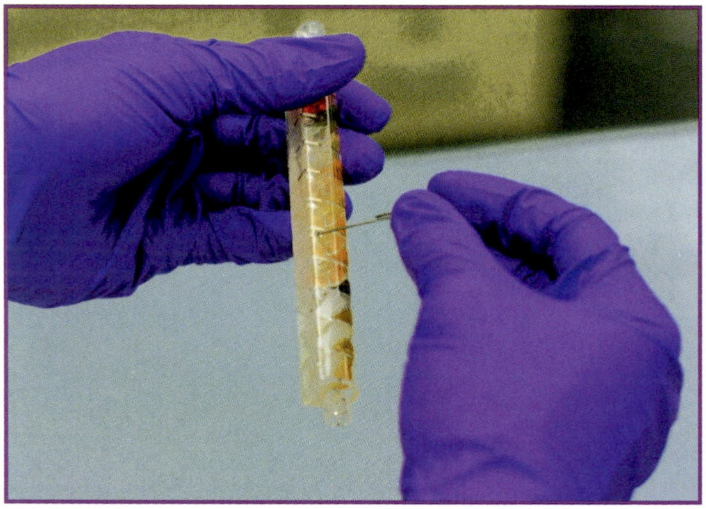

9-30 Puncture Air Inlets ◆ Using the broken wire, puncture the plastic covering the eight air inlets on the back side of the tube.

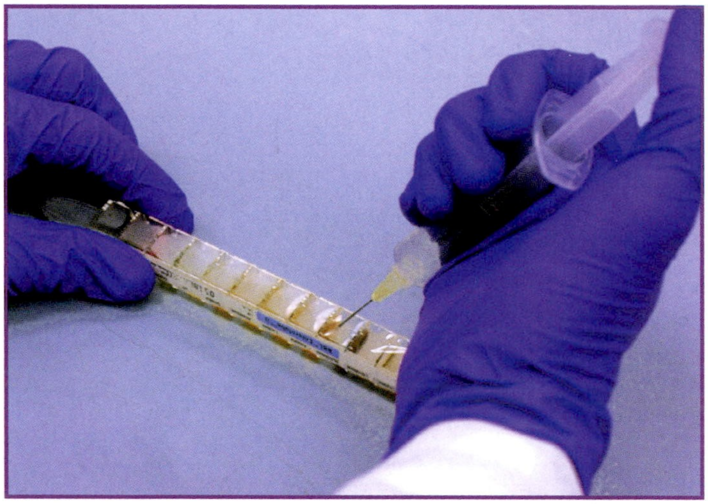

9-31 Add Reagents after Incubation ◆ Use a needle and syringe to add the reagents through the plastic film on the flat side of the tube. Do this with the tube in a test tube rack and the glucose end pointing down. The indole test should be done only after other tests have been read. Add the Voges-Proskauer reagents only if directed to do so by the CCIS.

The CCIS is a master list of all enterics and their assigned numeric codes. In most cases, the five-digit number applies to a single organism, but when two or more species share the same code, a confirmatory VP test is performed to further differentiate the organisms. An Enterotube® II before and after inoculation is shown in Figure 9-34.

Application

The Enterotube® II is a multiple test system used for rapid identification of bacteria from the family *Enterobacteriaceae*.

In This Exercise

You will use a multitest system to run an entire battery of tests designed to identify a member of the *Enterobacteriaceae*.

Materials

Per Class

- ✓ Enterotube® II Computer Coding and Identification System (CCIS) identification booklet from Becton Dickinson Microbiology Systems

Per Student Group

- ✓ Becton Dickinson Microbiology Systems' Enterotube® II Identification System for *Enterobacteriaceae*
- ✓ Kovacs' reagent
- ✓ VP reagents (KOH and α-naphthol)
- ✓ needles and syringes or disposable pipettes for addition of reagents
- ✓ streak plate containing colonies of an unknown enteric (appropriate to facilities and course level)

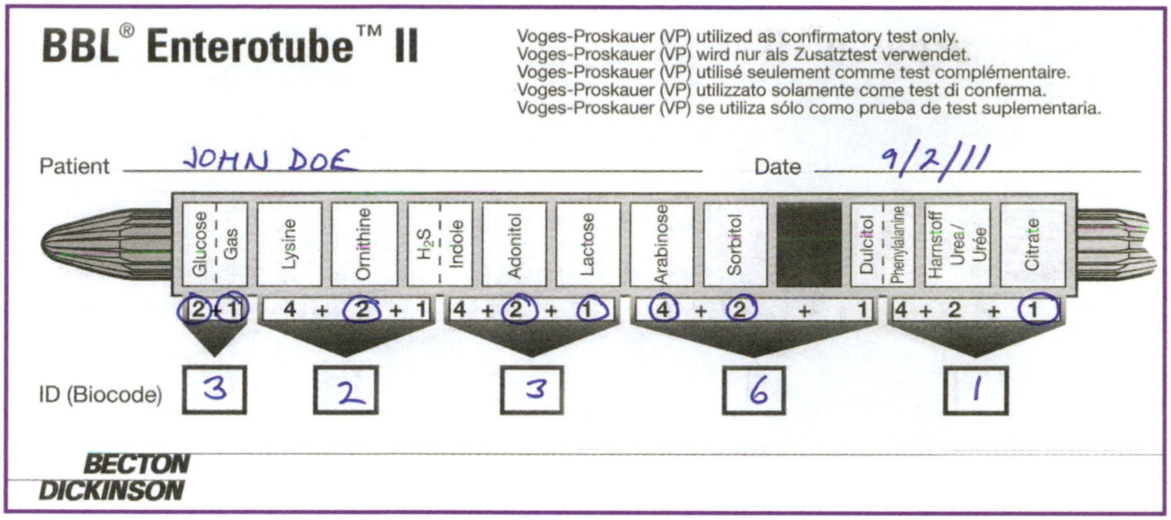

BBL® Enterotube™ II
Voges-Proskauer (VP) utilized as confirmatory test only.
Voges-Proskauer (VP) wird nur als Zusatztest verwendet.
Voges-Proskauer (VP) utilisé seulement comme test complémentaire.
Voges-Proskauer (VP) utilizzato solamente come test di conferma.
Voges-Proskauer (VP) se utiliza sólo como prueba de test suplementaria.
Patient JOHN DOE
Date 9/2/11
Glucose
Gas
Lysine
Ornithine
H_2S
Indole
Adonitol
Lactose
Arabinose
Sorbitol
Dulcitol
Phenylalanine
Harnstoff Urea/ Urée
Citrate
2 + 1
4 + 2 + 1
4 + 2 + 1
4 + 2 + 1
4 + 2 + 1
ID (Biocode)
3
2
3
6
1
BECTON DICKINSON

9-32 BBL™ Enterotube® II Result Sheet ◆ This is the result sheet for the tube in Figure 9-34B. The ID value obtained was 32361, which identifies the organism as *Enterobacter aerogenes*.

ID VALUE	ORGANISM IDENTIFICATION	ATYPICAL TESTS	CONFIRMATORY TESTS				
			MAL	TRE			
20406	PROTEUS VULGARIS	H2S-	+	V(30%)			
	MORGANELLA MORGANII	ORN-	–	–			
	PROVIDENCIA STUARTII	PAD-	–	+			
20407	PROVIDENCIA STUARTII	NONE					
			SHA				
20410	*SHIGELLA SEROGROUPS A, B OR C	DUL+	+				
			SHA	ADA			
20420	*SHIGELLA SEROGROUPS A, B OR C	SOR+	+	–			
	ESCHERICHIA COLI (AD)	ARA-	–	+			
20425	PROVIDENCIA STUARTII	SOR+					
20427	PROVIDENCIA STUARTII	SOR+					
20430	ESCHERICHIA COLI (AD)	ARA-					
			SHA	ADA	MOT	VP	YEL
20440	*SHIGELLA SEROGROUPS A, B OR C	NONE	+	–	–	–	–
	ESCHERICHIA COLI (AD)	SOR-	–	+	–	–	–
	ENTEROBACTER AGGLOMERANS	IND+	–	–	+	V(70%)	+(75%)
20441	ENTEROBACTER AGGLOMERANS	IND+					

9-33 A Sample of the Enterotube® II Computer Coding and Identification System ◆ The five-digit numbers on the Enterotube® II Results Pad are matched against the database in the CCIS to get identification.

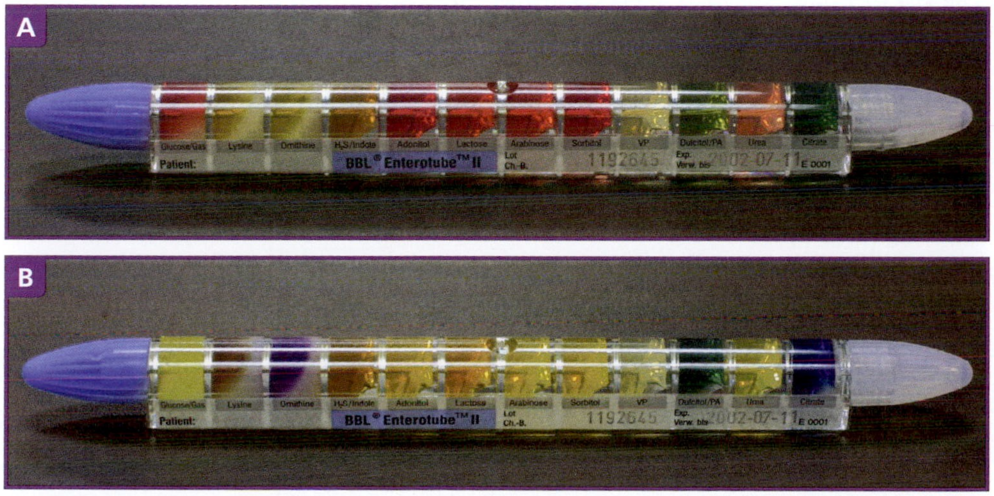

A
B
Glucose/Gas
Lysine
Ornithine
H₂S/Indole
Adonitol
Lactose
Arabinose
Sorbitol
VP
Dulcitol/PA
Urea
Citrate
Patient:
BBL® Enterotube™ II
1192645
2002-07-11

9-34 Enterotube® II Test Results ◆ **A** Uninoculated tube. **B** Tube inoculated with *Enterobacter aerogenes* after 24 hours incubation. This tube shows atypical negative results for lysine decarboxylase; after an additional 24 hours incubation the lysine medium turned purple (+).

Reagent Recipes

Kovacs' Reagent

♦ Amyl alcohol	75.0 mL
♦ Hydrochloric acid, concentrated	25.0 mL
♦ *p*-dimethylaminobenzaldehyde	5.0 g

KOH Reagent

♦ Potassium hydroxide	20.0 g
♦ Distilled water to bring volume to	100.0 mL

α-Naphthol Reagent

♦ α-naphthol	5.0 g
♦ Absolute Ethanol to bring volume to	100.0 mL

Procedure

Lab One

1 Perform a Gram stain on your isolate to verify that it is a Gram-negative rod.

2 Place a paper towel on the tabletop and soak it with disinfectant.

3 Remove the blue cap and then the white cap from the Enterotube® II, being careful not to contaminate the sterile wire tip. Place the caps open end down on the paper towel.

4 Aseptically remove a large amount of growth from one of the plated colonies (Figure 9-26). Try not to remove any of the agar with it.

5 Grasp the looped end of the wire and, while turning, gently pull it back through all of the Enterotube® II compartments (Figure 9-27). You do not have to completely remove the wire, but be careful to pull it back far enough to inoculate the last compartment.

6 Using the same turning motion as described above, slide the wire back into the tube (Figure 9-28). Push it in until the tip of the wire is inside the citrate compartment and the notch in the wire lines up with the opposite end of the tube.

7 Bend the wire at the notch until it breaks off (Figure 9-29).

8 Locate the air inlets on the side of the tube opposite the Enterotube® II label. Using the removed piece of inoculating wire, puncture the plastic membrane in the adonitol, lactose, arabinose, sorbitol, Voges-Proskauer, dulcitol/PA, urea, and citrate compartments (Figure 9-30). These openings will create the necessary aerobic conditions for growth. Be careful not to perforate the plastic film covering the flat side of the tube.

9 Discard the wire in disinfectant or a sharps container and replace the Enterotube® II caps.

10 Incubate the tube lying flat at 35 ± 2°C for 18 to 24 hours. (Note: If the tube cannot be read at 24 hours, have someone place it in a refrigerator until it can be read.)

Lab Two

1 Examine the tube and circle the appropriate positive results on the Enterotube® II Results Pad.

2 *After* circling all positive test results on the Enterotube® II Results Pad, place the tube in a rack with the glucose compartment on the bottom. Puncture the plastic membrane of the H_2S/Indole compartment and (with a needle and syringe or disposable pipette) add one or two drops Kovacs' reagent to the compartment (Figure 9-31).

3 Observe the compartment for the formation of a red color within 10 seconds. Record a positive result on the Enterotube® II Results Pad.

4 Add the numbers in each bracketed section on the Enterotube® II Results Pad to obtain a five-digit number. Find the number in the CCIS. If not directed to perform a VP test, record the name of your organism on the score sheet and skip to #7 below. If directed to run a confirmatory Voges-Proskauer test, proceed to #5.

5 Puncture the plastic membrane of the VP compartment and (with a needle and syringe or disposable pipette) add two drops of the KOH reagent and three drops of the α-Naphthol reagent. Observe for the formation of red color within 20 minutes. Interpret the result using the CCIS.

6 Discard the Enterotube® II in an appropriate autoclave container.

7 Enter all results on the Data Sheet, and answer the questions.

References

Enterotube® II Identification System for *Enterobacteriaceae* package insert.

Enterotube® II Computer Coding and Identification System.

TABLE 9-5 Enterotube® II Results and Interpretations

TABLE OF RESULTS				
Compartment	Test	Result	Interpretation	Symbol
1	Glucose	Red	Organism does not ferment glucose	–
		Yellow, wax not lifted	Organism ferments glucose to acid	+
		Yellow, wax not lifted	Organism ferments glucose to acid and gas	+
2	Lysine	Purple	Organism decarboxylates lysine	+
		Yellow	Organism does not decarboxylate lysine	–
3	Ornithine	Purple	Organism decarboxylates ornithine	+
		Yellow	Organism does not decarboxylate ornithine	–
4	H_2S	Beige	Organism does not reduce sulfur	–
		Black	Organism reduces sulfur	+
	Indole	Red after Kovacs' reagent	Organism produces indole	+
		No red after Kovacs' reagent	Organism does not produce indole	–
5	Adonitol	Red	Organism does not ferment adonitol	–
		Yellow	Organism ferments adonitol	+
6	Lactose	Red	Organism does not ferment lactose	–
		Yellow	Organism ferments lactose	+
7	Arabinose	Red	Organism does not ferment arabinose	–
		Yellow	Organism ferments arabinose	+
8	Sorbitol	Red	Organism does not ferment sorbitol	–
		Yellow	Organism ferments sorbitol	+
9	VP	Colorless	Organism does not produce acetoin	–
		Red	Organism produces acetoin	+
10	Dulcitol	Green	Organism does not ferment dulcitol	–
		Yellow	Organism ferments dulcitol	+
	PA	Green	Organism does not deaminate phenylalanine	–
		Black/smokey-gray	Organism deaminates phenylalanine	+
11	Urea	Beige	Organism does not hydrolyze urea	–
		Red-purple	Organism hydrolyzes urea	+
12	Citrate	Green	Organism does not utilize citrate	–
		Blue	Organism utilizes citrate	+

Name ______________________ Date ______________________

Lab Section ______________________ I was present and performed this exercise (initials) ________

Enterotube® II

OBSERVATIONS AND INTERPRETATIONS

1 Tape your BBL® Enterotube® II Result Sheet here.

2 Enter the results from the CCIS booklet below.

CCIS Five-Digit Code	Possible Organisms	VP Result + / −	Identified Organism

Questions

1 Fecal coliforms such as *Escherichia coli, Enterobacter aerogenes*, and *Klebsiella pneumoniae* are enterics that ferment lactose to acid and gas at 35°C within 48 hours. For most strains of these organisms, the chart below summarizes reactions in the Enterotube® II. Fill in the missing information and, using colored pencils, fill in the appropriate colors for each positive test.

	Glu	Gas	Lys	Orn	H_2S	Ind	Adon	Lact	Arab	Sorb	VP	Dulc	PA	Urea	Citrate
E. coli			+	−	−	+	−		+	+	−	−	−	−	−
E. aerogenes			+	+	−	−	+		+	+	+	−	−	−	+
K. pneumoniae			+	−	−	−	+		+	+	+	−	−	+	+

2 Based on the information above, enter the five-digit codes for the three organisms.

E. coli ________ *E. aerogenes* ________ *K. pneumoniae* ________

3 Examine the following chart. Fill in the color reactions for each test based on the ID value given.

ID Value	Glucose/ Gas	Lysine	Ornithine	H_2S/ Indole	Adonitol	Lactose	Arabinose	Sorbitol	VP	Dulcitol/ PA	Urea	Citrate
31122												
27345												
04004												
05511												

4 Which of the four organisms are not members of *Enterobacteriaceae*? Why?

5 Which of the four ID values is questionable? Why?

EXERCISE 9-6

Identification of *Streptococcus* and *Streptococcus*-like Organisms by Latex Agglutination: The Streptex® Rapid Test

Theory

The genus *Streptococcus* is composed of Gram-positive, catalase-negative cocci in pairs or chains. Most are facultative anaerobes and ferment carbohydrates to acid, but not gas. Historically, it was a heterogeneous group in which the species could be grouped according to functional characteristics (pyogenic streps, oral streps, enteric streps, lactic acid streps, and the others that don't fit), hemolytic reaction (α-, β-, or γ-hemolysis, Exercise 5-21), and finally, by antigenic characters (the **Lancefield groups**—more on these later).

Application of molecular methods has led to extensive modification of the genus since the 1st edition of *Bergey's Manual of Systematic Bacteriology*, Volume 2 (1986). Species that formerly were grouped as enterococci and lactococci now belong in their own genera, *Enterococcus* and *Lactococcus*, respectively. Other new genera have been created for Gram-positive, catalase-negative cocci, but these contain species that are not of great clinical significance or are not identifiable by this procedure.

In Exercise 9-2, you identified Gram-positive cocci based on hemolytic reactions and biochemical test results. In this exercise you will be identifying clinically significant "streptococci" based on cellular antigens originally classified by Rebecca Lancefield in 1933. The separation into Lancefield groups is still a useful method for presumptive identification of streptococci. The Streptex® Rapid Test kit (Remel) has six different antibody suspensions, each with a different antibody coating polystyrene latex particles and homologous to one of six different group-specific antigens (Figure 9-35).

In performing the test, first the antigen is extracted from a colony growing on an agar plate. Then each antibody suspension is added to one of six circles on the reaction card (Figure 9-36), followed by a known volume of extracted antigen. After mixing the antigen–antibody suspensions in each circle, the card is rocked gently for one minute. A positive test result shows agglutination. A negative result retains a homogeneous appearance. See Exercise 8-4 for more on agglutination reactions.

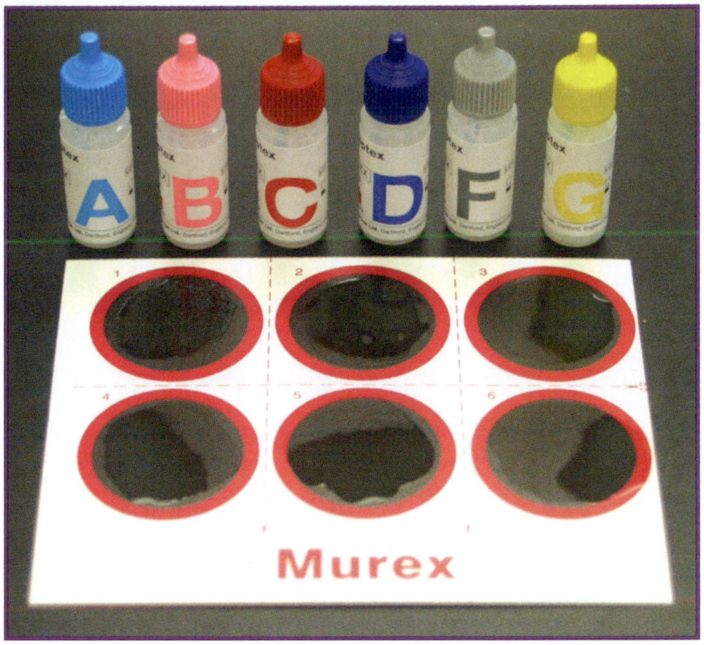

9-35 The Streptex® Rapid Test Kit (Remel) ◆ Shown are the six bottles of latex antibody suspensions (labeled with the group specific antigen with which each reacts) and a reaction card.

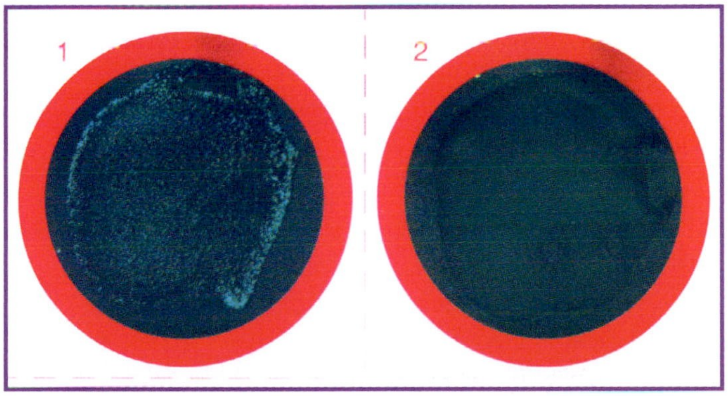

9-36 A Portion of the Reaction Card ◆ Two of the six circles are shown. The circle on the left shows agglutination and is positive. The one on the right is negative.

The Streptex® test is most reliable when grouping β-hemolytic organisms, but it can be used to identify some other groups—most notably the enterococci, which typically are α- or γ-hemolytic. The two most important streptococcal groups are Group A and Group B, but the kit is designed to identify Groups A, B, C, D, F and G. These are summarized below.

- Group A streptococci (GAS) is composed of *S. pyogenes* (Figure 9-37), the causative agent of bacterial pharyngitis ("strep throat"), impetigo, scarlet fever, necrotizing fasciitis (the "flesh eating bacterium"), streptococcal toxic shock syndrome (STSS), and many other superficial and deep infections. In most cases, penicillin is still effective in treating *S. pyogenes* infections. The Group A antigen is a branched polysaccharide or N-acetylglucosamine and the sugar rhamnose.

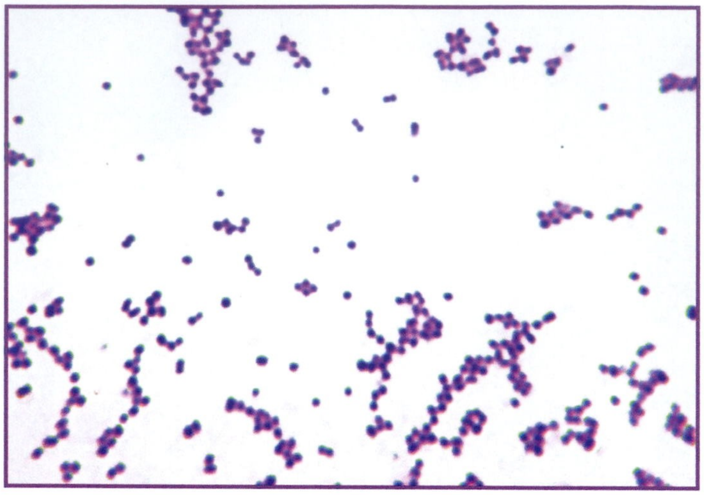

9-37 Gram Stain of a *Streptococcus pyogenes* Grown in Culture (X1000) ◆ Specimens isolated from patients usually are seen in pairs or chains.

- ◆ *S. agalactiae* is the only species of Group B streptococcus (GBS). It was first isolated from cows, where it causes bovine mastitis, but it also is a human pathogen, especially in neonates (neonatal sepsis and meningitis) and immunocompromised adults. The Group B antigen is a complex of N-acetylglucosamine and the sugars galactose and rhamnose.
- ◆ *S. bovis* group and some members of the genus *Enterococcus* comprise Group D streptococci. *S. bovis*[1] is associated with bacteremia, endocarditis, and meningitis. The two most frequently recovered enterococcal species from fecal material are *E. faecium* and *E. faecalis*. A major cause of nosocomial infections, they are of increasing concern because of their resistance to multiple antibiotics and possession of several intrinsic resistance factors. The Group D antigen is a teichoic acid composed of glycerol phosphate and glucose.
- ◆ Other β-hemolytic streptococci that possess group-specific antigens are the *S. anginosus* group, which have the A, C, F and G antigens, and *S. dysgalactiae* with either the C or G antigen. Although these species demonstrate antigen variability, each isolate will have only a single antigen. *S. anginosus* species cause abscesses, and *S. dysgalactiae* is a causative agent of pharyngitis and sometimes glomerulonephritis.
- ◆ Although they are not classified by the Lancefield system, two other streptococcal groups are important. *S. pneumoniae* is α-hemolytic and is a causative agent of bacterial pneumonia, meningitis, otitis media, and bacteremia. Viridans streptococci comprise more than 20 α- or γ-hemolytic species and are so-named because of the greening they often produce on blood agar plates. They are normal inhabitants of the oropharynx, digestive tract, and urogenital tract but are responsible for diseases as varied as dental caries, endocarditis, and abdominal infections.

[1] *S. bovis* has been recognized as a strain of *S. equinus*. To save confusion of changing a long-standing name, these are frequently referred to together as the "*S. bovis/S. equinus* complex" or "*S. bovis/S. equinus* group."

Application

Serological testing of Gram-positive, catalase-negative cocci in pairs or chains allows rapid identification of pathogenic and nonpathogenic strains.

In This Exercise

You will see a demonstration of how a latex agglutination test can be used to place an unknown streptococcal isolate in its appropriate Lancefield group. Alternatively, each student or student group may be assigned an unknown to classify using the agglutination test.

Materials

Per Class (Demonstration)

- ✓ one Streptex® Rapid Test Kit (Murex Biotec Limited, available from Remel Inc., Lenexa, KS 66215; http://www.remel.com/). The kit includes Groups A, B, C, D, F, and G latex suspensions, a polyvalent antigen positive control, extraction enzyme, mixing sticks, and reaction cards. Two kits are available, one sufficient for 50 tests, the other for 200 tests.
- ✓ streptococcal unknown (professor's choice) or a throat swab streaked on a blood agar plate.
- ✓ 37°C water bath or a beaker of water in a 37°C incubator
- ✓ inoculating loop
- ✓ pipettes to deliver 40 µL and 400 µL volumes and tips
- ✓ microtube(s)—one for each test to be run

Per Student or Student Group (Alternative Procedure)

- ✓ one Streptex® Rapid Test Kit (per class)
- ✓ 37°C water bath or a beaker of water in a 37°C incubator (per class)
- ✓ one streptococcal unknown pure culture streaked on blood agar and labeled with an identifying number. (Note: the number and identities of different organisms used as unknowns depend on the course level, facilities, number of students, and organisms available in the institution's culture collection.)
- ✓ one reaction card
- ✓ six mixing sticks
- ✓ inoculating loop
- ✓ pipettes to deliver 40 µL and 400 µL volumes and tips
- ✓ microtube

Procedure for Demonstration[2]

These instructions are for testing a single isolate. More isolates may be demonstrated by adjusting the numbers of items used.

1. Obtain two reaction cards.
2. Add 400 µL of extraction enzyme solution to a microtube. Number the colony type on the plate and label the microtube accordingly. The isolate used is "professor's choice." The isolate can be from a pure culture streak or from a real sample—e.g., a throat swab, where colony morphology gives a good indication of appropriate growth.
3. Scoop five or so colonies of identical morphology with an inoculating loop. Then add the growth to the enzyme solution to make a light suspension.
4. Incubate the microtube at 37°C. Shake the microtube after 5 minutes of incubation. Incubate for no less than 10 minutes and no more than an hour.
5. While the microtube is incubating, shake each of the latex suspensions to mix well, then add one drop (20 µL) of each to a different circle on one reaction card. Record the circle number next to each antiserum in the table on the Data Sheet.
6. After incubation, add 40 µL of polyvalent antigen solution to each circle on the card.
7. Use a different mixing stick to mix the two solutions in each circle. The mixture should cover each circle completely. Dispose of each mixing stick properly after use.
8. Gently rock the card for up to one minute.
9. Hold the card at a comfortable reading distance, and evaluate the results for obvious agglutination.
10. After incubation, repeat steps 6 through 10 for the extract instead of the polyvalent antigen solution. Use the same circle numbers for each antiserum.
11. If desired, prepare a third card in which antibody suspensions are mixed with extraction enzyme.

[2] Adapted from the Murex Biotec Ltd. package insert. Check the current package insert for any recent modifications.

Alternative Procedure for Student-run Test

1. Obtain a single reaction card, 6 mixing sticks, an inoculating loop, pipettors that dispense 40 µL and 400µL volumes, pipette tips, and one microtube. (Note: controls can be run by the instructor or by one student group not assigned an unknown. See "Procedure for Demonstration" for instructions.)
2. Obtain a numbered unknown organism plate from your instructor.
3. Add 400 µL of extraction enzyme solution to a microtube. Label the microtube with the unknown culture's number.
4. Scoop five or so colonies of identical morphology with an inoculating loop. Then add the growth to the enzyme solution to make a light suspension.
5. Incubate the microtube at 37°C. Shake the microtube after 5 minutes of incubation. Incubate for no less than 10 minutes and no more than an hour.
6. While the microtube is incubating, shake each of the latex suspensions to mix well, then add one drop (20 µL) of each to a different circle on one reaction card.
7. After incubation, add 40 µL of extract solution to each circle on the card. Record the circle number next to each antiserum in the table on the Data Sheet.
8. Use a different mixing stick to mix the two solutions in each circle. The mixture should cover each circle completely. Dispose of each mixing stick properly after use.
9. Gently rock the card for up to one minute.
10. Hold the card at a comfortable reading distance and evaluate the results for obvious agglutination.

References

Austrian, Robert. 2004. Chapter 184 in *Infectious Diseases*, 3rd ed., edited by Sherwood L. Gorbach, MD, John G. Bartlett, MD, and Neil R. Blacklow, MD. Lippincott Williams and Wilkins, Philadelphia, PA.

Murex Biotec Limited. 2001. Package insert for Streptex® Rapid Test Kit. Dartford, Kent, DA1 5LR, UK.

Murray, Barbara E., and John G. Bartlett. 2004. Chapter 18 in *Infectious Diseases*, 3rd ed., edited by Sherwood L. Gorbach, MD, John G. Bartlett, MD, and Neil R. Blacklow, MD. Lippincott Williams & Wilkins, Philadelphia, PA.

Murray, Patrick R., Ken S. Rosenthal, and Michael A. Pfaller. 2005. Chapters 23 and 24 in *Medical Microbiology*, 5th ed. Elsevier Mosby, Philadelphia, PA.

Spellerberg, Barbara, and Claudia Brandt. 2007. Chapter 29 in *Manual of Clinical Microbiology*, 9th ed., edited by Patrick R. Murray, Ellen Jo Baron, James H. Jorgensen, Marie Louise Landry, and Michael A. Pfaller. American Society for Microbiology Press, Washington, DC.

Stollerman, Gene H. 2004. Chapter 183 in *Infectious Diseases,* 3rd ed., edited by Sherwood L. Gorbach, MD, John G. Bartlett, MD, and Neil R. Blacklow, MD. Lippincott Williams and Wilkins, Philadelphia, PA.

Teixeira, Lúcia Martins, Maria da Glória Siqueira Carvalho, and Richard R. Facklam. 2007. Chapter 30 in *Manual of Clinical Microbiology*, 9th ed., edited by Patrick R. Murray, Ellen Jo Baron, James H. Jorgensen, Marie Louise Landry, and Michael A. Pfaller. American Society for Microbiology Press, Washington, DC.

Wessels, Michael R., and Dennis L. Kasper. 2004. Chapter 186 in *Infectious Diseases*, 3rd ed., edited by Sherwood L. Gorbach, MD, John G. Bartlett, MD, and Neil R. Blacklow, MD. Lippincott Williams and Wilkins, Philadelphia, PA.

Whiley, Robert A., and Jeremy M. Hardie. 2009. "*Streptococcus*" in *Bergey's Manual of Systematic Bacteriology,* 2nd ed., Vol 3: *The Firmicutes*. Springer, New York.

Winn Jr., Washington, Stephen Allen, William Janda, Elmer Koneman, Gary Procop, Paul Schreckberger, and Gail Woods. 2006. Chapter 13 in *Koneman's Color Atlas and Textbook of Diagnostic Microbiology*, 6th ed., Lippincott Williams & Wilkins, Philadelphia, PA.

NAME ______________________ DATE ______________________

LAB SECTION ______________________ I WAS PRESENT AND PERFORMED THIS EXERCISE (INITIALS) ______

The Streptex® Rapid Test

OBSERVATIONS AND INTERPRETATIONS

1 Record the results of each agglutination series in the spaces provided. Write "+" for agglutination and "−" for no agglutination.

Antiserum	Circle Number on Card	Polyvalent Control	Demonstration Unknown Number ____	Student Unknown Number ____ (Optional Procedure)
A				
B				
C				
D				
F				
G				

2 Based on the agglutination results in the demonstration, which organism(s) might the unknown be?

3 Based on the agglutination results for the optional student unknown (if done), which organism(s) might the unknown be?

Questions

1 What does "polyvalent" antigen solution mean?

2 What is the purpose of the polyvalent antigen solution? What results are desired? What results would invalidate the test?

3 What is the purpose of the third card (see Step 11, Demonstration Procedure) in which antibody suspensions are mixed with extraction enzyme? What results are desired? What results would invalidate the test?

4 Why must different mixing sticks be used for each circle on the reaction card?

Biochemical Pathways

APPENDIX A

So much of what is done in microbiology relies on an understanding of basic biochemical pathways. It's not as important to memorize them (although, with exposure they will become second nature) as it is to understand their importance in metabolism and to interpret diagrams of them when available. The following discussion is provided so you can see how the various biochemical tests presented in this manual fit into the overall scheme of cellular chemistry—that is, **metabolism**.

Oxidation of Glucose: Glycolysis, Entner-Doudoroff, and Pentose-Phosphate Pathways

Most organisms use **glycolysis** (also known as the "Embden-Meyerhof-Parnas pathway, Figure A-1) in energy metabolism. It performs the stepwise disassembly of glucose into two pyruvates, releasing some of its energy and electrons in the process. The exergonic (energy-releasing) reactions are associated with ATP synthesis by a process called **substrate phosphorylation**. Although a total of four ATPs are produced per glucose in glycolysis, two ATPs are hydrolyzed early in the pathway, leaving a net production of two ATPs per glucose. In one glycolytic reaction, the loss of an electron pair (oxidation) from a three-carbon intermediate occurs simultaneously with the reduction of NAD^+ to $NADH+H^+$. The $NADH+H^+$ then may be oxidized in an electron transport chain or a fermentation pathway, depending on the organism and the environmental conditions. The former yields ATP, and the latter generally does not. In summary, each glucose oxidized in glycolysis yields two pyruvates, 2 NADH+2 H^+, and a net of 2 ATPs (Table A-1).

Although the intermediates of glycolysis are carbohydrates, many are entry points for amino acid, lipid, and nucleotide catabolism. Many glycolytic intermediates also are a source of carbon skeletons for the synthesis of these other biochemicals. Some of these are shown in Figure A-1. Note: For clarity, many details have been omitted from these other pathways in Figure A-1. Single arrows may represent several reactions, and other carbon compounds not illustrated may be required to complete a particular reaction.

The **Entner-Doudoroff pathway** (Figure A-2) is an alternative means of degrading glucose into two pyruvates. This pathway is found exclusively among prokaryotes (e.g., *Pseudomonas* and *E. coli*, as well as other Gram-negatives and certain Archaea). It allows utilization of a different category of sugars (aldonic acids) than glycolysis and therefore improves the range of resources available to the organism. It is less efficient than glycolysis because only one ATP is phosphorylated and only one NADH is produced. Table A-2 summarizes this pathway.

The **pentose-phosphate pathway** is a complex set of cyclic reactions that provides a mechanism for producing five-carbon sugars (**pentoses**) from six-carbon sugars (**hexoses**). Pentose sugars are used in ribonucleotides and deoxyribonucleotides, as well as being precursors to aromatic amino acids. Further, this pathway produces NADPH, which is used as an electron donor in anabolic pathways. Unlike NADH, produced in glycolysis and Entner-Doudoroff, NADPH is not used as an electron donor in an electron transport chain for **oxidative phosphorylation** of ADP.

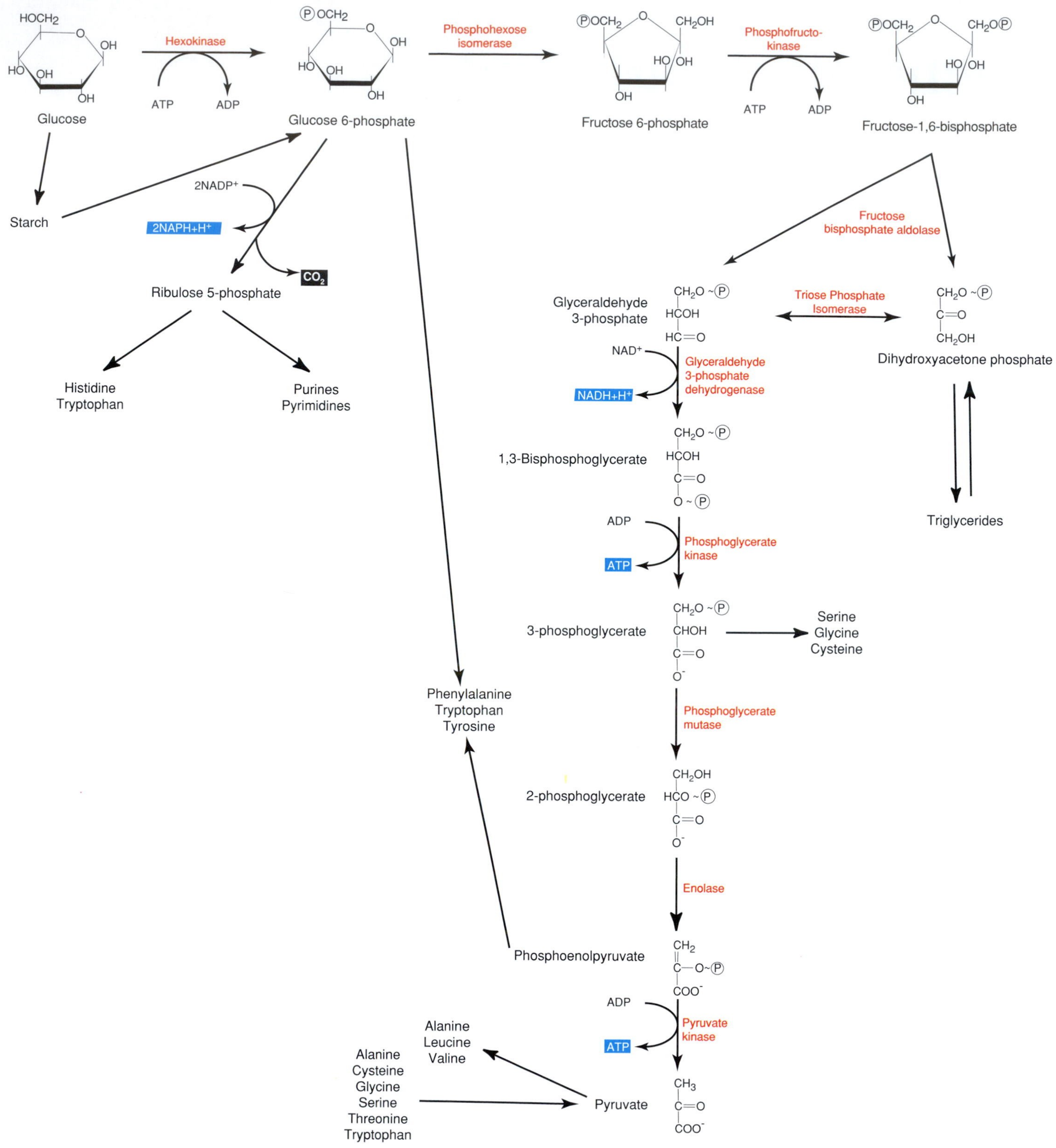

A-1 Glycolysis and Associated Pathways ◆ The names of glycolytic intermediates are printed in black ink; the enzyme names are in red. Reducing *power* (in the form of NADH+H$^+$) and ATP are highlighted in blue. The major key to getting product yields correct is to recognize that both C_3 compounds (Glyceraldehyde 3-phosphate and Dihydroxyacetone phosphate) produced from splitting Fructose 1,6-bisphosphate can pass through the remainder of the pathway because of the triose phosphate isomerase reaction. The conversion of each into pyruvate results in the formation of 2 ATPs and 1 NADH+H$^+$ (Table A-1).

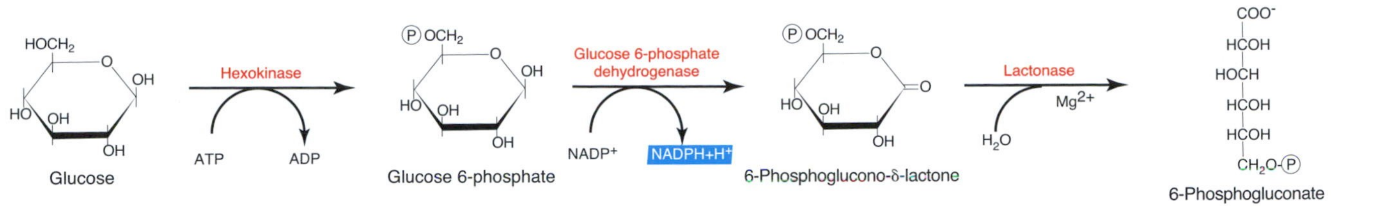

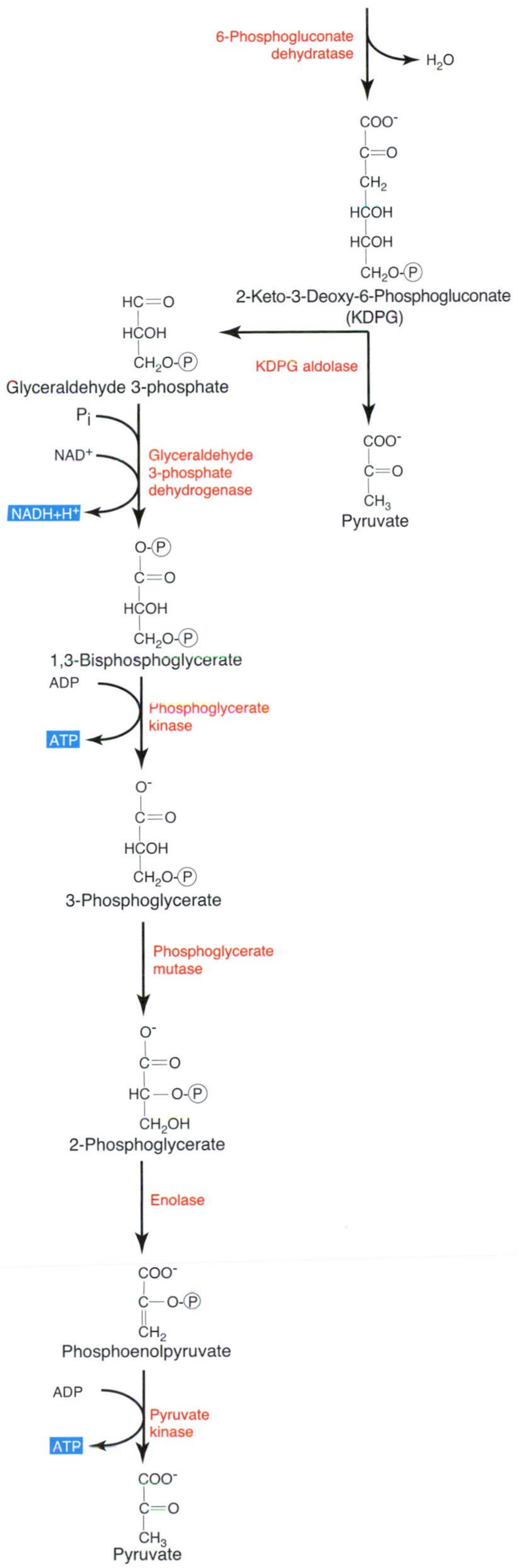

A-2 Entner-Doudoroff Pathway ◆

Notice the similarities between this pathway and glycolysis (Figure A-1). The main difference is in the six-carbon compound that is split into two three-carbon compounds. The result of this split is pyruvate and glyceraldehydes-3-phosphate, which is oxidized as in glycolysis to pyruvate. Because only one three-carbon compound goes through the sequence of reactions leading to pyruvate, the ATP and NADH yield is one-half that of glycolysis. But one NADPH is produced that is not made in glycolysis.

TABLE A-1 Summary of Glycolytic Reactants and Products per Glucose ◆

TABLE OF RESULTS

Reactant	Product
Glucose ($C_6H_{12}O_6$)	2 Pyruvates (2 $C_3H_3O_3$)
2 ATP	2 ADP
4 ADP	4 ATP
NET: 2 ADP	NET: 2 ATP
2 NAD^+	2 NADH + 2 H^+

TABLE A-2 Summary of Entner-Doudoroff Reactants and Products per Glucose ◆

TABLE OF RESULTS

Reactant	Product
Glucose ($C_6H_{12}O_6$)	2 Pyruvates (2 $C_3H_3O_3$)
1 ATP	1 ADP
2 ADP	2 ATP
NET: 1 ADP	NET: 1 ATP
1 NAD^+	1 NADH + $1H^+$
1 $NADP^+$	1 NADPH + $1H^+$

TABLE A-3 Summary of Pentose-Phosphate Reactants and Products per Glucose-6-phosphate ◆

TABLE OF RESULTS	
Reactant	**Product**
Glucose-6-phosphate (C_6)	$6CO_2 + 1\ P_i$
12 $NADP^+$	12 NADPH + 12 H^+

The pentose-phosphate reactants and products are listed in Table A-3, and the overall path is shown in Figure A-3. To completely oxidize one hexose to $6CO_2$, a total of six hexoses must enter the cycle as glucose-6-phosphate and follow one of three different routes (notice the symmetry of pathways as drawn). Notice in Figure A-3 that each hexose loses a CO_2 upon entry into the cycle, but at the end, five hexoses are produced. Thus, the net reaction is one hexose being oxidized to $6CO_2$. Notice also the reactions that transfer two-carbon and three-carbon fragments between the five-carbon intermediates. **Transketolase** catalyzes the two-carbon transfer, whereas **transaldolase** catalyzes the three-carbon transfer. Alternatively, the five-carbon intermediates can be redirected into pathways for synthesis of aromatic amino acids and nucleotides (not shown).

Oxidation of Pyruvate: The Krebs Cycle and Fermentation

Pyruvate represents a major crossroads in metabolism. Some organisms are able to further disassemble the pyruvates produced in glycolysis and Entner-Doudoroff and make more ATP and NADH+H^+ in the **Krebs cycle**. Other organisms simply reduce the pyruvates (or a pyruvate derivative) with electrons from NADH+H^+ without further energy production in **fermentation**.

The Krebs cycle (also known as the **tricarboxylic acid cycle** and the **citric acid cycle**) is a major metabolic pathway used in energy production by organisms that respire aerobically or anaerobically (Figure A-4). Pyruvate produced in glycolysis or other pathways is first converted to acetyl-coenzyme A during the **entry step** (also known as the **intermediate** or **gateway step**). Acetyl-CoA enters the Krebs cycle through a condensation reaction with oxaloacetate. Products for each pyruvate that enters the cycle via the entry step are: 3 CO_2, 4 NADH+H^+, 1 $FADH_2$, and 1 GTP. (Because two pyruvates are made per glucose, these numbers are doubled in Table A-4). The energy released from oxidation of reduced coenzymes (NADH+H^+ and $FADH_2$) in an electron transport chain is then used to make ATP. ATP yields are summarized in Table A-5.

Like glycolysis, many of the Krebs cycle's intermediates are entry points for amino acid, nucleotide and lipid catabolism, as well as a source of carbon skeletons for synthesis of the same compounds. These pathways are shown, but details have been omitted. Single arrows may represent several reactions, and other carbon compounds, not illustrated, may be required to complete a given reaction.

Figure A-5 illustrates some major fermentation pathways exhibited by microbes (though no single organism is capable of all of them). Pyruvate (shown in the blue box) is typically the starting point for each. End products of fermentation are shown in red. Fermentation allows a cell living under anaerobic conditions to oxidize reduced coenzymes (such as NADH+H^+ and shown in blue) generated during glycolysis or other pathways and makes them available as reactants in those same pathways. Some bacteria (aerotolerant anaerobes) rely solely on fermentation and do not use oxygen even if it is available. Table A-6 summarizes major fermentations and some representative organisms that perform each.

Notice that fermentation end products typically fall into three categories: acid, gas, or an organic solvent (an alcohol or a ketone). The specific fermentation performed is the result of the enzymes present in a species and often is used as a basis of classification.

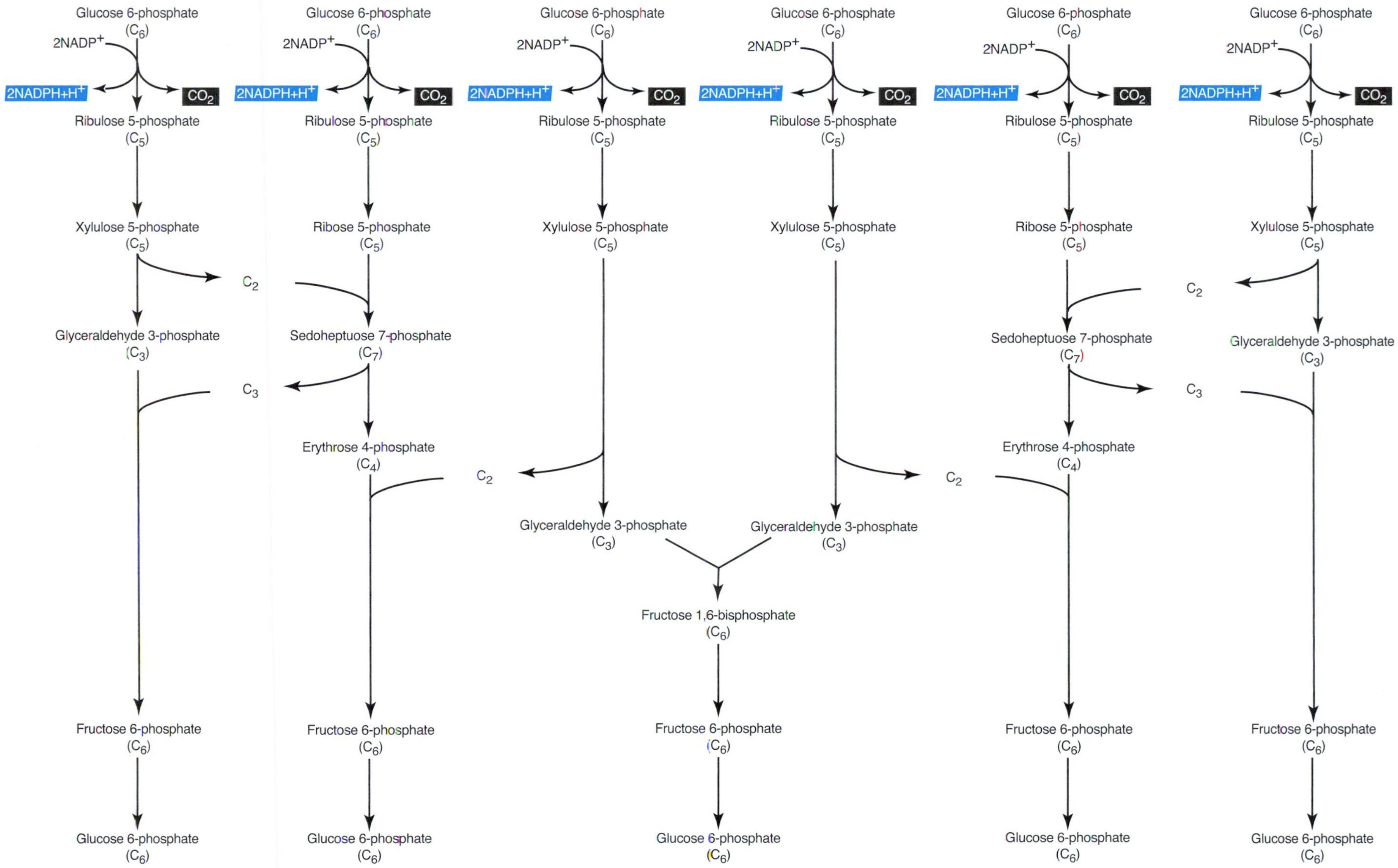

A-3 Pentose-Phosphate Cycle ◆ For every six glucose-6-phosphates that enter and complete the cycle, $6CO_2$ and 12 NADPH+H+ are produced. Some of the five-carbon intermediates, however, may be redirected into synthesis of aromatic amino acids and nucleotides. If the cycle is performed as shown, 36 carbons enter as six glucose 6-phosphate ($6 \times C_6 = 36C$). Six CO_2 are immediately lost, leaving a total of 30C to get shuffled around by the remaining reactions to form five glucose 6-phosphates ($5 \times C_6 = 30C$).

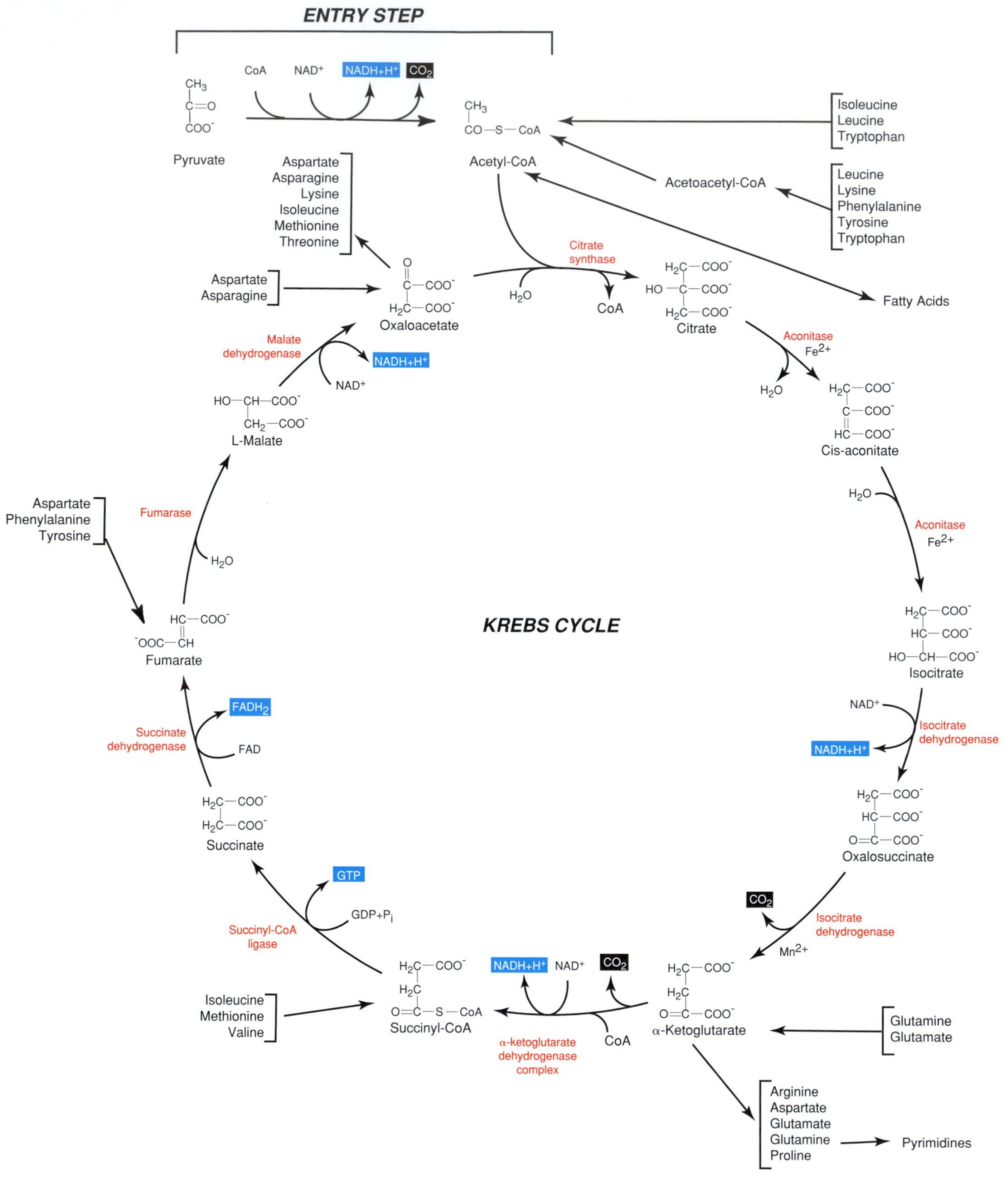

A-4 The Entry Step and Krebs Cycle ◆ The names of intermediates are printed in black ink; enzymes are in red. Reducing power (in the form of NADH+H^+ and $FADH_2$) and GTP are highlighted in blue. CO_2 produced from the oxidation of carbon is highlighted in black.

TABLE A-4 Summary of Reactants and Products per Glucose in the Entry Step and the Krebs Cycle ◆

Entry Step		Krebs Cycle	
Reactant	**Product**	**Reactant**	**Product**
2 Pyruvates	2 Acetyl CoA + 2 CO_2	2 Acetyl CoA	4 CO_2
2 Coenzyme A			2 Coenzyme A
2 NAD^+	2 NADH + $2H^+$	6 NAD^+	6 NADH + $6H^+$
		2 GDP + 2 P_i (=2 ADP + 2 P_i)	6 NADH + $6H^+$

TABLE A-5 ATP Yields from Complete Oxidation of Glucose to CO_2 by Prokaryote Using Glycolysis, Entry Step, and the Krebs Cycle with O_2 as the Final Electron Acceptor ◆

Compound	Number Produced	ATP Value*	Total ATPs per Glucose
NADH + H^+	10	3	30
$FADH_2$	2	2	4
ATP (by substrate phosphorylation)	4		4

* These values are estimates and are probably too high. Current evidence suggests each NADH is worth closer to 2.5 ATP and each $FADH_2$ is worth 1.5 ATP (White, David. [2000]. *The Physiology and Biochemistry of Prokaryotes*, 2nd ed. Oxford University Press, New York and Oxford)

TABLE A-6 Major Fermentations, Their End-Products, and Some Organisms That Perform Them ◆

Reactant	Product	Product
Alcoholic fermentation	Ethanol and CO_2	*Saccharomyces cerevisiae*
Homofermentation	Lactate	*Streptococcus* and some *Lactobacillus*
Heterofermentation	Lactate, ethanol, and acetate	*Streptococcus*, *Leuconostoc*, and *Lactobacillus*
Mixed acid fermentation	Acetate, formate, succinate, CO_2, H_2, and ethanol	*Escherichia*, *Salmonella*, *Klebsiella*, and *Shigella*
2,3-Butanediol fermentation	2,3-Butanediol	*Enterobacter*, *Serratia*, and *Erwinia*
Butyrate/butanol fermentation	Butanol, butyrate, acetone, and isopropanol	*Clostridium*, *Butyrivibrio*, and some *Bacillus*
Propionic acid fermentation	Propionate, acetate, and CO_2	*Propionibacterium*, *Veillonella*, and some *Clostridium*

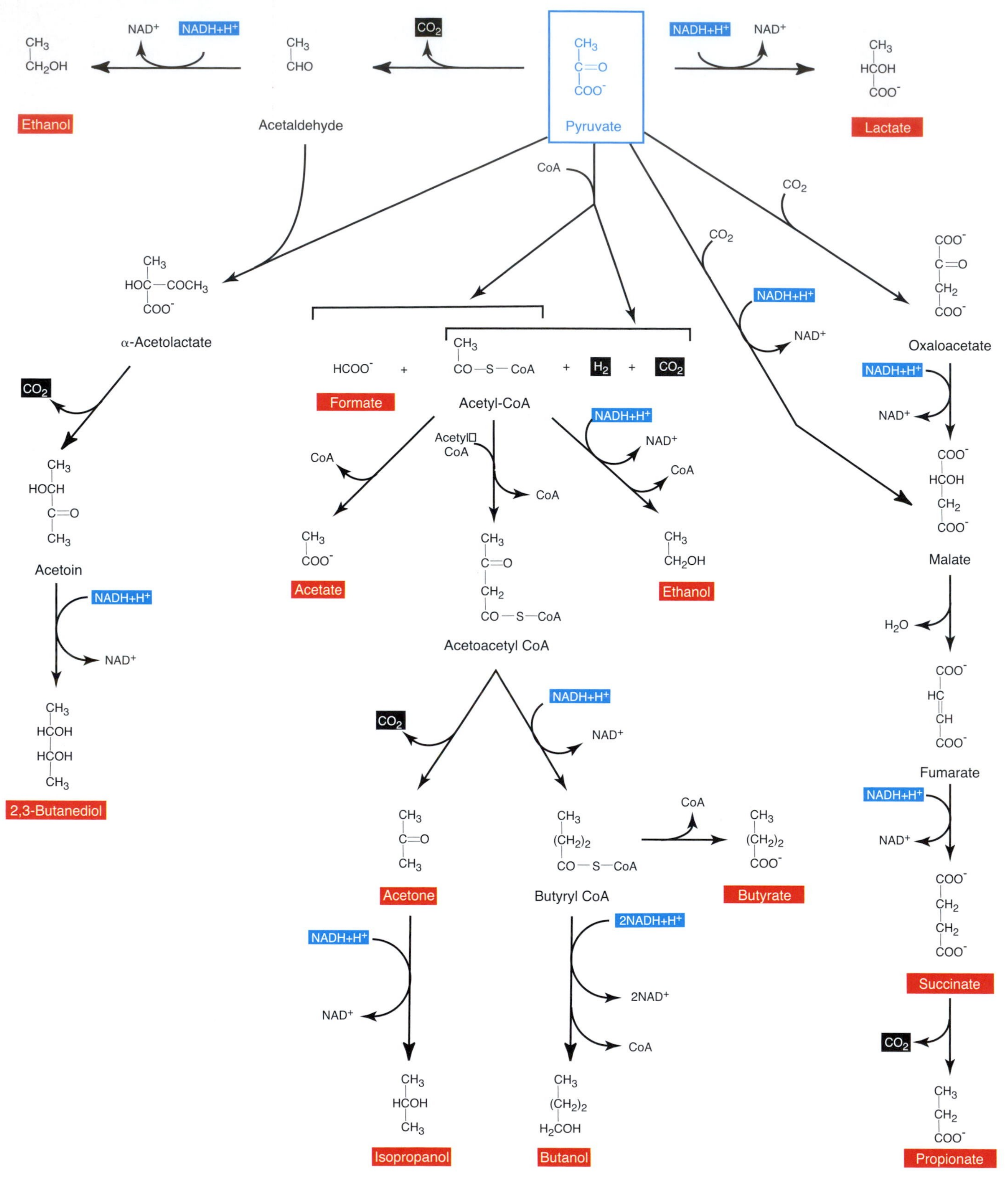

A-5 A Sampling of Fermentation Pathways ◆ Note that all pathways start with pyruvate, have a step(s) where $NADH + H^+$ (in blue) is oxidized to NAD^+, and produce end-products falling into one of three categories: acid, gas, or alcohol.

Miscellaneous Transfer Methods

APPENDIX

B

Following are instructions for transfer methods that are performed less routinely than those in Exercise 1-4.

Transfers Using a Sterile Cotton Swab

A sterile swab generally is used to obtain a sample from a primary source—either a patient or an environmental site. Occasionally, swabs are used to transfer pure cultures. Sterile swabs may be dry, or they may be in sterile water, depending on the source of the sample. In either case, care must be taken not to contaminate the swab by unintentionally touching other surfaces with it. Your instructor may provide specific instructions on collection of samples from sources other than the ones below.

Obtaining a Sample from a Patient's Throat with a Swab

1. Use a sterile tongue depressor and swab prepared in sterile water to obtain a sample from the throat. Have the patient open his/her mouth, then gently press down on the tongue (Figure B-1).
2. With the swab in your dominant hand, carefully sample the patient's throat with a swirling motion. Touching other parts of the oral cavity is likely to cause contamination. Also, avoid touching the soft palate or it may initiate a gag reflex!
3. Transfer the sample to an appropriate plated medium as quickly as possible. Use the swab to begin a quadrant streak (Exercise 1-4, Figure 1-33). Properly dispose of the swab, then complete the streaking with your loop.
4. Label the plate with the sample source, date, and medium.
5. If plating is to be done at a later time, place the swab in an appropriate sterile container (such as a sterile, capped test tube).
6. Incubate the plate in an inverted position at the appropriate temperature for the designated time.

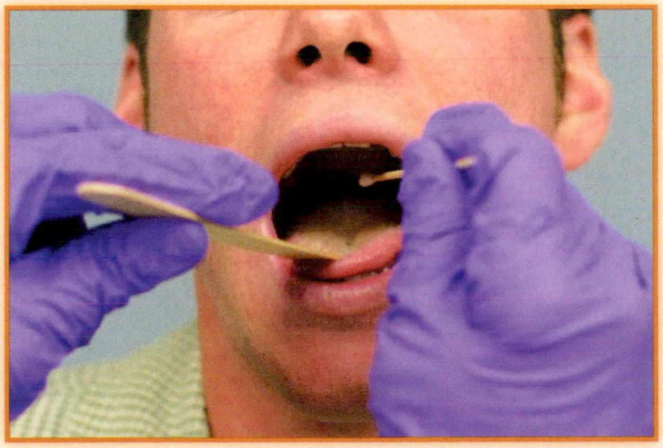

B-1 Taking a Throat Sample ◆ Use a sterile tongue depressor and swab to obtain a sample from a patient's throat. Be careful not to touch other parts of the oral cavity or the sample will get contaminated. Transfer the sample to a sterile medium as soon as possible.

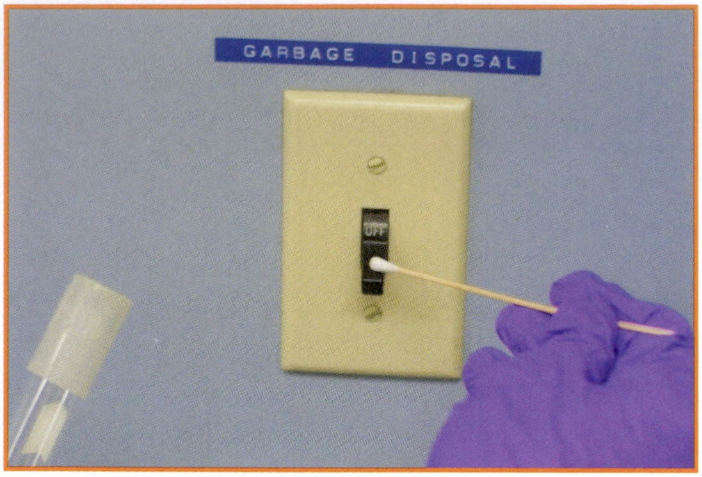

B-2 Taking an Environmental Sample ◆ Use a spinning motion of a sterile swab to sample inanimate objects in the environment. The swab may be placed in a sterile test tube until it is convenient to transfer the sample to a growth medium.

Obtaining an Environmental Sample with a Cotton Swab

1. Use a sterile swab prepared in sterile water to obtain a sample from an environmental source.
2. Rotate the swab to collect from the area to be sampled (Figure B-2).
3. Transfer the sample to an appropriate plated medium as quickly as possible. Depending on the anticipated cell density, use the swab to perform a zig-zag streak pattern (Exercise 1-4, Figure 1-32) or to begin a quadrant streak (Exercise 1-4, Figure 1-33). Properly dispose of the swab.
4. Label the plate with the sample source, date, and medium.
5. If plating is to be done at a later time, place the swab in an appropriate sterile container (such as a sterile, capped test tube).
6. Incubate the plate in an inverted position at the appropriate temperature for the designated time.

Stab Inoculation of Agar Tubes Using an Inoculating Needle

Stab inoculations of agar tubes are used for several types of differential media (usually to examine growth under anaerobic conditions or to observe motility). A stab is *not* used to produce a culture of microbes for transfer to another medium.

1. Label the tube with your name, date, and organism or sample sources.
2. Obtain the specimen with a sterile inoculating needle.
3. Remove the cap of the sterile medium with the little finger of your inoculating needle hand, and hold it there.

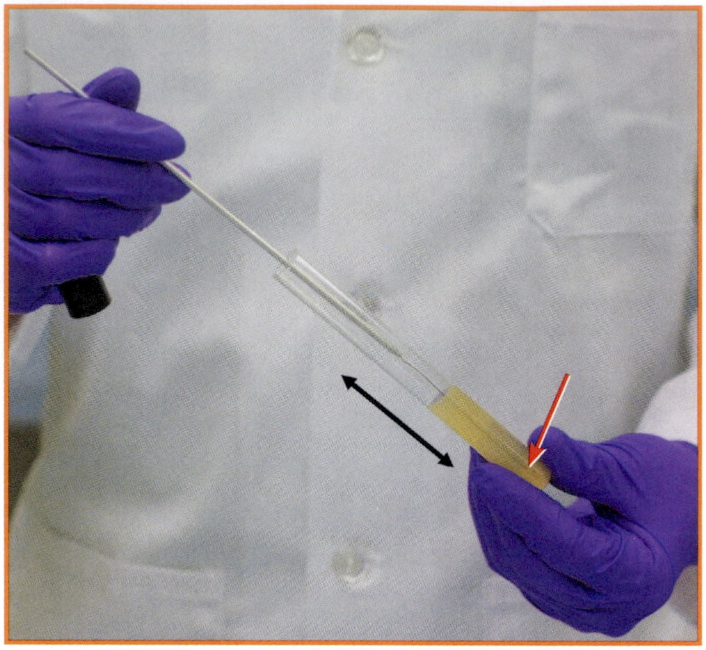

B-3 Agar Deep Stab ◆ Use the inoculating needle to stab the agar to a depth about 1 cm from the bottom (tip of red arrow). It is generally desirable to remove the needle along the original stab line and not create a new one. Upon completion, sterilize the needle.

4. Flame the tube by quickly passing it through the Bunsen burner flame a couple of times. Keep your needle hand still.
5. Hold the open tube on an angle to minimize airborne contamination. Keep your needle hand still.
6. Carefully move the agar tube over the needle wire (Figure B-3). Insert the needle into the agar to about 1 cm from the bottom.
7. Withdraw the tube carefully so the needle leaves along the same path it entered. (When removing the tube, be especially careful not to catch the needle tip on the tube lip. This springing action of the needle creates bacterial aerosols.)
8. Flame the tube lip as before. Keep your needle hand still.
9. Keeping the needle hand still (remember, it has growth on it), move the tube to replace its cap.
10. Flame the needle wire from base to tip to minimize aerosol production.
11. Incubate at the appropriate temperature for the assigned time.

Spot Inoculation of an Agar Plate

Sometimes an agar plate may be used to grow several different specimens at once. This is a typical practice with plated *differential* media (i.e., media designed to differentiate organisms based on growth characteristics). Prior to beginning the transfer, divide the plate into as many as four sectors, using a marking pen. (Some plates already have marks on the base for this purpose.) Then each may be inoculated with a different organism. Inoculation involves touching the loop to the agar surface once so growth is restricted to a single spot—hence the name "spot inoculation."

1. Label the base of the plate with your name, date, medium, and organism(s) inoculated.
2. With a marking pen, divide the base of the plate into the desired number of sectors. Figure B-4 shows a plate divided into four sectors.
3. Aseptically get growth from the source culture on the inoculating loop.
4. Lift the lid of the sterile agar plate and use it as a shield to prevent airborne contamination.
5. Touch the agar surface toward the periphery of the sector (Figure B-4).
6. Remove the loop and replace the lid.
7. Sterilize your loop as before. It is especially important to flame it from base to tip now because the loop has lots of bacteria on it.
8. Incubate the plate in an inverted position for the assigned time at the appropriate temperature.

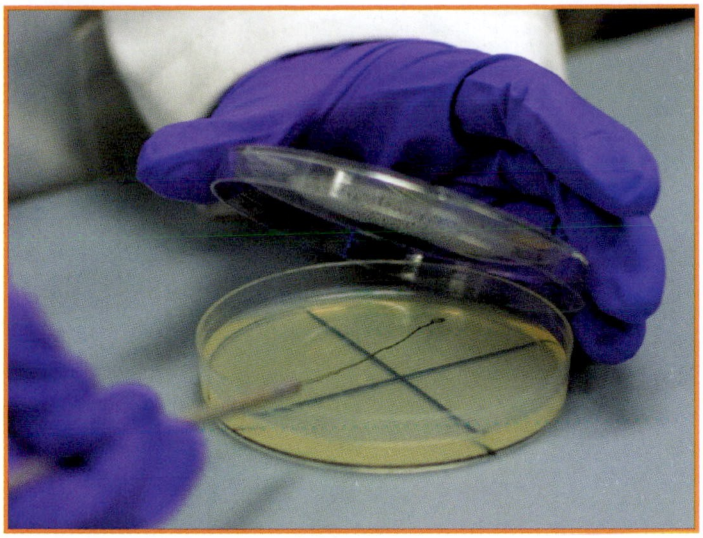

B-4 Spot Inoculation of a Plate ◆ Each of four sectors is spot inoculated with a different organism by touching the loop to the surface and making a mark about 1 cm in length. Generally, spot inoculations are done toward the edge (rather than the crowded middle) to prevent overlapping growth and/or test results.

APPENDIX

C

Transfer from a Broth Culture Using a Glass Pipette

Glass pipettes are used to transfer a known volume of liquid diluent, media, or culture. Originally pipettes were filled by sucking on them like a drinking straw, but mouth pipetting is dangerous and has been replaced by mechanical pipettors. Three examples are shown in Figure C-1, each with its own method of operation. Your instructor will show you how to properly use the type of pipettor available in your lab.

To use a pipette correctly, you must be able to correctly read the calibration. Examine Figure C-2. The numbers at the right indicate the pipette's *total volume* and its *smallest calibrated increments*. This is a 10.0 mL pipette divided into 0.1 mL increments.

When reading volumes, use the base of the meniscus (Figure C-3). The volume in the center pipette is read as exactly 3.0 mL because the meniscus is resting on the line. The left pipette is read as 2.9 mL and the right pipette is read as 3.1 mL (0 is always at the top of the pipette). Although the difference in volume between these three pipettes may seem negligible (1 part in 70,[1] about a 1.5% error), it may be enough to introduce significant error into your work.

Two pipette styles are used in microbiology (Figure C-4): the **serological pipette** and the **Mohr pipette**. A serological pipette is calibrated *to deliver* (TD) its volume by completely draining it and blowing out the last drop. The tip of a Mohr pipette is not graduated, so fluid

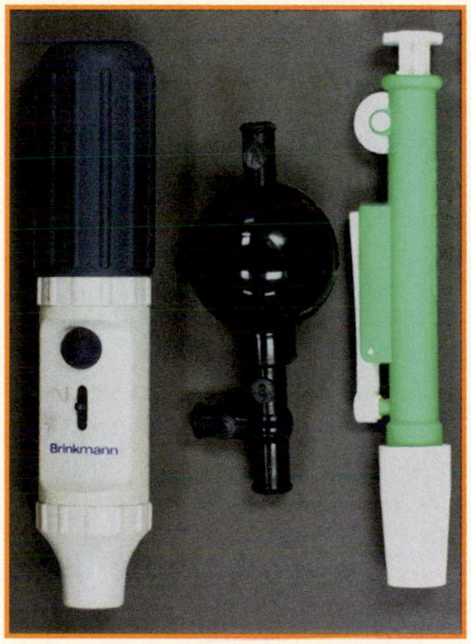

C-1 Mechanical Pipettors ◆ Three examples of mechanical pipettors are shown here, each with its own method of operation. Your instructor will show you how to properly use the style of pipettor available in your lab. From left to right: A pipette filler/dispenser, a pipette bulb, and a plastic pump.

[1] This number presumes the pipette shown has a total volume of 10.0 mL. If the meniscus is on the 3, the pipette will dispense 7.0 mL. If emptied completely, an error of 0.1 mL above or below the 3 would represent 0.1 mL in 7.0 mL or 1 part in 70.

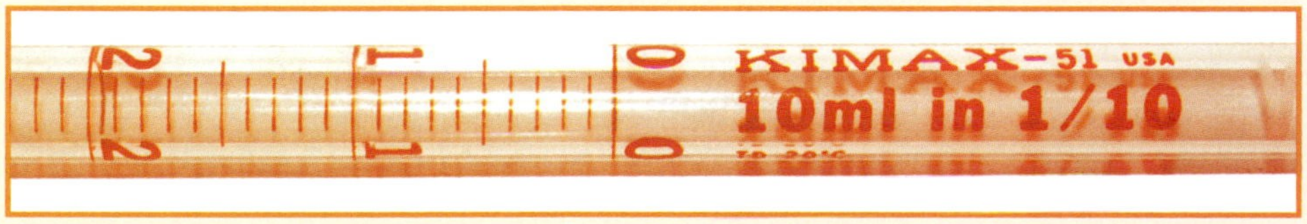

C-2 Pipette Calibration ◆ Prior to using a pipette, read its calibration. The numbers indicate the pipette's *total volume* and its *smallest calibrated increments*. This is a 10.0 mL pipette divided into 0.1 mL increments.

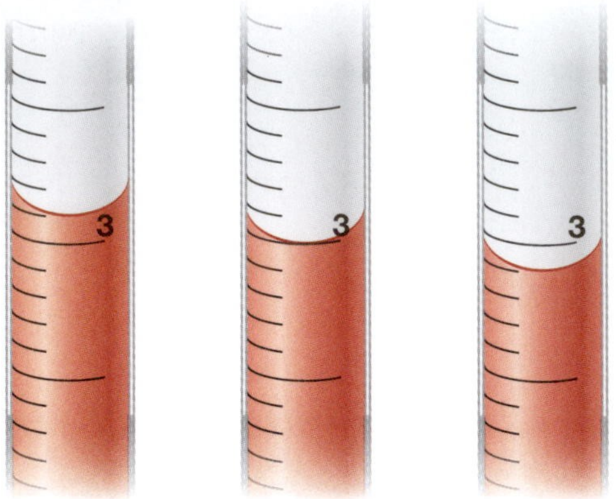

C-3 Read the Base of the Meniscus ◆ When reading volumes, use the base of the meniscus. The volume in the center pipette is read at exactly 3.0 mL because the meniscus is resting on the line. The left pipette is read as 2.9 mL and the right pipette is read as 3.1 mL (0 is always at the top of the pipette).

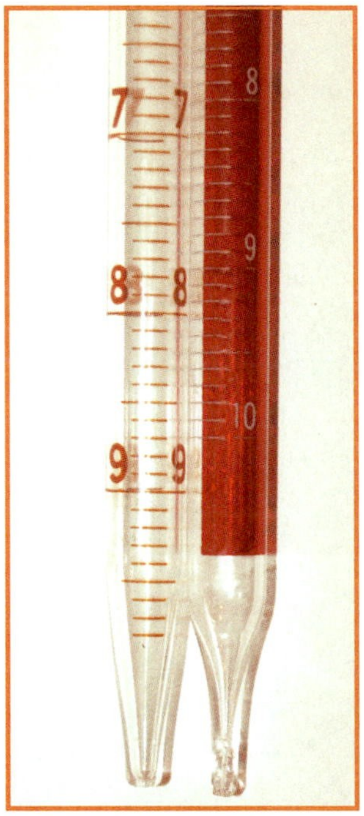

C-4 Two Types of Pipettes ◆ Two pipette styles—the *serological pipette* (left) and the *Mohr pipette* (right)—are used in microbiology. A serological pipette is calibrated *to deliver* (TD) its volume by completely draining it and blowing out the last drop. The tip of a Mohr pipette is not graduated, so fluid flow must be stopped at a calibration line. Stopping the fluid beyond the last line on a Mohr pipette results in an unknown volume being dispensed.

C-5 Getting the Sterile Pipette ◆ Pipettes of the same size are autoclaved in canisters, which then are opened and placed flat on the table. Pipettes are removed as needed.

flow must be stopped at a calibration line. Stopping the fluid beyond the last line on a Mohr pipette results in an unknown volume being dispensed. In either case, volumes are read at the bottom of the fluid meniscus.

Important: If pipetting a bacterial culture, be careful not to allow any to drop from the pipette before disposing of it in the autoclave container. Clean up any spills. Some labs recommend pipetting over a disinfectant-soaked towel.

Following are instructions for using a glass pipette.

Filling a Glass Pipette

1 Label the tube or plate you will be inoculating with your name, date, medium, and inoculum.

2 Bacteria should be suspended in the broth with a vortex mixer (Figure 1-16) or by agitating with your fingers (Figure 1-17). Be careful not to splash the broth into the cap or lose control of the tube.

3 Pipettes are sterilized in metal canisters, individually in sleeves, or as multiples in packages (if disposable). They typically are stored in groups of a single size (Figure C-5). *Be sure you know what volume your pipette will deliver*. Set the canister at the table edge and remove its lid. (Canisters should not be stored in an upright position, as they may fall over and break the pipettes or become contaminated.) If using pipettes in a package, open the end *opposite the tips*. Grasp *one pipette only* and remove it.

4 Carefully insert the pipette into the mechanical pipettor (Figure C-6). It's best to grasp the pipette near the end with your fingertips. This gives you more control and reduces the chance that you will break the pipette and cut your hand. *Do not touch* any part of the pipette that will contact the specimen or the medium or you will risk introducing a contaminant. Also, do not lay the pipette on the tabletop while you continue.

5 While keeping your pipette hand still, bring the culture tube toward it. Use your little finger to remove and hold its cap (Figure C-7).

6 Flame the open end of the tube by passing it through a Bunsen burner flame two or three times.

7 Hold the tube at an angle to prevent contamination from above.

8 Insert the pipette and withdraw the appropriate volume (Figure C-7). Bring the pipette to a vertical position briefly to read the meniscus accurately. (Remember: The volumes in the pipette are correct only if the meniscus of the fluid inside is resting *on* the line, not below it.) Then carefully remove the pipette from the tube.

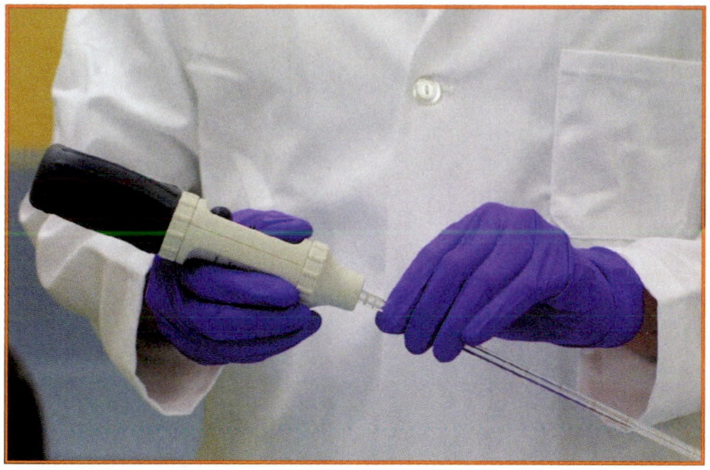

C-6 Assembling the Pipette ◆ Carefully insert the pipette into a mechanical pipettor. Notice that the pipette is held near the end with the fingertips. For safety, the hand is out of the way in case the pipette breaks.

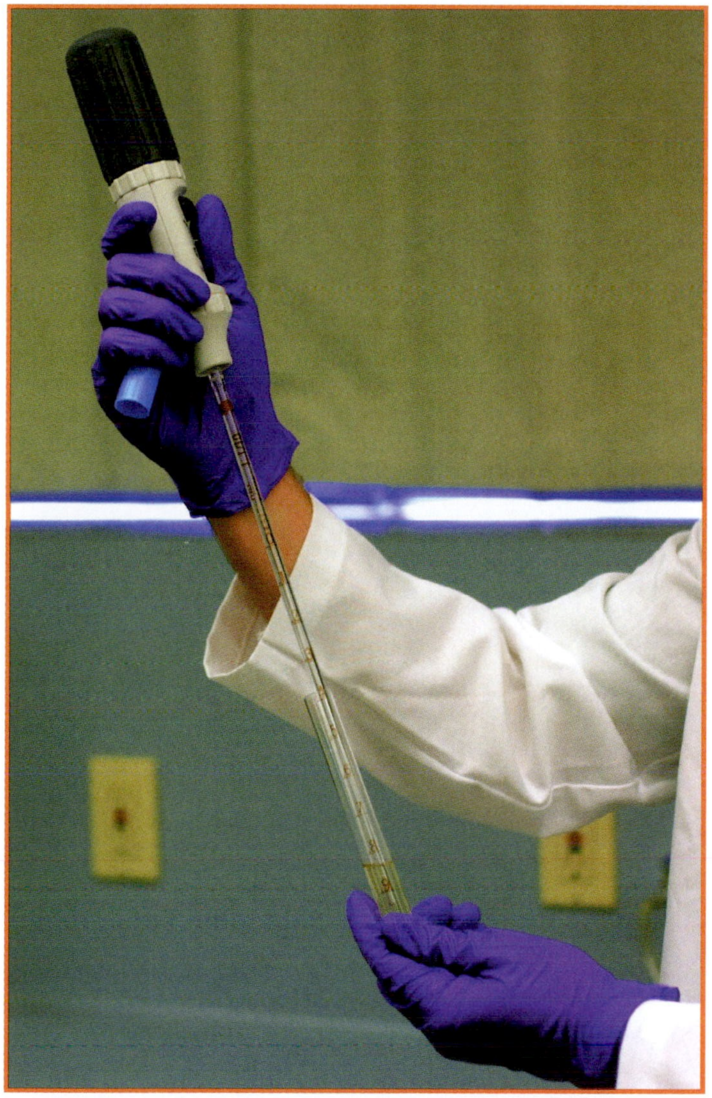

C-7 Filling the Pipette ◆ Carefully draw the fluid into the pipette. Briefly bring it to a vertical position and read the volume. Notice how the tube's cap is being held by the pinky of the pipettor hand.

9 Flame the tube lip as before. Keep your pipette hand still.

10 Keeping the pipette hand still (remember, it may contain fluid with microbes in it), move the tube to replace its cap.

11 What you do next depends on the medium to which you are transferring the liquid. Please continue with the appropriate inoculation section.

Inoculation of Broth Tubes with a Pipette

Pipettes are often used to inoculate a known volume of culture into a known volume of diluent during serial dilutions.

1 While keeping the pipette hand still, bring the broth culture tube toward it. Use your little finger to remove and hold its cap.

2 Flame the tube while you keep your pipette hand still.

3 Hold the open tube on an angle.

4 Insert the pipette tip and dispense the correct volume of inoculum.

5 Withdraw the tube from over the pipette. Before completely removing it, touch the pipette tip to the glass to remove any excess broth.

6 Completely remove the pipette, but avoid waving it around. This can create aerosols.

7 Flame the tube lip, and while keeping the pipette hand still, move the tube to replace its cap.

8 The pipette is contaminated with microbes and must be disposed of correctly. Each lab has its own specific procedures, and your instructor will advise you what to do. Glass pipettes typically are placed in a pipette disposal container containing a small amount of disinfectant until they are autoclaved and reused (Figure C-8). Disposable pipettes must be placed in an appropriate biohazard container. In either case, be careful when removing the pipette from the mechanical pipettor. There is danger of culture dripping from the pipette or of breaking the glass or plastic.

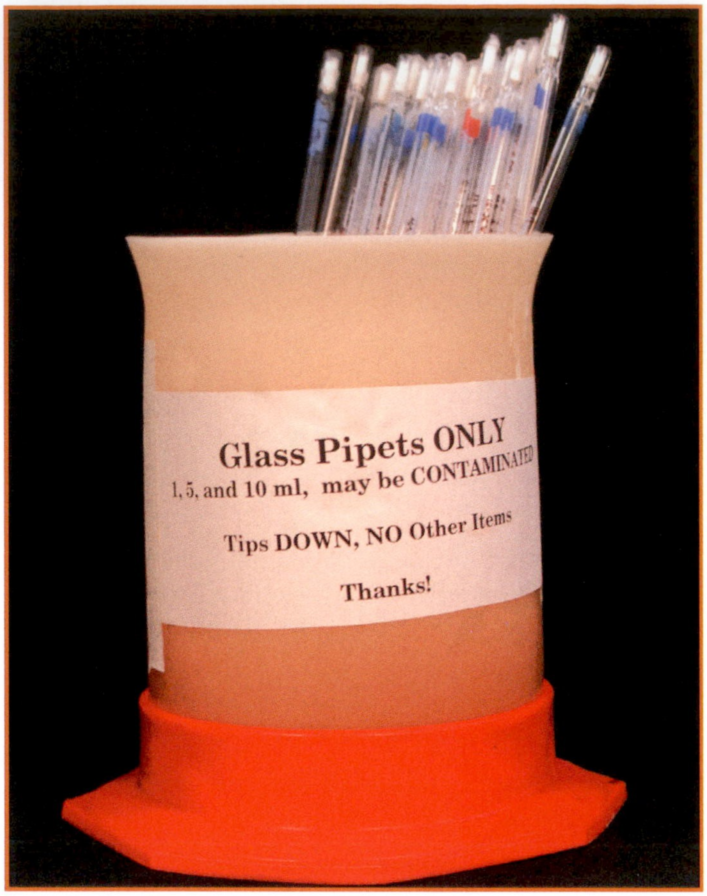

C-8 Dispose of the Pipette ◆ The pipette may be contaminated with microbes and must be disposed of correctly. Each lab has its own specific procedures, and your instructor will advise you what to do. Shown here is a canister used for glass pipettes. Disposable pipettes must be placed in an appropriate biohazard container. In either case, be careful when removing the pipette from the mechanical pipettor. There is danger of culture dripping from the pipette or of breaking the glass or plastic.

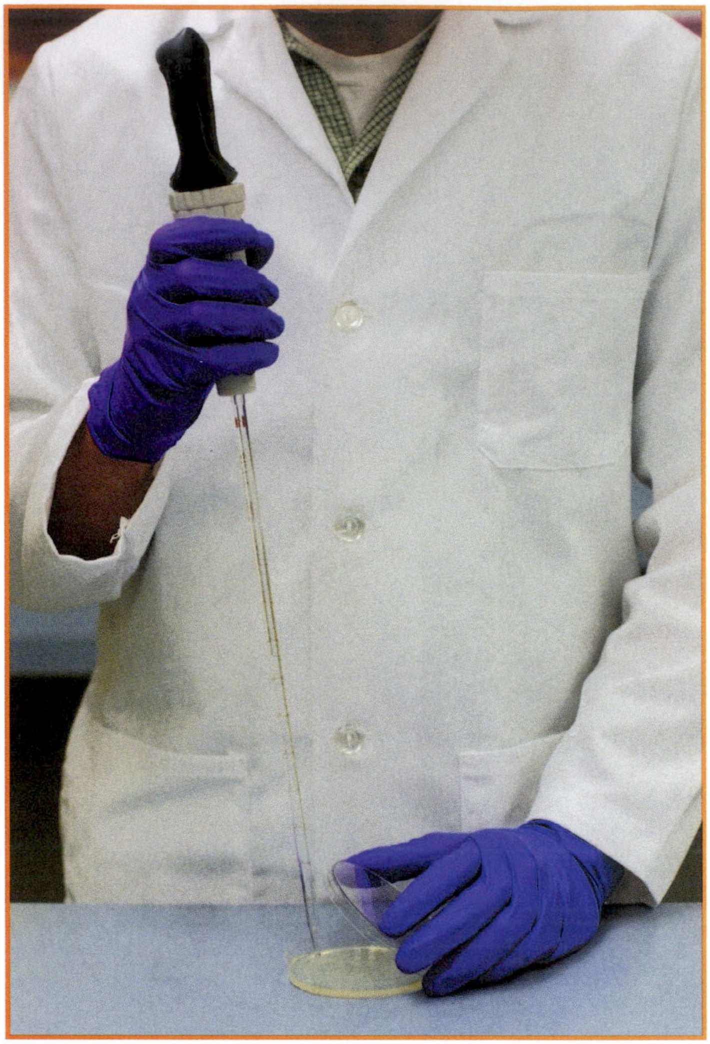

C-9 Inoculating the Plate with a Pipette ◆ Open the lid and dispense the inoculum onto the surface of the agar near the middle. Notice how the plate's lid is held to prevent aerial contamination.

Inoculation of Agar Plates with a Pipette

Pipettes also can be used to dispense a known volume of inoculum to an agar plate.

1. Lift the lid of the plate and use it as a shield to protect it from airborne contamination.
2. Hold the pipette over the agar and dispense the correct volume (often 0.1 mL) onto the center of the agar surface (Figure C-9). From this point, the remainder of steps should be completed within about 15 seconds to prevent the inoculum from soaking into the agar.
3. The pipette is contaminated with microbes and must be disposed of correctly. Each lab has its own specific procedures, and your instructor will advise you what to do.
4. Continue with the spread plate technique (Exercise 1-6).

APPENDIX

D

Transfer from a Broth Culture Using a Digital Pipette

Modern molecular biology procedures often involve transferring extremely small volumes of liquid with great precision and accuracy. This has led to the development of digital pipettors (Figure D-1) that can be set up to dispense microliter volumes through milliliter volumes (recall that 1.0 mL equals 1000 µL). Common digital pipettors are calibrated to dispense volumes of 1 to 10 µL, 10 to 100 µL, 100 to 1000 µL, or 1.0 to 5.0 mL.

Filling a Digital Pipettor

Many manufacturers make digital pipettors, but they all work basically the same way. The following instructions are for Eppendorf Series 2100 models.

1. Label the tube or plate you will be inoculating with your name, date, medium, and inoculum.
2. Growth may be suspended in the broth with a vortex mixer (Figure 1-16) or by agitating with your fingers (Figure 1-17). Be careful not to splash the broth into the cap or lose control of the tube.
3. Determine which digital pipettor should be used to dispense the desired volume.
4. Turn the setting ring to set the desired volume (Figure D-2). CAUTION: If you turn the dial past the volume range of the pipettor, you will damage it.
5. Hold the digital pipettor in your dominant hand.
6. Open the rack of appropriate pipette tips for your pipettor (these are often color-coded and match the pipettor) and push the pipettor into a sterile tip (Figure D-3). Close the rack. Never touch the pipette tip with your hands or leave the rack open. *Never use a digital pipettor without a tip.*
7. Remove the cap from the culture tube with the little finger of your dominant hand, and flame the tube.
8. Press down the control button with your thumb to the first stop. This is the measuring stroke.

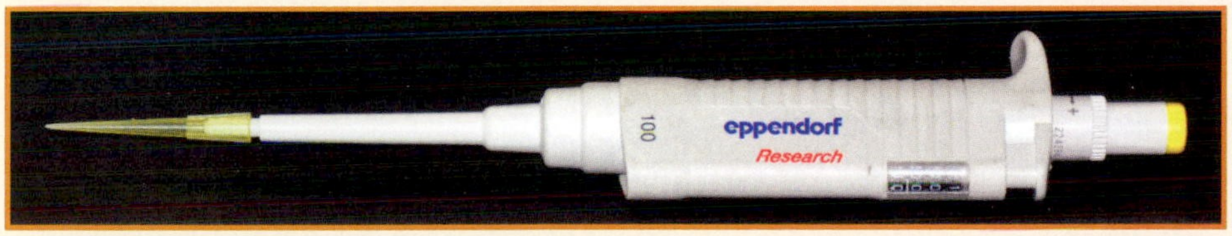

D-1 Digital Pipettor ◆ This is a digital pipettor to be used for dispensing volumes betwen 10 µL and 100 µL. Never set the pipettor to volumes above or below the intended volumes as damage may occur. Also, always use a digital pipettor with the appropriate tip. As shown here, tips and pipettors are often color-coded.

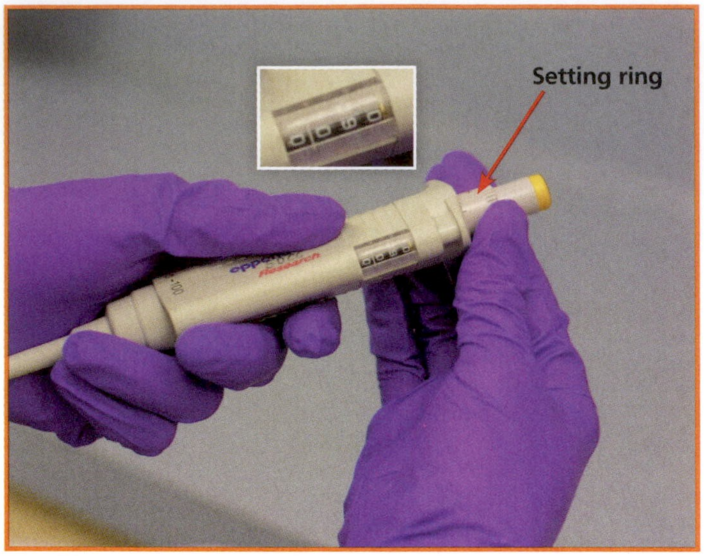

D-2 Setting the Volume ◆ Rotate the setting ring to adjust the volume on a digital pipettor. This pipettor has been set at 90.0 μL (see inset—the horizontal line between the third and fourth numerals is a decimal point). Never rotate the adjustment knob beyond the volume limits of the pipettor.

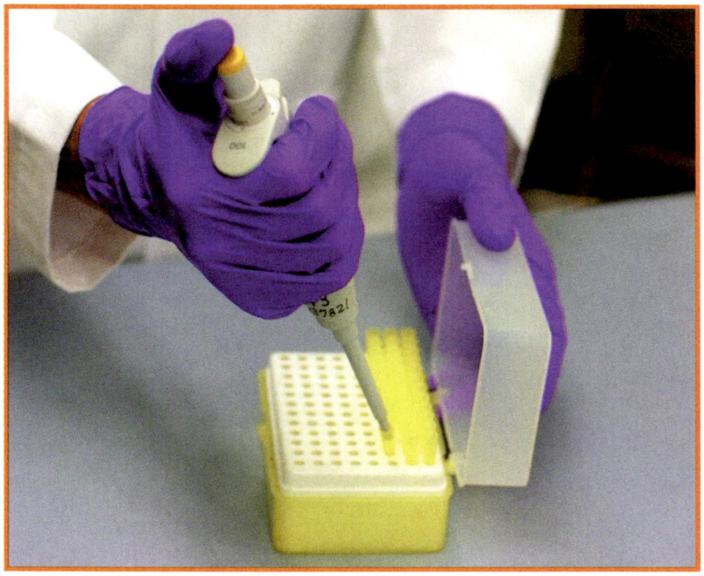

D-3 Digital Pipettor Tips ◆ Digital pipettors must be fitted with a sterile tip of appropriate size (often color-coded to match the pipettor). Open the case and press the pipettor into a tip, then close the case to maintain sterility. Do not touch the pipettor tip.

9 Insert the tip into the broth approximately 3 mm while holding it vertically (Figure D-4).

10 Slowly release pressure with your thumb to draw fluid into the tip. Be careful not to pull any air into the tip.

11 Remove the pipettor and hold it still as you flame the tube as before and replace the cap.

12 What you do next depends on the medium to which you are transferring the growth. Please continue with the appropriate inoculation section.

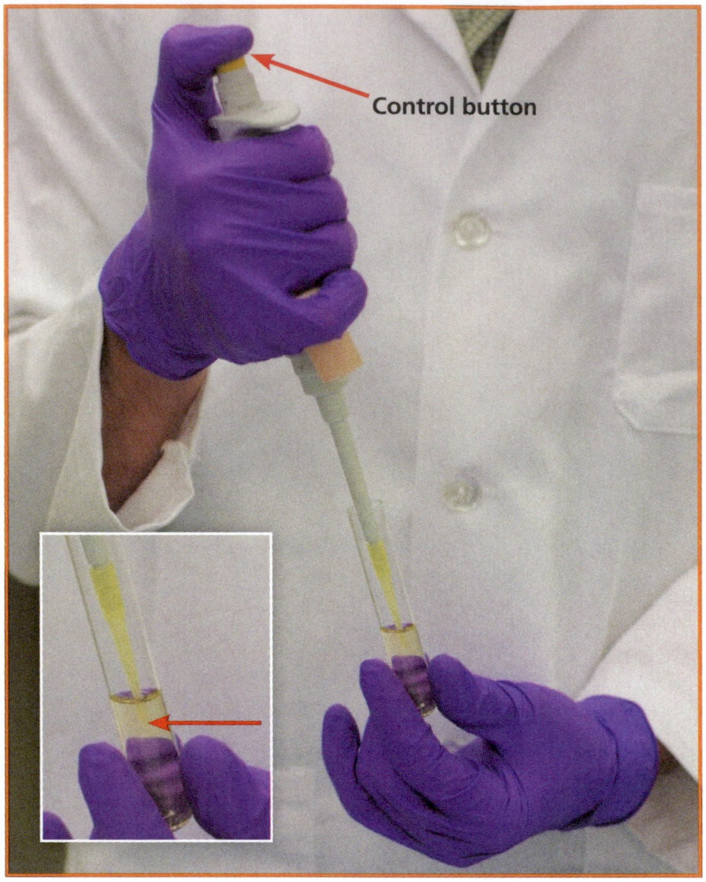

D-4 Filling the Pipette ◆ Depress the control button with your thumb to the first stop (measuring stroke). Holding the tube and pipettor in a vertical position, insert the tip into the fluid to a depth of 3 mm (see inset). Slowly release pressure with your thumb to fill the pipettor.

Inoculation of Broth Tubes with a Digital Pipettor

1 Remove the cap of the sterile medium with the little finger of your pipettor hand, and flame the tube.

2 Insert the pipette tip into the tube. Hold it at an angle against the inside of the glass (Figure D-5).

3 Depress the control button slowly with your thumb to the first stop and pause until no more liquid is dispensed. Then continue pressing to the second stop to deliver the remaining volume. This is the blow-out stroke.

4 While keeping pressure on the control button, carefully remove the pipettor from the tube by sliding it along the glass. Once it is out of the tube, slowly release pressure on the control button. (If you release pressure on the control button before the tip is out of the liquid, you will remove a portion of the measured volume dispensed.)

5 Flame the tube lip and replace the cap as before. Keep your pipette hand still.

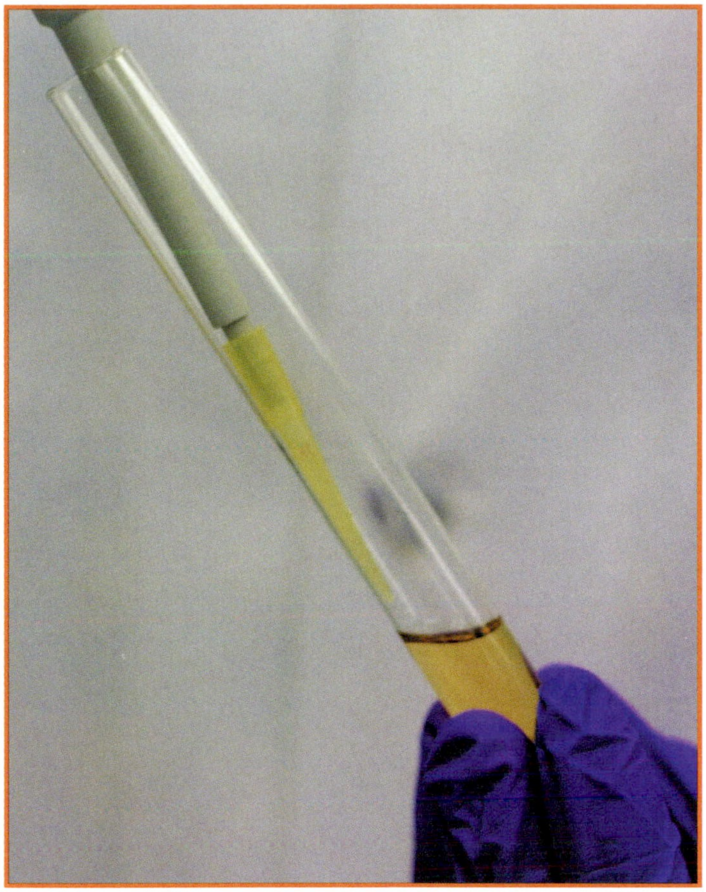

D-5 Dispensing to a Liquid ◆ Hold the pipettor at an angle with the tip against the glass. Press the control button to the first stop. When no more fluid comes out, continue pressing to the second stop. Remove the pipettor, then slowly release pressure on the control button. Flame the tube and replace its cap. Then, eject the tip into a biohazard container.

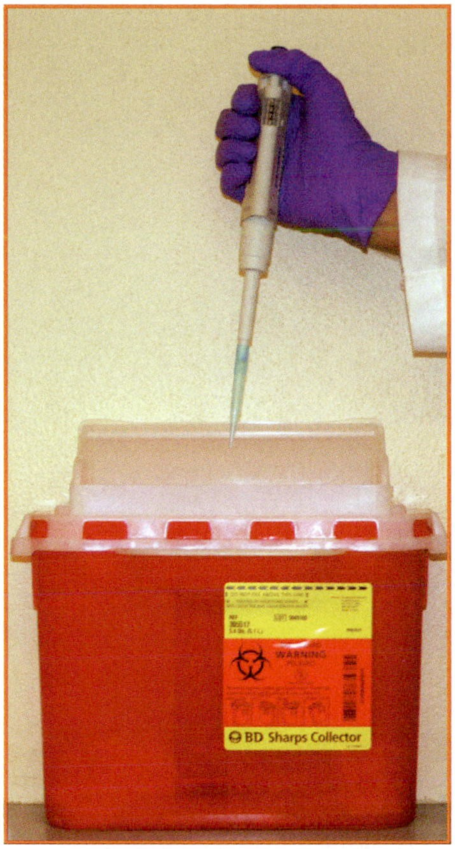

D-6 Ejecting the Tip ◆ Using the tip ejector button, eject the contaminated tip into an appropriate biohazard container.

6 The pipettor tip is contaminated with microbes and must be disposed of correctly. Use the ejector button to remove the tip into an appropriate biohazard container (Figure D-6). Each lab has its own specific procedures, and your instructor will advise you what to do.

Inoculation of Agar Plates with a Pipettor

1 Lift the lid of the plate and use it as a shield to protect from airborne contamination.

2 Place the pipette tip over the agar surface. Be sure to hold the pipettor in a vertical position (Figure D-7).

3 Depress the control button slowly with your thumb to the first stop, and pause until no more liquid is dispensed. Then continue pressing to the second stop to deliver the appropriate volume. This is the blow-out stroke. From this point, the remainder of steps should be completed within about 15 seconds to prevent the inoculum from soaking into the agar.

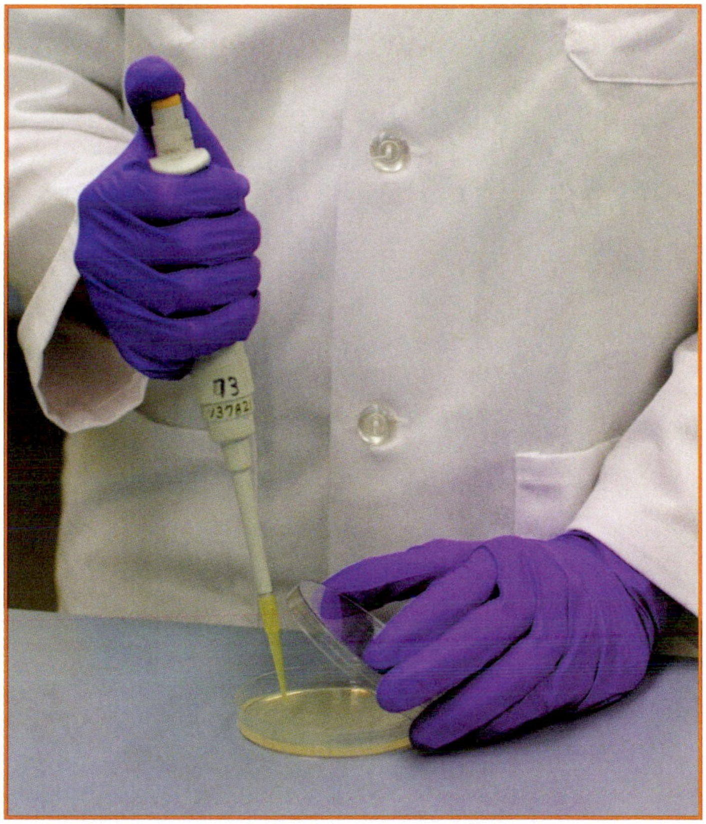

D-7 Dispensing to a Plate ◆ Hold the pipette in a vertical position and press the control button to the first stop. When no more fluid comes out, continue pressing to the second stop. Slowly release pressure on the control button, then eject the tip into a biohazard container.

4 Use the ejector button to remove the contaminated tip into an appropriate biohazard container (Figure D-6). Each lab has its own specific procedures, and your instructor will advise you what to do.

5 Continue with the Spread Plate Technique (Exercise 1-6).

Reference

Brinkman Instruments, *Eppendorf Series 2001 Pipette Instruction Manual.* Westbury, New York, NY.

APPENDIX

E

Medium, Reagent, and Stain Recipes

❖ Media

Unless otherwise specified, prepare all media as follows:

Broths

1 Suspend the dry ingredients in the water. Agitate and heat slightly (if necessary) to dissolve completely.

2 Dispense 7.0 mL portions into 16 X 150 mm test tubes and cap loosely.

3 Autoclave for 15 minutes at 121°C to sterilize the medium.

Slants and Stabs

1 Suspend the dry ingredients in the water, mix well, and boil until completely dissolved.

2 Dispense 7.0 mL (or 10.0 mL for deep stabs) portions into 16 X 150 mm test tubes, and cap loosely.

3 Autoclave for 15 minutes at 121°C to sterilize the medium.

4 Slant the tubes or leave upright as needed for stabs and cool to room temperature.

Plates

1 Suspend the dry ingredients in the water, mix well, and boil until completely dissolved.

2 Cover loosely and sterilize in the autoclave at 15 lbs. pressure (121°C) for 15 minutes.

3 Remove from the autoclave, allow the medium to cool to approximately 50°C, and aseptically pour into sterile 100 mm Petri dishes (20.0 mL/plate).

4 Allow the medium to cool to room temperature. Store upside down.

Bile Esculin Agar

♦ Pancreatic digest of gelatin	5.0 g
♦ Beef extract	3.0 g
♦ Oxgall	20.0 g
♦ Ferric citrate	0.5 g
♦ Esculin	1.0 g
♦ Agar	14.0 g
♦ Distilled or deionized water	1.0 L

pH 6.6–7.0 at 25°C

1 Suspend the dry ingredients in the water, mix well, and boil until completely dissolved.

2 Cover loosely and sterilize in the autoclave at 15 lbs. pressure (121°C) for 15 minutes.

3 Remove from the autoclave, allow the medium to cool slightly, and aseptically pour into sterile Petri dishes (20.0 mL/plate).

4 Allow the medium to cool to room temperature. Store upside down.

Blood Agar

♦ Infusion from beef heart (solids)	2.0 g
♦ Pancreatic digest of casein	13.0 g
♦ Sodium chloride	5.0 g
♦ Yeast extract	5.0 g
♦ Agar	15.0 g
♦ Defibrinated sheep blood	50.0 mL
♦ Distilled or deionized water	1.0 L

pH 7.1–7.5 at 25°C

1 Suspend the dry ingredients in the water, mix well, and boil until completely dissolved. This is blood agar base.

2 Cover loosely and sterilize in the autoclave at 15 lbs. pressure (121°C) for 15 minutes.

3 Remove from the autoclave and cool to 45°C.

4 Aseptically add the sterile, room-temperature sheep blood to the blood agar base and mix well.

5 Pour into sterile Petri dishes and allow the medium to cool to room temperature.

Brain-Heart Infusion Agar

♦ Calf brains, infusion from 200.0 g	7.7 g
♦ Beef heart, infusion from 250.0 g	9.8 g
♦ Proteose peptone	10.0 g
♦ Dextrose	2.0 g
♦ Sodium chloride	5.0 g
♦ Disodium phosphate	2.5 g
♦ Agar	15.0 g
♦ Distilled or deionized water	1.0 L

pH 7.2–7.6 at 25°C

Brain-Heart Infusion Broth

♦ Calf brains, infusion from 200.0 g	7.7 g
♦ Beef heart, infusion from 250.0 g	9.8 g
♦ Proteose peptone	10.0 g
♦ Dextrose	2.0 g
♦ Sodium chloride	5.0 g
♦ Disodium phosphate	2.5 g
♦ Distilled or deionized water	1.0 L

pH 7.2–7.6 at 25°C

Brilliant Green Lactose Bile Broth

♦ Peptone	10.0 g
♦ Lactose	10.0 g
♦ Oxgall	20.0 g
♦ Brilliant green dye	0.0133 g
♦ Distilled or deionized water	1.0 L

pH 7.0–7.4 at 25°C

1 Suspend the dry ingredients in the water, mix well, and boil until completely dissolved.

2 Dispense 10.0 mL portions into test tubes.

3 Place an inverted Durham tube in each broth and cap loosely.

4 Sterilize in the autoclave at 15 lbs. pressure (121°C) for 15 minutes.

5 Remove the medium from the autoclave and allow it to cool before inoculating.

Citrate Medium (Simmons)

♦ Ammonium dihydrogen phosphate	1.0 g
♦ Dipotassium phosphate	1.0 g
♦ Sodium chloride	5.0 g
♦ Sodium citrate	2.0 g
♦ Magnesium sulfate	0.2 g
♦ Agar	15.0 g
♦ Bromthymol blue	0.08 g
♦ Distilled or deionized water	1.0 L

pH 6.7–7.1 at 25°C

Columbia CNA with 5% Sheep Blood Agar

♦ Pancreatic digest of casein	12.0 g
♦ Peptic digest of animal tissue	5.0 g
♦ Yeast extract	3.0 g
♦ Beef extract	3.0 g
♦ Corn starch	1.0 g
♦ Sodium chloride	5.0 g
♦ Colistin	10.0 mg
♦ Nalidixic acid	10.0 mg
♦ Agar	13.5 g
♦ Sheep blood (defibrinated, sterile)	50.0 mL
♦ Distilled or deionized water	1.0 L

pH 7.1–7.5 at 25°C

1 Suspend the dry ingredients in the water, mix well, and boil until completely dissolved.

2 Cover loosely and sterilize in the autoclave at 15 lbs. pressure (121°C) for 15 minutes.

3 Remove from the autoclave and cool to 45 to 50°C.

4 Add the sheep blood to 950.0 mL of the mixture and aseptically pour into sterile Petri dishes (20.0 mL/plate).

5 Allow the medium to cool to room temperature. Store upside down.

Decarboxylase Medium (Møller)

♦ Peptone	5.0 g
♦ Beef extract	5.0 g
♦ Glucose (dextrose)	0.5 g
♦ Bromcresol purple	0.01 g
♦ Cresol red	0.005 g
♦ Pyridoxal	0.005 g
♦ L-Lysine, L-Ornithine, or L-Arginine	10.0 g
♦ Distilled or deionized water	1.0 L

pH 5.8–6.2 at 25°C

DNase Test Agar with Methyl Green

◆ Tryptose	20.0 g
◆ Deoxyribonucleic acid	2.0 g
◆ Sodium chloride	5.0 g
◆ Agar	15.0 g
◆ Methyl green	0.05 g
◆ Distilled or deionized water	1.0 L

pH 7.1–7.5 at 25°C

EC Broth

◆ Tryptose	20.0 g
◆ Lactose	5.0 g
◆ Dipotassium phosphate	4.0 g
◆ Monopotassium phosphate	1.5 g
◆ Sodium chloride	5.0 g
◆ Distilled or deionized water	1.0 L

pH 6.7–7.1 at 25°C

1. Suspend the dry ingredients in the water, and mix well until completely dissolved.
2. Dispense 10.0 mL portions into test tubes.
3. Place an inverted Durham tube in each broth and cap loosely.
4. Sterilize in the autoclave at 15 lbs. pressure (121°C) for 15 minutes.
5. Remove the media from the autoclave and allow it to cool before inoculating.

Endo Agar LES

◆ Yeast extract	1.2 g
◆ Casitone	3.7 g
◆ Thiopeptone	3.7 g
◆ Tryptose	7.5 g
◆ Lactose	9.4 g
◆ Dipotassium phosphate	3.3 g
◆ Monopotassium phosphate	1.0 g
◆ Sodium chloride	3.7 g
◆ Sodium desoxycholate	0.1 g
◆ Sodium lauryl sulfate	0.05 g
◆ Sodium sulfite	1.6 g
◆ Basic fuchsin	0.8 g
◆ Agar	15.0 g
◆ Distilled or deionized water	1.0 L

pH 7.3–7.7 at 25°C

Eosin Methylene Blue Agar (Levine)

◆ Peptone	10.0 g
◆ Lactose	10.0 g*
◆ Dipotassium phosphate	2.0 g
◆ Agar	15.0 g
◆ Eosin Y	0.4 g
◆ Methylene blue	0.065 g
◆ Distilled or deionized water	1.0 L

pH 6.9–7.3 at 25°C

Glucose Salts Medium

◆ Glucose	5.0 g
◆ NaCl	5.0 g
◆ $MgSO_4$	0.2 g
◆ $(NH_4)H_2PO_4$	1.0 g
◆ K_2HPO_4	1.0 g
◆ Distilled or deionized water	1.0 L

Hektoen Enteric Agar

◆ Yeast extract	3.0 g
◆ Peptic digest of animal tissue	12.0 g
◆ Lactose	12.0 g
◆ Sucrose	12.0 g
◆ Salicin	2.0 g
◆ Bile salts	9.0 g
◆ Sodium chloride	5.0 g
◆ Sodium thiosulfate	5.0 g
◆ Ferric ammonium citrate	1.5 g
◆ Bromthymol blue	0.064 g
◆ Acid fuchsin	0.1 g
◆ Agar	13.5 g
◆ Distilled or deionized water	1.0 L

pH 7.4–7.8 at 25°C

1. Suspend the dry ingredients in the water, mix well, and boil until completely dissolved.
2. Do not autoclave.
3. When cooled to 50°C, pour into sterile plates.
4. Cool to room temperature with lids slightly open.

* An alternative recipe replaces the 10.0 g of lactose with 5.0 g of lactose and 5.0 g of sucrose.

Kligler Iron Agar

◆ Pancreatic digest of casein	10.0 g
◆ Peptic digest of animal tissue	10.0 g
◆ Lactose	10.0 g
◆ Dextrose (glucose)	1.0 g
◆ Ferric ammonium citrate	0.5 g
◆ Sodium chloride	5.0 g
◆ Sodium thiosulfate	0.5 g
◆ Agar	15.0 g
◆ Phenol red	0.025 g
◆ Distilled or deionized water	1.0 L

pH 7.27.6 at 25°C

1 Suspend the dry ingredients in the water, mix well, and boil until completely dissolved.

2 Transfer 7.0 mL portions to test tubes and cap loosely.

3 Sterilize in the autoclave at 15 lbs. pressure (121°C) for 15 minutes.

4 Remove from the autoclave and slant in such a way as to form a deep butt.

5 Allow the medium to cool to room temperature.

Lauryl Tryptose Broth

◆ Tryptose	20.0 g
◆ Lactose	5.0 g
◆ Dipotassium phosphate	2.75 g
◆ Monopotassium phosphate	2.75 g
◆ Sodium chloride	5.0 g
◆ Sodium lauryl sulfate	0.1 g
◆ Distilled or deionized water	1.0 L

pH 6.6–7.0 at 25°C

1 Suspend the dry ingredients in the water until completely dissolved. Heat slightly if necessary.

2 Dispense 10.0 mL portions into test tubes.

3 Place an inverted Durham tube in each broth and cap loosely.

4 Sterilize in the autoclave at 15 lbs. pressure (121°C) for 15 minutes.

5 Remove the medium from the autoclave and allow it to cool before inoculating.

Luria-Bertani Agar

◆ Tryptone	10.0 g
◆ Yeast extract	5.0 g
◆ NaCl	10.0 g
◆ Agar	15.0 g
◆ Distilled or deionized water	1.0 L

pH 7.4 at 25°C

Luria-Bertani Broth

◆ Tryptone	10.0 g
◆ Yeast extract	5.0 g
◆ NaCl	10.0 g
◆ Distilled or deionized water	1.0 L

pH 7.4 at 25°C

MacConkey Agar

◆ Pancreatic digest of gelatin	17.0 g
◆ Pancreatic digest of casein	1.5 g
◆ Peptic digest of animal tissue	1.5 g
◆ Lactose	10.0 g
◆ Bile salts	1.5 g
◆ Sodium chloride	5.0 g
◆ Neutral red	0.03 g
◆ Crystal violet	0.001 g
◆ Agar	13.5 g
◆ Distilled or deionized water	1.0 L

pH 6.9–7.3 at 25°C

Mannitol Salt Agar

◆ Beef extract	1.0 g
◆ Peptone	10.0 g
◆ Sodium chloride	75.0 g
◆ D-Mannitol	10.0 g
◆ Phenol red	0.025 g
◆ Agar	15.0 g
◆ Distilled or deionized water	1.0 L

pH 7.2–7.6 at 25°C

(Skim) Milk Agar

◆ Beef extract	3.0 g
◆ Peptone	5.0 g
◆ Agar	15.0 g
◆ Powdered nonfat milk	100.0 g
◆ Distilled or deionized water	1.0 L

pH 7.0–7.4 at 25°C

1 Suspend the powdered milk in 500.0 mL of water in a 1-liter flask, mix well, and cover loosely.

2 Suspend the remainder of the ingredients in 500.0 mL of water in a 1-liter flask, mix well, boil to dissolve completely, and cover loosely.

3 Sterilize in the autoclave at 113–115°C for 20 minutes.

4 Remove from the autoclave and allow the mixtures to cool slightly.

5 Aseptically pour the agar solution into the milk solution. Mix *gently* (to prevent foaming).

6 Aseptically pour into sterile Petri dishes (15.0 mL/plate).

7 Allow the medium to cool to room temperature.

Motility Test Medium

◆ Beef extract	3.0 g
◆ Pancreatic digest of gelatin	10.0 g
◆ Sodium chloride	5.0 g
◆ Agar	4.0 g
◆ Triphenyltetrazolium chloride (TTC)	0.05 g
◆ Distilled or deionized water	1.0 L

pH 7.1–7.5 at 25°C

MR-VP Broth

◆ Buffered peptone	7.0 g
◆ Dipotassium phosphate	5.0 g
◆ Dextrose (glucose)	5.0 g
◆ Distilled or deionized water	1.0 L

pH 6.7–7.1 at 25°C

Mueller-Hinton II Agar

◆ Beef extract	2.0 g
◆ Acid hydrolysate of casein	17.5 g
◆ Starch	1.5 g
◆ Agar	17.0 g
◆ Distilled or deionized water	1.0 L

pH 7.1–7.5 at 25°C

1 Suspend the dry ingredients in the water, mix well, and boil until completely dissolved.

2 Cover loosely and sterilize in the autoclave at 121°C (15 lbs.) for 15 minutes.

3 Remove from the autoclave, allow to cool to approximately 50°C.

4 Aseptically pour into sterile Petri dishes to a depth of 4 mm.

5 Allow the medium to cool to room temperature.

Nitrate Broth

◆ Beef extract	3.0 g
◆ Peptone	5.0 g
◆ Potassium nitrate	1.0 g
◆ Distilled or deionized water	1.0 L

pH 6.6–7.0 at 25°C

1 Suspend the ingredients in the water; mix well, and warm until completely dissolved.

2 Transfer 8.0 mL portions to test tubes, and cap loosely. (Add inverted Durham tubes before capping, if desired.)

3 Sterilize in the autoclave at 15 lbs. pressure (121°C) for 15 minutes.

4 Remove from the autoclave and allow the medium to cool to room temperature.

Nutrient Agar

◆ Beef extract	3.0 g
◆ Peptone	5.0 g
◆ Agar	15.0 g
◆ Distilled or deionized water	1.0 L

pH 6.6–7.0 at 25°C

Nutrient Broth

◆ Beef extract	3.0 g
◆ Peptone	5.0 g
◆ Distilled or deionized water	1.0 L

pH 6.6–7.0 at 25°C

Nutrient Gelatin

◆ Beef extract	3.0 g
◆ Peptone	5.0 g
◆ Gelatin	120.0 g
◆ Distilled or deionized water	1.0 L

pH 6.6–7.0 at 25°C

1 *Slowly* add the dry ingredients to the water while stirring.

2 Warm to >50°C and maintain temperature until completely dissolved.

3 Dispense 7.0 mL volumes into test tubes and cap loosely.

4 Sterilize in the autoclave at 15 lbs. pressure (121°C) for 15 minutes.

5 Remove from the autoclave immediately and allow the medium to cool to room temperature in the upright position.

O-F Basal Medium (See O-F Carbohydrate Solution)

◆ Pancreatic digest of casein	2.0 g
◆ Sodium chloride	5.0 g
◆ Dipotassium phosphate	0.3 g
◆ Agar	2.5 g
◆ Bromthymol blue	0.03 g
◆ Distilled or deionized water	1.0 L

pH 6.6–7.0 at 25°C

O-F Carbohydrate Solution (See O-F Basal Medium)

- Carbohydrate (glucose, lactose, sucrose) 1.0 g
- Distilled or deionized water to total 10.0 mL

1. Suspend the dry ingredients, *without the carbohydrate*, in the water, mix well, and boil to dissolve completely. This is basal medium.
2. Divide the medium into 10 aliquots of 100.0 mL each.
3. Cover loosely and sterilize in the autoclave at 121°C for 15 minutes.
4. Prepare carbohydrate solution, cover loosely, and autoclave at 118°C for 10 minutes.
5. Allow both solutions to cool to 50°C.
6. Aseptically add 10.0 mL sterile carbohydrate solution to a basal medium aliquot, and mix well.
7. Aseptically transfer 7.0 mL volumes to sterile test tubes, and allow the medium to cool to room temperature.

Phenol Red (Carbohydrate) Broth

- Pancreatic digest of casein 10.0 g
- Sodium chloride 5.0 g
- Carbohydrate (glucose, lactose, sucrose) 5.0 g
- Phenol red 0.018 g
- Distilled or deionized water 1.0 L

pH 7.1±7.5 at 25°C

1. Suspend the dry ingredients in the water; mix well, and warm slightly to dissolve completely.
2. Dispense 7.0 mL volumes into test tubes.
3. Insert inverted Durham tubes into the test tubes and cap loosely.
4. Sterilize in the autoclave at 116–118°C for 15 minutes.
5. Remove from the autoclave and allow the medium to cool to room temperature.

Phenylalanine Deaminase Agar

- DL-Phenylalanine 2.0 g
- Yeast extract 3.0 g
- Sodium chloride 5.0 g
- Sodium phosphate 1.0 g
- Agar 12.0 g
- Distilled or deionized water 1.0 L

pH 7.1–7.5 at 25°C

1. Suspend the dry ingredients in the water, mix well, and boil until completely dissolved.
2. Dispense 7.0 mL volumes into test tubes and cap loosely.
3. Sterilize in the autoclave at 15 lbs. pressure (121°C) for 10 minutes.
4. Remove from the autoclave, slant, and allow the medium to cool to room temperature.

Phenylethyl Alcohol Agar

- Tryptose 10.0 g
- Beef extract 3.0 g
- Sodium chloride 5.0 g
- Phenylethyl alcohol 2.5 g
- Agar 15.0 g
- Distilled or deionized water 1.0 L

pH 7.1–7.5 at 25°C

SIM (Sulfur-Indole-Motility) Medium

- Pancreatic digest of casein 20.0 g
- Peptic digest of animal tissue 6.1 g
- Ferrous ammonium sulfate 0.2 g
- Sodium thiosulfate 0.2 g
- Agar 3.5 g
- Distilled or deionized water 1.0 L

pH 7.1–7.5 at 25°C

Skim Milk Agar—See (Skim) Milk Agar

Snyder Test Medium

- Pancreatic digest of casein 13.5 g
- Yeast extract 6.5 g
- Dextrose 20.0 g
- Sodium chloride 5.0 g
- Agar 16.0 g
- Bromcresol green 0.02 g
- Distilled or deionized water 1.0 L

pH 4.6–5.0 at 25°C

1. Suspend the dry ingredients in the water, mix well, and boil until completely dissolved.
2. Transfer 7.0 mL portions to test tubes and cap loosely.
3. Sterilize in the autoclave at 118–121°C for 15 minutes.
4. Remove from the autoclave and place in a hot water bath set at 45–50°C. Allow at least 30 minutes for the agar temperature to equilibrate before beginning the exercise.

Soft Agar

◆ Beef extract	3.0 g
◆ Peptone	5.0 g
◆ Sodium chloride	5.0 g
◆ Tryptone	2.5 g
◆ Yeast extract	2.5 g
◆ Agar	7.0 g
◆ Distilled or deionized water	1.0 L

1 Suspend the dry ingredients in the water, mix well, and boil until completely dissolved.

2 Transfer 2.5 mL portions to test tubes and cap loosely.

3 Sterilize in the autoclave at 15 lbs. pressure (121°C) for 15 minutes.

4 Remove from the autoclave and place in a hot water bath set at 45°C. Allow 30 minutes for the agar temperature to equilibrate.

Sporulating Agar

◆ Pancreatic digest of gelatin	6.0 g
◆ Pancreatic digest of casein	4.0 g
◆ Yeast extract	3.0 g
◆ Beef extract	1.5 g
◆ Dextrose	1.0 g
◆ Agar	15.0 g
◆ Manganous sulfate	0.3 g
◆ Distilled or deionized water	1.0 L

Final pH 6.6±0.2 at 25°C

Starch Agar

◆ Beef extract	3.0 g
◆ Soluble starch	10.0 g
◆ Agar	12.0 g
◆ Distilled or deionized water	1.0 L

pH 7.3–7.7 at 25°C

Thioglycollate Medium (Fluid)

◆ Yeast extract	5.0 g
◆ Casitone	15.0 g
◆ Dextrose (glucose)	5.5 g
◆ Sodium chloride	2.5 g
◆ Sodium thioglycolate	0.5 g
◆ L-Cystine	0.5 g
◆ Agar	0.75 g
◆ Resazurin	0.001 g
◆ Distilled or deionized water	1.0 L

pH 7.1–7.5 at 25°C

1 Suspend the dry ingredients in the water, mix well, and boil until completely dissolved.

2 Dispense 10.0 mL into sterile test tubes.

3 Autoclave for 15 minutes at 15 lbs. pressure (121°C) to sterilize. Allow the medium to cool to room temperature before inoculating.

Tributyrin Agar

◆ Beef extract	3.0 g
◆ Peptone	5.0 g
◆ Agar	15.0 g
◆ Tributyrin oil	10.0 mL
◆ Distilled or deionized water	1.0 L

pH 5.8–6.2 at 25°C

1 Suspend the dry ingredients in the water, mix well, and boil until completely dissolved.

2 Cover loosely, and sterilize together with the tube of tributyrin oil in the autoclave at 15 lbs. pressure (121°C) for 15 minutes.

3 Remove from the autoclave, and aseptically pour agar mixture into a sterile glass blender.

4 Aseptically add the tributyrin oil to the agar mixture and blend on "High" for 1 minute.

5 Aseptically pour into sterile Petri dishes (20.0 mL/plate). Allow the medium to cool to room temperature.

Triple Sugar Iron Agar (TSIA)

◆ Pancreatic digest of casein	10.0 g
◆ Peptic digest of animal tissue	10.0 g
◆ Dextrose (glucose)	1.0 g
◆ Lactose	10.0 g
◆ Sucrose	10.0 g
◆ Ferrous ammonium sulfate	0.2 g
◆ Sodium chloride	5.0 g
◆ Sodium thiosulfate	0.2 g
◆ Agar	13.0 g
◆ Phenol red	0.025 g
◆ Distilled or deionized water	1.0 L

pH 7.1–7.5 at 25°C

1 Suspend the dry ingredients in the water, mix well, and boil until completely dissolved.

2 Transfer 7.0 mL portions to test tubes and cap loosely.

3 Sterilize in the autoclave at 15 lbs. pressure (121°C) for 15 minutes.

4 Remove from the autoclave and slant in such a way as to form a deep butt.

5 Allow the medium to cool to room temperature.

Tryptic Soy Agar

◆ Tryptone	15.0 g
◆ Soytone	5.0 g
◆ Sodium chloride	5.0 g
◆ Agar	15.0 g
◆ Distilled or deionized water	1.0 L

pH 7.1–7.5 at 25°C

Tryptic Soy Agar (Enriched with Yeast Extract)

◆ Tryptone	15.0 g
◆ Soytone	5.0 g
◆ Sodium chloride	5.0 g
◆ Yeast extract	5.0 g
◆ Agar	15.0 g
◆ Distilled or deionized water	1.0 L

pH 7.1–7.5 at 25°C

1. Suspend the dry ingredients in the water, mix well, and boil until completely dissolved.
2. Transfer 10.0 mL portions to test tubes and cap loosely.
3. Sterilize in the autoclave at 121°C for 15 minutes.
4. Allow the medium to cool to room temperature with the tubes in an upright position (to be used for agar stabs).

Tryptic Soy Broth

◆ Tryptone	17.0 g
◆ Soytone	3.0 g
◆ Sodium chloride	5.0 g
◆ Dipotassium phosphate	2.5 g
◆ Distilled or deionized water	1.0 L

pH 7.1–7.5 at 25°C

Tryptic Soy Broth Plus 1% Glucose

◆ Tryptic soy broth (see above)	
◆ Glucose	10 g

Urea Broth (Rustigian and Stuart)

◆ Yeast extract	0.1 g
◆ Potassium phosphate, monobasic	9.1 g
◆ Potassium phosphate, dibasic	9.5 g
◆ Urea	20.0 g
◆ Phenol red	0.01 g
◆ Distilled or deionized water	1.0 L

pH 6.6–7.0 at 25°C

1. Suspend the dry ingredients in the water and mix well.
2. Filter sterilize the solution. *Do not autoclave.*
3. Aseptically transfer 1.0 mL volumes to small sterile test tubes and cap loosely.

❖ Reagents

MR-VP (Methyl Red–Voges-Proskauer) Test Reagents

Methyl Red

◆ Methyl red dye	0.1 g
◆ Ethanol	300.0 mL
◆ Distilled water to bring volume to	500.0 mL

1. Dissolve the dye in the ethanol.
2. Add water to bring the total volume up to 500.0 mL.

VP Reagent A (Barritt's)

◆ β-naphthol	5.0 g
◆ Absolute Ethanol to bring volume to	100.0 mL

1. Dissolve the β-naphthol in approximately 95.0 mL of ethanol.
2. Add ethanol to bring the total volume up to 100.0 mL.

VP Reagent B (Barritt's)

◆ Potassium hydroxide	40.0 g
◆ Distilled water to bring volume to	100.0 mL

1. Dissolve the potassium hydroxide in approximately 60.0 mL of water. (***Caution:*** This solution is highly concentrated and will become hot as the KOH dissolves. It should be prepared in appropriate glassware on a stirring hot plate.) Allow it to cool to room temperature.
2. Add water to bring the volume up to 100.0 mL.

Methyl Red (See MR-VP)

Methylene Blue Reductase Reagent

◆ Methylene blue dye	8.8 mg
◆ Distilled or deionized water	200.0 mL

Nitrate Test Reagents

Reagent A

◆ Sulfanilic acid	1.0 g
◆ 5N Acetic acid	125.0 mL

Reagent B

◆ Dimethyl-β-naphthylamine	1.0 g
◆ 5N Acetic acid	200.0 mL

Oxidase Test Reagent

◆ Tetramethyl-*p*-phenylenediamine dihydrochloride	1.0 g
◆ Distilled or deionized water	100.0 mL

Phenylalanine Deaminase Test Reagent

◆ Ferric chloride	10.0 g
◆ Distilled or deionized water	≅90.0 mL

1. Dissolve the ferric chloride in approximately 90.0 mL of distilled or deionized water.
2. Add water to bring the total volume up to 100.0 mL.

10X Tris-Acetate-EDTA buffer

◆ Trisma base	48.4 g
◆ Glacial acetic acid	11.42 mL
◆ 0.5 *M* EDTA, pH 8.0	20.0 mL
◆ Distilled or deionized water	1.0 L

Voges-Proskauer Reagents A and B (See MR-VP)

Solutions

10x Phosphate Buffered Saline (PBS) Stock Solution

♦ Na_2HPO_4, anhydrous, reagent grade	12.36 g
♦ $NaH_2PO_4 \bullet H_2O$, reagent grade	1.80 g
♦ NaCl, reagent grade	85.00 g

Dissolve ingredients in distilled water to a final volume of 1.0 L.

1x PBS Working solution (0.01 M phosphate, pH 7.6)

♦ Stock solution	100.0 mL
♦ dH_2O	900.0 mL

Stains and Staining Solutions

Acid-Fast, Cold Stain Reagents (Modified Kinyoun)

Carbolfuchsin

♦ Basic fuchsin	1.5g
♦ Phenol	4.5 g
♦ Ethanol (95%)	5.0 mL
♦ Isopropanol	20.0 mL
♦ Distilled or deionized water	75.0 mL

1 Dissolve the basic fuchsin in the ethanol and add the isopropanol.

2 Mix the phenol in the water.

3 Mix the solutions together and let stand for several days.

4 Filter before use.

Decolorizer

♦ H_2SO_4	1.0 mL
♦ Ethanol (95%)	70.0 mL
♦ Distilled or deionized water	29.0 mL

Brilliant Green

♦ Brilliant green dye	1.0 g
♦ Sodium azide	0.01g
♦ Distilled or deionized water	100.0 mL

Acid-Fast, Hot Stain Reagents (Ziehl-Neelson)

Carbolfuchsin

♦ Basic fuchsin	0.3 g
♦ Ethanol	10.0 mL
♦ Distilled or deionized water	95.0 mL
♦ Phenol	5.0 mL

1 Dissolve the basic fuchsin in the ethanol.

2 Dissolve the phenol in the water.

3 Combine the solutions and let stand for a few days.

4 Filter before use.

Decolorizer

♦ Ethanol	97.0 mL
♦ HCl (concentrated)	3.0 mL

Methylene Blue Counterstain

♦ Methylene blue chloride	0.3 g
♦ Distilled or deionized water	100.0 mL

Azotobacter Cyst Stain

♦ Glacial acetic acid	8.5 mL
♦ Sodium sulfate	3.25 g
♦ Neutral red	200.0 mg
♦ Light green	200.0 mg
♦ Ethanol	50.0 ml
♦ Distilled or deionized water	100.0 mL

Basic Fuchsin Stain (0.5%)

♦ Basic fuchsin	0.5 g
♦ dH_2O	99.5 mL

Capsule Stain

Congo Red

♦ Congo red dye	5.0 g
♦ Distilled or deionized water	100.0 mL

Maneval's Stain

♦ Phenol (5% aqueous solution)	30.0 mL
♦ Acetic acid, glacial (20% aqueous solution)	10.0 mL
♦ Ferric chloride (30% aqueous solution)	4.0 mL
♦ Acid fuchsin (1% aqueous solution)	2.0 mL

Carbolfuchsin (See Acid Fast, Hot)

Crystal Violet (See Gram Stain)

Gram Stain Reagents

Gram Crystal Violet (Modified Hucker's)

Solution A

♦ Crystal violet dye (90%)	2.0 g
♦ Ethanol (95%)	20.0 mL

Solution B

♦ Ammonium oxalate	0.8 g
♦ Distilled or deionized water	80.0 mL

1 Combine solutions A and B. Store for 24 hours.

2 Filter before use.

Gram Decolorizer

- ♦ Ethanol (95%)
- ♦ Can use 75% ethanol and 25% acetone, but decolorization time must be reduced.

Gram Iodine

♦ Potassium iodide	2.0 g
♦ Iodine crystals	1.0 g
♦ Distilled or deionized water	300.0 mL

1 Dissolve the potassium iodide in the water *first.*

2 Dissolve the iodine crystals in the solution.

3 Store in an amber bottle.

Gram Safranin

◆ Safranin O	0.25 g
◆ Ethanol (95%)	10.0 mL
◆ Distilled or deionized water	100.0 mL

1 Dissolve the safranin O in the ethanol.

2 Add the water.

Methylene Blue (See Acid Fast, Hot)

Nigrosin Stain

◆ Nigrosin	10.0 g
◆ Distilled or deionized water	100.0 mL

Safranin (See Gram Stain)

Spore Stain

Malachite Green

◆ Malachite green dye	5.0 g
◆ Distilled or deionized water	100.0 mL

Safranin (See Gram Stain)

References

Baron, Ellen Jo, Lance R. Peterson, and Sydney M. Finegold. 1994. *Bailey & Scott's Diagnostic Microbiology*, 9th ed. Mosby–Year Book, St. Louis.

Eisenstadt, Bruce, C. Carlton, and Barbara J. Brown. 1994. *Methods for General and Molecular Bacteriology*, edited by Philipp Gerhardt, R. G. E. Murray, Willis A. Wood, and Noel R. Krieg. American Society for Microbiology, Washington, DC.

Farmer, J. J. III. 2003. Chapter 41 in *Manual of Clinical Microbiology*, 8th ed. Edited by Patrick R. Murray, Ellen Jo Baron, James H. Jorgensen, Michael A. Pfaller, and Robert H. Yolken. ASM Press, American Society for Microbiology, Washington, DC.

Forbes, Betty A., Daniel F. Sahm, and Alice S. Weissfeld. 2002. Chapter 19 in *Bailey & Scott's Diagnostic Microbiology*, 11th ed. Mosby, St. Louis.

Koneman, Elmer W., et al. 1997. *Color Atlas and Textbook of Diagnostic Microbiology*, 5th ed. Lippincott-Raven Publishers, Philadelphia.

MacFaddin, Jean F. 1980. *Biochemical Tests for Identification of Medical Bacteria*, 2nd ed. Williams & Wilkins, Baltimore.

Zimbro, Mary Jo, and David A. Power, eds. 2003. *Difco™ and BBL™ Manual —Manual of Microbiological Culture Media*. Becton Dickinson and Co., Sparks, MD.

Glossary

A

absorbance A measurement, using a spectrophotometer, of how much light entering a substance is *not* transmitted.

acetoin A four-carbon intermediate in the conversion of pyruvic acid to 2,3-Butanediol; the compound detected in the Voges-Proskauer test.

acetyl-CoA Compound that enters the Krebs cycle by combining with oxaloacetate to form citrate; may be produced from pyruvate in carbohydrate metabolism or from oxidation of fatty acids.

acid clot Pink clot formed in litmus milk from the precipitation of casein under acidic conditions; an indication of lactose fermentation.

acidic stain Staining solution with a negatively charged chromophore.

acidophil *See* eosinophil.

acidophile Microorganism adapted to a habitat below pH 5.5.

acid reaction Pink color reaction in litmus milk from acidic conditions produced by lactose fermentation.

adenosine triphosphate (ATP) In hydrolysis of ATP to adenosine diphosphate (ADP), supplies the energy necessary to perform most work in the cell.

aerobe Microorganism that requires oxygen for growth.

aerobic respiration A type of energy releasing metabolism in which oxygen is the final electron acceptor in the pathway. It also has the highest potential ATP yield of all catabolic and exergonic reactions.

aerosol Droplet nuclei that remain suspended in the air for a long time.

aerotolerance Designation of an organism based on its ability to grow in the presence of oxygen.

aerotolerant anaerobe Anaerobe that grows in the presence of oxygen but does not use it metabolically.

agar overlay Used in plaque assays, the soft agar containing a mixture of bacteriophage and host poured over Nutrient Agar as an inoculum.

agglutination Visible clumping produced when antibodies and particulate antigens react.

agranulocyte Category of white blood cells characterized by the absence of prominent cytoplasmic granules; includes monocytes and lymphocytes.

akinete A thick-walled resting spore produced by some cyanobacteria. It is derived from a vegetative cell.

alkaliphile Microorganism adapted to a habitat above pH 8.5.

alpha (α) hemolysis The greening around a colony on blood agar as a result of partial destruction of red blood cells; typical of certain members of *Streptococcus*.

ammonification The metabolic production of ammonia. *See* deamination.

amphitrichous Describes a flagellar arrangement with flagella at both ends of an elongated cell.

amylase Family of enzymes that hydrolyze starch.

anaerobe Microorganism that cannot tolerate oxygen.

anaerobic respiration A type of energy releasing metabolism in which an inorganic compound other than oxygen is the final electron acceptor. These include NO_3, NO_2, SO_4, and CO_2. The ATP yield is higher than fermentation, but not as high as aerobic respiration.

annealing The attachment of a primer to a complimentary sequence of DNA. Annealing is done on each strand of the denatured DNA in positions flanking the desired sequence (template) to be replicated. Annealing is the second of three phases in a PCR cycle, following denaturation and prior to extension.

antibiotic Refers to an antimicrobial substance produced by a microorganism such as a bacterium or fungus.

antibody A glycoprotein produced by plasma cells, in response to an antigen, that reacts with the antigen specifically.

antigen A molecule of high molecular weight and complex three-dimensional shape that stimulates the production of antibodies and reacts specifically with them.

antimicrobic(al) Describes any substance that kills microorganisms, natural or synthetic.

antiporter A system that simultaneously transports substances in opposite directions across the cytoplasmic membrane.

antiserum General term applied to a serum that contains antibodies of a specific type.

Ascomycetes A division of fungi that produces sexual spores in a saclike structure called an ascus.

aseptic Describes the condition of being without contamination.

assimilation (nitrogen) The incorporation of NH_4 or NO_3 into organic form (amino acids) by plants.

assimiliatory sulfate reduction The conversion of sulfate (SO_4^{2-}) to a sulfhydryl group (–SH) in the biosynthesis of amino acids cysteine and methionine.

autoinducer Substance secreted by bioluminescent bacteria that, upon reaching sufficient concentration, triggers the bioluminescent reaction; believed to help conserve energy by synchronizing individual cellular reactions.

autotroph An organism that is able to make all of its organic molecules using CO_2 as the source of carbon. *See also* heterotroph.

auxotroph A nutritional mutant; a cell incapable of synthesizing the nutrient that is acting as the marker in a specific genetic experiment; will grow only on a complete medium. *See also* prototroph.

B

bacillus Rod-shaped cell.

back mutation Mutation in a gene that reverses the effect of an original mutation.

bacteriophage Virus that attacks bacteria; usually host-specific.

basic stain Staining solution with a positively charged chromophore.

Basidiomycetes Division of fungi characterized by club-like appendages (basidia) that produce haploid spores; include common mushrooms and rusts.

basophil Category of white blood cells; one of three types of granulocyte (along with eosinophils and neutrophils) characterized by a cytoplasm with dark-staining granules and a lobed nucleus (often difficult to see because of the dark cytoplasmic granules).

β-hemolysis The clearing around a colony on blood agar resulting from complete destruction of red blood cells; typical of certain members of *Streptococcus*.

β-lactamase Enzyme found in some bacteria that breaks a bond in the β-lactam ring of β-lactam antibiotics, rendering the antibiotic ineffective; the source of resistance against these antibiotics.

β-lactam antibiotic A group of structurally related antimicrobial chemicals that interfere with cross-linking of peptidoglycan subunits, which results in a defective wall structure and cell lysis; examples are penicillins and cephalosporins.

biofilm A layer of bacterial cells adhering to and reproducing on a surface; used commercially for acetic acid production and water purification.

bioluminescence Process by which a living organism emits light.

biosafety level (BSL) One of four sets of minimum standards for laboratory practices, facilities, and equipment to be used when handling organisms at each level; BSL-1 requires the least care, BSL-4 the most.

bound coagulase An enzyme (also called clumping factor) bound to the bacterial cell wall, responsible for causing the precipitation of fibrinogen and coagulation of bacterial cells.

C

capnophile Microaerophile that requires elevated CO_2 levels.

capsule Insoluble, mucoid, extracellular material surrounding some bacteria.

carbohydrate One of four families of biochemicals; characterized by containing carbon, hydrogen, and oxygen in the ratio 1:2:1.

carcinogen Any substance that causes cancer.

cardinal temperatures Minimum, optimum, and maximum temperatures for an organism.

casease A family of hydrolytic enzymes that break down casein, often extracellularly.

casein Milk protein that gives milk its white color.

catalase Enzyme produced by some bacteria that catalyzes the breakdown of metabolic H_2O_2.

cestode The class of parasitic worms commonly called tapeworms; characterized by having a head (scolex) with suckers and hooks and numerous segments (proglottids) that are little more than sex organs.

chemoheterotroph Any organism that uses chemicals for energy and organic compounds as a carbon source. Usually the same organic compounds serve both purposes.

chemolithotroph Any organism that gets its energy from chemicals and its electrons from an inorganic source. No eukaryotes are chemolithotrophic, only some prokaryotes such as sulfur and ammonia oxidizing bacteria.

chromogenic reducing agent A substance that produces color when it gives up electrons (becomes oxidized).

chromophore The charged region of a dye molecule that gives it its color.

coagulase *See* bound coagulase; free coagulase.

coagulase-reacting factor Plasma component that reacts with free coagulase to trigger the clotting mechanism.

coccus Spherical cell.

coenzyme Nonprotein portion of an enzyme that aids the catalytic reaction (usually by accepting or donating electrons).

coliform Member of *Enterobacteriaceae* that ferments lactose with production of gas within 48 hours at 37°C.

colony Visible mass of cells produced on culture media from a single cell or single CFU (cell forming unit); a pure culture.

colony forming unit Term used to define the cell or group of cells (e.g., *staphylococci, streptococci*) that produces a colony when transferred to plated media.

commensal Describes a synergistic relationship between two organisms in which one benefits from the relationship and the other is affected neither negatively nor positively.

common source epidemic An epidemic in which the disease is transmitted from a single source (such as water supply) and is not transmitted person-to-person.

compatible solutes Compounds such as amino acids that function both as metabolic constituents and solutes necessary to maintain osmotic balance between the internal and external environments.

competent cell Cell capable of picking up DNA from the environment; some cells are naturally competent, in other cases the cells are made to be competent artificially.

competitive inhibition Process whereby a substance attaches to an enzyme's active site, thereby preventing attachment of the normal substrate.

complete medium A medium used in genetic experiments that supplies all nutrients for growth required by prototrophs and auxotrophs.

complex medium A medium in which at least one nutritional ingredient is of unknown composition or amount. Media with any plant or animal extract are always complex.

confirmatory test One last test after identification of an unknown, to further support the identification.

conjugate acid A compound (one of a conjugate pair) that donates a proton in solution.

conjugate base A compound (one of a conjugate pair) that accepts a proton in solution.

conjugate pair A compound that alternates between acid and base forms by losing or gaining one proton. Example: HCl/Cl^-.

constitutive Term used to describe an enzyme that is produced continuously, as opposed to an enzyme produced only when its substrate is present. *See also* induction.

cos site (cohesive end sites) In Lambda phage (linear) DNA, the single stranded tails at the 5' ends of both strands are complimentary to each other and can form a circular molecule. These "sticky" ends can be a source of error in DNA fragment size determinations if not properly heated before running gel electrophoresis.

counterstain Stain applied after decolorization to provide contrast between cells that were decolorized and those that weren't.

culture A liquid or solid medium with microorganisms growing in or on it.

cyst "Resting stage" in life cycle of certain protozoans. *See also* trophozoite.

cytochrome A class of iron-containing enzyme in electron transport chains. Characterized by a porphyrin ring structure.

cytochrome oxidase An electron carrier in the electron transport chain of aerobes, facultative anaerobes, and microaerophiles that makes the final transfer of electrons to oxygen.

cytoplasm The semifluid component of cells inside the cellular membrane in which many chemical reactions take place.

D

deaminase Enzyme that catalyzes the removal of the amine group (NH_2) from an amino acid.

deamination The removal of an amine from an organic molecule, resulting in the production of ammonia and whatever is left of the organic molecule.

death phase Closed system microbial growth phase immediately following stationary phase; characterized by population decline usually resulting from nutrient deficiencies or accumulated toxins.

decarboxylase Enzyme that catalyzes the removal of the carboxyl group (COOH) from an amino acid.

decimal reduction value Amount of time at a specific temperature to reduce a microbial population by 90% (one log cycle).

defined medium A growth medium in which the exact amount and chemical formula (and thus identity) of each ingredient are known.

denaturation The first of three phases in a PCR cycle, followed by annealing and extension. The separation of the strands in double stranded DNA.

denitrification The bacterial process in which nitrate is reduced to various forms of nitrogen in its role as final electron acceptor of anaerobic respiration.

deoxyribonuclease A family of hydrolytic enzymes that depolymerize DNA into polynucleotides. The precise location of hydrolysis within the sugar-phosphate backbone depends on which DNase is acting. These are usually secreted enzyme.

Deuteromycetes An unnatural grouping of fungi in which the sexual stages are either unknown or are not used in classification.

differential medium Growth medium that contains an indicator (usually color) to detect the presence or absence of a specific metabolic activity.

differential stain Staining procedure that allows distinction between cell types or parts of cells; often involves more than one stain, but not necessarily.

dilution blank A test tube containing a measured volume of sterile diluent (water, saline, or buffer) used to serially dilute a concentrated solution or broth.

dilution factor Proportion of original sample present in a new mixture after it has been diluted; calculated by dividing the volume of the original sample by the total volume of the new mixture. (Subsequent dilutions in a serial dilution must be multiplied by the dilution factor of the previous dilution.)

diploid Defines a cell that has two complete sets of genetic information; one stage in the life cycle of all eukaryotic microorganisms (which also have a haploid [one set] stage).

direct agglutination Serological test in which the combination of antibodies and *naturally* particulate antigens, if positive, form a visible aggregate. *See also* indirect agglutination.

disaccharide A sugar made of two monosaccharide subunits; e.g., the monosaccharides glucose and galactose can combine to make the disaccharide lactose.

disk diffusion test (Kirby-Bauer Test) This is one way to check susceptibility and resistance of a microbe to an antibiotic. A paper disk containing an antibiotic is placed on a plate inoculated to produce confluent growth of a bacterium. Susceptibility is determined by the size of the zone of inhibition of growth around the disk.

disproportionation reaction A redox reaction of some sulfur reducers whereby a substrate is split into two molecules, one that is more reduced and one that is more oxidized than the original compound. Example: $S_2O_3^{2-} \rightarrow SO_4^{2-} + H_2S$.

dissimilatory sulfate reduction A respiratory reaction in which sulfate is reduced exclusively for the purpose of gaining energy.

DNA polymerase A group of enzymes that catalyze the addition of deoxyribonucleotides to the 3' end of an existing polynucleotide chain.

DNase *See* deoxyribonuclease.

Durham tube A small, inverted test tube used in some liquid media to trap gas bubbles and indicate gas production.

E

electromagnetic energy The energy that exists in the form of waves (including X-rays, ultraviolet, visible light, and radio waves).

electron donor A compound that can be oxidized to transfer electrons to another compound, which in turn becomes reduced.

electron transport chain A series of membrane-bound electron carriers that participate in oxidation-reduction reactions whereby electrons are transferred from one carrier to another until given to the final electron acceptor (oxygen in aerobes, other inorganic substances in anaerobes) in respiration.

endergonic Any metabolic reaction in which the products have more potential energy than the reactants. These are usually associated with anabolic (synthesis) reactions.

endospore Dormant, highly resistant form of bacterium; produced only by species of *Bacillus*, *Clostridium*, and a few others.

enteric bacteria Informal name given to bacteria that occupy the intestinal tract.

Enterobacteriaceae A group of Gram-negative rods that also are oxidase-negative, ferment glucose to acid, have polar flagella if motile, and usually are catalase-positive and reduce nitrate to nitrite.

enzyme A protein that catalyzes a metabolic reaction by interacting with the reactant(s) specifically; each metabolic reaction has its own enzyme.

eosinophil A category of white blood cells; one of three types of granulocyte (along with basophils and neutrophils) characterized by a cytoplasm with red (in typical stains) granules and lobed nucleus.

epitope That portion of an antigen that stimulates the immune system and reacts with antibodies.

eukaryote Type of cell with membranous organelles, including a nucleus, and having 80S ribosomes and many linear molecules of DNA.

exergonic Any metabolic reaction in which the reactants have more potential energy than the products. These are usually associated with catabolic (degradative) reactions.

exoenzyme Enzyme that operates in the external environment after being secreted from the cells.

exponential phase Closed-system microbial growth phase immediately following the lag phase; characterized by constant maximal growth during which population size increases logarithmically.

extension The third of three phases in PCR, following denaturation and annealing, whereby a polymerase (usually *Taq*I) catalyzes the replication of a DNA template.

extracellular Pertains to the region outside a cell.

extreme halophile Organism that grows best at 15% or higher salinity.

extreme thermophile Organism that grows best at temperatures above 80°C.

F

facultative anaerobe Microorganism capable of both fermentation and respiration; grows in the presence or absence of oxygen.

facultative thermophile Microorganism that prefers temperatures above 40°C but will grow in lower temperatures.

false negative Test result that is negative when the sample is actually positive; usually a result of lack of sensitivity in the test system.

false positive Test result that is positive when the sample is actually negative; usually a result of lack of specificity of the test system.

fastidious (microorganism) Describes a microorganism with strict physiological requirements; difficult or impossible to grow unless specific conditions are provided.

fatty acid A long chain, organic molecule with a carboxylic acid at one end with the rest of the carbons being bonded to hydrogens.

fermentation Metabolic process in which an organic molecule acts as an electron donor and one or more of its organic products act as the final electron acceptor, marking the end of the metabolic sequence (differs from respiration in that an inorganic substance is not needed to act as final electron acceptor).

final electron acceptor (FEA) The molecule receiving electrons (it becomes reduced) at the end of a metabolic sequence of oxidation/reduction reactions.

flavin adenine dinucleotide (FAD) Coenzyme used in oxidation/reduction reactions that acts as an electron acceptor or donor, respectively.

flavoprotein A flavin-containing protein in the electron transport chain, capable of receiving and transferring electrons as well as entire hydrogen atoms; sometimes bypasses normal route of transfer and reduces oxygen directly, forming hydrogen peroxide and superoxide radicals.

free coagulase An enzyme produced and secreted by some microorganisms that initiates the clotting mechanism in plasma; seen as a solid mass in the coagulase tube test or clumps in the slide test.

free energy (of formation) The energy released by a molecule in a chemical reaction that is able to do work (e.g., produce ATP). It is equivalent to the energy required to form the molecule from its individual elements.

free living A term used to describe nonparasitic organisms.

G

gametangium A structure that produces gametes; used in describing fungi.

gamete Reproductive cell that must undergo fusion with another gamete to continue the life cycle; usually are haploid. *See also* spore.

gelatinase Enzyme secreted by microorganisms, which catalyzes the hydrolysis of gelatin.

generation time The time required for a population to produce offspring (i.e., in bacteria that reproduce by binary fission, the time needed for the population to double); inverse of mean growth rate.

germ theory Theory holding that diseases and infections are caused by microorganisms; first hypothesized in the 16th century by Girolamo Fracastoro of Verona.

glycolysis Metabolic process by which a glucose molecule is split into two 3-carbon pyruvic acid molecules, producing two ATP (net) and two $NADH_2$ molecules.

gradient bacteria Microaerophiles with very narrow nutritional requirements living in habitats characterized by nutrients diffusing upward from sediments below and oxygen diffusing downward from air above.

granulocyte A category of white blood cells characterized by prominent cytoplasmic granules; includes neutrophils, eosinophils, and basophils.

group A streptococci α-hemolytic members of the genus *Streptococcus*, characterized by possessing the Lancefield group A antigen; belong to the species *S. pyogenes*, an important human pathogen.

group B streptococci β-hemolytic members of the genus *Streptococcus*, characterized by possessing the Lancefield group B antigen; belong to the species *S. agalactiae*.

group D streptococci Streptococci possessing the Lancefield group D antigen; include *S. bovis* and species of the genus *Enterococcus*.

H

halophile Microorganism that grows best at 3% or higher salinity.

haploid Describes a cell that has one complete set of genetic information; all eukaryotic microorganisms have a haploid stage and a diploid (two sets) stage in their life cycle; all prokaryotes are haploid.

hemagglutination General term applied to any agglutination test in which clumping of red blood cells indicates a positive reaction.

hemoglobin Iron-containing protein of red blood cells that is responsible for binding oxygen.

hemolysin A class of chemicals produced by some bacteria that break down hemoglobin and produce hemolysis.

heterocyst This is a hollow-appearing cell found in some cyanobacteria specialized to perform nitrogen fixation.

heterotroph An organism that requires carbon in the form of organic molecules. *See also* autotroph.

host An organism that serves as a habitat for another organism such as a parasite or a commensal.

hydrolysis The metabolic process of splitting a molecule into two parts, adding a hydrogen ion to one part and a hydroxyl ion to the other. (One water molecule is used in the process.)

hyperosmotic (solution) A term used to describe extracellular solute concentration relative to the cell; a solution that contains a higher concentration of solutes than the cell, such that water tends to move down its concentration gradient and diffuse out of the cell.

hypha A filament of fungal cells.

hyposmotic A term used to describe extracellular solute concentration relative to the cell; a solution that contains a lower concentration of solutes than the cell, such that water tends to move down its concentration gradient and diffuse into the cell.

I

IMViC Acronym representing the four tests used in identification of *Enterobacteriaceae:* Indole, Methyl Red, Voges-Proskauer, and Citrate.

incomplete medium *See* minimal medium.

incubation The process of growing a culture by supplying it with the necessary environmental conditions.

index case First occurrence of an infection or disease that results in an epidemic.

indirect agglutination Serological test in which artificially produced particulate antigens or antibodies are used to form a visible aggregate if positive. *See also* direct agglutination.

induction Process by which a substrate (inducer) causes the transcription of the genes used in its digestion.

infectious disease A transmissible illness or infection.

inoculum The organisms used to start a new culture or transferred to a new place.

inorganic molecule A molecule that does *not* contain carbon and hydrogen.

intracellular Within the cell; as an intracellular enzyme catalyzing reactions inside the cell.

iron-sulfur protein Iron-containing electron carrier in electron transport chains. Differentiated from cytochromes by the lack of porphyrin ring structure.

isolate (*v.*) The process of separating individual cell types from a mixed culture; (*n.*) the group of cells resulting from isolation.

isosmotic Describe extracellular solute concentration relative to the cell; a solution that contains the same concentration of solutes as the cell, such that water tends to move equally into and out of the cell.

K

karyogamy The process of nuclear fusion that occurs after fertilization (plasmogamy) in sexual life cycles.

Krebs cycle A cyclic metabolic pathway found in organisms that respire aerobically or anaerobically.

L

lag phase Closed-system microbial growth phase immediately preceding the exponential phase; characterized by a period of adjustment in which no growth takes place.

limit of resolution The closest two points can be together for the microscope lens to make them appear separate; two points closer than the limit of resolution will blur together.

lipase A family of hydrolytic enzymes that break down fats into their component parts: glycerol and up to three fatty acids.

lipid A fat.

lophotrichous Describes a flagellar arrangement with a group of flagella at one end of an elongated cell.

lymphocyte A category of white blood cells; one of two types of agranulocyte (along with monocytes); characterized by a large nucleus and little visible cytoplasm; involved in specific acquired immunity as T-cells and B-cells.

lysozyme A naturally occurring bactericidal enzyme in saliva, tears, urine, and other body fluids; functions by breaking peptidoglycan bonds.

lytic cycle The viral life cycle from attachment to lysis of the host cell.

M

mean growth rate constant The number of generations produced per unit time; the inverse of generation time.

medium A substance used for growing microbes; may be liquid (usually a broth) or solid (usually agar).

meiosis The process in which the nucleus of a diploid eukaryotic cell nucleus divides to make four haploid nuclei.

mesophile A microorganism that grows best at temperatures between 15°C and 45°C.

methylation A protective enzymatic mechanism in many organisms whereby methyl groups ($–CH_3$) attach to DNA, thus blocking attachment and destruction by restriction endonucleases.

microaerophile A microorganism that requires oxygen, but can't tolerate atmospheric levels of it.

microbial growth curve Graphic representation of microbial growth in a closed system, consisting of lag phase, exponential (log) phase, stationary phase, and death phase.

minimal medium A medium used in genetic experiments, supplying all nutrients for growth *except* the one required by auxotrophs, and thus supporting growth of only prototrophs.

minimum inhibitory concentration The lowest concentration of an antimicrobial substance required to inhibit growth of all microbial cells it contacts; on an agar plate, typically the outer edge of the zone of inhibition where the substance has diffused to the degree that it no longer inhibits growth.

mitosis The process in which a nucleus divides to produce two identical nuclei.

mixed acid fermentation Vigorous fermentation producing many acid products including lactic acid, acetic acid, succinic acid, and formic acid, and subsequently lowering the pH of the medium to pH 4.4 or below.

mold Informal grouping of filamentous fungi. *See also* yeast.

monocyte A category of white blood cells; one of two types of agranulocyte (along with lymphocytes); characterized by large size and lack of cytoplasmic granules; the blood form of macrophages.

monotrichous Describes a flagellar arrangement consisting of a single flagellum.

morbidity Epidemiological measurement of incidence of a disease; typically accompanied by "incidence rate," referring to the incidence of a disease over time.

morphology The shape of an organism.

mortality Epidemiological measurement of death caused by a disease; typically accompanied by "incidence rate," referring to the incidence of death from a disease over time.

mutagen A substance that causes mutation in DNA; most mutagens are carcinogens.

mutation Alteration in a cell's DNA.

mutualistic Describes a synergistic relationship between two organisms in which both benefit from the interaction.

mycelium A mass of fungal filaments (hyphae).

N

nematode A class of roundworms; environmentally abundant in some parasitic species.

neutrophil Category of white blood cells; one of three types of granulocyte (along with eosinophils and basophils) characterized by a granular cytoplasm and lobed nucleus; also known as "polymorphonuclear granulocytes" or "PMNs."

neutrophile Microorganism adapted to a habitat between pH 5.5 and 8.5.

nicotinamide adenine dinucleotide (NAD) A coenzyme used in oxidation/reduction reactions that acts as an electron acceptor or donor, respectively.

nisin Antibiotic produced by *Lactococcus lactis.*

nitrate A highly oxidized form of nitrogen; NO_3.

nitrate reductase An enzyme produced by all members of *Enterobacteriaceae* (and others) that catalyzes the reduction of nitrate (NO_3) to nitrite (NO_2).

nitrification The chemical process in which nitrogen compounds, such as ammonia and nitrite are oxidized to form nitrate. Typically, different bacteria perform these two steps in which ammonia is first oxidized to nitrite, which is subsequently oxidized to nitrate.

nitrite NO_2, an oxidized form of nitrogen.

nitrogen fixation The bacterial process in which gaseous nitrogen (N_2) becomes reduced to NH_4. This is ecologically important because N_2, while abundant in the atmosphere, is not in a form usable by most organisms, whereas NH_4 is usable by plants and can thus enter food chains.

nitrogenase The enzyme involved in nitrogen fixation, that is, converting gaseous N_2 to NH_4. This is performed by free-living bacteria (e.g., *Azotobacter*), symbiotic bacteria (e.g., *Rhizobium*), and many cyanobacteria.

noninfectious diseases Conditions not caused by microorganisms; examples are stroke, heart disease, and emphysema.

O

objective lens The microscope lens that first produces magnification of the specimen in a compound microscope.

obligate (strict) aerobe Microorganism that requires oxygen to survive and grow.

obligate (strict) anaerobe Microorganism for which oxygen is lethal; requires the complete absence of oxygen.

obligate thermophile Microorganism that grows only at temperatures above 40°C.

ocular lens The lens the microscopist looks through; produces the virtual image by magnifying the real image.

ocular micrometer A uniformly graduated linear scale placed in the microscope ocular used for measuring microscopic specimens.

oligonucleotide A short nucleic acid molecule.

operon A prokaryotic structural and functional genetic unit consisting of two or more structural genes that code for enzymes in the same pathway and that are regulated together.

opportunistic pathogen A microorganism not ordinarily thought of as pathogenic (i.e., most enterics) that will cause infection when out of its normal habitat.

organic molecule A molecule made of reduced carbon—that is, containing at least carbon and hydrogen.

osmosis Diffusion of water across a semipermeable membrane.

osmotolerant Microorganism that will grow outside of its preferred salinity range.

oxidation-reduction Chemical reaction in which electrons are transferred. (The molecule losing the electrons becomes oxidized; the molecule gaining the electrons becomes reduced.)

oxidative deamination The metabolic process in which an amino acid has its amine removed, producing ammonia and an organic acid.

oxidative phosphorylation The process by which an electron transport chain is used to add a phosphate to ADP to make ATP.

oxidizing agent A substance that removes electrons from (oxidizes) another. *See also* reducing agent.

P

parasite An organism that lives symbiotically with another, but to the detriment of the other organism (called a host).

parthenogenesis A process in which females produce offspring from an unfertilized egg.

peptidoglycan The insoluble, porous, cross-linked polymer comprising bacterial cell walls; generally thick in Gram-positive organisms and thin in Gram-negative organisms.

peptone A digest of protein used in formulating some bacteriological media.

peritrichous Describes a flagellar arrangement in which flagella arise from the entire surface of the cell.

pH The measure of a solution's alkalinity or acidity; the negative logarithm of the hydrogen ion concentration.

phage *See* bacteriophage.

phage host Bacteria attacked by a virus.

phototaxis Phototaxis is a response to light stimuli. Positive phototaxis is movement toward a light stimulus, whereas negative phototaxis is movement away from light.

plaque The clearing produced in a bacterial lawn as a result of cell lysis by a bacteriophage; used to calculate phage titer (PFU/mL) when accompanied by a serial dilution.

plaque forming unit (PFU) Term that replaces "viral particle" or "single virus" when referring to phage titer; accounts for multiple particle arrangements in which more than one virus is responsible for initiating a plaque.

plasma The noncellular (fluid) portion of blood; consists of serum (including serum proteins) and clotting proteins.

plasmid Small, circular, extrachromosomal piece of DNA found in prokaryotic cells; often carries genes for antibiotic resistance.

plasmogamy The cytoplasmic fusion of gametes at the time of fertilization. *See also* karyogamy.

plasmolysis Shrinking of cell membrane (pulling away from the rigid cell wall) because of loss of water to the environment and reduced turgor pressure.

polar flagellum A single flagellum at one end of an elongated cell.

pour plate technique Method of plating bacteria in which the inoculum is added to the molten agar prior to pouring the plate.

preadsorption period When performing a plaque assay, the time given to allow a bacteriophage to attach to the host before adding the mixture to the molten agar being plated.

precipitate An insoluble material that comes out of a solution during a precipitation reaction.

presumptive identification Tentative identification of an isolate based on one or more key test results.

primary stain The first stain applied in many differential staining techniques; usually subjected to a decolorization step that forms the basis for the differential stain.

primer A short nucleotide sequence (typically 10–20 nucleotides) used in PCR to attach to DNA for the purpose of replicating a desired sequence of nucleotides. Typically, two primers attach to sites on opposite strands, flanking the area (template) to be replicated.

proglottid Tapeworm segments posterior to the scolex, used for absorption of nutrients and containing reproductive organs.

prokaryote Type of cell lacking internal compartmentalization (membranous organelles, including a nucleus) and having 70S ribosomes and a circular molecule of DNA; more primitive than eukaryotes.

promoter site The patch of DNA upstream from the structural gene(s), which binds RNA polymerase to begin transcription.

propagated transmission Conveying a disease person-to-person.

proteolytic Refers to catabolism of protein; e.g., a *proteolytic* enzyme.

prototroph A strain that is capable of synthesizing the nutrient that is acting as the marker in a particular genetic experiment; prototrophs will grow on complete and minimal medium. *See also* auxotroph.

pseudohypha A chain of fungal cells produced by budding (rather than by cytokinesis that produces two equally sized cells) and which is characterized by constrictions at cell junctions.

psychrophile A microorganism that grows only at temperatures below 20°C, with an optimum around 15°C.

psychrotroph A microorganism that grows optimally at temperatures between 20 and 30°C but will grow at temperatures as low as 0°C and as high as 35°C.

pure culture Microbial culture containing only a single species.

purine nucleotide Purines are made of a sugar (ribose or deoxyribose), phosphate, and a nitrogen-containing base composed of two joined rings. Examples are adenine (A) and guanine (G). These are found in DNA and RNA.

putrefaction The process of digesting dead organic material; decay.

pyrimidine nucleotide Pyrimidines are made of a sugar (ribose or deoxyribose), phosphate, and a nitrogen-containing base composed of a single ring. Examples are cytosine (C) in DNA and RNA, thymine (T) in DNA, and uracil (U) in RNA.

pyruvic acid (pyruvate) A three-carbon compound produced at the end of glycolysis that may enter a respiration or a fermentation pathway; also serves as a starting point for synthesis of certain amino acids and an entry point for their digestion.

Q

quinoidal (compound) A color-producing compound containing quinone as its central structure.

quorum sensing The phenomenon in bioluminescing bacteria whereby the light-emitting reaction of all cells takes place simultaneously when a threshold concentration of secreted autoinducer is reached.

R

real image Magnified image of a specimen produced by the objective lens of a microscope; the real image is magnified again by the ocular lens to produce the virtual image.

recognition site A short DNA sequence where a restriction enzyme attaches. Each enzyme recognizes and attaches to its own specific sequence.

reducing agent A substance that donates electrons to (reduces) another. *See also* oxidizing agent.

reductase An enzyme that catalyzes the transfer of electrons from donor molecule to acceptor molecule, thereby reducing the acceptor.

reduction *See* oxidation-reduction.

refraction The bending of light as it passes from a medium with one refractive index into another medium with a different refractive index.

reservoir A nonhuman host or other site in nature serving as a perpetual source of pathogenic organisms.

resolution The clarity of an image produced by a lens; the ability of a lens to distinguish between two points in a specimen; high resolution in a microscope is desirable.

resolving power *See* limit of resolution.

respiration Metabolic process by which an organic molecule acts as an electron donor and an inorganic substance—such as oxygen, sulfur, or nitrate—acts as the final electron acceptor in an electron transport chain, marking the end of the metabolic sequence (differs from fermentation, which uses one of its own organic products as the final electron acceptor).

restriction enzyme Enzymes involved in cutting DNA at specific recognition sites, unique to the enzyme. Involved in removing damaged DNA and destroying foreign (e.g., phage) DNA.

reticuloendothelial system Combination of macrophages and associated cells located in the liver, spleen, bone marrow, and lymph nodes.

reverse citric acid cycle An energy consuming process used by green sulfur bacteria whereby molecules ordinarily associated with the oxidative respiratory reactions of the Krebs cycle perform reactions in the opposite direction to reduce carbon dioxide and form pyruvate.

reverse electron flow An energy consuming process used by purple sulfur bacteria whereby electrons transferred down the ETC of the photosystem are boosted upward against their thermodynamic gradient to reduce $NADP^+$ needed by the Calvin cycle.

reversion In carbohydrate fermentation tests, the phenomenon of a microorganism fermentively depleting the carbohydrate and reverting to amino acid metabolism, thereby neutralizing acid products with alkaline products; produces a false negative.

rhizoid A root-like structure used for attachment of some fungi to the substrate.

RNA polymerase A group of enzymes that catalyze the addition of ribonucleotides to the 3' end of an existing polynucleotide chain.

S

saltern Salterns are low pools of saltwater. Evaporation of the water leaves salt, which can then be harvested.

saprophyte A heterotroph that digests dead organic matter; a decomposer.

scolex The "head" of a tapeworm, often with suckers and hooks for attachment.

selective medium Growth medium that favors growth of one group of microorganisms and inhibits or prevents growth of others.

sensitivity This is a measure of a test's ability to detect small amounts of the item being tested for. The better the sensitivity, the fewer the false negatives (due to smaller amounts triggering a positive reaction), and the more useful the test is.

serial dilution Series of dilutions used to reduce the concentration of a culture and thereby produce between 30 and 300 colonies when plated, providing a means of calculating the original concentration.

serology A discipline that utilizes a serum containing antibodies (antiserum) to detect the presence of antigens in a sample; also refers to identification of antibodies in a patient's serum.

serum Fluid portion of blood minus the clotting factors.

soft agar A semisolid growth medium containing a reduced concentration of agar; used in plaque assay to allow diffusion of bacteriophage while arresting movement of the bacteriophage host.

solute The dissolved substance in a solution.

solution The mixture of dissolved substance (solute) and solvent (liquid).

solvent The liquid portion of a solution in which solute is dissolved.

specificity This is a measure of a test's ability to produce a positive response only when reacting with the particular item being tested for. The better the test's specificity, the fewer the false positives (due to better discrimination), and the more useful it is.

spirillum Spiral-shaped cell.

sporangium Structure that produces spores.

spore In bacteria, a dormant form of a microbe protected by specialized coatings produced under conditions of, and resistant to, adverse conditions; also known as an endospore; in fungi and plants, spores are specialized reproductive cells; frequently a means of dissemination.

spread plate technique Method of plating bacteria in which the inoculum is transferred to an agar plate and spread with a sterile bent-glass rod or other spreading device.

stage micrometer A microscope ruler used to calibrate an ocular micrometer.

standard curve A graph constructed from data obtained using samples of known value for the independent variable; once made, can be used to experimentally determine the value of the independent variable when the dependent variable is measured on an unknown.

stationary phase Closed system microbial growth phase immediately following exponential phase; characterized by steady, level growth during which death rate equals reproductive rate.

stolons Surface hyphae of some molds (e.g., *Rhizopus*) that attach to the substrate with rhizoids.

stormy fermentation Vigorous fermentation produced in litmus milk by some species that produce an acid clot but subsequently break it up because of heavy gas production (members of *Clostridium*).

streptolysin Hemolysin (blood hemolyzing exotoxin) produced and secreted by members of *Streptococcus*.

superoxide dismutase Enzyme produced by some bacteria that catalyzes the conversion of superoxide radicals to hydrogen peroxide.

syntrophy A situation where two or more organisms, having differing metabolic capabilities, derive mutual benefit by providing and/or receiving essential nutrients otherwise not available to them.

T

thermal death time The amount of time required to kill a population of a specific size at a specific temperature.

thermophile A microorganism that grows best at temperatures above 40°C.

titer A measurement of concentration of a substance or particle in a solution; used in measurements of phage concentration.

transformant cells Cells that have undergone transformation by picking up foreign DNA in a genetic engineering experiment.

transformation A form of genetic recombination performed by some bacteria in which DNA is picked up from the environment and incorporated into its genome.

trematode A class of parasitic flatworms; also known as "flukes."

trend line A line drawn on a graph to show the general relationship between X and Y variables; also known as a regression line.

triacylglycerol *See* triglyceride.

tricarboxylic acid cycle *See* Krebs cycle.

trichome A filament of cyanobacterial cells. It does not include the sheath, if present.

triglyceride A molecule composed of glycerol and three long chain fatty acids.

trophozoite "Feeding" stage in the life cycle of certain protozoans. *See also* cyst.

tryptophan An amino acid.

tryptophanase Enzyme that catalyzes the hydrolysis of tryptophan into indole and pyruvic acid; detected in the indole test.

turgor pressure The pressure inside a cell that is required to maintain its shape, tonicity, and necessary biochemical functions.

2,3-butanediol fermentation The end-product of a metabolic pathway leading from pyruvate through acetoin with the associated oxidation of NADH; detected in the Voges-Proskauer test.

U

undefined medium *See* complex medium.

urease Enzyme that catalyzes the hydrolysis of urea into two ammonias and one carbon dioxide.

utilization medium A differential medium that detects the ability or inability of an organism to metabolize a specific ingredient.

V

variable A factor in a scientific experiment that is changed; the control and experimental groups in a good experiment differ in only one variable, the one being tested.

vector In genetic engineering, a means, often a plasmid or a virus, of introducing DNA into a new host.

vegetative cell An actively metabolizing cell.

virtual image The image produced when the ocular lens of a microscope magnifies the real image; appears within or below the microscope.

Voges-Proskauer test A differential test used to identify organisms that are capable of performing a 2,3-butanediol fermentation.

W

wavelength Measurement of a wave from crest to crest, usually in nanometers—as in electromagnetic energy.

whey Watery portion of milk as seen upon coagulation of casein in the production of a curd.

X

X-Y scatter plot Graph presenting the relationship between two variables.

Y

yeast An informal grouping of unicellular fungi. *See also* mold.

Z

zone of inhibition On an agar plate, the area of nongrowth surrounding a paper disc containing an antimicrobial substance. (The zone typically ends at the point where the diffusing antimicrobial substance has reached its minimum inhibitory concentration, beyond which it is ineffective.)

zygospore The product of fertilization and the site of meiosis in some molds.

zygote The product of gamete fusion (plasmogamy) and nuclear fusion (karyogamy); a fertilized egg.

Index